Structure and Dynamics of Biomolecules

HERCULES Volume IV

Structure and Dynamics of Biomolecules:

Neutron and Synchrotron Radiation for Condensed Matter Studies

Edited by

Eric Fanchon
Institut de Biologie Structurale Jean-Pierre Ebel Grenoble

Erik Geissler
Laboratoire de Spectrométrie Physique, Université Joseph Fourier de Grenoble I

Jean-Louis Hodeau
Laboratoire de Cristallographie, CNRS, Grenoble

Jean-René Regnard
Université Joseph Fourier de Grenoble I and CEA-Grenoble

Peter A. Timmins
Institut Laue Langevin, Grenoble

This book has been printed digitally in order to ensure its continuing availability

OXFORD
UNIVERSITY PRESS

Great Clarendon Street, Oxford OX2 6DP
Oxford University Press is a department of the University of Oxford.
It furthers the University's objective of excellence in research, scholarship,
and education by publishing worldwide in
Oxford New York
Auckland Bangkok Buenos Aires Cape Town Chennai
Dar es Salaam Delhi Hong Kong Istanbul Karachi Kolkata
Kuala Lumpur Madrid Melbourne Mexico City Mumbai Nairobi
São Paulo Shanghai Singapore Taipei Tokyo Toronto
with an associated company in Berlin

Oxford is a registered trade mark of Oxford University Press
in the UK and in certain other countries

Published in the United States by Oxford University Press Inc., New York

© Oxford University Press, 2000

The moral rights of the author have been asserted
Database right Oxford University Press (maker)

First published 2000
Reprinted 2002

All rights reserved. No part of this publication may be reproduced,
stored in a retrieval system, or transmitted, in any form or by any means,
without the prior permission in writing of Oxford University Press,
or as expressly permitted by law, or under terms agreed with the appropriate
reprographics rights organization. Enquiries concerning reproduction
outside the scope of the above should be sent to the Rights Department,
Oxford University Press, at the address above

You must not circulate this book in any other binding or cover
and you must impose this same condition on any acquirer

A catalogue record for this book is available from the British Library

Library of Congress Cataloging in Publication Data
(Data available)

ISBN 0 19 950453 (Hbk)
ISBN 0 19 850452 7 (Pbk)

Foreword

HERCULES is a training course both for PhD students and for postdoctoral and senior scientists who require neutron scattering or synchrotron radiation techniques in their research. Its name is an acronym for the Higher European Research Course for Users of Large Experimental Systems. Since its inception in 1991 it is organized by the Université Joseph Fourier de Grenoble, the Institut National Polytechnique de Grenoble and the Université de Paris Sud XI, under the auspices of the Institut d'Etudes Scientifiques Avancées de Grenoble.

The principal emphasis of HERCULES is upon laboratory practicals. These are carried out at four large installations, two of which, the European Synchrotron Radiation Facility and the Institut Laue Langevin, are located in Grenoble, while the other two, the Laboratoire pour l'Utilisation du Rayonnement Electromagnétique and the Laboratoire Léon Brillouin, are in the Paris region. The practicals are accompanied by tutorials and lectures, the latter of which have been collected into a series entitled *Neutron and synchrotron radiation for condensed matter studies*.

In the last few years, requirements of biologists in the field of large facilities and their contributions to the development of these techniques have become increasingly significant. For this reason the need has arisen for a further volume in the HERCULES series that is devoted entirely to recent biological applications. That is the purpose of the present volume. It has been a pleasure for us to collaborate with Oxford University Press in its preparation.

We are grateful to the following organizations for continued financial support to HERCULES:

- Ministère de l'Education Nationale, de la Recherche et de la Technologie
- Commission of the European Communities
- Centre National de la Recherche Scientifique
- Commissariat à l'Energie Atomique
- European Synchrotron Radiation Facility
- Institut Laue Langevin
- Laboratoire pour l'Utilisation du Rayonnement Electromagnétique
- Laboratoire Léon Brillouin
- Institut des Etudes Scientifiques Avancées de Grenoble

We would like to express our gratitude to the contributors of this volume for their willing cooperation and solidarity in this undertaking. We are particularly grateful to Marie-Claude Simpson for her invaluable help and her unstinting contribution to the smooth operation of HERCULES.

Eric Fanchon
Erik Geissler
Jean-Louis Hodeau
Jean-René Regnard
Peter A. Timmins

Previous volumes in the HERCULES series *Neutrons and synchrotron radiation for condensed matter studies*:

Volume I—*Theory, instruments and methods*, eds J. Baruchel, J.L. Hodeau, M.S. Lehmann, J.R. Regnard and C. Schlenker, Les Editions de Physique and Springer Verlag, Les Ulis, France 1993.

Volume II—*Applications to solid state physics and chemistry*, eds J. Baruchel, J.L. Hodeau, M.S. Lehmann, J.R. Regnard and C. Schlenker, Les Editions de Physique and Springer Verlag, Les Ulis, France 1994.

Volume III—*Applications to soft condensed matter and biology*, eds J. Baruchel, J.L. Hodeau, M.S. Lehmann, J.R. Regnard and C. Schlenker, Les Editions de Physique and Springer Verlag, Les Ulis, France 1994.

Introduction

Molecular science has come to rely on increasingly large and expensive instruments, where techniques of the highest precision can be brought together and used efficiently. This applies particularly to the intense beam sources for photons and neutrons which become essential for more demanding observations of biological samples. Experimental measurements are made by scientists who bring their samples to the equipment. Many of them, although familiar with their samples, are strangers to the equipment they have to use. At the same time, the cost of beam time, and the difficulties of obtaining it, place a high premium on using the facility for appropriately designed experiments which are executed correctly.

The HERCULES courses were designed to address these problems. They offer anyone, from starting doctoral students onwards, a detailed background to the various experimental techniques employed at synchrotron and neutron beam facilities. The lecturers are all leaders in their fields, expert in both the theoretical background and the practical implementation, and fully aware of current developments.

The courses have now been running with tremendous success for nine years. The march of science is inexorable, and the courses develop rapidly. Almost every year needs new, different lectures; and even when one is repeated, we all find that last year's lecture is already out of date and needs some changes. Students receive a manuscript prepared by each speaker for the particular lecture, but in 1993–94 a series of volumes was produced, documenting the whole HERCULES course. Five years later, the Editors have decided that a new edition is needed. When I compare the current volume with its predecessor from 1994, I see that decision was essential. 'Biology' has expanded from part of a volume to a full volume. Its 10 chapters have extended to 19. Very few authors have survived from one volume to the next, and where they have done so, their contribution is a very different one.

The tremendous burgeoning of protein crystal structures in the last few years had led to a popular perception that when you know the structure, you understand the molecule. This volume emphasizes that crystal structure determination is only one among many physical techniques that help us to describe the complex properties of macromolecules, and the way these may combine to make organisms. Molecular science must study the dynamics of systems, the steps involved in transitions, and the energy changes in these transitions. The environment in a crystal is nothing like that of a molecule tumbling in the cytoplasm, and it is essential to get as much information as possible about the behaviour of molecules in solution. Some molecules have never been crystallized; others may form two-dimensional arrays in membranes, or one-dimensional arrays in fibres; yet others are still too large for conventional crystallography to be feasible. These are some of the tough problems for the 21st century, and all of them are discussed here.

The HERCULES courses are rooted in the dedication and enthusiasm of a group of Grenoble scientists who identified a need, and created a way to meet it. The original initiative from the Grenoble universities and from Grenoble's large facilities was strongly supported by other groups, notably from Orsay and Saclay, and the programme was set up with the help of an international advisory committee. But the heavy task of maintaining an organization that responds to advances in knowledge and to changes in students' needs, has fallen on a few people. Outstanding among these are Jean-René Regnard, currently Director of the HERCULES programme, and Claire Schlenker, who was his co-director up to the end of 1996. They have coordinated the whole course since its inception, and created its strong international flavour by attracting lecturers and students from many countries. We are all indebted to the Editorial Board who have put together this volume. Another person sometimes overlooked is Marie-Claude Simpson, whose efficiency and friendliness contribute so much to the pleasant atmosphere of the courses, and which help the lecturers to relate closely to the students.

David Blow
Imperial College of Science, Technology and Medicine,
London SW7 2BZ, England

Contents

The plates section falls between pages 208–209

List of abbreviations   xi

List of contributors   xiii

Protein structure

1   Challenges in structural molecular biology: genomics in three dimensions   3
  J. Janin
2   Data collection and reduction methods   14
  A.G.W. Leslie
3   Experimental determination of structure factor phases in biocrystallography   36
  R. Fourme, W. Shepard, M. Schiltz, M. Ramin and R. Kahn
4   Electron density—calculation, modification and interpretation   79
  M. Kjeldgaard
5   Neutron crystallography of biological molecules   102
  E. Pebay-Peyroula and D. Myles
6   Reactions in crystals and time-resolved crystallography   116
  I. Schlichting
7   Local structure of metalloproteins at atomic resolution by XAFS   124
  S.S. Hasnain

Dynamics of proteins

8   Molecular dynamics   141
  M.J. Field
9   Inelastic and quasielastic neutron scattering: complementarity with biomolecular simulation   161
  J.C. Smith
10   Diffuse X-ray scattering from molecular crystals   181
  D.S. Moss, G.W. Harris and A. Wostrack

Solutions and partially ordered systems

11   Small-angle X-ray scattering by solutions of biological macromolecules   199
  P. Vachette and D. Svergun

12	Small-angle neutron scattering G. Zaccai	238
13	Structure and dynamics of biological membranes G. Büldt, R. Schlesinger, E. Pebay-Peyroula and H.J. Sass	251
14	Fibre and muscle diffraction J.M. Squire	272

Selected topics

15	Biological spectroscopy using low energy (VUV/UV) synchrotron radiation G.R. Jones and I.H. Munro	305
16	X-ray microscopy C.J. Buckley	338
17	Crystallography of viruses and very large macromolecules D.I. Stuart, J.M. Grimes and E.E. Fry	353
18	The quest for high resolution phasing for large macromolecular assemblies exhibiting severe non-isomorphism, extreme beam sensitivity and no internal symmetry A. Yonath	367
19	Nuclear spin contrast variation studies on macromolecular complexes H.B. Stuhrmann	390

Index   407

Abbreviations

ADC	amplitude-to-digital converter	MASC	multiple wavelength anomalous solvent contrast
AFP	adiabatic fast passage		
aPP	avian pancreatic polypeptide	MBP	mannose binding protein
AS	anomalous scattering	MIR	multiple isomorphous replacement
ATCase	aspartate transcarbamylase		
BPTI	bovine pancreatic trypsin inhibitor	MWPC	multiwire proportional chamber/counter
bR	bacteriorhodopsin	NCS	non-crystallographic symmetry
BTV	bluetongue virus		
CASP	critical assessment of methods of protein structure	NEXAFS	near-edge X-ray absorption fine structure
		OD	optical density
CCD	charge coupled device	ORFs	open reading frames
CD	circular dichroism	PCR	polymerase chain reaction
DNP	dynamic nuclear spin polarization	PDB	protein data bank
EISF	elastic incoherent structure factor	PM	purple membranes
		PMF	potential of mean force
EPMA	electron probe microanalysis	PPE	porcine pancreatic elastase
ESRF	European Synchrotron Radiation Facility	PRCs	photosynthetic reaction centres
EXAFS	extended X-ray absorption fine structure	RH	relative humidity
		rms	root mean square
FFT	fast Fourier transform	SAD	single anomalous diffraction
FTIR	Fourier transform infrared	SAS	small-angle scattering
FWHM	full width at half maximum	SAXS	small-angle X-ray scattering
GUI	graphical user interfaces	SCOP	structural classification of proteins
HEWL	hen egg white lysozyme		
LacR	lac repressor of *E. coli*	SIR	single isomorphous replacement
LADI	Laue diffractometer		
LEED	low energy electron diffraction	SIRAS	single isomorphous replacement with anomalous scattering
MAD	multiple wavelength anomalous diffraction		
		SR	synchrotron radiation

STEM	scanning transmission electron microscope	TFXA	time focusing crystal analysis spectrometer
TAC	time-to-amplitude converter	TIS	triple isotropic substitution
TBSV	tomato bushy stunt virus	XAFS	X-ray absorption fine structure
TDS	thermal diffuse scattering	XANES	X-ray absorption near-edge structure
TEM	transmission electron microscope	wARP	automated refinement protocol

Contributors

C.J. Buckley Department of Physics, King's College London, Strand, London WC2R 2LS, UK. E-mail: chris.buckley@kcl.ac.uk

Georg Büldt Forschungszentrum Jülich, IBI-2: Biologische Strukturforschung, D-52425 Jülich, Germany. E-mail: g.bueldt@kfz-juelich.de

Martin J. Field Laboratoire de Dynamique Moléculaire, Institut de Biologie Structurale Jean-Pierre Ebel, 41 Avenue des Martyrs, F-38027 Grenoble Cedex 01, France. E-mail: mjfield@ibs.fr

Roger Fourme LURE (CNRS, CEA, MENRT), Université Paris-Sud, Bât. 209D, B.P. 34, F-91898 Orsay, France. E-mail: fourme@lure.u-psud.fr

Elizabeth E. Fry Wellcome Trust Centre for Human Genetics, Division of Structural Biology, Roosevelt Drive, Headington, Oxford OX3 7BN, UK. E-mail: liz@strubi.ox.ac.uk

Jonathan M. Grimes Wellcome Trust Centre for Human Genetics, Division of Structural Biology, Roosevelt Drive, Headington, Oxford OX3 7BN, UK. E-mail: jonathan@strubi.ox.ac.uk

G.W. Harris Birkbeck College, University of London, London, UK.

S.S. Hasnain CLRC Daresbury Laboratory, Warrington WA4 4AD, Cheshire, UK. E-mail: s.hasnain@dl.ac.uk

Joël Janin Laboratoire d'Enzymologie et de Biochimie Structurales, UPR 9063 CNRS, Bât. 34, F-91198-Gif-sur-Yvette, France. E-mail: janin@lebs.cnrs-gif.fr

G.R. Jones CCLRC Daresbury Laboratory, Warrington WA4 4AD, Cheshire, UK. E-mail: g.r.jones@dl.ac.uk

Richard Kahn Institut de Biologie Structurale (CEA, CNRS), 41 Avenue des Martyrs, F-38027 Grenoble, France. E-mail: kahn@ibs.fr

Morten Kjeldgaard Institute of Molecular and Structural Biology, Aarhus University, Gustav Wieds Vej 10C, DK-8000 Århus C, Denmark. E-mail: mok@imsb.au.dk

A.G.W. Leslie MRC Laboratory of Molecular Biology, Hills Rd, Cambridge CB2 2QH, UK. E-mail: andrew@mrc-lmb.cam.ac.uk

D.S. Moss Birkbeck College, University of London, London, U.K. E-mail: d.moss@mail.cryst.bbk.ac.uk

I.H. Munro Department of Physics, UMIST, Manchester, M60 1QD, UK. E-mail: ian.h.munro@umist.ac.uk

Dean Myles European Molecular Biology Laboratory, 156X, F-38042 Grenoble Cedex, France. E-mail: myles@embl-grenoble.fr

Eva Pebay-Peyroula Institut de Biologie Structurale, CEA-CNRS, Université Joseph Fourier, 41 Avenue des Martyrs, F-38027 Grenoble Cedex 1, France. E-mail: pebay@ibs.fr

Michel Ramin LURE (CNRS, CEA, MENRT), Université Paris-Sud, Bât. 209D, B.P. 34, F-91898 Orsay, France. E-mail: ramin@lure.u-psud.fr

HansJürgen Sass Forschungszentrum Jülich, IBI-2: Biologische Strukturforschung, D-52425 Jülich, Germany.

Marc Schiltz LURE (CNRS, CEA, MENRT), Université Paris-Sud, Bât. 209D, B.P. 34, F-91898 Orsay, France. E-mail: schiltz@lure.u-psud.fr

Ramona Schlesinger Forschungszentrum Jülich, IBI-2 Biologische Strukturforschung, D-52425 Jülich, Germany.

Ilme Schlichting Max Planck Institut für Molekulare Physiologie, Abteilung Physikalische Biochemie, Otto Hahnstr. 11, D-44227 Dortmund, Germany.

William Shepard LURE (CNRS, CEA, MENRT), Université Paris-Sud, Bât. 209D, B.P. 34, F-91898 Orsay, France. E-mail: shepard@lure.u-psud.fr

Jeremy C. Smith Lehrstuhl für Biocomputing, IWR der Universität Heidelberg, Im Neuenheimer Feld 368, D-69120 Heidelberg, Germany. E-mail: biocomputing@iwr.uni-heidelberg.de

John M. Squire Biological Structure and Function Section, Biomedical Sciences Division, Imperial College of Science, Technology and Medicine, London SW7 2AZ, UK.

David I. Stuart Wellcome Trust Centre for Human Genetics, Division of Structural Biology, Roosevelt Drive, Headington, Oxford OX3 7BN, UK. E-mail: dave@strubi.ox.ac.uk

Heinrich B. Stuhrmann GKSS Forschungszentrum, D-21502 Geesthacht; Institut de Biologie Structurale, Jean-Pierre Ebel, CEA, CNRS, F-38027 Grenoble Cedex 01, France. E-mail: stuhrmann@godot.ibs.fr

Dmitri Svergun European Molecular Biology Laboratory, c/o DESY, Notketstr. 85, D-22603 Hamburg, Germany; Institute of Crystallography, Russian Academy of Sciences, Leninsky pr. 59, 117333 Moscow, Russia, E-mail: svergun@embl-hamburg.de

Patrice Vachette LURE, Bâtiment 209d, Université Paris-Sud, BP 34, F-91898 Orsay Cedex, France. E-mail: vachette@lure.u-psud.fr

A. Wostrack Birkbeck College, University of London, London, UK.

Ada Yonath Department of Structural Biology, Weizmann Institute, Rehovot, Israel. E-mail: csyonath@weizmann.weizmann.ac.il; The Max-Planck Research Unit at DESY, Hamburg, Germany. E-mail: yonath@mpgars.desy.de

Giuseppe Zaccai Institut de Biologie Structurale CEA-CNRS, 41 Avenue des Martyrs, F-38027 Grenoble Cedex 1, France. E-mail: zaccai@ibs.fr

Protein structure

1

Challenges in structural molecular biology: genomics in three dimensions

Joël Janin

Challenges are the daily bread of science. Research is all about creating new knowledge, new ways of understanding and mastering the world around us. Structural molecular biology is even more so, as a relatively new and fast moving field of investigation. Its specific aim is to give a description of biological macromolecules and their assemblies at the atomic level, and to relate their structure to the many elaborate functions they perform in cells and organisms. The founding example, arguably still the best example of a structure that explains a function, is the DNA double helix discovered by Jim Watson and Francis Crick in 1953 (Watson and Crick 1953). Their discovery entirely changed biology, and the consequences are still far from exhausted. A few years after the double helix, Max Perutz and John Kendrew found the haemoglobin structure (Perutz *et al.* 1959) and established X-ray crystallography as the major tool for determining macromolecular structures at the atomic level. At the end of this century, crystallography has been joined by electron microscopy and high resolution nuclear magnetic resonance (NMR). The three techniques have different capacities and limitations, they complement each other and, together, they strive to meet new challenges. One of these is to tackle larger and larger objects and to make structural biology progress upwards from the angstrom (Å) scale of the atom to the micrometre (µm) scale of cells, where optical microscopy can take over. In specific cases, X-ray crystallography almost reaches there, as it has succeeded in solving the atomic structure of viruses nearly 1000 Å, or 0.1 µm, in diameter. Still, viruses are best studied by the very powerful combination of crystallography and electron microscopy pioneered by Aaron Klug (1983) and now extended to macromolecular assemblies such as the ribosome or the cell membrane. In general, large structures are being approached by crystallography in parallel with other techniques, among which near-field microscopies may become important contributors in coming years.

Size has been a challenge from the very beginning of structural molecular biology, and methods of dealing with it are described in other chapters of this book. Here, I will deal with a novel challenge arising from an entirely new field of biology, genomics, which deals with whole genomes containing all the information necessary to make an organism. Genomics was born in 1995 when the first bacterial genome sequence was determined, thus defining amino acid sequences for all proteins in the bacterium. Can we attach three-dimensional structures and functions to all these sequences? Not at present, but we must in the future. This will not be the task of structural molecular biologists alone, but we trust that they will contribute their share and collaborate with their colleagues in fields ranging from genetics and evolution biology to physical chemistry. Do they not stand right in the middle after all?

1.1 The sequence explosion

Ever since DNA sequencing became feasible, which was the great achievement of the late 1970s (Sanger 1981), the gap between the number of known primary (sequence) and tertiary (three-dimensional) structures has been increasing fast, and it is now enormous. As a result, far too many biologists now view proteins as one-dimensional objects and look for functional clues in amino acid sequences considered as chains of characters rather than chemical entities. Structural molecular biologists have a duty to provide their colleagues with the right picture, remind them that proteins are molecules made of atoms, not letters, and, above all, find out how the biological function relates to the three-dimensional structure.

The era of gene sequencing has been now succeeded by the era of whole genome sequencing, which progresses even faster. It began only in 1995, yet already at the end of 1997, a dozen microbial genomes and one eukaryote genome, baker's yeast, have been entirely sequenced (Table 1.1). The number doubled in 1998, and we have the complete sequence of a simple animal, a nematode. Many more will follow, including of course that of *Homo sapiens sapiens*. The

Table 1.1. Completed genome sequences 1995–97

Organism	Genome size (megabases)	Tentative genes	Year
Saccharomyces cerevisiae[1]	12.1	6034	1997
Escherichia coli[2]	4.6	4288	1997
Bacillus subtilis[3]	4.2	4100	1997
Synechocystis sp.[4]	3.6	3168	1996
Archaeoglobus fulgidus[5]	2.2	2471	1997
Hemophilus influenzae[6]	1.8	1740	1995
Methanobacterium thermoautotrophicum[7]	1.8	1855	1997
Methanococcus jannaschii[8]	1.7	1692	1996
Helicobacter pylori[9]	1.7	1590	1997
Borrelia burgdorferi[10]	1.3	863	1997
Mycoplasma pneumoniae[11]	0.8	677	1996
Mycoplasma genitalium[12]	0.6	470	1995

References: (1) Goffeau, A et al. (1997). *Nature*, **387**, S5; (2) Blattner, FR et al. (1997). *Science*, **277**, 1453; (3) Kunst, F et al. (1997). *Nature*, **390**, 249; (4) Kaneko, T et al. (1996). *DNA Res.*, **3**, 109; (5) Klenk, HP et al. (1997). *Nature*, **390**, 364; (6) Fleischmann, RD et al. (1995). *Science*, **269**, 496; (7) Smith, DR et al. (1997). *J. Bacteriol.*, **179**, 7135; (8) Bult, CJ et al. (1996). *Science*, **273**, 1058; (9) Tomb, JF et al. (1997). *Nature*, **388**, 539; (10) Fraser, CM et al. (1997). *Nature*, **390**, 580; (11) Himmelreich, R et al. (1996). *Nucleic Acid Res.*, **24**, 4420; (12) Fraser, CM et al. (1995). *Science*, **270**, 397.

huge task of sequencing our 46 chromosomes is well under way, and it should be completed early in the next century, providing biology and medical sciences with an access to *all* proteins in our body. We do not know yet how many genes there are in the human genome—the current guess is 100 000—but it will be a huge amount of information compared to the 4000 odd human genes that have been analysed up to now. We can already see that the bacterium *Escherichia coli* and the yeast *Saccharomyces cerevisiae* each contain about 5000 genes. At present, only half of these can be associated with a known biological function, even though these two mono-cellular organisms have been favourites with geneticists and molecular biologists during the last fifty years. We know them much better than we know our own cells and we should expect that in the human genome, most of the genes will have no recognizable function.

Albeit less publicized than the DNA sequence explosion, three-dimensional structure determination by X-ray diffraction and high resolution nuclear magnetic resonance has also undergone rapid progress in the last ten years. This is apparent in the number of atomic coordinate sets deposited at the PDB (Protein Data Bank, Brookhaven, New York, with mirror sites in Europe and Asia, see Table 1.2 for the Web addresses). The PDB remained very small, two or three per month, from 1973, the year where it was created, until about 1987. Then, it began to grow faster and faster, and in 1997, five data sets were deposited every day (Fig. 1.1). The PDB contained about 7000 entries at the end of 1997 and it will have well over 10 000 by year 2000—about as many as there are genes in *E. coli* and *S. cerevisiae* together. Does that mean that we could have three-dimensional structures for all proteins in these two organisms if we concentrated on them? Of course not, because the PDB is highly redundant. Many coordinate sets refer to the same protein, perhaps with different bound ligands, to variant proteins obtained by site-directed mutagenesis, or to homologous proteins with closely related sequences prepared from different organisms. In 1994, the PDB

Table 1.2. Some major databases and their URL addresses

Database	Content	URL
1. Swiss-Prot	Protein sequences	http://www.expasy.ch/
2. PDB	Protein data bank, Brookhaven Mirror at EBI-EMBL, Cambridge	http://www.pdb.bnl.gov/pdb/ http://www.ebi.ac.uk/pdb/
3. SCOP	Structure classification	http://scop.mrc-lmb.cam.ac.uk/scop/
4. CATH	Structure classification	http://www.biochem.ucl.ac.uk/bsm/cath/
5. DALI	Structure classification	http://www.ebi.ac.uk/dali/
6. ENTREZ	Sequences and structures	http://www.ncbi.nlm.nih.gov/Entrez/

References: (1) Appel, RD, Bairoch, A and Hochstrasser, DF (1994). *TIBS* **19**, 258–260; (2) Bernstein, FC et al. (1977). *J. Mol. Biol.*, **112**, 535–542; (3) Murzin, A, Brenner, SE, Hubbard, T and Chothia, C (1995). *J. Mol. Biol.*, **247**, 536–540; (4) Orengo, C, Mitchie, A, Jones, S, Swindells, M and Thornton, J (1996). *Protein Data Bank Quart. Newsletter*, **78**, 8–9; (5) Holm, L and Sander, C (1997). *Nucl. Acids Res.*, **25**, 231–234; (6) Hogue, CWV, Ohkawa, H and Bryant, SH (1996). *TIBS*, 226–229.

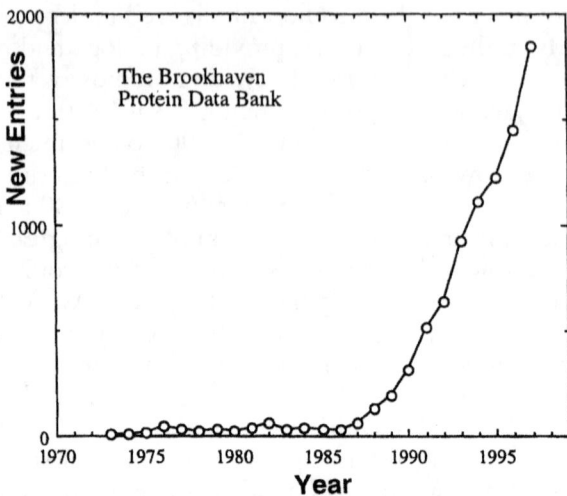

Fig. 1.1. New entries in the PDB. The number of coordinate data sets deposited each year since 1973.

contained 2854 entries which concerned 752 different proteins, a redundancy factor of 3.8, and 30% of the new entries contained novel tertiary structures, the remaining 70% being variants of these 752 proteins. These proportions were maintained in following years (Brenner et al. 1997). Of course, the deposition of these data sets was still fully justified; it brought structural information on the way an enzyme binds a substrate or an inhibitor, or on how a transcription factor binds DNA, and this information may be just as important in interpreting the function as the original structure itself. Yet, the fact is that in 1998 we had three-dimensional structures for only some 2000 proteins counting close homologues as a single molecular species, and this was a very small fraction of the 70 000 proteins for which sequence information was accessible through the Swiss-Prot depository and the ExPASy Web server (Table 1.2).

1.2 From ORF to PDB

Even in *E. coli* and yeast, a majority of the proteins found in sequenced genomes are no more than chains of characters called 'open reading frames' (ORFs) and covering segments of DNA sequence comprised between a start and a stop codon. A putative protein remains an ORF until subjected to functional, biochemical and structural analysis. There is an imperious necessity to perform that analysis, and first of all, to convert the DNA sequence into an actual protein molecule, indeed into milligrams of it for crystallization or NMR structure analysis. Except for very small proteins which can be synthesized chemically, this implies (a) cloning a functional version of the gene; (b) choosing an expression system; (c) expressing and purifying the protein product of the gene. Then, steps (d) crystallization and (e) structure solution can be attempted. Steps (a) and (b) rely on well-established techniques of molecular biology, step (c) on chromatography and other preparative methods of biochemistry. Though unexpected difficulties may be encountered, cloning and expression can be efficiently performed on most bacterial or yeast genes. With higher organisms, the genome sequence is not directly useful, because most genes are split into pieces (often many of them) by introns. Cloning must be performed on the cDNA rather than on the genome DNA. The cDNA is a copy of a messenger RNA (mRNA) that can be made by the enzyme reverse transcriptase, and it is collinear with the protein sequence, just like in bacteria. The genome sequence nevertheless provides the necessary information to make primer oligonucleotides for the polymerase chain reaction (PCR), which is a powerful way to select by specific amplification the right DNA molecules among all those produced by reverse transcription of all mRNA present in a cell extract.

To yield protein, a cloned gene must be associated with the necessary elements for efficient expression in cell culture. This is done by inserting the coding sequence in an expression vector, a plasmid or viral DNA. Expression in bacteria (usually *E. coli*) or yeast has the advantage that these microorganisms are grown easily and cheaply in large quantity. The protein yield can be very high, and then purification becomes very easy: some proteins are produced in tens or hundreds of milligrams per litre of culture. Not surprisingly, bacterial genes perform best in bacteria (and yeast genes in yeast), whereas many genes of higher organisms are either poorly expressed or expressed as insoluble aggregates that form inclusion bodies in the bacterial cell. This is the rule rather than the exception with genes that code for very long polypeptide chains of over 1000 amino acid residues. They seem not to fold properly in bacteria, perhaps because they need the help of the chaperone machinery, which is different in eukaryotes and bacteria. Long polypeptide chains usually fold into a number of distinct domains, which are independent structural and functional units of moderate size (50–200 residues). If domain boundaries can be located in the DNA sequence, the corresponding segments can be subdivided and expressed separately with a better chance of achieving correct folding. In the absence of three-dimensional structure information, the subdivision relies on trial and error, and many genetic constructions may be needed before efficient expression is obtained in bacteria, mammalian cell cultures, or in insect cell cultures using an insect virus as a carrier.

1.3 Homology in one and three dimensions

This procedure is labour intensive and prone to failure. Major efforts are under way to increase its efficiency and automate some of its steps, and a systematic analysis of proteins coded by the archaebacterium *Methanococcus jannaschii* is being attempted in the USA as a collaboration between the Institute for Genomic Research and a crystallography laboratory in Berkeley (*Science* 1998). The results are promising, but a systematic analysis of this type is unlikely to be carried out on the tens of thousands of ORFs identified in complete genomes of higher organisms. However, a majority of these putative proteins are clearly related. Gene homology is the result of biological evolution and it indicates that two DNA sequences derive from a common ancestor by duplication. It is found between genomes (orthologous genes), and also within genomes (paralogous genes) as the larger bacterial genomes and the yeast genome show clear indications of internal repetition. Gene duplication is even more extensive in most genomes of higher organisms. Homologous genes are expected to produce proteins with identical functions and closely similar structures. Therefore, we really need to know only one in detail. How many proteins are there in nature that warrant a separate study? Over two decades ago, it was proposed that the number of non-homologous protein families is small, of the order of 1000 (Chothia 1992). The 'one-thousand families' hypothesis was based on a comparison of the relatively few sequences known at the time, yet a much more recent analysis of a much larger body of sequences suggests that it still holds (Brenner et al. 1997). It is based on statistical arguments rather than on just homology as detected by classical sequence comparison algorithms. Alignment algorithms implemented in programmes BLAST or FASTA easily identify the evolutionary relationship of two proteins that share 30–40% identical residues, but they generally assign statistically non-significant scores to more divergent sequences. Homology at lower levels of sequence identity can nevertheless be detected when three-dimensional structures are available, for folds are much better conserved throughout evolution than the sequences themselves.

Three-dimensional structures may be compared and classified. Examples are the SCOP, CATH, Entrez/MMDB and DALI/FSSP data bases (Table 1.2). SCOP (Structural Classification Of Proteins) was created and maintained by A. Murzin, C. Chothia and collaborators in Cambridge. SCOP groups together into *families* proteins with related sequences, and into *super-families*, proteins that have no significant sequence homology but sufficiently similar structures to be nevertheless considered as evolutionarily related. Because super-families are larger than families, there are more homologous structures in SCOP than indicated by sequence alignment alone. Above the super-families, the classification is based on the presence of secondary structure (α-helices and β-sheets). Three major *classes* group all-α, all-β and α-and-β proteins, and in each class, *folds* describe the topology and the way secondary structure is arranged.

According to SCOP, the June 1996 PDB contained representatives of some 460 non-homologous families and 300 super-families of proteins or protein domains. Representatives of new families still appear at the high rate of more than 100 a year (Fig. 1.2). Thus, if the 'one-thousand families' hypothesis is correct, three-dimensional structures are presently known for about half of the proteins existing in nature. We may have the impression that it will be possible to assign a three-dimensional structure to most protein sequences by the time the human genome is completed, early in the next century. This impression is over-optimistic for at least two reasons. One is the law of diminishing returns, which will slow the process in coming years due to saturation. In 1994, three-quarters of the estimated total was still to be discovered, yet only one in ten of the new PDB entries represented a new family, and one in thirty a new fold (Fig. 1.2 bottom). This year, the ratio may be less than one in twenty for families. The other reason is that the families are far from equal. The PDB contains many proteins

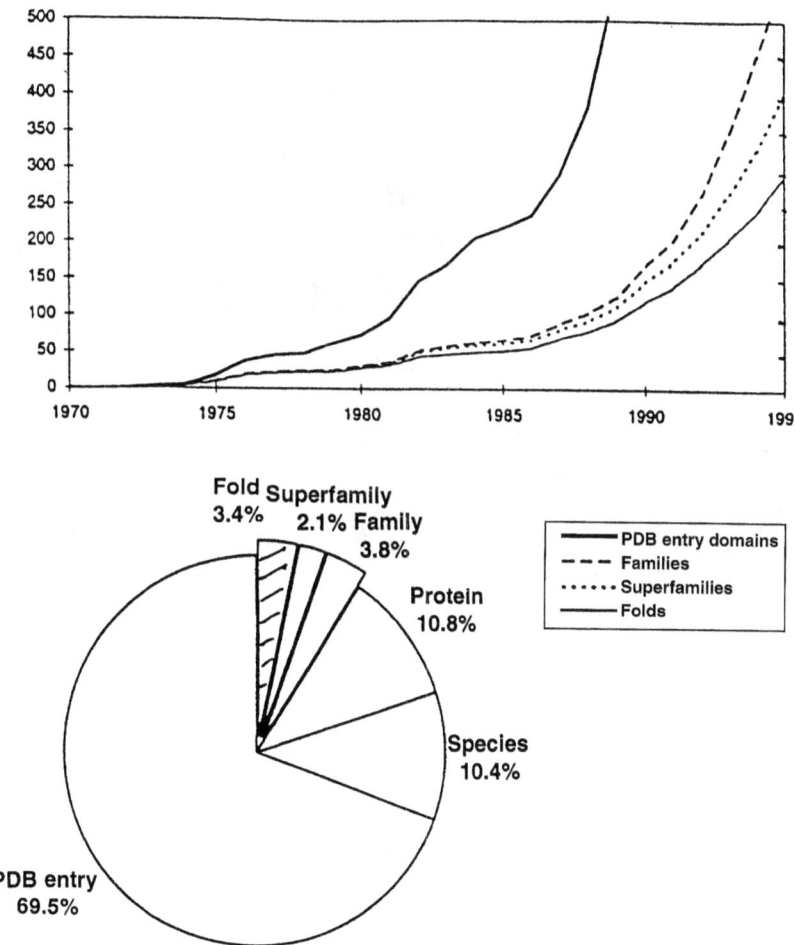

Fig. 1.2. Structures, families and folds in the PDB. (Top) The cumulative growth of the PDB is described as a function of time for each level of the classification used in the SCOP database: all entries (heavy line), families of homologous proteins (dashes), super-families of distant evolutionary relatives (dots), and folds (light line). (Bottom) Of 1392 new PDB entries in 1994, 70% were for proteins already in the PDB, another 20% for proteins with related amino acid sequences. Only 9% (exploded sector) had no detectable homology to a known structure, and therefore a potential to have a new fold, which occurred in 3.4% of cases (shaded sector). [Adapted from Brenner et al. (1997) with permission.]

related to immunoglobulins and to haemoglobins, all with the same characteristic fold. Other very common folds are the α/β barrel or triose-phosphate isomerase fold and the $\beta\alpha\beta$ sandwich or ferredoxin fold. Unlike the immunoglobulin and haemoglobin folds, they occur in many proteins with unrelated sequences, and they may be the result of convergent evolution. As much as 60% of the proteins in the PDB contain one (or several) of the nine folds represented in Fig. 1.3 and called 'superfolds' by Orengo et al. (1994). In contrast, some classes

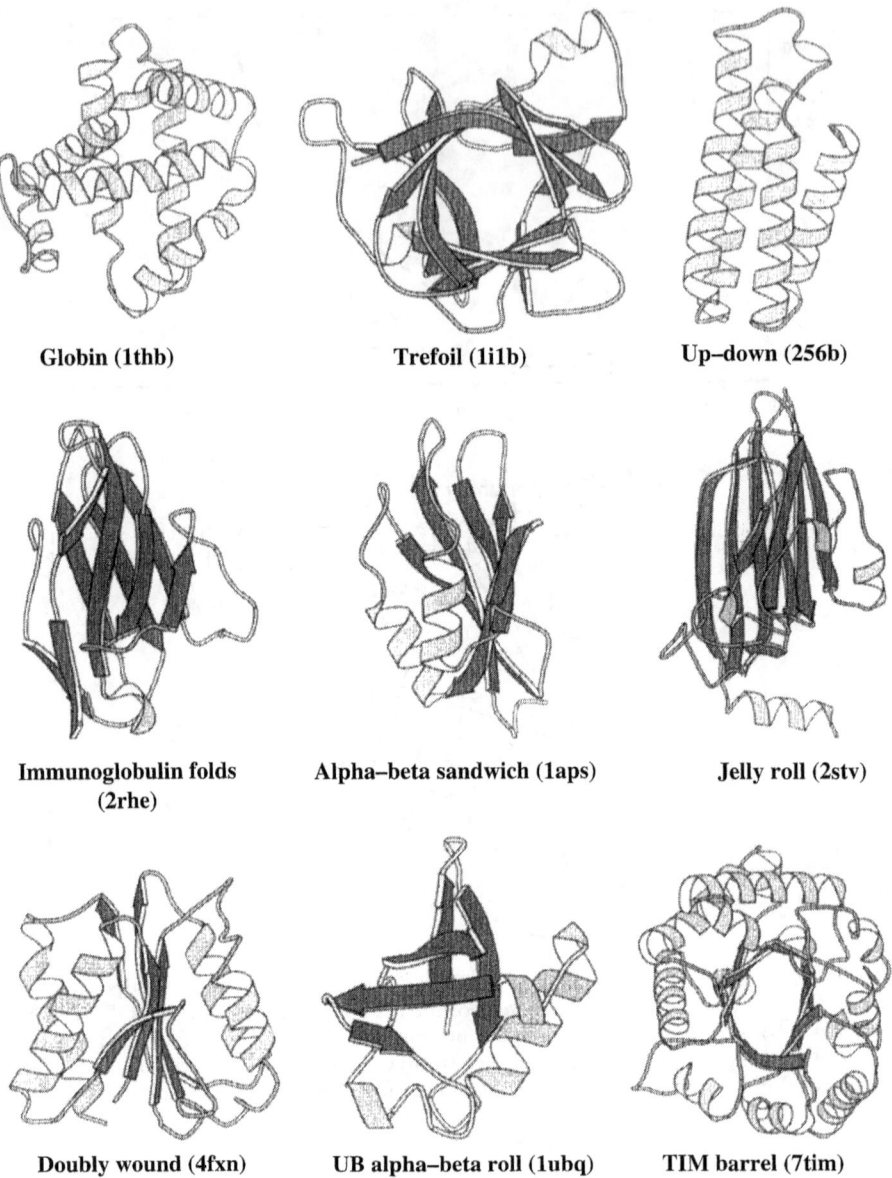

Fig. 1.3. Superfolds. Nine protein folds that are present in over half of the known protein structures [Orengo *et al.* (1994), with permission].

of proteins are very poorly represented at present, and they are likely to remain incompletely surveyed in the coming years. The most obvious case is membrane proteins, which represent a large fraction of the gene products and of which we have few examples in the PDB.

1.4 Homology modelling

Homology modelling is so cheap and fast compared to the standard five-step (a–e) procedure described above, that it must become more and more popular—several of the Web servers in Table 1.2 will perform it for you if you give it an amino acid sequence and atomic coordinates for a known homologous protein structure. Existing methods have been tested by blind structure prediction, followed with a comparison with the experimental structure, in two 'Critical assessment of methods of protein structure' (CASP) experiments, conducted in 1994 (CASP1) and 1996 (CASP2). A third one was completed in 1998 (CASP3). For each experiment, predictors were given a set of target sequences for proteins which were studied by crystallography or NMR, their predictions were collected by a team of assessors and they were carefully compared with the experimental structures as these came out. The results of the assessment have been published together with reports by the predictors themselves who describe the methods they use.

In these experiments, some targets had homologues in the PDB that could be identified by sequence comparison, and others did not. The first type of target was suitable for homology modelling. The results confirmed that it yields useful models when the sequences have over 35% identity, but that the models remain inaccurate even at high levels of identity. The accuracy of main chain prediction can be judged from the root-mean-square (rms) value of the distance D between the $C\alpha$ atoms of the model and the experimental structure. At the 35% sequence identity level, the expected rms distance between two X-ray structures is less than 1.5 Å, implying that the chain fold is similar over the whole sequence. Very few models built by homology achieved that accuracy, and most had rms distances of 2–6 Å. These discrepancies were due to large errors at places where the alignment required a gap due to an insertion or deletion in the sequences. Insertion–deletion events generally concern loops of polypeptide chain on the protein surface, and nearly all of these loops were incorrectly modelled. Loop prediction, which has been the object of much effort, is still a major problem, and it seems to have made little progress from CASP1 to CASP2. One or two target sequences were so close (85%) to a known protein that the main chain should be nearly identical. On those, side-chain prediction could be tested separately from main chain prediction. Values of the $\chi 1$ dihedral angle governing the side-chain orientation had a rather high 45° rms error, and they were essentially random with more divergent targets. Thus, side-chain prediction is very sensitive to the main-chain conformation and by no means as easy as is often assumed.

When the target sequences are not obviously related to proteins in the PDB or when identity is less than 25%, homology modelling breaks down altogether, and yields no useful result. Then, prediction is based on a search for adequacy between the target sequence and an already known fold. The technique is called fold recognition or *threading*, for it amounts to threading the target sequence through a set of three-dimensional structures representing the whole

PDB. Each threading is attributed a score using empirical potentials derived from a previous statistical study of real structures. High scores point to folds that fit the target sequence. With about one-half of the possible folds already known, fold recognition should already be a powerful way of assigning structures to sequences, and it can only become more so in the future. Moreover, sequence comparison methods miss many homologies that structure-based methods reveal, and fold recognition should be able to identify them. A recent study shows that unsuspected homologies occur widely in the PDB (Brenner et al. 1997). A sample of 1323 proteins (or protein domains) covering the whole PDB in June 1996 yielded 9044 pairs of homologous proteins based on structure, some with levels of sequence identity as low as 15%. Fewer than 10% of these could be detected by sequence comparison methods, whereas 95% of the pairs with at least 30% identity were assigned correctly. In CASP2, the best models derived by threading had an rms distance near 3 Å, and therefore they were as good as those based on homology modelling at the 35% identity level. All these models were for target proteins that did have homologues in the PDB, but at identity levels of 10–15% which sequence alignment cannot identify. On other targets, fold recognition was still successful in the sense that predictors did identify the closest relative of each target in the PDB, but models derived from the prediction had rms distances greater than 10 Å, which is very poor and useless for most purposes. The main problem appears to be that threading may identify the right structure in the PDB, yet fail to align it correctly on the target sequence unless there are clues in the sequence itself due to unsuspected homology. As a consequence, threading is more an extension of homology modelling at present than a truly independent way of generating three-dimensional structures from sequences.

1.5 Conclusion

Prediction methods will not take the burden of assigning three-dimensional structures to amino acid sequences from the shoulders of X-ray crystallographers and NMR spectroscopists in the foreseeable future. They will only relieve them from the obligation of solving many very similar structures for proteins with close homology. Even in these cases, unpredictable changes sometimes happen as a point mutation is introduced or a ligand binds. There is therefore no alternative to experiment. The community of molecular structural biologists presently provides the PDB with thousands of new entries per year. This is a large fraction of the estimated number of non-homologous protein families, and molecular structural biology would soon catch up with genomics if the redundancy of the new entries were not so high. To cover all protein families systematically, we need large scale concerted endeavour and the three-dimensional equivalent of the effort that has led to the determination of whole genome sequences and opened the era of one-dimensional genomics. Determining structures for new families involves setting up expression systems for a very large number of genes, developing efficient purification procedures and automating many steps including homogeneity tests and, of course, crystallization. This is being attempted by some of the same laboratories that started genomics in the USA and elsewhere. Beyond the critical step of crystallization, the powerful tools of X-ray crystallography, first of all the use of synchrotron radiation, can be set in movement to produce reliable and accurate structures, making three-dimensional genomics a reality.

References

Brenner, SE, Chothia, C and Hubbard, TJP (1997). *Current Opinion in Structural Biology*, **7**, 369–376.
CASP1 (1995). *Proteins*, **23**, 295–462.
CASP2 (1997). *Proteins*, Suppl. 1.
CASP3 (1999). *Proteins*, Suppl. 3.
Chothia, C (1992). *Nature*, **357**, 543–544.
Klug, A (1983). *Biosci. Reports*, **3**, 395–430.
Orengo, C, Jones, S and Thornton, J (1994). *Nature*, **372**, 631–634.
Perutz, MF, Rossmann, MG, Cullis, AF, Muirhead, H, Will, G and North, ACT (1959). *Nature*, **185**, 416–422.
Sanger F (1981). Nobel lecture. *Biosci. Reports*, **1**, 3–18.
See *Science*, (1998). **279**, 979.
Watson, JD and Crick, FC (1953). *Nature*, **171**, 737–738.

2
Data collection and reduction methods

A.G.W. Leslie

This chapter describes modern methods for collecting and processing X-ray diffraction data from macromolecules. The first five sections deal with practical aspects of data collection, with Section 2.1 focusing primarily on laboratory sources. Complementary information on synchrotron radiation instrumentation is given in the next chapter by Fourme *et al.* (Chapter 3). The final section deals with the computational aspects of integrating diffraction intensities and merging and scaling the data. More detailed accounts of many of these topics can be found in Carter and Sweet (1997) and Sawyer *et al.* (1993).

2.1 X-ray optics

2.1.1 General principles

To obtain the highest possible data quality, it is important to minimize the general X-ray background scatter and also to minimize the size of the diffraction spots on the detector, since larger spots will inevitably have a greater background component, which in turn leads to greater error in estimating the intensity (see Section 2.6.3). There are a number of ways to reduce the X-ray background:

(1) Match the beam size to the sample size. If the beam is made larger than the sample, scattering from the air, glass capillary and mother liquor or cryoprotectant and loop will increase without any increase in the strength of diffraction.
(2) Minimize the distance from the end of the collimator to the crystal and the crystal to the backstop to reduce air scatter.
(3) Use radiation with a high spectral purity—radiation at wavelengths other than the characteristic wavelength will not contribute to the measured diffraction intensities but will contribute to the X-ray background.
(4) Maximize the crystal-to-detector distance, as the X-ray background falls off as the inverse square of this distance, while the intensity of the diffracted beams falls off much more slowly (depending on the degree of collimation of the incident beam).

The size of the diffraction spots on the detector is determined primarily by the X-ray beam divergence, although other factors (the crystal mosaicity and the characteristics of the detector) also make a contribution. It is therefore desirable to have a beam that has a very low divergence, 0.1° or less. It is always possible to reduce the divergence, but this can lead to an unacceptable loss in intensity. *In practice, a compromise between beam divergence and beam intensity always has to be made.* For large unit cells, a low beam divergence may be essential in order to resolve adjacent spots on the detector.

The absorption of CuKα radiation in air is about 1% per cm, so for low resolution work with crystal-to-detector distances greater than 30 cm it is worth considering the use of a helium path between the crystal and the detector.

2.1.2 *Pinhole collimation*

The simplest type of collimation consists of two small pinholes, between 0.1 and 0.5 mm in diameter, separated by a distance of several cm. Providing the distance from the focal spot to the first pinhole is several cm, this will give a beam of the same size as the pinhole, with a divergence of a few tenths of a degree. A nickel filter is necessary to discriminate against CuKβ radiation, but even with a filter a significant amount of white radiation will reach the sample, giving a higher X-ray background and increased radiation damage.

2.1.3 *Monochromator*

A single crystal, usually graphite, is positioned close to the X-ray source and oriented to produce Bragg reflection from the (001) planes. This gives a monochromatic X-ray beam (apart from a fairly weak $\lambda/2$ component) which is collimated by a single pinhole. The resulting radiation has a much higher spectral purity than with pinhole collimation, leading to a significant reduction in the level of the X-ray background. However, the beam divergence is rather high (e.g. 0.3°), and this can only be reduced at the cost of a loss in beam intensity.

2.1.4 *Double mirrors*

X-ray mirrors for laboratory use are made of glass coated with nickel (or less commonly gold or platinum). X-rays incident at an angle below the critical angle (0.38° for Ni) are totally reflected at the air/mirror interface. Bending the glass substrate allows the reflected beam to be focused onto the sample or the detector. Two orthogonal mirrors are used to provide focusing in both the horizontal and vertical planes. When used with a small focal spot size (100–200 µm) mirrors can provide a more brilliant and more highly collimated beam than either pinholes or a graphite monochromator. They are therefore preferred when collecting data from crystals with long unit cell axes (e.g. > 100 Å) (Harrison 1968). Although Ni mirrors provide a degree of discrimination against CuKβ, it is still desirable to have an additional Ni filter although this does not need to be as thick as for pinhole collimation. Because of the variation in critical angle with wavelength, mirrors act as a good filter for shorter wavelengths, but longer wavelengths (which are thought to be more damaging to the sample) are not attenuated significantly. Until relatively recently, double mirrors (6 cm in length) were used in conjunction with a 100 µm focal spot size for problems where a very long cell edge demanded the use of focusing optics. More recently, longer mirror systems (8 cm for the first mirror and 16 cm for the second) used with a 300 μm focal spot size have become popular, as this combination produces a high intensity with a modest (~0.2°) beam divergence and can be used effectively for many crystallographic problems.

A double mirror system is more difficult to set up than other types of collimator, and the use of an 'X-ray eye' (an image intensifier/CCD coupled to a TV monitor) greatly facilitates alignment.

2.1.5 Novel collimation devices

A number of novel collimating systems have been developed recently, although none of these is yet in widespread use. A single tapered capillary or a bundle of tapered capillaries can be used to provide a very small (1–10 μm) focal spot size, as used on the microfocus beamline at the ESRF (Engström and Riekel 1996). A multilayer optical assembly is available from Osmic which produces five times the intensity of a conventional double mirror system. Finally, a miniature ellipsoidal mirror has been developed for use in conjunction with a microfocus X-ray tube (Arndt *et al.* 1998).

2.2 Goniostat

The goniostat is used to orient the crystal and subsequently rotate the crystal to allow the collection of a complete data set in an efficient manner.

Crystals are normally mounted on a goniometer head, which has two orthogonal translation stages to allow the crystal to be centred on the rotation axis. Most goniometers also have two orthogonal arcs that allow rotation of the crystal by ±20°, and can be used to align crystals of well-defined morphology to get a particular axis approximately parallel to the rotation axis of the goniostat.

Many commercially available detectors are supplied with the simplest possible goniostat, that is, a single rotation axis. Others provide more complex goniostats which have three independent axes of rotation which allow the crystal to be placed in any desired orientation. With the advent of modern area detectors, particularly image plates with their large active area, it is no longer as important as it once was to be able to orient the crystal precisely. Particularly for laboratory work, a single axis rotation device is perfectly adequate. However, for collecting multiwavelength anomalous data at a synchrotron, where it is crucial to minimize the systematic errors in data collection, there is an advantage in being able to orient the crystal so that Friedel mates are measured at the same or very similar times.

2.3 Crystal mounting

Macromolecular crystals are very sensitive to dehydration and frequently to temperature, so it is crucial to have an appropriate sample environment. Typically, this is achieved by sealing the crystal in a glass capillary which also contains some of the crystallization medium (mother liquor) to maintain high humidity. The crystals are normally mounted slightly 'wet', that is, in physical contact with a small amount of mother liquor. The glass capillary can be placed in a cold air stream to provide cooling down to about −20 °C (the actual temperature used depends on the freezing point of the mother liquor), in order to reduce radiation damage.

2.3.1 Cryocooling

In recent years an alternative method of crystal mounting known as flash freezing or cryocooling has become increasingly popular (Rogers 1997). The crystal is either grown in or transferred to a suitable cryoprotectant (e.g. 20% glycerol), from which it is picked up with small fibre loop attached to a copper pin. The crystal in its loop is then rapidly moved into a stream of nitrogen gas at ~ 100 K or alternatively frozen in liquid N_2. The presence of the cryoprotectant and the very rapid cooling prevent the formation of crystalline ice and the diffraction properties of the crystal should be preserved, although the mosaic spread may increase slightly. This procedure has the very great advantage that radiation damage on exposure to X-rays is dramatically reduced, to the point at which it is usually possible to collect a complete data set from one crystal. It has the second advantage that it requires much less physical manipulation of the crystal than the standard mounting technique, which, for mechanically fragile crystals, can result in an improvement in the quality of the diffraction relative to mounting the crystal in a capillary. Usually, some experimentation is required to find the best cryoprotectant (i.e. that which best preserves the diffraction properties of the crystal), and it may be necessary to increase the cryoprotectant concentration gradually over a period of hours. In some instances it has proved possible to reduce crystal mosaicity by thawing and re-freezing a flash-frozen crystal (Harp et al. 1997). The very high X-ray flux available at a synchrotron makes cryocooling essential for many samples.

2.3.2 Tapered glass capillaries

A variant of the standard mounting technique is to mount crystals in a glass capillary that has been drawn out to a very fine tip (20–50 μm in diameter) that is completely filled with mother liquor (Richmond et al. 1984). The crystal is transferred from the crystallization well or harvesting solution and placed in the wide end of a vertical tapered capillary (e.g. using a Pasteur pipette). The crystal will then fall under gravity until it becomes wedged in the narrow end of the capillary. The wide end is sealed (with plasticine) and the capillary mounted on a goniometer head. The capillary should be oriented so that it is always pointing slightly downwards during data collection to minimize crystal slippage. This procedure works very well for fragile crystals as it minimizes crystal handling.

Tapered capillaries have also been used for cryocooling particularly sensitive crystals such as the nucleosome core particle (Luger et al. 1997) and viruses

(D. Stuart, personal communication) using a protocol that involves slow cooling to about −70 °C prior to flash-freezing.

2.4 Detectors

There have been substantial improvements in X-ray detector technology in the last 15 years, which have had a dramatic impact on both the quality of diffraction data and the ease of data collection. The principles of operation of the three types of detector currently in widespread use are considered below.

2.4.1 *Multiwire proportional chambers (MWPCs, e.g. Xentronix and Xuong Hamlin)*

In a multiwire chamber (see Kahn and Fourme (1997) for an overview) the incident X-ray is absorbed by a gas, typically xenon. The resulting electrons are accelerated by a potential of several kV towards a set of closely spaced thin wires (the anode wires). The electrons gain sufficient energy to cause further ionization events leading to an avalanche effect which produces a multiplication factor of about 10^5. The event is detected as a pulse on at least two anode wires and this induces a signal on two sets of cathode wires, one set running parallel and the other perpendicular to the anode wires. These pulses can be decoded electronically to give the coordinates of the ionization event. This system allows the detection of single photons, and sophisticated electronics can discriminate between an ionization event and electronic noise. So, these detectors are very low noise devices. However, because of the finite time taken for the positional decoding, there is a maximum global count rate above which counts will be 'lost'. In addition to this global count rate limitation there is a maximum local count rate which is due to space charge effects caused by the accumulation of positive ions around the anode wires. These effects can limit the data collection rate for strongly diffracting samples, even with laboratory sources, and make these detectors unsuitable for use at synchrotrons except in very special circumstances (e.g. the wire chamber detector at LURE).

2.4.2 *Image plate detectors (Mar Research, R-Axis, MacScience)*

An image plate consists of a thin layer of phosphor (Eu^{2+} doped BaFBr) deposited on a plastic backing. Absorbed X-ray photons create lattice defects (known as F centres), resulting in a latent image in the phosphor. The latent image can be read out by scanning the plate with a helium−neon laser ($\lambda \sim 633$ nm). The laser light restores electrons in the F centres to the ground state, resulting in the emission of photoluminescence at 390 nm with an intensity proportional to the number of absorbed X-ray photons. The photoluminescence is measured with a photomultiplier and converted to a digital signal which is stored in computer memory. Image plates are integrating devices, as distinct from multiwire proportional counters which are photon counting detectors. An overview of the properties of image plates is given in Amemiya (1997).

2.4.3 CCD detectors

The latest generation of detectors are based on the use of charge coupled devices (CCDs), as described in Westbrook and Naday (1997).

CCD detectors are, like image plates, integrating rather than photon counting devices. The X-rays are absorbed by a thin phosphor layer to produce visible light photons. The phosphor is linked by a reducing fibre optic bundle to a CCD, either directly or via an image intensifier. The latent image is stored in the CCD and then read out in a serial manner to give a digitized image. Recent results using directly coupled CCD detectors (with no image intensifier) at synchrotrons in Europe and the USA have given very promising results. The disadvantage of CCD detectors is the limited size of the active area (typically about 100 mm square), which is dictated by the size of commercially available CCDs. However, detectors are now available consisting of several modules in a 'tiled' arrangement (2×2 or 3×3) to give a more useful active area.

2.4.4 Characteristics of an ideal detector

An ideal detector for protein crystallography should have the following characteristics:

(1) High absorption efficiency so that all incident photons are detected.
(2) Very low intrinsic noise so that weak signals are not lost in 'instrument noise'.
(3) Unlimited global and local counting rates.
(4) A very large dynamic range.
(5) A large active area so that the detector-to-crystal distance is large while still collecting a large solid angle of diffraction data.
(6) Excellent spatial resolution so that closely spaced diffraction spots can be resolved.
(7) Very rapid readout.
(8) Stable and easily determined spatial distortion and non-uniformity of response.
(9) Rugged, reliable and easy to use.

Some of these criteria (e.g. 3, 4, 7) assume particular importance at synchrotron sources. In practice, there is no detector that satisfies all of these criteria, so compromises have to be made.

For laboratory work, image plates are excellent detectors, and satisfy most of the criteria except for the readout time, which is between 36 s (Mar Research Mar 345 scanner, 18 cm diameter scan) and 8 min (R-Axis II fine scan). In the case of the R-Axis II the disadvantage of the long readout time is partially offset by the fact that this detector has two image plates, so that one plate can be scanned while the other is being exposed.

Multiwire chambers also give data of a high quality, and because they are photon counting devices they will be superior to image plates for *very* weak diffraction. However, they are inferior to image plates in terms of maximum count rates, the size of the active area, spatial resolution, long-term stability of calibration and arguably their reliability and ease of use. Their main advantage over image plate detectors is that there is essentially no readout time, which means they can collect data in very fine phi rotation slices, which improves the data quality because it minimizes the background that is included in each Bragg diffraction spot (see Section 2.5.1).

CCD detectors are particularly suitable for use at synchrotron sources because the rapid readout time (5–20 s) allows more efficient use of beamtime. In principle they could also be used to collect data in fine phi slices, although they are not typically used in this way. At present the main disadvantage is the relatively small active area, which is a circle of diameter 135 mm for the Mar Research CCD and a 188 mm square for the ADSC 2×2 tiled detector, which are the two commercial detectors currently in use on synchrotron beamlines. To some extent this disadvantage is offset by the very small beam sizes ($< 100 \mu m$) and low beam divergence that is available at a synchrotron source. The spatial resolution of the CCD is better than an image plate, and so the same number of orders of diffraction can be resolved in spite of its smaller size. The dynamic range of the CCD can also be better than an image plate, which is particularly important for synchrotron data.

2.5 Collecting the images

2.5.1 *Fine vs coarse phi slicing*

There is a basic difference in practice between the way that diffraction images are collected using detectors with a relatively long readout time, such as image plates, compared to those with a short readout time, such as MWPCs and CCDs. In order to avoid excessive dead-time, image plate data is collected using a crystal rotation of typically 0.5–2.0° per image (coarse phi slices), while for wire chambers (and CCDs) a much smaller crystal rotation of 0.1–0.3° is more common (fine phi slices). The disadvantage of the larger rotation range is that, for any given reflection, Bragg diffraction will only be taking place for a fraction of the total rotation angle. (The actual fraction depends on the beam divergence, crystal mosaic spread and geometric factors.) For the remainder of the rotation, X-ray background scatter will accumulate leading to an increase in the noise component of the total intensity, with a corresponding decrease in the signal-to-noise ratio. For weak reflections, this can have a significant effect on data quality.

Fortunately, the large active area of the image plate means that it can be placed at a large distance from the crystal and so the background is correspondingly lower, which to some extent offsets the disadvantage of having to use a larger rotation angle per image.

For MWPC detectors the rotation angle per image is often dictated by the software, which will not integrate reflections with a width of more than 10 images, but disk space limitations can be another consideration.

For image plates, the rotation angle per image must be small enough to avoid spatial overlap of adjacent reflections in reciprocal space. This is dictated by the cell dimensions which lie approximately normal to the rotation axis and can be checked by the software. The other consideration is to maintain an efficient duty cycle, so that the exposure time lost because the plate is being scanned is a small percentage of the total time, which argues in favour of larger rotation angles. It is now possible to process images in which all the reflections are partially recorded, so there is no practical limitation imposed by the software on the minimum rotation angle.

2.5.2 Data collection strategy

When collecting data it is important to consider the strategy required to collect a complete data set [reviewed in Dauter (1997)]. With image plate detectors, the large active area means that it is normally possible to collect all the data with a single detector position with the beam in the centre of the detector, rotating the crystal around a single axis. The total rotation required depends on the Laue group and the orientation of the crystal. However, with smaller detectors it is often necessary to offset the detector, usually on a two-theta arm, and in these cases more thought is required to collect a complete data set, and the scaling of the images can also be problematic. For three-circle goniostats there are additional mechanical restrictions which limit the rotation range on some axes that also need to be taken into account. A number of software tools are available (e.g. MOSFLM, RSPACE, LATTICEPATCH) to aid in the design of the data collection strategy.

2.6 Data processing

Data processing falls naturally into three quite distinct steps:

(1) Determination of crystal cell parameters, space group and orientation.
(2) Integration of the images (with concurrent refinement of crystal, beam and detector parameters).
(3) Data reduction, that is placing all data on a common scale, merging multiple observations to give a unique data set while rejecting outliers and reducing intensities to amplitudes for use in heavy atom phasing, Fourier syntheses, model refinement, etc.

Further details of processing fine phi-slice data (MADNES) are given in Pflugrath (1997), and coarse phi-slice data (DENZO) in Otwinowski and Minor (1997).

2.6.1 Autoindexing

With modern software packages it is possible to determine the crystal orientation and unit cell parameters using data from a very small phi rotation—a few degrees of data from fine phi slice images or a single image for coarse phi slices is usually sufficient. The basic principles of autoindexing procedures are described

2.6.1.1 Definition of crystal orientation

The crystal orientation can be defined as

$$X = \Phi UBh, \qquad (2.1)$$

where

X is a vector in the laboratory frame giving the position of the reciprocal lattice vector with indices h;

B is an orthogonalization matrix, which defines a set of orthogonal axes based on the crystal axes (this matrix depends only on the crystal cell parameters);

U is a pure rotation matrix describing the orientation of the crystal in the laboratory frame in the initial or standard setting;

Φ is the rotation around the spindle axis for a single axis device, or more generally the goniostat matrix.

For convenience, the product of the U and B matrices is often denoted as the 'setting matrix' A,

$$A = UB. \qquad (2.2)$$

The orthogonalization matrix B is given by (Busing and Levy 1967)

$$\begin{pmatrix} a^* & b^* \cos \gamma^* & c^* \cos \beta^* \\ 0 & b^* \sin \gamma^* & -c^* \sin \beta^* \cos \alpha \\ 0 & 0 & 1/c \end{pmatrix} \qquad (2.3)$$

which, as mentioned earlier, depends only on the crystal cell parameters. If the cell parameters of the crystal are known with reasonable accuracy, then autoindexing requires a determination of the matrix U. When the cell parameters are unknown, both U and B have to be determined.

2.6.1.2 Autoindexing when cell parameters are known

To illustrate the principle underlying autoindexing it is useful to consider the relatively straightforward task of indexing a zero level precession photograph (Fig. 2.1A). This could be done as follows:

(1) Choose two vectors which together define the lattice of spots in the precession photograph. In Fig. 2.1A the obvious choice is a horizontal and a vertical vector.
(2) Measure the lengths of these seed vectors (by measuring the distance between many pairs of spots and taking an average) and the angle between them.
(3) Generate, using the known cell parameters, a list of calculated reciprocal lattice vectors with low indices (001, 010, 100, 110, 101, 011, etc.).

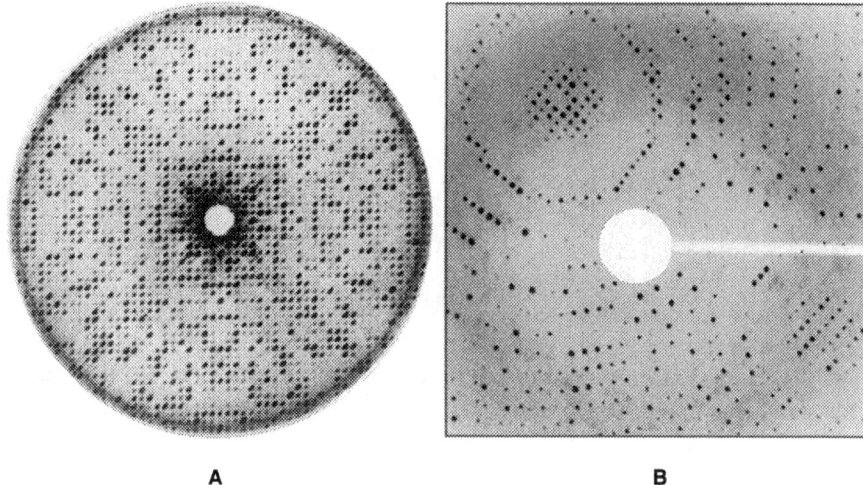

Fig. 2.1. (A) A precession photograph gives an undistorted image of the reciprocal lattice, from which it is easy to identify two vectors that define the lattice. (B) In an oscillation image the reciprocal lattice is distorted, but a lattice is still clearly visible in parts of the image.

(4) Identify the seed vectors chosen in (1) by comparing their lengths and relative orientation with the list of calculated lattice vectors from (3). Usually, it will be possible to identify the two seed vectors uniquely (ignoring solutions related by crystallographic symmetry). The identification of these two vectors then determines the orientation of the crystal.

The case of indexing a precession photograph is made simpler because the photograph represents an undistorted representation of the reciprocal lattice. In the case of a rotation image, it is also possible to identify sections of the reciprocal lattice, particularly when a crystal is almost aligned on a principal zone (Fig. 2.1B). In this case the lattice is distorted, but using the Ewald sphere construction the observed spot positions (X_d^i, Y_d^i, Φ^i) can be mapped back into reciprocal space to give a set of scattering vectors S_i^Φ. In vector notation

$$S_i^\Phi = \begin{pmatrix} D/r - 1 \\ X_d^i/r \\ Y_d^i/r \end{pmatrix} \qquad (2.4)$$

where

$$r = \sqrt{X_d^{i2} + Y_d^{i2} + D^2}, \qquad (2.5)$$

D being the crystal-to-detector distance and X_d, Y_d are the spot coordinates relative to the direct beam position on a flat, untilted detector.

This gives the scattering vector when the reciprocal lattice point lies *exactly* on the Ewald sphere. To place all scattering vectors in the same coordinate frame, we must correct for the fact that different spots will arise at different Φ values, giving

$$S_i = \Phi_i^{-1} S_i^{\Phi}. \qquad (2.6)$$

This creates a problem when a large rotation angle per image has been used (e.g. 0.5–2°) because there is normally no information on the true Φ values for different spots. In practice, all spots are assigned a Φ equal to the midpoint of the rotation range, but this is of course only an approximation and will inevitably give rise to errors in the derived vectors S_i. When fine phi slicing is used (e.g. 0.1–0.25° rotation per image) the spots will usually extend over several images and an accurate Φ centroid can be determined empirically, and consequently the scattering vectors S_i will be far better determined.

A list of difference vectors

$$S_{ij} = S_i - S_j \qquad (2.7)$$

is constructed from the set of scattering vectors S_i. The resulting difference vectors are then sorted on increasing length $|S_{ij}|$. This list will contain 'clusters' of vectors corresponding to short reciprocal lattice vectors (clusters rather than points because of errors in the vectors S_i which will be proportionately greater in the difference vectors S_{ij}). The vectors within each cluster are averaged to improve accuracy. The shortest vectors (which will also be those with the highest multiplicity) should correspond to reciprocal lattice vectors with small Miller indices (e.g. 100, 010, 001, etc.). The orientation of the crystal can be determined by identifying two (or more) of these short difference vectors (the seed vectors) and this can be done in exactly the same way as for the precession photograph providing the cell dimensions are known and the seed vectors are non-collinear.

Once an initial orientation has been determined (using either two or three seed vectors) the crystal orientation and cell parameters can be refined by minimizing

$$\Omega = \sum_i (h_i - [h_i])^2, \qquad (2.8)$$

where

$$h_i = A^{-1} S_i, \qquad (2.9)$$

$[h_i]$ is the closest integer to h_i.

Because of the relatively large errors in the scattering vectors S_i, the initial estimate of the orientation will not be very precise and attempts to index the actual scattering vectors S_i or even the longer difference vectors S_{ij} using eqn (2.9) will fail. To prevent this, only relatively short difference vectors S_{ij} are used in the first stages of refinement, and the number of difference vectors included is gradually increased and finally the scattering vectors themselves are used. Clearly, the inclusion of the longer scattering vectors will improve the accuracy of the refined parameters but, conversely, it is the use of very short difference vectors that is crucial to the success of the initial indexing.

2.6.1.3. *Autoindexing when cell parameters are unknown*

The first stages of the procedure are identical up to the calculation of the list of difference scattering vectors S_{ij}. At this point three short, non-coplanar seed vectors are selected with a minimum angle between them (typically 40°). Since the unit cell is unknown, arbitrary indices can be assigned to the three seed vectors [typically (100), (010) and (001)]. This assignment is used to generate an initial estimate of the setting matrix A, which can be used to index other scattering vectors (initially using difference vectors and then going on to use scattering vectors). The initial orientation and cell parameters can be refined in the same way as described for the case of a known cell [eqns (2.8) and (2.9)].

An additional step is required in this case to produce a final cell, as the cell chosen by the initial assignment of indices to the seed vectors will not necessarily correspond to the optimum choice of unit cell. In particular, it may not reflect the symmetry of the reciprocal lattice (as there is no information available on reflection intensities at this stage). The normal procedure is then to try to fit the 14 distinct Bravais lattices to the deduced lattice, and assign a 'quality index' to each based on how well the Bravais lattice matches. The user is finally presented with a list of alternative solutions from which one is selected. The true symmetry can, of course, only be determined when integrated intensities are available.

2.6.2 *Parameter refinement*

Once an orientation matrix and cell parameters have been derived from the autoindexing, these parameters (and others) are refined further using a different algorithm. The parameters to be refined can conveniently be grouped into three classes:

(a) Crystal parameters: cell parameters, crystal orientation and mosaic spread (isotropic or anisotropic).
(b) Detector parameters: the detector position and orientation and (if appropriate) distortion parameters (e.g. the radial and tangential offsets for the Mar image plate scanner).
(c) Beam parameters: the orientation of the primary beam and beam divergence (isotropic or anisotropic).

The refinement of these parameters is achieved by least-squares minimization of two residuals: a positional residual

$$\Omega_1 = \sum_i w_{ix}\left(X_i^{\text{calc}} - X_i^{\text{obs}}\right)^2 + w_{iy}\left(Y_i^{\text{calc}} - Y_i^{\text{obs}}\right)^2, \qquad (2.10)$$

where X and Y are the spot coordinates on the detector, and w_{ix} and w_{iy} are appropriate weights; and an angular residual

$$\Omega_2 = \sum_i w_i \left[\left(R_i^{\text{calc}} - R_i^{\text{obs}} \right) / d_i^* \right]^2, \tag{2.11}$$

where R_i^{calc}, R_i^{obs} are the calculated and observed distances of the reciprocal lattice point d_i^* from the centre of the Ewald sphere (OP and OP' in Fig. 2.2) and again w_i is a weighting term. R_i^{calc} is determined from the current values for the cell parameters and crystal orientation. R_i^{obs} is obtained from the Φ centroid if fine Φ slices have been used. For coarse Φ slices, the reciprocal lattice point is either assumed to lie exactly on the Ewald sphere at the midpoint of the rotation, or for partially recorded reflections its position is estimated from the degree of partiality of the reflection (i.e. the way in which the total intensity is distributed between the two abutting images). This latter approach, known as post-refinement (Winkler et al. 1979; Rossmann et al. 1979) because it depends on a knowledge of the integrated intensities, requires a model for the rocking curve, and permits refinement of either crystal mosaicity or beam divergence.

The effective radius of the reciprocal lattice point (see Fig. 2.2) is given by

$$\varepsilon = \frac{\gamma d^*}{2} \cos \theta, \tag{2.12}$$

where γ is the combined mosaic spread and beam divergence, d^* is the reciprocal lattice spacing and θ is the Bragg angle. The distance of the reciprocal lattice

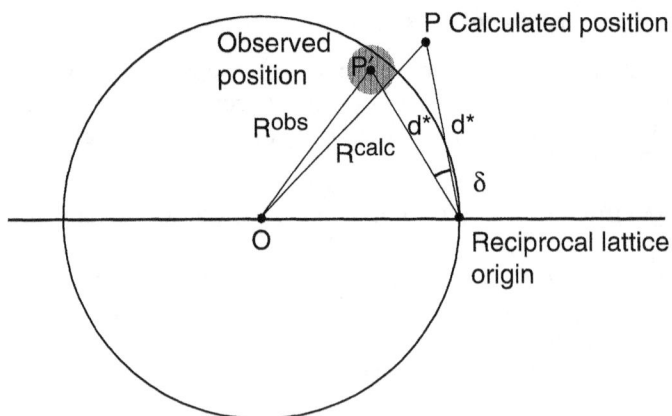

Fig. 2.2. The shaded circle represents the position of a reciprocal lattice point at the end of an oscillation. A fraction of the total intensity corresponding to the volume of the reciprocal lattice point that has already passed through the Ewald sphere will be recorded on this image. The remaining intensity will be recorded on the next image as P' rotates clockwise.

point from the Ewald sphere, Δr, is then given by

$$P = \tfrac{1}{2}[1 + \sin(\pi \Delta r/2\varepsilon)], \tag{2.13}$$

where

$$P = \frac{I_1}{I_1 + I_2} \tag{2.14}$$

and I_1, I_2 are the intensities recorded on the two abutting images. Knowing P from the measured intensities, Δr can be calculated from eqn (2.13), and thus R^{obs} can be determined. Rocking curve models other than the simple sine model in eqn (2.13) have also been used. Because ε depends on the combined mosaic spread and beam divergence, this parameter can also be refined. (For fine Φ slices the mosaic spread or beam divergence is estimated from the observed reflection width in Φ.)

The refinement strategy can depend on how the data has been collected. If fine Φ slices have been used, accurate Φ centroids and coordinates (X, Y) are available for most strong reflections (excluding those very close to the rotation axis) and both residuals (Ω_1, Ω_2) can be minimized simultaneously by using a suitable selection of reflections (strong and evenly distributed over the detector and in Φ). Problems arising due to correlations of different parameters can be avoided either by fixing some parameters or by the use of eigenvalue filtering. These problems can be particularly serious for low resolution data, where there is a strong correlation between crystal-to-detector distance and the cell parameters, or for an offset detector where there is a high correlation between the detector swing angle and the (horizontal) primary beam coordinate. If only a narrow Φ range of reflections is used in the refinement then some unit cell parameters will be defined poorly and may be correlated with the crystal setting angles, and there will also be a strong correlation between the detector orientation around the X-ray beam and the crystal setting angle around the beam. In such circumstances the refined parameters may assume physically unrealistic values, but this will not necessarily impair the accuracy of the prediction of reflection positions and widths.

When the data are collected with coarse Φ slices, only fully recorded reflections will give accurate spot positions (X, Y), and accurate Φ centroids can only be determined for partially recorded reflections. In some programs (e.g. MOSFLM) the two residuals are therefore minimized independently. Only the detector parameters are refined when minimizing the positional residual, and only cell, crystal orientation and optionally beam parameters are refined against the angular residual. This approach does have the advantage that the accuracy of the refined cell parameters does not depend on the accuracy of the crystal-to-detector distance or direct beam position, providing these are known sufficiently well to allow correct indexing of the reflections.

2.6.3 Integration of the images

Once an estimate of the crystal cell parameters and orientation have been obtained, the images can be integrated. Stated in the simplest way, this procedure involves predicting the position in the digitized image of each Bragg reflection, and then estimating its intensity (after subtracting the X-ray background) and an error estimate of the intensity. In practice, this apparently simple task is quite complex.

2.6.3.1 Predicting reflection positions

A knowledge of the crystal cell and orientation will allow the prediction of spot positions on a 'virtual detector'—that is, a detector whose position and orientation are exactly known. These positions must then be mapped onto the digitized image, and this mapping must take into account any spatial distortions introduced by the detector, either using a pre-determined calibration table or by refining the distortion parameters for each image.

2.6.3.2 Defining the peak/background mask

Because it is physically impossible to measure the X-ray background actually under the diffraction spot (which strictly is what is required to obtain the background subtracted intensity) the background is measured in a region around the spot either in two dimensions (X, Y, the detector coordinates) for coarse Φ slices or in three dimensions (X, Y and Φ) for fine Φ slices. A background plane is fitted to these background pixels, and this plane is then used to estimate the background under the spot. To do this, it is necessary to define a pixel mask, which, when centred on the predicted position of the spot, will define which pixels are to be considered as part of the peak and which are to be used to determine the background (see Fig. 2.3). This mask is usually defined by the user after visual inspection of the spot shapes, but some programs (MOSFLM or MADNES) will automatically optimize the peak/background definition. It is clearly important that pixels are not misclassified, as this will lead to systematic

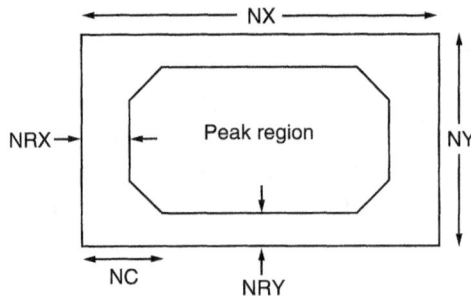

Fig. 2.3. The peak/background mask definition used in MOSFLM. The overall mask size (in pixels) is defined by NX and NY, and the differentiation between peak and background pixels is defined by a background rim in X and Y (NRX, NRY) and a corner cutoff (NC).

errors in the integrated intensity. The presence of strong diffuse scattering, which is quite commonly observed with synchrotron data, can lead to difficulties in differentiating between peak and background pixels. Unfortunately, there is no simple way of dealing with this problem.

2.6.3.3 *Summation integration and profile fitting*

Having determined the background plane, the simplest way to obtain an estimate of the integrated intensity is to sum the pixel values of all pixels in the peak area of the mask, and then subtract the sum of the background values calculated from the background plane for the same pixels. This is known as summation integration and, for spots where the background level is very low compared to the intensity of the spot, this will give as accurate an estimate of the intensity as it is possible to get. (In such cases the accuracy is determined by counting statistics, so for a total count of N photons the standard deviation is $\sqrt{N}$.) For weaker reflections, it is possible to get a more accurate estimate of the integrated intensity by using a procedure known as profile fitting (Diamond 1969; Ford 1974). In this procedure, it is assumed that the shape or profile (in two or three dimensions) of the spots is known. The background plane is determined in the same way as for summation integration, but the intensity is derived by determining the scale factor which, when applied to the *known* spot profile, gives the best fit to the *observed* spot profile. This scale factor is then proportional to the profile fitted intensity for the reflection. In practice, the fitting is done by least squares methods, to minimize the residual

$$R = \sum_{\text{peak pixels}} w_i(X_i - KP_i)^2, \qquad (2.15)$$

where X_i is the background subtracted intensity at pixel i, P_i the value of the standard profile at the corresponding pixel, w_i a weight, derived from the expected variance of X_i, and K the scale factor to be determined.

The improvement gained by profile fitting depends on the spot intensity relative to background and the spot shape, but typically it can provide a reduction in variance by a factor of 2.0 (1.4 in the standard deviation) for weak reflections (see Appendix). This is a significant gain, and all modern software packages employ profile fitting, although the implementation differs in detail.

The procedure assumes a knowledge of the *true* reflection profile. In practice, this is determined from the observed reflection profiles of a number of reflections in the immediate vicinity of the reflection being integrated. An appropriate weighted sum of the individual profiles is used to form the 'true' or standard profile. The reflection shape will vary with position on the detector (due to changes in obliquity of incidence and other factors) and it is important to allow for this. Some programs find a 'standard' profile for several defined areas and then calculate the best profile for each reflection as a weighted mean of the closest 'standard' profiles. Other programs evaluate a separate profile for every reflection using only reflections within a circle of defined radius centred on the reflection being integrated.

Profile fitting is a powerful technique for reducing the random error in weak diffraction data, but equally an error in determining the standard profiles will lead to systematic errors in all measured intensities. Modern software packages go to some lengths to minimize the magnitude of the systematic errors introduced by the use of non-ideal standard profiles.

2.6.3.4 Standard deviation estimates

It is important to obtain reasonable estimates of the standard deviations of the integrated intensities, since these are used as weights when merging multiple observations, and in subsequent steps of the structure determination (e.g. identification of heavy atom derivatives, heavy atom parameter refinement and possibly model refinement). For summation integration, a standard deviation can be obtained based on Poisson statistics, while for profile fitted intensities the goodness of fit of the scaled standard profile to the true reflection profile can be used. These will generally underestimate the true errors, as they take no account of systematic errors arising from effects such as absorption, beam instability, detector non-linearity or errors in non-uniformity corrections. The standard deviation estimates should therefore be modified when the data are merged, making use of the *observed* agreement between multiple observations (see Section 2.6.4.2).

2.6.4 Data reduction

2.6.4.1 Scaling

Corrections are applied to all intensities to allow for the polarization of the incident X-ray beam and geometric factors relating to the speed of the reciprocal lattice point through the Ewald sphere (the Lorentz factor). The intensities from different images then need to be put on a common scale. This is to allow for variations in incident beam intensity (e.g. due to beam decay or re-injection at a synchrotron) and those in effective diffracting volume (or different crystals) and can also provide a partial correction for absorption and radiation damage.

Scaling is commonly performed by applying a scale factor K and temperature factor B to each image (Kabsch 1988; Evans 1997), and these parameters are refined to minimize the residual

$$R = \sum_h \sum_i w_{hi}(I_{hi} - \langle I_h \rangle / K_{hi})^2, \qquad (2.16)$$

where I_{hi} is the ith measurement of reflection h, w_{hi} the weight for that observation (the inverse of the variance), $\langle I_h \rangle$ the weighted mean intensity for reflection h and

$$K_{hi} = K_j \exp(-2B_j \sin^2 \theta_h / \lambda^2), \qquad (2.17)$$

K_j and B_j being the scale and temperature factors for image j on which I_{hi} was measured. Furthermore, θ_h is the Bragg angle for reflection h and λ the radiation wavelength.

The success of this scaling procedure depends on the presence of multiple (symmetry related) observations on different images. For large area detectors where data are collected on both sides of the X-ray beam this condition is usually met, but difficulties in scaling can arise with offset detectors where data have only been collected on one side of the X-ray beam. In such cases it is sometimes necessary to collect data about a second rotation axis (preferably orthogonal to the first rotation axis) in order to provide the observations required for scaling.

2.6.4.2 Merging data

Once all the observations have been placed on a common scale, multiple observations are reduced to a weighted mean intensity and standard deviation. At this stage it should also be possible to detect 'rogue' observations or outliers, providing the multiplicity is at least three-fold, and these observations can be rejected. A common cause of outliers arises when one or more of the observations lie behind the backstop shadow or in a defective area of the detector. Ideally, only those observations for which there is an identifiable physical reason for the error should be rejected, and only in exceptional cases should the proportion of observations rejected exceed a fraction of 1%.

Statistics on the agreement between multiple observations, data completeness, evidence for systematic errors such as partial bias (which arise from errors in modelling the reflection width) are also calculated at this stage.

The level of agreement between multiple observations can be used to modify the standard deviations of the intensities. Providing the multiplicity is high, then the standard deviation ratio

$$\text{SDRATIO} = \left\langle \frac{I_{hi} - \langle I_h \rangle}{\sigma(I_{hi})} \right\rangle, \tag{2.18}$$

where I_{hi}, $\sigma(I_{hi})$ are the intensity and standard deviation of the ith observation of reflection h, and $\langle I_h \rangle$ is the weighted mean intensity, should equal unity when averaged over a significant number of reflections. A mean greater than unity suggests the standard deviations are underestimated. This ratio is evaluated as a function of reflection intensity, and the standard deviations are modified to give

$$\sigma(I_{hi}) = A\sqrt{\sigma^2(I_{hi}) + BI_{hi}^2}, \tag{2.19}$$

where the values of the parameters A and B are chosen to get SDRATIO of unity for all intensity ranges.

2.6.4.3 Reducing intensities to amplitudes

The simplest way to determine the structure factor amplitudes (F) from the observed intensities (I) is simply to take

$$F = \sqrt{I}. \tag{2.20}$$

However, this cannot be applied to observations for which the observed intensity is negative (statistically, a certain percentage of the data will be expected to have negative intensities). To overcome this problem, Bayesian statistics have been applied (French and Wilson 1978), making use of the prior knowledge that the true intensity must be greater than zero, and the distribution of intensities obeys Wilson statistics, to derive the most likely (positive) intensity for those reflections with an observed negative intensity. This approach is implemented in the CCP4 program TRUNCATE.

The TRUNCATE program also produces statistics on the cumulative intensity distribution, known as the $N(z)$ test. The theoretical distribution can be calculated using Wilson statistics, and protein data normally follow the theoretical values to within 1–2%. Deviations greater than this can indicate problems with the integration, or that the crystals are in fact twinned. This is usually the only indication of twinning from the processing statistics, and so the $N(z)$ distribution should always be checked. Procedures are available to 'detwin' data in favourable cases, and molecular replacement and refinement are possible even with perfectly twinned data (Yates 1997).

Appendix: profile fitting equations

Profile fitting assumes a knowledge of the spot profile (shape), and uses this knowledge to reduce the error in the estimated integrated intensity.

As stated in eqn (2.15), the best estimate of the spot intensity is obtained by minimizing

$$R = \sum w_i (X_i - KP_i)^2, \qquad (2.A1)$$

where X_i is the background subtracted counts at pixel i, P_i the value of the standard profile at pixel i, w_i the weight and K the scale factor to be determined. Then the profile fitted intensity I_P is given by

$$I_P = K \sum P_i. \qquad (2.A2)$$

From standard least squares, the value of K that minimizes R is given by

$$K = \frac{\sum w_i X_i P_i}{\sum w_i P_i^2}. \qquad (2.A3)$$

The weights w_i should be the inverse of the variance of X_i.

From Poisson statistics, the variance is equal to the total number of counts,

$$\text{Var}(X_i) = N_{\text{bg}} + KP_i, \qquad (2.A4)$$

where N_{bg} is the background counts at pixel i.

In practice, the summation integration intensity can be used to estimate the scale factor K to allow the weights to be calculated.

Substituting for K from eqn (2.A3) in the equation for the profile fitted intensity (2.A2) gives

$$I_P = K \sum P_i = \frac{\sum w_i X_i P_i}{\sum w_i P_i^2} \sum P_i. \tag{2.A5}$$

Profile fitted intensity of strong reflections

For strong reflections the background level N_{bg} is negligible and so using eqn (2.A4) we get

$$w_i = 1/\mathrm{Var}(X_i) \simeq 1/KP_i. \tag{2.A6}$$

Substituting in eqn (2.A5) gives

$$I_P = \sum X_i. \tag{2.A7}$$

Thus profile fitting and summation integration are equivalent if the background is weak compared to the reflection intensity.

Profile fitted intensity of weak reflections

For very weak reflections, the background is much greater than the Bragg intensity so that using eqn (2.A4),

$$w_i = 1/\mathrm{Var}(X_i) \simeq 1/N_{bg}. \tag{2.A8}$$

Assuming a constant background, on substituting in eqn (2.A5) this gives

$$I_P = \sum X_i P_i \frac{\sum P_i}{\sum P_i^2}. \tag{2.A9}$$

The second term depends only on the standard profile, not on the intensity of the reflection.

In this case, I_P is a weighted sum of X_i, with higher weights for the stronger pixels in the spot.

Improvement afforded by profile fitting weak reflections

For very weak reflections, from eqn (2.A9) the variance in I_P is given by

$$\mathrm{Var}(I_P) = \sum \mathrm{Var}(X_i) \cdot P_i^2 \cdot \left(\frac{\sum P_i}{\sum P_i^2}\right)^2. \tag{2.A10}$$

Assuming a flat background and a very weak intensity,

$$\text{Var}(X_i) = N_{\text{bg}}; \qquad (2.\text{A}11)$$

so

$$\text{Var}(I_P) = N_{\text{bg}} \cdot \frac{\left(\sum P_i\right)^2}{\sum P_i^2}. \qquad (2.\text{A}12)$$

For summation integration the variance is given by

$$\text{Var}(I_S) = \sum \text{Var}(X_i) = m N_{\text{bg}}, \qquad (2.\text{A}13)$$

where m is the number of pixels in the spot.

Thus the ratio of the variances in the summation integration and profile fitted estimates of the intensity is

$$\frac{\text{Var}(I_S)}{\text{Var}(I_P)} = \frac{m \sum P_i^2}{\left(\sum P_i\right)^2}. \qquad (2.\text{A}14)$$

For a typical spot profile, this has a value of about 2, illustrating that profile fitting can reduce the variance by a factor of two for very weak reflections on a high background.

References

Amemiya, Y (1997). *Methods in Enzymology*, **A276**, Carter Jr, CW and Sweet, RM, eds, 233–243.

Arndt, UW, Duncumb, P, Long, JVP, Pina, L and Inneman, A (1998). *Journal of Applied Crystallography*, **31**, 733–41.

Busing, WR and Levy, HA (1967). *Acta Crystallographica*, **22**, 457–464.

Carter Jr, CW and Sweet, RM (eds) (1997). *Methods in Enzymology*, **A276**.

Dauter, Z (1997). *Methods in Enzymology*, **A276**, Carter Jr, CW and Sweet, RM, eds, 326–344.

Diamond, R (1969). *Acta Crystallographica*, **A25**, 43–54.

Engström, P and Riekel, C (1996). *Journal of Synchrotron Radiation*, **3**, 97–100.

Evans, PR (1997). *Scaling of MAD Data, Proceedings of CCP4 Study Weekend*, Wilson, KS, Davies, G, Ashton, AW, and Bailey, S, eds, Daresbury Laboratory, U.K.

Ford, GC (1974). *Journal of Applied Crystallography*, **7**, 555–564.

French, S and Wilson, K (1978). *Acta Crystallographica*, **A34**, 517–525.

Harp, JM, Timm, DE and Bunick, GJ (1997). *American Crystallographic Association Annual Meeting*, Meeting Abstract P205.

Harrison, SC (1968). *Journal of Applied Crystallography*, **1**, 84–90.

Higashi, T (1990). *Journal of Applied Crystallography*, **23**, 253–257.

Kabsch, W (1988). *Journal of Applied Crystallography*, **21**, 916–924.

Kabsch, W (1993). *Journal of Applied Crystallography*, **26**, 795–800.

Kahn, R and Fourme, R (1997). *Methods in Enzymology*, **A276**, Carter Jr, CW and Sweet, RM, eds, 268–286.

Luger, K, Mader, AW, Richmond, RK, Sargent, DF and Richmond, TJ (1997). *Nature*, **389**, 251–260.
Otwinowski, Z and Minor, W (1997). *Methods in Enzymology*, **A276**, Carter Jr, CW and Sweet, RM, eds, 307–326.
Pflugrath, W (1997). *Methods in Enzymology*, **A276**, Carter Jr, CW and Sweet, RM, eds, 286–306.
Richmond, TJ, Finch, JT, Rushton, B, Rhodes, D and Klug, A (1984). *Nature*, **311**, 532–537.
Rogers, DW (1997). *Methods in Enzymology*, **A276**, Carter Jr, CW and Sweet, RM, eds, 183–203.
Rossmann, MG, Leslie, AGW, Abdel-Meguid, SS and Tsukihara, T (1979). *Journal of Applied Crystallography*, **12**, 570–581.
Sawyer, L, Isaacs, N and Bailey, S (eds) (1993). *Proceedings of the CCP4 Study Weekend*, Daresbury Laboratory, U.K.
Westbrook, EM and Naday, I (1997). *Methods in Enzymology*, **A276**, Carter Jr, CW and Sweet, RM, eds, 244–268.
Winkler, FK, Schutt, CE and Harrison, SC (1979). *Acta Crystallographica*, **A35**, 901–911.
Yates, TO (1997). *Methods in Enzymology*, **A276**, Carter Jr, CW and Sweet, RM, eds, 344–358.

3
Experimental determination of structure factor phases in biocrystallography

*Roger Fourme, William Shepard, Marc Schiltz,
Michel Ramin and Richard Kahn*

3.1 Introduction

X-ray crystallography plays a central role in the elucidation of three-dimensional structures of biological macromolecules as it can reach, in favourable conditions, atomic or near-atomic resolution. The sample is a single crystal that diffracts X-ray radiation. In optical microscopy, the recombination of scattered rays (which is expressed mathematically as a Fourier synthesis) is performed by the lens system of the microscope; in the case of crystallography, the synthesis is performed by computation. Unfortunately, measurements allow us to derive quantities proportional to the amplitudes of diffracted rays, but the phases are lost. This difficulty, known as the phase problem, is familiar to all crystallographers. In the case of macromolecular crystals, special procedures have been designed to get estimates of lacking phases. The present chapter is devoted to this question.

3.1.1 *The phase problem*

Consider a sample of volume V with a structure described by the electron density $\rho(r)$ expressed in the number of electrons per Å^3 (r is the position vector in direct space, with O as the origin). The sample is illuminated by (ideally) a plane wave of X-rays with wave vector k_0. The elastic scattering of X-rays by the electron density is observed at a distance from the sample that is very large with respect to the X-ray wavelength, so that the total wave scattered in a given direction can be considered as plane and thus expressed mathematically as a complex number. Instead of k, the wave vector of scattered radiation, the direction of observation is rather specified by, $S = (k - k_0)/2\pi$, the so-called scattering vector in reciprocal space with I as the origin (Fig. 3.1). For a certain value of S, let us define as $\zeta(S)$ and $\zeta_0(S)$ waves scattered, respectively, by the sample and by a hypothetical free electron placed at O, and the structure factor $F(S)$ as the ratio $\zeta(S)/\zeta_0(S)$. It can be shown that

$$F(S) = \int_V \rho(r)\, dv \exp(i2\pi S \cdot r), \qquad (3.1)$$

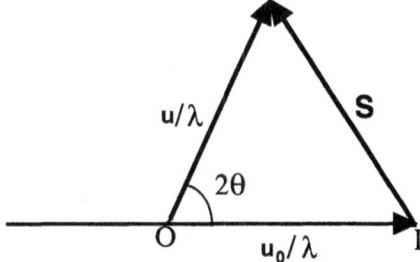

Fig. 3.1. Basic geometry of elastic scattering. O: origin of real space (crystal); I: origin of reciprocal space (scattering); θ: Bragg angle; u_0 and u: unit vectors along the X-ray beam and the direction of scattering respectively; S: scattering vector, with $S = (k - k_0)/2\pi = (u - u_0)/\lambda$. The resolution is defined either by S, the modulus of, S, with $S = 2\sin\theta/\lambda$ (Å^{-1}) or by the Bragg spacing $d = 1/S$ (Å).

so that $F(S)$ is the inverse Fourier transform of $\rho(r)$. Conversely, $\rho(r)$ is the Fourier transform of $F(S)$.

We now take into account the fact that $\rho(r)$ can be decomposed into non-overlapping atoms. The integral eqn (3.1) restricted to a single atom j is called the scattering factor f_j of this atom. As the atomic electron cloud has spherical symmetry, the scattering factor is a real number dependent on the modulus S of S. From eqn (3.1), $f_j(0)$ is obviously equal to the atomic number Z_j. The atomic electron density has an extent that is not negligible with respect to the X-ray wavelength; accordingly, scattered waves interfere more and more destructively and $f_j(S)$ decreases as S increases. For each atomic species, the scattering factor can be calculated from first principles and tabulated. The structure factor value at S is then a discrete sum of complex numbers over all atoms contained in the sample. In the simplified assumption of atoms at rest,

$$F(S) = \sum_j f_j \exp(i2\pi S \cdot r_j), \qquad (3.2)$$

where r_j is the position vector of the jth atom.

In the case of a crystal, it turns out that the amplitude of scattered waves may be different from zero only for values of S that are reciprocal lattice vectors (denoted h). Further, the structure factor value relates to scattering by the building block of the crystal, i.e. the content of the unit cell,

$$F(h) = \sum_j f_j \exp(i2\pi h \cdot r_j). \qquad (3.3)$$

$F(h)$ is a complex number obtained by summing the n atomic contributions in the unit cell with moduli f_j and phases $2\pi h \cdot r_j$. The representation of $F(h)$ in the complex plane is shown in Fig. 3.2, where each small vector contributes to the resultant $F(h)$.

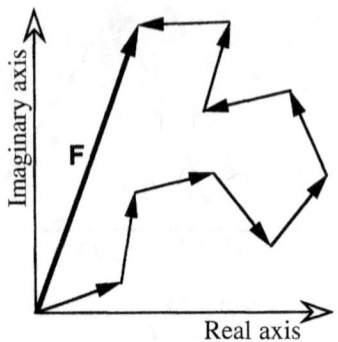

Fig. 3.2. Representation, in the complex plane, of a structure factor $F(\mathbf{h})$ as the sum of contributions from atoms with scattering factor f_j located at $\mathbf{r}_j$ in the crystal unit cell.

Finally, the expression of the electron density at $\mathbf{r}$ is given by

$$\rho(\mathbf{r}) = \frac{1}{V} \sum_{\mathbf{h}} F(\mathbf{h}) \exp(-i2\pi \mathbf{h} \cdot \mathbf{r}). \tag{3.4}$$

The structure factor values are complex numbers:

$$F(\mathbf{h}) = |F(\mathbf{h})| \exp[i\phi(\mathbf{h})]. \tag{3.5}$$

The intensity of each diffracted beam $\mathbf{h}$, a quantity that is proportional to the square of the modulus of the associated structure factor value, can be measured with a detector:

$$I(\mathbf{h}) \propto |F(\mathbf{h})|^2 = F(\mathbf{h})F^*(\mathbf{h}), \tag{3.6}$$

where $F^*(\mathbf{h})$ is the complex conjugate of $F(\mathbf{h})$. The detector is not sensitive to phase, and this loss of information precludes a direct calculation of the electron density. This is the origin of the so-called 'phase problem', which remains, once appropriate crystals have been grown, the central problem in macromolecular crystallography.

It is interesting to point out the fundamental difference between X-ray diffraction and optical microscopy. In a microscope, all rays scattered in the same direction interfere at a point in the image focal surface of the lens system. This surface contains points which are the inverse Fourier transform (analysis) of the sample electron density and act as secondary sources. Rays emitted by these sources interfere in the image plane of the lens system to produce an image (synthesis). Here, the lens system acts as an analog computer. In an X-ray diffraction experiment, since large aperture X-ray lenses are not available, the object plane is at infinity and the image recombination does not occur (Fig. 3.3).

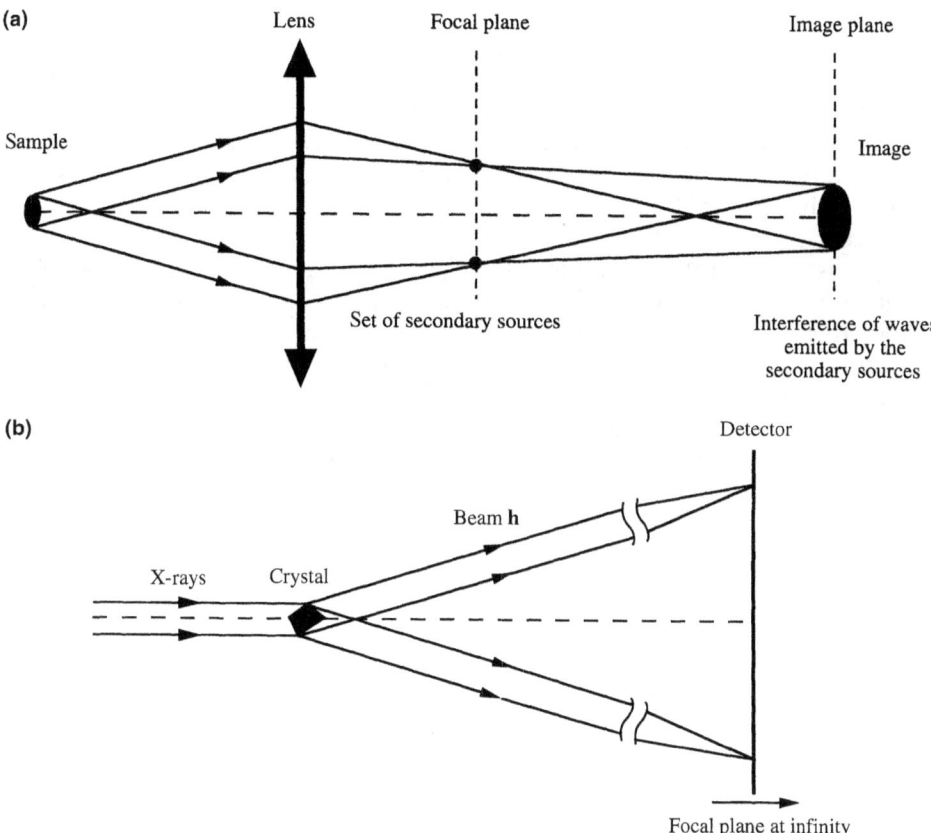

Fig. 3.3. Comparison of (a) optical microscopy and (b) single crystal X-ray diffraction.

Prior knowledge of a structural model for at least a fragment of the macromolecule can be used to initiate the phase determination. As the number of three-dimensional structures in the databases is increasing at a fast pace, this 'molecular replacement' method becomes more frequently used. When there is no solid prior information on the model, initial phases are estimated by *de novo* methods based on measurements of diffracted intensities and intensity differences. These phases are generally of a limited accuracy, and they need improvement through various procedures, such as density modification, histogram matching or non-crystallographic symmetry (if any).

The first, and still very important, *de novo* phasing method is multiple isomorphous replacement (MIR), which exploits the diffraction from heavy (i.e. high-Z) atoms. Special diffraction characteristics can also be obtained by a physical effect, namely that the scattering of X-rays is modified by resonance of the incoming radiation with natural angular frequencies of bound electrons in atoms. This 'anomalous' dispersion (dispersion means frequency dependence, similar to the dispersion of light of different frequencies by a glass prism) of scattering is always present to some extent at wavelengths of interest for diffraction experiments, but it is possible to design experiments in which the contribution of specific atoms can be isolated from many other atoms which scatter in a quasi-'normal' way. This contribution can be used to derive independent phase information. As the full exploitation of anomalous dispersion for the phase problem requires wavelength tunability, the development of these methods has been closely associated with the progress of X-ray synchrotron radiation sources and related instrumentation.

The distribution of amplitudes of diffracted beams contains information of a statistical nature on phases. 'Direct' phasing methods have been developed to extract this information [for a review see e.g. Woolfson (1987)]. They are of crucial importance for the solution of reference structures in *de novo* methods, especially for MAD (an acronym for multiple wavelength anomalous diffraction) applications, as discussed later. The direct solution of macromolecular structures is restricted to cases where very high resolution data ($d < 1.3$ Å) are available. Finally, the use of direct methods to improve *de novo* phases has been the subject of many articles, but applications have been marginal up to now. In a broader context, we mention the 'Bayesian approach' (Bricogne 1993, 1997), which is likely soon to play a crucial role in this improvement process, as a general tool encompassing all sources of phase information.

3.1.2 *Principle of* de novo *phasing methods*

Suppose that we want to determine the phase ϕ of a plane wave expressed conveniently with the complex notation

$$\xi = \xi_0 \exp i(\omega t - \boldsymbol{k} \cdot \boldsymbol{r} + \phi). \tag{3.7}$$

The first step might be the determination of the amplitude ξ_0 by measurement (#1) of wave intensity I_1.

Assume now that we have another plane wave ζ_1 with the same angular frequency and wave vector, but with known amplitude ζ_{01} and known phase α, so that it can be considered as a reference wave,

$$\zeta_1 = \zeta_{01} \exp i(\omega t - \boldsymbol{k} \cdot \boldsymbol{r} + \alpha). \tag{3.8}$$

A measurement (#2) of intensity I_2 of interference between the two waves restricts ϕ to two possible values. In effect,

$$I_2 = \xi_0^2 + \zeta_{01}^2 + 2\xi_0\zeta_{01}\cos(\phi - \alpha); \tag{3.9}$$

then

$$\cos(\phi - \alpha) = (I_2 - \xi_0^2 - \zeta_{01}^2)/2\xi_0\zeta_{01}. \tag{3.10}$$

There are two values of $(\phi - \alpha)$, and accordingly of ϕ, that satisfy eqn (3.10). The ambiguity in ϕ can be relieved by measurement (#3) of intensity I_3 of

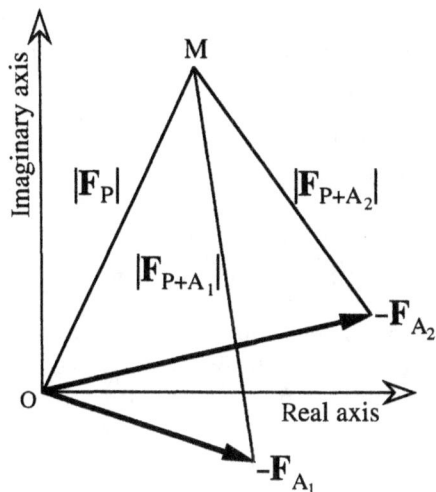

Fig. 3.4. Principle of the determination of the phase of $F_P(h)$ by *de novo* phasing methods in the ideal case (perfect isomorphism and error-free measurements). P is the native structure; A_1 and A_2 are partial structures; $P + A_1$ and $P + A_2$ are derivative structures. The construction is based on $F_P = F_{P+A_1} + (-F_{A_1}) = F_{P+A_2} + (-F_{A_2})$. $(-F_{A_1})$ and $(-F_{A_2})$ are reference vectors calculated from the known partial structures. Measurements provide amplitudes of F_P, F_{P+A_1} and F_{P+A_2}. M is determined without ambiguity, and so is the phase of F_P.

interference of ξ with a second reference wave $\zeta_2 = \zeta_{02} \exp i(\omega t - \boldsymbol{k} \cdot \boldsymbol{r} + \beta)$. Two new values are obtained for ϕ. The correct value is common to both sets of results.

In general, at least three intensity measurements are necessary to determine a phase by this method. The transposition of this principle to the crystallographic case is the following. Each diffracted beam $\boldsymbol{h}$ is a plane wave, with intensity proportional to the square of the structure factor amplitude $|F(\boldsymbol{h})|$. From eqn (3), this plane wave is produced by interference of n plane waves scattered by the n atoms in the unit cell. The ξ, ζ_{01} and ζ_{02} waves may be the contribution to the total structure factor $F(\boldsymbol{h})$ of three subsets of atoms, respectively the main structure P and substructures (or partial structures) A_1 and A_2. The structure of P is unknown. The structures of A_1 and A_2 are assumed to be known, so that their *calculated* contributions to the total structure factor can be used as references (Fig. 3.4).

It is important to note that ξ—and accordingly the crystal structure of P— should ideally remain invariant during the various intensity measurements.

3.2 Methods based on chemical variation

3.2.1 *Phasing based on partial structures of ordered heavy atoms*

This principle is exploited in the MIR method, which was historically the first to be used (Green *et al.* 1954) and is still the major *de novo* phasing method in macromolecular crystallography. Here, each partial structure A_i consists of a few ordered heavy atoms. The intensity measurements required for the determination of the phase of $F_P(\mathbf{h})$ are performed on the native crystal P and at least two heavy atom derivatives $P + A_1$ and $P + A_2$.

The expected relative variation in structure factor amplitudes for the native and a particular derivative labelled with a single atomic species, which quantifies the strength of the isomorphous signal, is

$$\frac{\langle \Delta F_{iso} \rangle}{\langle |F_P| \rangle} \approx \frac{1}{\sqrt{2}} \frac{\sqrt{N_A}\, f_A}{\sqrt{N_P}\, f_P}, \qquad (3.11)$$

where $\langle x \rangle$ denotes the root mean square (rms) value of x, $\Delta F_{iso} = \||F_{P+A}| - |F_P|\|$, f_A is the scattering factor of the heavy atom, f_P is the rms value of scattering factors in the native structure (≈ 6.7 electrons for a protein at low resolution), N_P is the number of atoms in the asymmetric unit of the native structure, N_A is the effective (i.e. taking into account the occupancy factor) number of atoms in the asymmetric unit of the partial structure. The factor $1/\sqrt{2}$ is due to the fact that experimental quantities are magnitude differences between vectors. Since f_A and f_P are resolution-dependent, this ratio, called the isomorphous ratio for the native–derivative pair, is calculated in shells of resolution.

A major advantage of the MIR method is the large magnitude of the isomorphous signal. As an example, the isomorphous ratio for a native 25 kDa protein and a derivative with a full occupancy mercury site per molecule is about 0.20. Indeed, with appropriate heavy atom labelling, this method is applicable to the determination of very large macromolecules and assemblies. A limitation of MIR is that, in most circumstances, measurements are performed on different crystals, which is the source of substantial intercrystal absorption and scaling errors that limit the accuracy. Even more serious is the fact that the chemical fixation of heavy atoms nearly always modifies, to some extent, the native structure, so that the P structure is not exactly the same in the native and derivative samples. This non-isomorphism between the native and derivative structures also degrades the accuracy of phases, and this effect is more pronounced as resolution increases.

The expertise on the fabrication of heavy atom derivatives is obviously very important, and databases are available.[1] Macromolecules can be purposely engineered, for instance by introducing cysteines as potential sites for the binding of mercury in proteins or bromine in nucleic acids. Also, we point out the use of selenium ($Z = 34$). The use of selenium incorporated in proteins was developed primarily, as described later, for the application of the MAD method (Hendrickson 1985; Hendrickson et al. 1990). In fact, selenium derivatives are also useful for producing isomorphous effects and as pointers that help to trace the polypeptide chain in the initial electron density. Noble gases xenon ($Z = 54$) and, to a lesser extent, krypton ($Z = 36$) are also commonly used since a few years. The idea of using xenon for *de novo* phasing was introduced by Schoenborn et al. (1965). The method, which consists of equilibrating a native protein crystal under a compressed xenon-gas atmosphere, had not found

---

[1] For instance: http://www.bmm.icnet.uk/had/heavyatom.html

widespread application until advances associated with the overcoming of practical and technical difficulties involved with the pressurization procedure were proposed (Schiltz et al. 1994). Under gas pressures between 2 and 15×10^5 Pa, xenon is able to bind at discrete sites in hydrophobic cavities, ligand and substrate binding pockets, and into the pore of channel-like structures (Schiltz et al. 1994; Prangé et al. 1998). As the interaction of xenon in proteins is the result of weak van der Waals forces, xenon binding is a reversible process and, in most cases, derivatives exhibit a high degree of isomorphism with the native crystal. Further, if crystals withstand two successive irradiations without substantial degradation, the same crystal can be used to collect both native and derivative data. In such a case, the ideal scheme described in Section 3.1.2 is essentially realized and high quality phasing is achieved. Xenon is now routinely used for determining protein phases by means of isomorphous replacement, often completed by anomalous scattering, as discussed later. Krypton also binds to crystallized proteins in a way similar to xenon, although noticeable substitution is achieved at substantially higher pressure (above 50×10^5 Pa) (Schiltz et al. 1997b).

The task of structure solution proceeds in two steps: the solution of partial structures A_i, which is a prerequisite, then the solution of the unknown structure P. Partial structures are solved by Patterson or direct methods based on estimates of $|F_A|$ magnitudes. The problem is complicated by the fact that, in contrast to small molecule crystallography, these magnitudes are not measured directly, but derived from intensity variations of the h beam diffracted by the native and derivatives. In the MIR method, only crude estimates of $|F_A|$ are obtained [see e.g. Drenth (1994) for a detailed analysis of the solution of heavy atom structures], which reduces contrast in Patterson maps and complicates further unravelling of these maps.

3.2.2 Phasing based on variations of solvent electron density

Macromolecular crystals include 30–80% of the solvent. In a simple model [Fig. 3.5(a)], the crystal can be described by two phases: one ordered phase

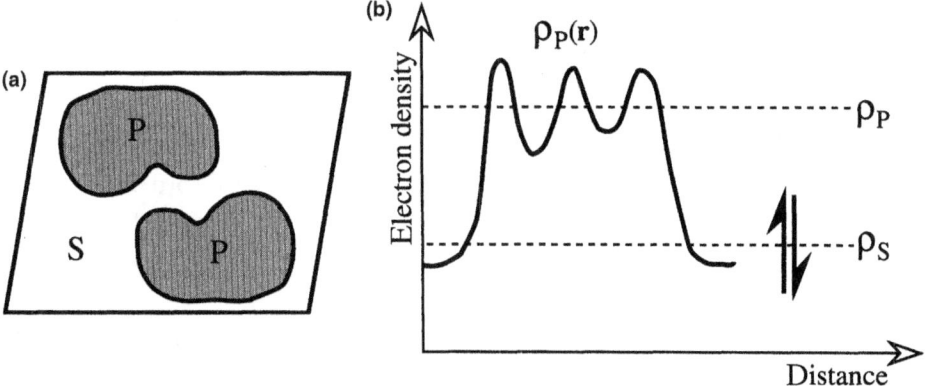

Fig. 3.5. (a) Binary model of the ordered (P) and solvent phases (S) in a macromolecular crystal. (b) Principle of contrast variation. The contrast variation is produced by modifications of the solvent electron density, ρ_S, either by a chemical change or by a physical change.

(macromolecule and ordered water molecules) with density $\rho_P(r)$; and a disordered phase (solvent) with a uniform electron density ρ_S and a large Debye–Waller-like factor B reflecting the lack of periodicity of solvent atoms.

Let us introduce the indicator χ_U (Bricogne 1974) of the ordered part of the structure (i.e. the function is equal to 1 in the ordered domain U of the unit cell and 0 elsewhere) and the complementary indicator χ_{V-U} of the solvent region $[\chi_U(r)+\chi_{V-U}(r)=1]$. The inverse Fourier transform at h of χ_U, the structure factor of the molecular envelope, is denoted $G(h)$. It can be shown (Bricogne, unpublished; Carter et al. 1990) that the structure factor value at $h \neq 0$ of the total structure (including solvent) is

$$F(h) = \Delta(h) + [\bar{\rho}_P - \rho_S \exp(-BS^2/4)]G(h). \qquad (3.12)$$

In this expression, the first term $\Delta(h)$ is the inverse Fourier transform (at h) of the electron density fluctuation $[\rho_P(r) - \bar{\rho}_P]$ in the ordered phase. It is just the structure factor of ordered atoms calculated as usual, i.e. with atoms in a vacuum. The second term, proportional to $G(h)$, describes the solvent effects. Simulations and experiments show that the average amplitude of the envelope terms $\{G(h)\}$ is quite large at very low resolution and decreases rapidly as resolution increases. This explains why standard crystallographic calculations are performed assuming implicitly that atoms are in a vacuum and why low resolution reflections, when available, are generally discarded in these calculations.

In principle, it is possible to use the low resolution part of the diffraction pattern to get some phase information. The solvent phase, reflected by the second term in eqn (3.12), acts as a partial structure [Fig. 3.5(b)]. By changing the average solvent density ρ_S by chemical modification, the amplitude, but not the phase, of the solvent contribution can be changed. The situation is equivalent to a single heavy atom derivative in which the occupancy, but not the position, of the heavy atom can be modified. Acentric reflections have two possible values, and additional information is necessary to lift this ambiguity. Further, the determination of the $G(h)$ coefficients is a major difficulty. It is much more difficult to determine a molecular envelope than the position of a few ordered atoms, and, to our knowledge, a generally applicable method is not yet available.

3.2.3 *From chemical to physical contrast variation*

Approaches using partial structures of ordered atoms or disordered solvent atoms can both be described in terms of contrast variation between a part of the total structure and the invariant remainder of this structure. In conventional phasing methods that have just been presented, the contrast variation is obtained by chemical modifications.

Nature offers us a way called anomalous dispersion (i.e. the wavelength dependence of atomic scattering) to perform contrast variation in a more subtle way, by a purely physical process that affects scattering properties of selected atoms without any change in the crystal structure.

The use of anomalous dispersion (at a single or at multiple wavelengths) for phasing structure factors is the central topic in the rest of this chapter.

3.3 Introduction to anomalous dispersion

3.3.1 Normal and anomalous scattering

The atomic scattering factor as discussed previously is the inverse Fourier transform of the atomic electron density. This number, denoted ^0f, is real, as the various electrons of the atomic cloud are assumed to behave as free electrons ('normal' scattering). In reality, electrons are in bound orbitals and vibrations induced by the incident X-rays resonate with the natural frequencies of bound electrons. The resulting scattering includes a phase shift with respect to that from free electrons, and the total scattering factor is thus a complex number which is dependent upon the wavelength λ of incident X-rays (or the energy E of X-ray photons). Separating the normal and the resonant scattering, then,

$$^\lambda f = {}^0f + {}^\lambda \delta, \tag{3.13}$$

where $^\lambda\delta$ is a complex number with real and imaginary components denoted by $^\lambda f'$ and $^\lambda f''$, respectively. Accordingly,

$$^\lambda f = {}^0f + {}^\lambda f' + i {}^\lambda f'' = {}^0f + {}^0f \left(\frac{^\lambda f'}{^0f} + i \frac{^\lambda f''}{^0f} \right). \tag{3.14}$$

The imaginary component $^\lambda f''$ of the scattering factor corresponds to a wave delayed by $\pi/2$ with respect to the normal wave, and thus to absorption of X-rays. It is expressed as

$$f'' = \frac{1}{2r_0 N} M \frac{\mu}{\rho} \frac{1}{\lambda}, \tag{3.15}$$

where N is the Avogadro constant, r_0 is the classical electron radius (2.8873 × 10^{-15} m), μ/ρ and M are the mass attenuation coefficient and the atomic weight of the element, respectively and λ the X-ray wavelength. Thus, f'' values can be derived from X-ray absorption or fluorescence measurements. f' values are derived by numerical integration from f'' values, as f' and f'' are interrelated by the Kramers–Kronig dispersion relation (James 1958),

$$f'(E) = \frac{2}{\pi} \int_0^\infty \frac{E' f''(E') \, dE'}{E^2 - E'^2}, \tag{3.15'}$$

where E is the energy of the incoming photons.

3.3.2 Some properties of f' and f''

As seen previously, anomalous scattering (instead of 'normal' scattering) is the rule, but this effect is, as the resonance phenomenon suggests, most pronounced in the immediate vicinity of an absorption edge. The normal scattering factor ^0f decreases with increasing scattering angle or resolution. In contrast, f' and f'' are generally assumed to be independent of resolution because the anomalous dispersion is due to interactions of photons with the innermost electronic shells which, for atoms used as anomalous scatterers, have radii much smaller than the X-ray wavelength. Consequently, the strength of anomalous scattering relative to normal scattering *increases* with resolution. The wavelength dependence of f' and f'' calculated from first principles for isolated atoms of a few elements is shown in Fig. 3.6. These examples illustrate several important points:

(i) L_{III} edges, which are associated with the six $^2p_{3/2}$ electrons, have anomalous scattering factor magnitudes, about three times greater than those for K edges, which are associated with the two 1s electrons. The M edges, especially M_{IV} and M_V, are associated with even larger values.
(ii) The experimental atomic absorption spectra (and hence the anomalous scattering profiles) are generally different from the spectra calculated for isolated atoms. The spectra in the vicinity of the edge are generally more strongly modulated and the edge position may be shifted by several eV. Features very close to the edge (about ± 10 eV) are due to the electronic structure and its perturbation by atoms bound to the probe. Features 10 eV from the edge, on the high energy side, are due to scattering of photoelectrons by neighbouring atoms. Between about 10 and 50–100 eV (X-ray absorption near edge structure, XANES), multiple scattering of low energy electrons must be taken into account. From 50–100 eV to 800–1000 eV (Extended X-ray absorptions fine structure, EXAFS region), weaker modulations are due essentially to single scattering of more energetic photoelectrons.

Let us give selected examples at some K, L and M edges that may be of particular interest in view of the phase problem. The f'' profile (Fig. 3.7) at the K-absorption edge of selenium in ammonium selenate shows a sharp resonance ('white line'). A stronger white line is also found at the L_{III} edge of elements from $Z = 55$ to 78, including lanthanides (Fig. 3.8); it is associated with a transition from the 2p core level to a spatially very compact 5d final state. Giant resonances have been observed at M edges of uranium (Liu *et al.* 1995), with $f' = -70\,e^-$ and $f'' = 80\,e^-$ at the M_{IV} edge; $f' = -90\,e^-$ and $f'' = 105\,e^-$ at the M_V edge.
(iii) With polarized X-rays such as those produced by synchrotron radiation sources, absorption spectra in the energy range close to the absorption edge depend on the orientation of the local environment of the anomalous scatterer with respect to the polarization direction of the X-ray beam. This effect, which is averaged out in the case of a powder sample, can be observed with a single crystal and may affect substantially both the edge

position and the magnitude of the absorption. There is a related anisotropy in the anomalous scattering. L.K. Templeton and D.H. Templeton (1982) have developed a formalism in which f' and f'' are described as second rank tensors instead of the usual scalar quantities.

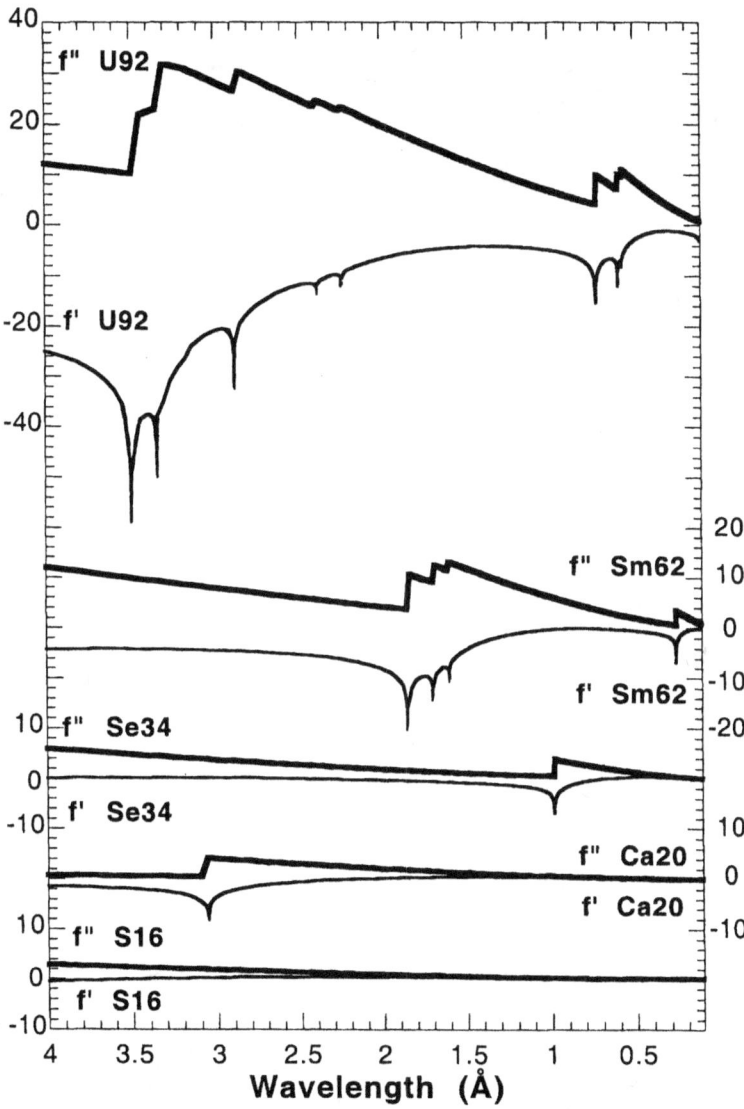

Fig. 3.6. Calculated spectra for certain isolated atoms. For each element, the imaginary component f'' is drawn in the upper curve and the real component (f') is in the lower curve. Origins for the five elements are displaced vertically as indicated. [Figure reproduced from Hendrickson (1991).]

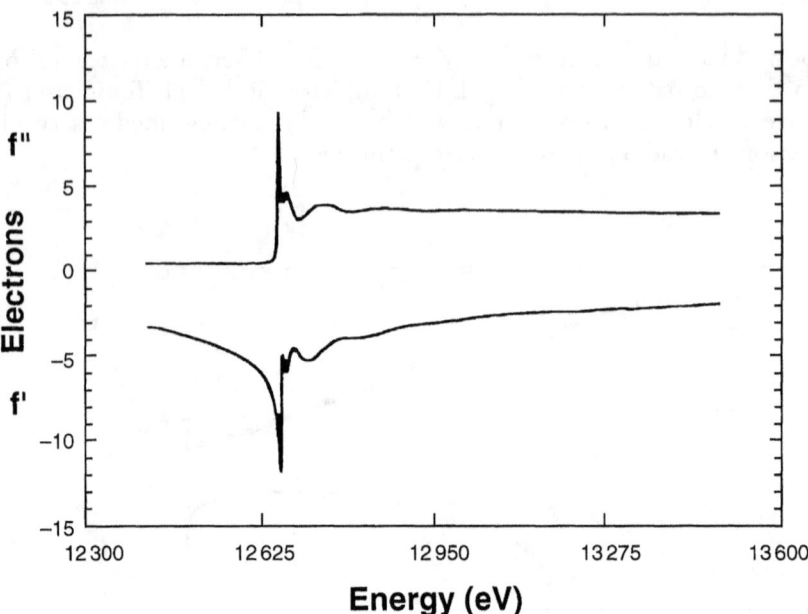

Fig. 3.7. Anomalous scattering factor spectra of selenium. Experimental values derived from an absorption spectrum of 0.2 M ammonium selenate recorded with a high energy resolution ($\Delta E/E = 3 \times 10^{-5}$). The narrow profile of the resonance associated with the K transition is very sensitive to the energy resolution of the monochromator. [Data analysis performed with MADSYS (Hendrickson 1991) on the basis of results communicated by F. Villain and M. Ramin (LURE).]

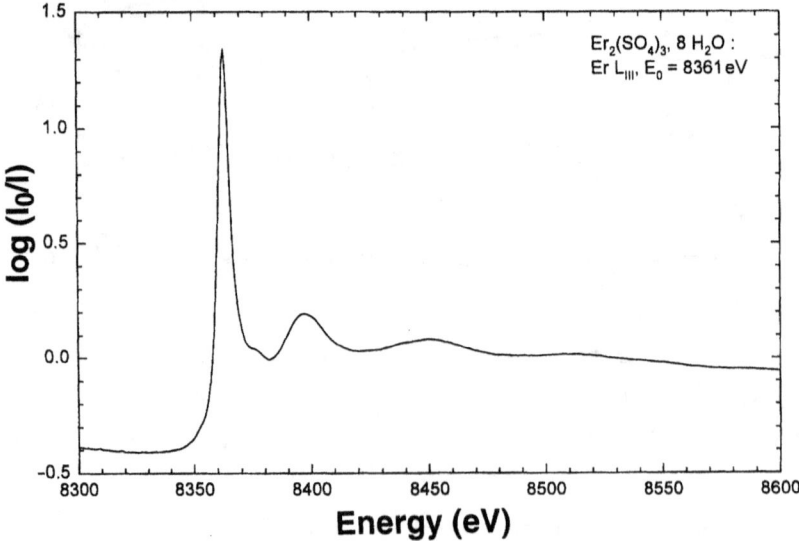

Fig. 3.8. Experimental transmission curve of a powder of erbium sulphate near the L_{III} absorption edge of erbium. $\Delta E/E = 2 \times 10^{-4}$. [Results communicated by C. Giorgetti, E. Dartyge and F. Baudelet (LURE).]

3.3.3 Accessible spectral range

The accessible wavelength range for experiments using anomalous diffraction is ≈ 0.3–3.5 Å. The lower limit (Schiltz et al. 1997a) can be reached with an appropriate X-ray source (wiggler or undulator on a high energy storage ring) and instrumentation (optics and detector). The upper limit is set by absorption of X-rays by the sample and its environment. Experiments have been performed with softer X-rays at the phosphorus and sulphur K edges, but this requires special procedures and instrumentation including vacuum chambers, very thin crystals and thin window detectors (Stuhrmann et al. 1991; Lehmann et al. 1993). The 0.3–3.5 Å window includes K edges from $Z=19$ (K) to $Z=58$ (Ce); L_{III} edges from $Z=48$ (Cd) to $Z=92$ (U); M_{IV} edges of $Z=91$ (Pa); M_{IV} and M_V edges of $Z=92$ (U).

3.3.4 Effects of resonant scattering on the diffraction pattern

In the case of normal scattering, the weighted reciprocal lattice (one with nodes weighted by structure factor amplitudes) has an inversion centre at the origin I. Structure factors $F(h)$ and $F(-h)$ have equal moduli and opposite phases and the associated vectors in the complex plane are symmetrical with respect to the real

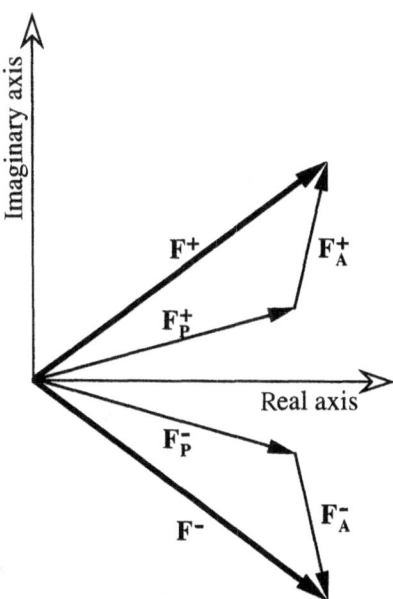

Fig. 3.9. Case of a derivative structure with a parent structure P and a partial structure A of normal scatterers. F_P, F_A and F are the structure factors at (h) of the parent, partial and total structures, respectively. For each of these structure factors, the two mates of the Friedel pair $(h, -h)$ are denoted $^+$ and $^-$ respectively. Representation in the complex plane of F^+ and F^-. In this case F^+ and F^- are symmetrical with respect to the real axis of the complex plane, so that F^+ and the complex conjugate of F^- are superimposed.

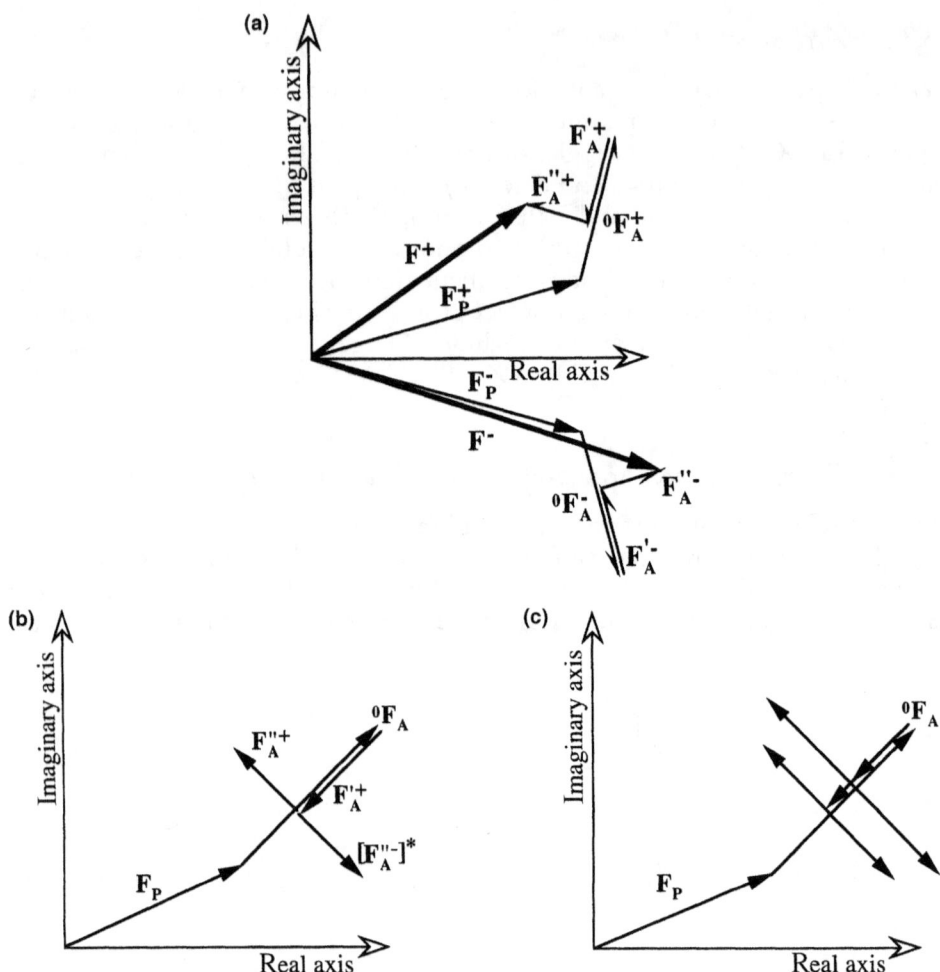

Fig. 3.10. Derivative structure with a partial structure of a single species of anomalous scatterers. Notations are as in Fig. 3.9. (a) Representation in the complex plane of the Friedel pair F^+ and F^-. As the imaginary component of F_A^- is still phase-advanced by $+\pi/2$ with respect to the real component, F^+ and F^- are no longer symmetrical with respect to the real axis. (b) Representation of F^+ and of the complex conjugate of F^-. The two vectors are superimposed, except for the imaginary components of F_A^+ and of the complex conjugate of F_A^-, which are opposite. (c) Same as (b), but at two wavelengths chosen in order to modify the resonant scattering.

axis (Fig. 3.9). Assume now that the structure contains a partial structure A consisting of a single anomalous scattering species with scattering factor $^\lambda f = {^0f} + {^\lambda f'} + i\,{^\lambda f''}$. Then $F_A(h)$, the complete structure factor of the anomalous species, has two components with moduli proportional to $^0f + {^\lambda f'}$ and $^\lambda f''$, respectively, the latter being phase advanced by $+\pi/2$. For $F_A(-h)$, the phase is still advanced by $+\pi/2$. Then $F_A(h)$ and $F_A(-h)$ have identical moduli but are not symmetrical with respect to the real axis. Hence, the total structure factors

$F(\mathbf{h})$ and $F(-\mathbf{h})$ have different moduli and are no longer symmetrical with respect to the real axis [Fig. 3.10(a) and (b)]. Figure 3.10(c) illustrates the two-wavelength case.

Mirror-related reflections (Bijvoet pairs) are also affected by anomalous scattering. Thus, when anomalous scattering is present, reflections which are normally equivalent to reflection $\mathbf{h}$ by inverse or mirror symmetry are split into two families with different structure factor amplitudes. Any pair of reflections with one mate in each family will be called a Bijvoet pair (with no further distinction between true Bijvoet and true Friedel pairs), keeping the notation used for Friedel pairs ($\pm\mathbf{h}$ or simply $\pm$).

The absolute value of the difference between the structure factor moduli of the two mates of a Bijvoet pair at a single wavelength is called a Bijvoet difference. The absolute value of the difference between the structure factor moduli of a given reflection $\mathbf{h}$ (or $-\mathbf{h}$) at two wavelengths is called a dispersive difference. The strength of signals due to anomalous dispersion may be quantified by anomalous ratios (Hendrickson 1985) in shells of resolution:

- At a given wavelength λ, the Bijvoet ratio is

$$\frac{\langle \Delta F_{\pm h}\rangle}{\langle |\tilde{F}|\rangle} \tag{3.16}$$

where $\Delta F_{\pm h} = ||{}^\lambda F(\mathbf{h})| - |{}^\lambda F(-\mathbf{h})||$ and $|\tilde{F}| = \frac{1}{2}(|{}^\lambda F(\mathbf{h})| + |{}^\lambda F(-\mathbf{h})|)$.
- At two wavelengths λ_i and λ_j, and assuming that the variation of $\langle |\tilde{F}|\rangle$ between the two wavelengths can be neglected, the dispersive ratio is

$$\frac{\langle \Delta F_{\Delta\lambda}\rangle}{\langle |\tilde{F}|\rangle}, \tag{3.17}$$

where $\Delta\lambda = \lambda_i - \lambda_j$ and $\Delta F_{\Delta\lambda} = ||{}^{\lambda_i}\tilde{F}| - |{}^{\lambda_j}\tilde{F}||$.

Expressions of expected values for these anomalous ratios have been derived for the various phasing techniques, and will be presented below.

In the 0.3–3.5 Å window, light atoms in macromolecules behave, to a good approximation, as normal scatterers since their resonant component is either very weak (H, C, N, O) or small (S, P). Assume that, in addition to these atoms, the crystal contains another species with significant resonant scattering ${}^\lambda\delta$. Then, by tuning the wavelength of the incoming X-ray beam, it is possible to change the amplitude and phase of ${}^\lambda\delta$ with a negligible effect on the scattering from other atoms [Fig. 3.10(c)]. Accordingly, resonant scattering is a physical way to produce effects which are equivalent to *in situ* multiple isomorphous replacements in an invariant crystal structure.

3.4 Multiple wavelength methods

The production of multiple isomorphous replacement by resonant scattering is completely exploited in multiwavelength methods, which can produce initial estimates for most phases without recourse to anything but anomalous dispersion. These methods are called MAD and MASC (multiple wavelength anomalous solvent contrast). The status of these methods is quite different. MAD is a mature technique which is now widely used for solving structures. MASC has a more limited domain of application and is still being developed. As in other *de novo* phasing methods, the structure solution proceeds in two steps: the determination of the (single) relatively simple substructure A; and the determination of the complex macromolecular structure. The basic advantages of these methods are the following:

(i) All experiments can be performed with a single sample. This is achieved in practice with a cryocooled crystal. Accordingly, the structural isomorphism is perfect, although some non-isomorphism can result from more subtle effects, such as the weak anomalous scattering from other atoms in the structure or changes in f' and f'' values due to unwanted wavelength drifts during data collection.
(ii) Normal diffraction from the substructure and from the rest of the atoms can be separated completely. Accordingly, at least in the MAD case, the solution of the substructure can be tackled with methods developed for small molecule crystallography, in particular direct methods.
(iii) There are general methods for incorporating anomalous scatterers in macromolecules, so that MAD is a rather systematic route to solving macromolecular structures.

3.4.1 The MAD method

The principle of MAD phasing was given four decades ago by Okaya and Pepinsky (1956). Various steps in the development of this method are retraced in Fourme and Hendrickson (1990). Other review articles on MAD have been published: Hendrickson (1991, 1994), Smith (1991), Fourme *et al.* (1996), Hendrickson and Ogata (1997).

3.4.1.1 Solution of the substructure of anomalous scatterers

The partial structure can be solved in various ways. The first approach is based on Bijvoet differences $\Delta F_{\pm h} = \||^{\lambda}F(h)| - |^{\lambda}F(-h)\|$ at a single wavelength. The peak wavelength is obviously the best choice in order to have largest differences. A Patterson map can be calculated with the $|\Delta F_{\pm h}|^2$ as coefficients (Rossmann 1961). Both abnormally large (outliers) and weak differences (noise) are discarded from this calculation. As shown by Karle (1980), such maps are in fact sharpened Patterson maps with a signal-to-noise ratio reduced by about 1/2 with respect to Patterson maps with $|^0F_A|^2$ as coefficients. Dispersive Patterson maps, although often noisy, are also useful. The similarity of dispersive and Bijvoet difference maps is often the first clear indication that the MAD experiment was successful. Dispersive Patterson difference maps can be clearer than the

Bijvoet map, in particular by using a very remote wavelength to give a genuine optimal dispersive difference. It is also possible to exploit direct methods. One of the largest substructures solved to date from a single wavelength or from a MAD analysis of protein data in the absence of any previously determined phases is 30 Se atoms (Turner et al. 1998) (see Section 3.7.1).

The parameters of the substructure may be refined using a limited subset of the largest anomalous differences, also excluding outliers (Hendrickson and Teeter 1981).

It is also possible to exploit the information at multiple wavelengths. Following a seminal idea of Karle (1980), the total structure factor is separated into λ-independent and λ-dependent contributions (which are denoted by superscripts 0 and λ respectively). The most popular decomposition is the following (Hendrickson 1985):

$$^\lambda F_T(h) = {}^0F_T(h) + {}^\lambda F_A(h), \quad (3.18)$$

where $^0F_T(h)$ is the total normal structure factor (i.e. including the normal contribution $^0F_A(h)$ from anomalous scatterers, and not only the P structure) of reflection h at wavelength λ, which is equivalent to the structure factor of the native crystal in MIR; $^\lambda F_A(h)$ is the resonant structure factor that is used as a tunable reference wave (Fig. 3.11). From eqn (3.14), $^\lambda F_A(h)$ can be written at a given wavelength λ, in the case of a single anomalous scattering species, as

$$^\lambda F_A(h) = \left(\frac{^\lambda f'}{^0f} + i \frac{^\lambda f''}{^0f} \right) {}^0F_A(h). \quad (3.19)$$

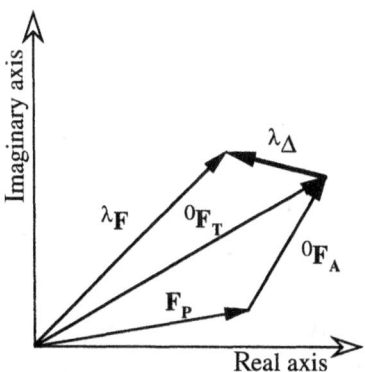

Fig. 3.11. Vector decomposition of $^\lambda F(h)$, the total structure factor value at λ, as used in the algebraic method. F_P is the (normal) structure factor of the P structure. ^0F_A and $^\lambda\Delta$ are the normal and resonant structure factors, respectively, of substructure A. ^0F_T is the total normal structure factor, i.e. including the normal contribution from A.

Combining eqns (3.18) and (3.19) gives

$$^\lambda F_T(h) = {}^0F_T + \left(\frac{^\lambda f'}{^0f} + i\frac{^\lambda f''}{^0f}\right){}^0F_A(h), \qquad (3.20)$$

and the complex conjugate of the structure factor for the Bijvoet mate $(-h)$,

$$[^\lambda F_T(-h)]^* = {}^0F_T(h) + \left(\frac{^\lambda f'}{^0f} - i\frac{^\lambda f''}{^0f}\right){}^0F_A(h). \qquad (3.20')$$

Multiplication of $^\lambda F_T(\pm h)$ by the complex conjugate gives algebraic equations

$$|^\lambda F(\pm h)|^2 = |^0F_T|^2 + a(\lambda)|^0F_A|^2 + b(\lambda)|^0F_T||^0F_A|\cos(^0\varphi_T - {}^0\varphi_A)$$
$$\pm c(\lambda)|^0F_T||^0F_A|\sin(^0\varphi_T - {}^0\varphi_A), \qquad (3.21)$$

where $^0\varphi_A$ and $^0\varphi_T$ are the phases of ^0F_A and ^0F_T, respectively, $a(\lambda)$ and $b(\lambda)$ and $c(\lambda)$ are known quantities which encode the wavelength dependence:

$$a(\lambda) = \frac{(^\lambda f'^2 + {}^\lambda f''^2)}{^0f^2}, \quad b(\lambda) = \frac{2^\lambda f'}{^0f}, \quad c(\lambda) = \frac{2^\lambda f''}{^0f}. \qquad (3.21')$$

Denote $\Delta\varphi = {}^0\varphi_T - {}^0\varphi_A$. With an appropriate choice of the unknowns $(|^0F_T|^2, |^0F_A|^2, |^0F_T||^0F_A|\cos\Delta\varphi$ and $|^0F_T||^0F_A|\sin\Delta\varphi)$ the system of eqn (3.21) F_A^+ is linear, with the non-linear constraint $\cos^2\Delta\varphi + \sin^2\Delta\varphi = 1$. This system can be solved for the desired parameters $|^0F_A|$, $|^0F_T|$ and $\Delta\varphi$ with the program MADLSQ (Hendrickson 1985).

The derivation of $|^0F_A|$ by least-squares method can yield estimates that are unrealistically high, particularly if the data contain large errors in measurements (Pähler et al. 1990). These large values, which can be a substantial fraction of all the structure factors, must be identified and eliminated (Yang et al. 1990). Terwilliger (1994a) has proposed a different procedure which explicitly incorporates information available a priori on the likely magnitudes $\{|^0F_A|\}$; he uses weighted-average estimates of these magnitudes and incorporates estimates of errors in the data that are not represented in the instrumental uncertainties.

As normal structure factor amplitudes for the A and T structures have been obtained, respective diffraction effects are now totally separated. This is a remarkable result, specific to multiwavelength methods.

As structure factor amplitudes $|^0F_A|$ are produced by the algebraic method, the solution and refinement of the A structure can be performed using methods and programs that are standard in small molecule crystallography. Patterson and direct methods are applicable for the structure solution. The entire set of $|^0F_A|$, after appropriate rejections, is valid data for the refinement of the initial model, in contrast to the single wavelength case.

3.4.1.2 Phasing the main structure

In the algebraic method, the phase of the total normal structure factor is calculated from $^0\varphi_T = \Delta\varphi + {}^0\varphi_A$. If A is non-centrosymmetric, then it is necessary to determine the absolute configuration. Using the incorrect enantiomorph of A to calculate $^0\varphi_A$ leads to wrong $^0\varphi_T$ values. In practice, Fourier syntheses based on $\{^0F_T\}$ are calculated for both hands, and one of them is selected on the basis of chemical reasonableness and symmetry considerations. The algebraic MAD approach, including preliminary steps such as the determination of f' and f'' values from fluorescence measurements, has been coded in the system of programs MADSYS (Hendrickson 1991).

Once the correct enantiomorph has been identified, the result of the algebraic analysis is a single numerical value. An estimate of the reliability of this value is given by the fitting residual and the standard deviation estimated from least-squares calculations. The objection is that, with data limited to a few wavelengths, it is difficult to make reliable estimates of the accuracy by the least-squares algebraic method. Subsequently, a method was developed to use the calculated structure factors from the partial structure to improve estimates of $\Delta\varphi$ and $^0\varphi_T$ and to calculate phase-probability distributions for the normal scattering from the total structure (Pähler et al. 1990). These distributions provide a means for obtaining estimates of reliability and ambiguity. They can be encoded in a compact form by 'ABCD' coefficients (Hendrickson and Lattman 1970), which provides a convenient means for combining information from diverse sources such as solvent flattening, molecular averaging, isomorphous replacement and molecular replacement.

The alternative is the use of programs that both refine the parameters of the substructure and calculate the phases by probabilistic methods. The use of probabilistic methods is reminiscent of procedures developed since the seminal work of Blow and Crick (1959) for the treatment of errors in MIR data. The principle of the transfer of this approach to MAD is found in Phillips and Hodgson (1980). Then, 3λ-MAD data taken on a crystal of the parvalbumin from *Opsanus tau* were analysed in a quasi-MIR mode by Kahn et al. (1985, 1986a,b). The $\{F^+\}$ set at one wavelength was treated as pseudo-native data, other sets as derivative data; 'best' (centroid) phases were obtained for the pseudo-native. Later, Chiadmi et al. (1993) completed this procedure by the calculation of best phases and moduli instead of best phases only. The MIR formalism was also used by Korszun (1987).

Since the principles underlying MAD and MIR are very similar, the question arose as to whether standard heavy atom refinement programs—which have been developed over decades—could be used to analyse MAD data. The structure of GH5, the globular domain of histone H5, was solved by MAD (Ramakrishnan et al. 1993); phases were obtained with the phasing program MLPHARE (Otwinowski 1991). The treatment of MAD data as a special case of MIR and the specificities of this treatment have been discussed by Terwilliger (1994b), Brünger et al. (1995) and Ramakrishnan and Biou (1997). Bella and Rossmann (1998) have addressed the question of the artificial 'native' role given to one of the wavelengths, which is assumed to be error-free. A number of structures have now been solved along these lines (e.g. Dyda *et al.* 1994; Glover *et al.* 1995; Müller *et al.* 1995; Lee *et al.* 1995). In addition to MLPHARE, other heavy atom refinement programs have been successfully used, including PHASIT in the PHASES package (Furey and Swaminathan 1990), HEAVY (Terwilliger and Eisenberg 1983) and X-PLOR (Brünger *et al.* 1995).

Refinement of the parameters of the partial structure remained a troublesome issue in macromolecular crystallography for decades. These difficulties have been traced back to inadequacies in conventional statistical procedures. The least-squares method is not suitable for this type of refinement, and recourse to maximum-likelihood (ML) is necessary to obtain a bias-free solution. Least-squares is a special case of ML when errors are normally distributed with fixed (co)variances. Two-dimensional statistical phasing (probability distribution on the phase and on the modulus of the 'native' structure factor) has been considered by Raiz and Andreeva (1970), leading to the first mention of likelihood in this context by Einstein (1977). The first mention of parameter estimation by ML, in a very limited context, is found in Green (1979). ML refinement for heavy atom refinement was then advocated by Bricogne (1985, 1988, 1991). Some features of ML have been incorporated in existing phasing programs such as MLPHARE (Otwinowski 1991) and MADPRB (Friedman *et al.* 1994). A full implementation of ML has been incorporated in SHARP (La Fortelle and Bricogne 1997). This program can handle MIR, SIRAS (single isomorphous replacement with anomalous scattering) and MAD data or any mixture of them. It can refine all parameters pertinent to the partial structure including f'' and f' values as a function of wavelength and for various blocks of data. The initial version of SHARP treated f' and f'' as scalar quantities. In the forthcoming version (Schiltz and Bricogne, private communication), they can be optionally treated as atomic tensors; in this case, the program uses goniometric information from unmerged reflection data. The tensor description allows one to take into account the anisotropy of the anomalous scattering (L.K. Templeton and D.H. Templeton 1982; Fanchon and Hendrickson 1990), thus giving a more accurate description of the reference structure and, finally, better phases for the total structure. SHARP calculates the probability distribution function of the modulus and phases of the 'native' structure factor and the centroid structure factor which are used to calculate the electron density map. Extensive tests of SHARP with simulated data have demonstrated that bias is effectively removed (La Fortelle and Bricogne 1997). Tests on phasing high quality MAD data from IF3-C, a small protein, have shown that SHARP performed somewhat better than other programs (Ramakrishnan and Biou 1996). These authors note that in more difficult cases where the data are weak, ML programs such as SHARP may produce the best initial set of phase probability distributions to serve as a starting point for density modification procedures. Indeed, the procedure currently used by many SHARP users, following Bricogne and co-workers, is to produce initial phases with SHARP and then to use the procedure of density modification as implemented in the program SOLOMON (Abrahams and Leslie 1996).

In a broader perspective, SHARP is designed in order to be interfaced to BUSTER (Bricogne 1988), a program that implements the 'Bayesian approach' (Bricogne 1993, 1997). BUSTER uses phase relationships between reflections through maximum-entropy extrapolation and likelihood scoring. It is able to make use of the phase information extracted from a MAD data set by SHARP in the form of 'ABCD' coefficients and performs 'mode permutation' on bimodal phase distributions, so as to select the correct mode in the case of a low signal-to-noise experiment. Results derived from MAD and other sources (SIR, MIR, density modifications, ...) are combined and the phase information is extracted in a statistically optimal way through a unified formalism. This treatment is especially beneficial to those reflections with weak amplitudes for the partial structure factor, and which would be poorly phased by the MAD method alone.

ML methods are also appropriate to determine the locations of atoms in the partial structure, as advocated by Bricogne (1993, 1997) and there are plans to incorporate this option in future upgrades of SHARP. The prospect is thus a unified treatment of all *de novo* phasing methods, from the solution of the reference structure(s) to the phase determination of the complex structure, which is consistent with the fact that these methods are based on a common principle.

3.4.1.3 *Anomalous ratios, choice of wavelengths and data collection strategy*

From eqn (3.21), the value expected for a Bijvoet ratio in a shell of resolution, as defined in Section 3.3.4, is given by (Hendrickson 1985)

$$\frac{\langle \Delta F_{\pm h} \rangle}{\langle |\tilde{F}| \rangle} \approx \frac{1}{\sqrt{2}} \frac{\sqrt{N_A}}{\sqrt{N_P}} \frac{2f''}{f_P} \qquad (3.22)$$

and the value expected for a dispersive ratio by

$$\frac{\langle \Delta F_{\Delta \lambda} \rangle}{\langle |\tilde{F}| \rangle} \approx \frac{1}{\sqrt{2}} \frac{\sqrt{N_A}}{\sqrt{N_P}} \frac{\Delta f'}{f_P}, \qquad (3.22')$$

where N_A and N_P are the number of anomalous scatterers and normal atoms, respectively, in the asymmetric unit and $\Delta f' = |{}^{\lambda_i}f' - {}^{\lambda_j}f'|$.

As there are three independent unknowns ($|{}^0F_A|$, $|{}^0F_T|$ and $\Delta \varphi$), at least three measurements are required for each h, and, in practical terms, at least two-λ data sets with measurements of Bijvoet pairs for at least one of those sets. For a given crystal containing anomalous scatterers with known f' and f'', it is essential to determine the best data collection strategy in order to get the most accurate phases: How many data sets and at which wavelengths? Is it necessary to measure Bijvoet pairs? Computer simulations were performed by Phillips and Hodgson (1980) in order to plan an optimal strategy in a MAD experiment. Narayan and Ramaseshan (1981) have given an analytical theory from which the expected rms phase error in a given MAD experiment can be calculated, assuming a given rms error in intensity measurements. For instance, in a two-λ

measurement with Bijvoet pairs (four equations for each $\boldsymbol{h}$), the quantity to be maximized for a minimum phase error is $|{}^{\lambda_1}f' - {}^{\lambda_2}f'|[({}^{\lambda_1}f'')^2 + ({}^{\lambda_2}f'')^2]$. More generally, large values of both f'' and $|{}^{\lambda_i}f' - {}^{\lambda_j}f'|$ are required.

With these considerations in mind, the strategy generally adopted in MAD studies is the following. Anomalous pair data are taken at several wavelengths in the vicinity of an absorption edge of the anomalous scattering species. These wavelengths always include the peak value of f'', for which the anomalous signal is largest; this data set should be recorded with particular care in terms of accuracy, redundancy and resolution. In the two-λ case, the second wavelength should be a remote wavelength on the high energy side of the absorption edge. In the three-λ case, the third wavelength is in general selected at the point of inflection of the absorption edge, giving the most negative value of f'. If a fourth measurement is performed, the additional point is chosen on the low energy side of the edge. One should note that, in the case of anomalous scatterers showing a 'white line', the choice of the peak wavelength is obvious; for scatterers such as Hg, whose compounds have often no structure at L edges, the inflection point is at the L_{III} edge and the peak value corresponds to a 'remote' wavelength at the peak of the L_I edge.

Data collection strategies in a MAD experiment have been investigated by Peterson et al. (1996) on an oligonucleotide crystal labelled with bromine.

3.4.1.4 Choice of the anomalous scattering species

A wide range of atoms and absorption edges can be used for MAD phasing. Considerations which guide the choice of the anomalous scattering species are the following:

(i) The anomalous scatterer, if not present in the wild-type macromolecule (intrinsic label), must be introduced into it (extrinsic label). The phasing power is improved by increasing the number of resonant centres; on the other hand, the solution of the partial structure must remain tractable by direct methods (the current limit is 50–70 atoms per asymmetric unit in the absence of other phase information.).
(ii) For a given atom, the edge position occurs at increasing energy (shorter wavelength) as the atomic number Z increases, but the edge profiles (for isolated atoms) are essentially alike: atoms that are not so heavy are quite usable for MAD phasing.
(iii) Anomalous signals calculated for isolated atoms increase from K to L and from L to M edges. Specific resonance features observed in many cases close to an absorption edge may be exploited to enhance these signals.
(iv) The energy of the selected absorption edge must be located in a range convenient for data collection and accessible with the particular set-up and X-ray source used for data collection.

In the case of nucleic acids, bromine ($Z = 35$, K edge at 0.9202 Å), introduced via brominated nucleic acid bases, is used as an anomalous scatterer for MAD (Ogata et al. 1989).

In the case of proteins, natural labels which have been used are transition metals (Cu, Fe, Zn, ...). The most commonly used extrinsic anomalous scatterers are currently selenium as seleno-methionine, common heavy atoms (Hg, Pt, Au, Pb, U, ...) and lanthanides (Yb, Ho, Tb, ...). The K edge of selenium ($Z=34$), at a wavelength (0.9796 Å) quite suitable for data collection, features a sharp resonance. As suggested by Hendrickson (1985), selenium can be substituted for natural sulphur in methionine by recombinant DNA techniques in which the target protein is expressed in a methionine auxotroph that is grown with selenomethionine as a nutrient (Hendrickson et al. 1990). Soluble proteins have on average one methionine for every 59 amino acids (Klapper 1977). Maximum Bijvoet and dispersive ratios expected for a Se-met protein with the average concentration of methionine would be about 4% and 3.5%, respectively (Smith 1991). If necessary, the number of methionines can be increased by substitution of methionine for other amino acids by site-directed mutagenesis. Krypton ($Z=36$, K edge at $\lambda=0.8655$ Å) is a potential candidate for MAD. Nevertheless, the ease of making krypton derivatives, the fact that krypton minimally perturbs the native structure and the isomorphous signal resulting from this binding are strong arguments for using krypton for SIRAS rather than MAD (Schiltz et al. 1997b). The same remarks apply even more to xenon. Although we have demonstrated that data collection at ultra-short wavelengths is quite feasible (Schiltz et al. 1997a), which opens the possibility of MAD experiments at the xenon K edge ($\lambda=0.3585$ Å), SIRAS is obviously a better choice than MAD. In effect, xenon is a fairly heavy atom ($Z=54$), and further the anomalous signal increases at longer wavelengths (for instance $f''=7.4\,e^-$ at 1.54 Å).

All of the commonly used substituting elements for conventional heavy atom derivatives have their L_{III} edges at wavelengths around 1 Å. Lanthanides can replace group II ions Ca^{2+} and Mg^{2+} and yield strong anomalous scattering at the L_{III} edges well located for MAD experiments.

The use of anomalous scattering at the M edges of heaviest atoms is another exciting possibility. Experiments by Liu et al. (1995) have revealed very large values of f'' and f' at the M_{IV} and M_V edges of uranium (at 3.3276 and 3.4908 Å, respectively). Although accurate data collection at such long wavelengths requires specific procedures, the use of M edges presents a new opportunity for structure determination, in particular of high molecular weight systems.

3.4.1.5 Conclusion

MAD is now a well-established technique in the crystallographers toolbox. Instruments with characteristics appropriate for MAD studies are available at most synchrotron radiation centres. MAD is in fact at the crossroads of much recent instrumental and methodological progress in biocrystallography, and it has stimulated a number of these developments and their application to the whole field: improvements in cryocooling methods, X-ray optics and detectors; data collection and data analysis methods; new ways to label macromolecules; unification of *de novo* phasing methods through a more rigorous treatment of information. Macromolecules are now often designed, based on the specific purpose, for the application of MAD (e.g. incorporation of selenium in proteins, of bromine in nucleic acids). With data accurate to high or very high resolution, more complete sets of $\{^0F_A\}$ can be derived, which may be crucial for unravelling a complicated partial structure by direct methods. Partial structures with several tens of anomalous centres in the asymmetric unit have already been solved by direct methods, so that the domain of MAD is extending toward more ambitious problems, such as high molecular weight macromolecules or assemblies.

Finally, the potential of strong anomalous scatterers for the derivation of accurate, model bias-free phases, should be underlined. As discussed in Section 3.7.1, MAD can indeed provide accurate phases up to very high resolution, allowing a detailed and unbiased description of the structure under study, even of the solvent phase.

3.4.2 The MASC method

As pointed out by Bricogne (1993), the anomalous scattering at several wavelengths by the solvent phase of a macromolecular crystal can be used to get phase information. A theory of MASC assuming a binary density model [Fig. 3.5(a)] of the protein and solvent volume in the crystal and a report on first MASC experiments was given by Fourme et al. (1995).

It is now assumed that anomalous scatterers of a single species are randomly dispersed in solvent. Formally, these atoms, with a scattering factor $^\lambda f = {}^0f + {}^\lambda f' + i{}^\lambda f''$ which can be assumed as constant in the low resolution domain, contribute to the solvent electron density by a complex, wavelength-dependent density denoted $^\lambda\rho_{SA}$:

$$^\lambda\rho_{SA} = {}^0\rho_{SA}\left(1 + \frac{{}^\lambda f'}{{}^0f} + i\frac{{}^\lambda f''}{{}^0f}\right), \tag{3.23}$$

where $^0\rho_{SA}$ is the normal part of the electron density of anomalous scatterers.

The total solvent electron density can be decomposed into two components,

$$^\lambda\rho_S = {}^0\rho_S + {}^0\rho_{SA}\left(\frac{{}^\lambda f'}{{}^0f} + i\frac{{}^\lambda f''}{{}^0f}\right), \tag{3.24}$$

where $^0\rho_S$ is the total normal electron density of the solvent.

Replacing ρ_S in eqn (3.12) by $^\lambda\rho_S$, it can be shown that the total structure factor of reflection $\boldsymbol{h}$ and the complex conjugate of the anomalous mate $-\boldsymbol{h}$, respectively, can be decomposed into two terms,

$$^\lambda F(\boldsymbol{h}) = {}^0F_T(\boldsymbol{h}) + \left(\frac{{}^\lambda f'}{{}^0f} + i\frac{{}^\lambda f''}{{}^0f}\right){}^0\Gamma_A(\boldsymbol{h}), \tag{3.25}$$

$$[^\lambda F(-\boldsymbol{h})]^* = {}^0F_T(\boldsymbol{h}) + \left(\frac{\lambda f'}{{}^0f} - i\frac{\lambda f''}{{}^0f}\right) {}^0\Gamma_A(\boldsymbol{h}), \tag{3.25'}$$

where

$$^0\Gamma_A(\boldsymbol{h}) = -{}^0\rho_{SA} G(\boldsymbol{h}) \exp(-BS^2/4) \tag{3.26}$$

and $G(\boldsymbol{h})$ and B have been defined in Section 3.2.2.

Multiplication of $^\lambda F(\pm\boldsymbol{h})$ by the complex conjugate gives equations which have exactly the same form as eqn (3.21).

$$|^\lambda F(\pm\boldsymbol{h})|^2 = |{}^0F_T|^2 + a(\lambda)|{}^0\Gamma_A|^2 + b(\lambda)|{}^0F_T||{}^0\Gamma_A|\cos(\Delta\varphi)$$
$$\pm c(\lambda)|{}^0F_T||{}^0\Gamma_A|\sin(\Delta\varphi), \tag{3.27}$$

where $a(\lambda)$, $b(\lambda)$ and $c(\lambda)$ are defined as in eqn (3.21') and $\Delta\varphi$ is the difference between $^0\varphi_T$ and $^0\varphi_A$, the phases of ^0F_T and $^0\Gamma_A$, respectively. Accordingly, the set of equations pertaining to the pair ($\pm\boldsymbol{h}$) can be solved for $|{}^0F_T|$, $|{}^0\Gamma_A|$ and $\Delta\varphi$ using MADLSQ. The set $\{|G|\}$ of is derived from the $\{|{}^0\Gamma_A|\}$ with eqn (3.26).

The determination of the $\{|G|\}$ phases is still in progress (Ramin 1999). An approach based on spherical harmonics has been developed by one of the authors and is being applied to test cases (R. Kahn, unpublished). The ordered ($\chi = 1$) fraction of the structure is described by an expansion in spherical harmonics. First, the domain is approximated by a sphere which is moved randomly in the unit cell. For each trial, the coordinates of the centre and a scale factor are refined. Selected solutions are expanded in spherical harmonics of higher rank and coefficients are refined. The best solutions are selected according to the agreement between the amplitudes of G_{obs} and G_{calc}, the rate of overlap between domains generated by space group operations and the speed of convergence. During the whole procedure, the volume of the ordered domain is kept constant, taking into account the estimated solvent content. Another possibility is the Bayesian approach (Bricogne 1993, 1997). Likelihood tests can be applied to models generated for the envelope and their refinement. An expression has been derived for the likelihood function attached to a putative envelope or to a parametrized modification of an existing envelope, incorporating either native data or a series of chemical and anomalous contrast variation measurements.

MASC experiments have been performed on three protein structures of different molecular weight (Ramin 1999; Ramin et al. 1999). In all cases, ordered anomalous scatterers were found in addition to disordered atoms (mixed MAD/MASC case). Equations (3.27) have to be modified accordingly, with $^0\Gamma_A$ being replaced by $(^0\Gamma_A + {}^0F_A)$ in order to incorporate the contribution from ordered anomalous scatterers. When only MASC effects are taken into account, the agreement between experimental and model values is satisfactory at very low resolution ($d > 20$ Å); it is also satisfactory at low and medium resolution when ordered atoms are included in the model.

As in the MAD method, the expected accuracy in phase determination is related to the magnitudes of Bijvoet and dispersive ratios (Fourme et al. 1995). Accordingly, the choice of wavelengths is made as described in Section 3.4.1.3. The choice of the anomalous scattering species is similar to that of the MAD method as described in Section 3.4.1.4, with the addition that the chemical compound that contains this species must be soluble at molar or multimolar concentrations and that the crystals must be stable in such conditions.

3.5 Single wavelength methods

3.5.1 SAD

In a single anomalous diffraction (SAD) experiment, Bijvoet pairs are measured at a single wavelength chosen in order to maximize the Bijvoet differences without native data collection. SAD is attractive as only one data set is recorded. But, as there are only two intensity measurements for each phase, additional information is necessary to remove the two-fold ambiguity. How to overcome the ambiguity inherent to SAD has been discussed in a number of articles, e.g. Okaya et al. (1955), Ramachandran and Raman (1956), Fan (1965), Hauptman (1982), Giacovazzo (1983), Karle (1984, 1989), Fan and Gu (1985), Hao and Woolfson (1989) and Ralph and Woolfson (1991).

A milestone in SAD was the solution of the structure of crambin (Hendrickson and Teeter 1981), a small protein which contains six native sulphur atoms (three Cys–Cys bridges). The structure was solved from the anomalous scattering of sulphur, using Friedel pairs to 1.5 Å resolution. The partial structure of sulphur atoms was solved from a Patterson map calculated with $|\Delta F_{\pm h}|^2$ as coefficients (Rossmann 1961). The parameters of the partial structure were refined using a limited subset of the largest Friedel differences (excluding outliers). Then phases for the non-anomalous scattering portion of the structure were derived by a probabilistic combination of the phase information from the anomalous scattering and that from the partial structure.

Recent developments in SAD phasing combine high accuracy data collection, direct methods of solving the partial structure of anomalous scatterers and statistical methods giving unbiased phases for solvent flattening. The result is that interpretable electron density maps have been obtained from SAD data with very small anomalous signals (see Section 3.7.2).

SAD achieves the goal of *de novo* phasing with one data set, and its recent developments are promising.

3.5.2 SIRAS

The SIRAS method combines single isomorphous replacement and anomalous dispersion. It requires a native crystal and a single derivative crystal in which the partial structure A consists of atoms with significant anomalous scattering. Data are collected at a single wavelength chosen in order to maximize Bijvoet differences. Three structure factor amplitudes are exploited for phasing: $|F_P(h)|$, $|F_{P+A}(h)|$ and $|F_{P+A}(-h)|$.

The solution of the partial structure and its refinement proceeds as described for SAD.

If the arrangement of anomalous scatterers is not centro-symmetric, then it is necessary to find their absolute configuration. A process for doing this, based on an article by Okaya *et al.* (1955), has been described by Woolfson and Yao (1994). This problem has also been discussed by Wang (1995). When the configuration is incorrect, the electron density map, phased with SIRAS data, is a negative density map where the observed ridges of density actually correspond to the deep valleys in the correctly phased map; this negative map will show well-defined electron density with excellent connectivity at a relatively correct location. Theoretical aspects and practical procedures used to know if the handedness is correct prior to examining the Fourier map have been discussed by the same author.

If the native and derivative structures are sufficiently isomorphous, SIRAS is a powerful and convenient method. The isomorphous difference is generally much larger than the largest dispersive difference in a multiwavelength experiment. Data collection of two data sets at a fixed wavelength is simpler and faster than with MAD. In the MAD method, wavelengths are selected close to absorption edges in order to maximize both dispersive and Bijvoet differences. In SIRAS, only Bijvoet differences are used, which makes the choice of the single wavelength more flexible.

3.6 Measurement and processing of anomalous dispersion data

The phasing power of anomalous dispersion methods resides in Bijvoet and dispersive differences. Accordingly, high precision of the measurement of intensity differences is essential. Anomalous signals (defined by Bijvoet and dispersive ratios) give expectations for the level of accuracy that is required in each case. In macromolecular crystallography, these signals may be as strong as 25% in specific cases, such as in MAD experiments on small proteins incorporating several lanthanide atoms per molecule and exploiting the 'white line' features at the L_{III} edges of lanthanides (see Section 3.7). In general, these signals are much smaller. However, successful experiments have been performed with signals as low as 2–3% (Hendrickson *et al.* 1989). High precision has been achieved by the interplay between careful experimental design, a strategy for efficient acquisition of data with appropriate counting statistics, and care in minimizing or correcting systematic errors at various steps from sample preparation, data acquisition and reduction of raw data (Hendrickson *et al.* 1988; Fourme and Hendrickson 1990).

3.6.1 *Instrumentation*

The main components of an instrument optimized for measurements of anomalous dispersion data are: the X-ray source, the X-ray optics, a goniometer provided with a device for cooling the sample to 100–120 K, a system for the detection of the X-ray fluorescence emitted by the sample and an area detector. The software components are an integral part of this equipment.

3.6.1.1 X-ray sources

Anomalous dispersion experiments require sources capable of producing X-rays at various suitable wavelengths. Although bremsstrahlung or characteristic emission lines from conventional X-ray tubes have been occasionally used (Hendrickson 1991), nearly all experiments are now performed with synchrotron radiation. Both bending magnets from the basic storage ring lattice and devices (wigglers and undulators) inserted in this lattice are used as X-ray sources. The spectrum from bending magnets and wigglers is a smooth continuum. Bending magnets of storage rings with an energy of at least ≈ 1.8 GeV provide adequate flux for many experiments. Multipole wigglers give enhanced flux and shift the spectrum to higher energies, thus improving the medium energy second-generation sources (such as DCI, Orsay, France and SRS, Daresbury, UK) and making third-generation, lower energy sources (such as LNLS, Campinas, Brazil; Elettra, Trieste, Italy; ALS, Berkeley, USA and MAX II, Lund, Sweden) suitable for anomalous dispersion studies. In an undulator, a periodic magnetic field is produced by an array of permanent magnets. Interferences between rays emitted by each electron travelling in the periodic field have two effects: the synchrotron emission is concentrated in a very narrow cone and the energy spectrum features bright lines. The spectrum of the on-axis emission includes the fundamental wavelength λ and odd harmonics. These wavelengths can be shifted by changing the magnetic field strength, which is performed by adjusting the magnetic gap. In order to keep a high brilliance for higher order harmonics, the emittance must be very low and field errors must be corrected accurately, which is done by shimming magnet blocks. The ESRF was the first high energy third-generation synchrotron radiation source and gave an outstanding contribution to the progress of undulators. Currently, the storage ring incorporates, more than 50 insertion devices, mostly undulators which are better sources than multipole wigglers down to about 0.2 Å. The X-ray beam stability in position and angle required to fully exploit these high brilliance sources has been achieved. That the undulator source can be used successfully for multi-wavelength data collection was first demonstrated in a MAD experiment by Shapiro et al. (1995) and later in a MASC experiment by Ramin et al. (1999), both using the 'Troika' beam line at ESRF. In addition to bending magnet beam lines, there are two MAD stations on undulator lines, ID14-EH4 (operational) and ID29 (under construction). MAD experiments at the other third-generation high energy facilities, the APS (Argonne, USA) and SPring-8 (near Kobe, Japan) are also installed on undulator lines. Undulators are without doubt the best sources for MAD and other techniques exploiting anomalous dispersion. In the current state of the art, the minimum energy required to get 1 Å radiation from an undulator is 2.4–2.5 GeV. This condition is just fulfilled by the SLS (Switzerland) and, more comfortably, by planned facilities in France (SOLEIL) and Great Britain (DIAMOND).

3.6.1.2 X-ray optics

X-ray optics are a crucial component of the experiment, which must be designed very carefully. Main specifications are: rapid and highly reproducible wavelength changes; a bandpass ($\delta\lambda/\lambda$) of 10^{-4}–10^{-5}, in order to avoid smearing of narrow resonance features at some absorption edges; fixed exit slit; good spectral purity (harmonic content $< 10^{-3}$); and a high flux for rapid data collection, which, combined with the requirement of a narrow bandpass, requires double focusing. The energy calibration of the monochromator must be efficient and convenient. An apparatus that provides the required wavelength information and which is used to stabilize the wavelength is in operation on the X31 beam line of the EMBL Outstation at Hamburg. This instrument provides monochromator-independent measurements of the incident X-ray energy and ensures that drift due to temperature changes and/or beam movements are eliminated (Evans et al. 1996).

Various optical configurations have been used. Monochromatization is commonly ensured by fixed-exit, double-crystal Si monochromator. Focusing and beam shaping is provided by double curvature mirrors or multilayers (ensuring also the rejection of harmonics) or the combination of a single curvature mirror or multilayer with a sagittally curved crystal. In some beam lines (e.g. BM14 at ESRF), focusing and monochromatization are separate functions, which is the simplest solution. In other beam lines, e.g. DW21 at LURE or the French CRG beam line FIP at the ESRF, the second crystal of the monochromator is also a focusing device.

Switching between different wavelengths is time consuming and puts great demands on the stability and reproducibility of the monochromator and the synchrotron beam. To eliminate the mechanical motion of a single monochromator, it has been suggested to use several pre-adjusted monochromators. In effect, very thin diamond plates are both efficient monochromators and transparent enough for the mounting, on the same undulator beam, of several plates reflecting different wavelengths ('Troika' concept as used at the ESRF). Using this concept, a cascade of three double crystal diamond monochromators could be installed, with switching from one wavelength to another obtained by moving shutters placed on the monochromatic beam paths (Yamamoto et al. 1998). The principle of simultaneous measurements at various wavelengths using polychromatic synchrotron radiation was first discussed by Arndt et al. (1982). The ability to measure reflections at six different wavelengths and their Bijvoet pairs at the same time was demonstrated by Lee and Ogata (1995), using an overbent crystal monochromator with an energy selecting grid-plate. Similar experiments have been performed at the ESRF by two authors of the present article (W. Shepard and R. Kahn, unpublished results).

3.6.1.3 Area detector

The detector is another critical component. The highest signal-to-noise ratio of intensity measurements are obtained with a detector that has very low intrinsic noise, a large sensitive area placed at a large distance from the crystal and data collection subdivided into many images (frames) corresponding to small (0.05–0.1°) $\Delta\omega$ rotations of the crystal (Fourme et al. 1991). These features are even more important in the case of MASC data collection, because the high concentration of anomalous scatterers in the solvent produces background fluorescence that accumulates on frames.

Successful anomalous dispersion experiments have been carried out with a variety of detectors including scintillation detectors, photographic films, MWPC, imaging plates, and CCD-based systems. Circular imaging plate devices with a built-in scanner (X-ray Research, Hamburg, Germany) are popular, as they are fully automated, simple to use and do not require frequent calibrations. Their main drawback is a dead time between two images, which, in the latest version, is about 45 or 95 s for a scanned area of diameter 180 or 345 mm, respectively. Thus, with the most brilliant sources (undulators), the ratio of exposure time to total elapsed time is generally poor. CCD-based detectors have emerged recently and are gradually superseding image plates for anomalous dispersion diffraction. In effect, imaging plates and CCD detectors are integrating devices producing data of similar quality, but the readout time of a CCD is much reduced, thus speeding up the data collection and making the use of small $\Delta\omega$ practical. MWPCs are fast readout, noise-free photon counters. They behave nearly as ideal Poissonian detectors and can produce highly accurate data (Kahn et al. 1986b; Kahn and Fourme 1996). But, due to count rate limitations and relative complexity, they are no longer used for biocrystallography with synchrotron radiation. An advanced gas chamber, combining microstrips (Oed 1988) instead of wires and pixel encoding instead of a single encoder, might be an alternative to integrating detectors. This is clearly a fairly long term project with challenging technological issues. The same remark applies to detectors consisting of monolithic arrays of reverse-biased semiconductor diodes (made of silicon or CdZnTe) hybridized with a specific circuit instrumenting each diode (Millau et al. 1995).

The data reduction programs, which extract structure factor amplitudes from frames, can be considered as an essential component of any detector, as they take into account various parameters and corrections which are more or less system-specific. Packages currently being most frequently used by the crystallographic community are DENZO (Otwinowski 1993), MOSFLM (Leslie 1994) and XDS (Kabsch 1988).

3.6.1.4 Other items

Although most instruments feature a single axis goniometer (Φ-axis), full flexibility is obtained with a multicircle goniometer, preferably with open geometry (kappa-type). A vertical equatorial plane is optimal as the synchrotron beam is polarized in the orbit plane of the storage ring. Slewing speeds on the Φ-axis should be as large as possible, in order to minimize dead times when using inverse beam geometry.[2]

---

[2] A data collection mode in which Friedel pairs are recorded by taking frames at Φ and $\Phi + 180°$. Rotating the crystal by 180° is equivalent to reversing the direction of propagation of X-rays while keeping the crystal fixed.

The goniometer should be provided with additional equipment:

(a) A beam intensity monitor placed between the collimation system of the goniometer and the sample. This monitor is necessary to correct intensity variations due to the finite lifetime of particles orbiting in the storage ring and displacements of the tightly focused X-ray beam.
(b) A device for cooling the crystal at 100–120 K during data collection. In effect, the last step towards the sample invariance during multiwavelength data collection has been achieved the past few years, with the generalized use of crystal-freezing techniques at cryogenic temperatures. Cooling to cryogenic temperatures minimizes the radiation damage during data collection so that, in many cases, an entire diffraction data set can be taken with one crystal.
(c) A system to scan absorption edges of the anomalous scattering species, permanently installed on the goniometer. This is done by detecting the fluorescence signal I_f from the crystal with a scintillation counter or a silicon diode. The reference signal I_0 may be supplied by the beam monitor of the goniometer.
(d) A small beam catcher placed at a fairly large distance from the sample and a helium path between the detector and the crystal. This equipment is required for MASC experiments, in which very low resolution data are collected.

Software for aligning crystals, assessing their quality and making data acquisition is closely associated with instrumentation to provide the flexibility that is required to accommodate a variety of experiments whatever the crystal morphology, symmetry and unit cell size. Software must simplify the task of the experimentalist through graphical user interfaces (GUI) and supervise the automatic execution of repetitive experimental protocols.

With undulator sources, the brilliance of (unattenuated!) beams is so high that 0.1–100 images (depending on the crystal) with 1–10 million pixels (depending on the detector) could potentially be recorded every second. The upper practical limit is currently in the range 0.3 images/s, but will move rapidly. Further, users wish to check and process their data as quickly as possible. Accordingly, the importance of the computing equipment (including large data storage and high speed networking) cannot be overestimated.

3.6.2 *From sample mounting to structure amplitudes*

The various steps in an anomalous dispersion diffraction experiment are as follows:

First, fluorescence X-ray spectra are recorded and exploited. As the edge structure depends not only on the environment and chemical bonding of the anomalous scatterers embedded in the sample, but also on the bandpass of the X-ray beam, recording fluorescence spectra from the actual sample is highly recommended. Spectra are recorded in the near-edge region and reduced to f''

values. The f' values are obtained by a Kramers–Kronig transformation. These calculations are made subject to constraints to values obtained from theoretical calculations at points fairly remote from the edge (Sasaki 1984). The choice of wavelengths is then performed.

The second step is data collection and data processing. Exposure times are adjusted so that the counting statistics is sufficient in view of the expected Bijvoet and dispersive ratios. The benefit of high redundancy to improving a weak difference signal in diffraction data is underlined. The data collection software takes into account standard factors, including detector corrections (such as flat field and positional corrections, etc), beam intensity, Lorentz and polarization factors. All data should be taken from the same sample (where possible), in order to eliminate the intercrystal absorption and scaling errors. The strategy of data collection is organized in order to reduce the remaining systematic errors in the determination of small intensity differences. Crystals are mounted so as to reduce absorption errors arising from different path lengths for the mates of a Bijvoet pair, or a reference data set with minimal absorption errors is collected. The most favourable configuration is obtained when Bijvoet pairs at each wavelength can be recorded simultaneously; if not, Friedel pairs are recorded close together in time using the inverse beam geometry. Data at the various wavelengths are also measured close in time. Absorption differences at different wavelengths can be taken into account in experimental transmission factors. Many of the residual errors in differences can be reduced numerically by local scaling of adjacent reflections (Hendrickson and Teeter 1981). Afterwards, the various data sets from separate wavelengths are brought into least-squares agreement by local scaling. Then, as the total scattering strength of the sample is dependent on the wavelength, additional scale factors accounting for this variation are applied. Finally, all data are placed on quasi-absolute scale by Wilson's statistics. Merging of equivalent and redundant data is generally not performed on structure amplitudes, but rather on phases determined for separate data sets.

In the last few years, anomalous dispersion experiments have been somewhat simplified. In many cases, it is now possible to collect all data sets with one cryo-cooled crystal. The beam stability and beam lifetime have generally improved. More brilliant sources and better detectors make experiments shorter and stability requirements are more easily achieved. Programs (in particular ML phasing programs) can refine efficiently the f' and f'' values, correcting approximate initial values as well as small changes due to wavelength drifts during data collection. Especially with very brilliant sources, full data sets are often recorded at each wavelength, instead of sectors of reciprocal space recorded at the various wavelengths. Good results have been obtained by merging equivalent reflections before phasing (Glover et al. 1995; Ramakrishnan et al. 1993).

Finally, we underline the importance of high completeness both overall and of Bijvoet mates (including low resolution reflections) and high redundancy. Instead of making experiments shorter, the higher intensity should primarily be used to collect better and more redundant data. In particular, it is recommended to record a peak wavelength data set at the maximum possible resolution; this data set may be extremely useful for SAD phasing or for extension of MAD phases at a higher resolution using solvent flattening. In several cases, an interpretable MAD electron density map has been obtained in less than 24 h after the beginning of data collection at the synchrotron radiation facility.

3.7 Examples

3.7.1 MAD

3.7.1.1 Phasing with selenium

The general method of incorporating selenium is via selenomethionines, as described previously. This technique has been successfully used to solve many structures. The use of selenium might become the method of choice in structural genomics, where the structures of many proteins with molecular weight around 30 kDa will have to be solved very efficiently.

A typical example of the solution of a medium size protein structure by three-λ MAD using the Se-met technique followed by solvent flattening is the C3d: AC3 fragment and ligand for complement receptor 2 (Nagar et al. 1998). The 294 residue selenomethionyl-labelled protein contains eight selenium atoms. Atomic positions for the Se atoms were obtained by the Patterson methods. The three data sets were treated as a multiple isomorphous replacement problem, using the remote wavelength data as 'native' and the peak and edge wavelength data as 'derivatives'. The resulting experimental map was of very high quality with an overall figure of merit of 0.8 (0.9 after solvent flattening).

Considerable difficulty has been encountered in the determination of the anomalous scattering substructure when the number of such atoms exceeds 18–20. This was the case for N-myristoyl transferase (Weston et al. 1998). The asymmetric unit contains 3×52 kDa monomers with six Se atoms per monomer. The 18 Se atoms could not be located directly. Eventually, a xenon derivative was prepared. The Xe atoms were located, and statistical heavy atom refinement of the Xe positions with SHARP was used to generate the phase probability distributions that allowed a determination of the Se positions. As mentioned previously, the largest Se-met substructure (30 atoms) solved to date from a single wavelength or from a MAD analysis of protein data in the absence of any previously determined phases belongs to the structure of Se-met S-adenosylhomocysteine hydrolase (Turner et al. 1998). The asymmetric unit contains a dimer of 2×432 residues with 2×15 Se atoms. Using single wavelength data (peak data set), the atomic positions of Se atoms were determined by the 'Shake-and-Bake' method using program Snb1v2.0 (Miller et al. 1994) in conjunction with anomalous difference E-magnitudes (diffE) (Smith et al. 1998). The Se positions were then refined by treating the data as a pseudo-MIR case.

3.7.1.2 Phasing with mercury

The *lac* repressor (LacR) is a DNA-binding protein which is an example of the negative control of transcription, a central mechanism of metabolism and development in all organisms. The crystal structure of the tryptic core fragment of the *LacR* of *Escherichia coli* complexed with the inducer isopropyl-β-D-thiogalactosidose was determined at 2.6 Å resolution (Friedman *et al.* 1995). This 128 kDa tetramer was determined on the basis of a MAD electron density map at 4.5Å. The anomalous scatterer was mercury (4 atoms/tetramer), incorporated prior to crystallization by reacting the core fragment with ethylmercury phosphate. The MAD data were collected on imaging plates using cryocooled crystals at the F2 beam line at CHESS (Cornell University, USA). Due to a lack of beam time, only two wavelength data sets were collected. The Bijvoet differences were measured with the inverse beam geometry at $\lambda = 0.9840$ Å which is on the high energy side of the Hg L_{III}-absorption edge. The dispersive differences were measured by collecting data at the inflection point of the measured fluorescence spectrum. Each set of data used for MAD determination thus consisted of only three measurements of the same reflection matched in time and absorption path, a single measurement at the absorption edge and a Bijvoet pair on the high energy side of the edge. This choice of measurements efficiently records the anomalous signal for elements such as mercury which lack a strong f'' peak. The scaling of the MAD data and phase determination were done with a suite of programs, MADPRB derived from the original MADSYS package (Hendrickson 1991). MADPRB incorporates local block scaling algorithms for both Bijvoet and cross-wavelength scaling, a reformulation of the structure factor algebra with the use of $|F|$ instead of $|F|^2$, a χ^2 and Bayesian-based analysis of phase probabilities. The fairly large (61.9°) mean phase difference $\langle \Delta(\Delta\phi) \rangle$ between multiple independent MAD determinations of the same phase is the consequence of small anomalous signals (Bijvoet ratio 4.6%, noise contribution 3.2%). Initial MAD phases were improved by solvent flattening and four-fold symmetry averaging at 4.5 Å, and subsequent phase extension to 2.6 Å. The structure was refined to a crystallographic *R*-factor of 0.222.

The determination of this large structure by MAD, solvent flattening and symmetry averaging extends the usefulness of this combination to significantly smaller anomalous diffraction signals.

3.7.1.3 *Phasing with lanthanides*

The first crystal structure solved by the MAD technique was parvalbumin from *Opsanus tau* (Kahn *et al.* 1985, 1986a), a protein from a family with a known folding, and took advantage of the strong anomalous scattering at the L_{III} edge of Tb^{3+} ions substituted for native Ca^{2+} ions. Several proteins have since been solved using a similar approach. As an example, the structure of a fragment of N-cadherin, a specific adhesive molecule on the surface of special cells that recognizes and binds to cells of its own type, was solved by MAD using Yb^{3+} at the L_{III} edge (1.3878 Å) substituted for an intrinsic Ca^{2+} ion (Shapiro *et al.* 1995). The experiment, performed at the ESRF, demonstrated the feasibility of using an undulator beam for MAD experiments.

A textbook example is the structure solution of the Ca-dependent lectin domain from a rat mannose-binding protein (MBP-A-F2), where four lanthanides in the 230-residue asymmetric unit produced exceptionally large

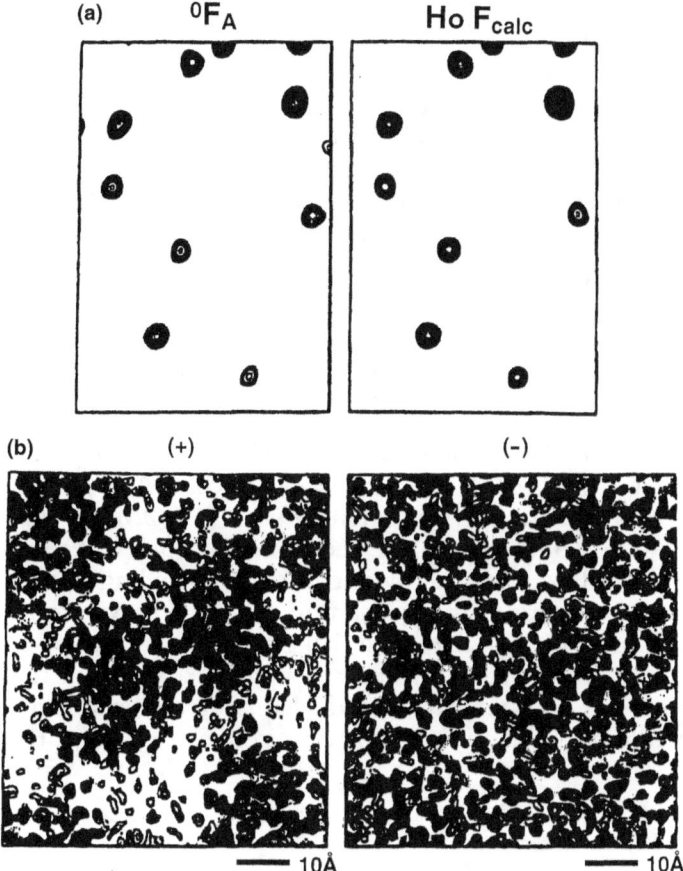

Fig. 3.12. MAD phasing of MBP-A-F2. Determination of Ho positions and choice of hand. (a) The $u\,1/2\,w$ section of the $|^0F_A|$ Patterson map, contoured in increments of 0.5σ, starting at 2.5σ. In addition to four self-vectors, there are numerous cross-vectors on this section because three of the four Ho sites have y coordinates near 0 or 1/2 (left). The same section calculated from the refined Ho model (right). (b) Comparison of 2.5 Å electron density maps computed from the two enantiomorphic choices of Ho partial structure. The Ho^{3+} appear as black circles. Each panel shows the equivalent 10 Å slab along the c-axis of one unit cell. The correct enantiomer is shown on the left. Both maps are contoured in 0.5σ intervals, starting at 1.0σ. [Figure reproduced from Weis et al. (1991).]

anomalous signals. The structure of the Ho^{3+} derivative was solved in a three-λ MAD experiment at room temperature (Weis et al. 1991; Fig. 3.12). A few years later, another four-λ MAD experiment at 1.8 Å resolution was performed at low temperature with the Yb^{3+} derivative (Temple-Burling et al. 1995). The redundantly determined phases for 99% of possible reflections had average agreement of 13°. Solvent flattening was not used. The correlation coefficients between the electron density map calculated with the refined (multiconformer) model and the experimental map are respectively, 0.93 and 0.91 for the two protomers in the asymmetric unit cell. A detailed image of both protein and solvent structures could thus be obtained without resort to phases calculated from a model. A similar experiment has been performed recently on human psoriasin, a small protein belonging to the EF hand class of calcium-binding proteins. Native Ca^{2+} ions were exchanged by Ho^{3+} ions and phasing at 1.9 Å resolution was performed exclusively from MAD data collected around the Ho L_{II} absorption edge. The structure was refined using another data set at 1.05 Å. These results illustrate the potential of MAD for producing, in favourable cases, very accurate model bias-free phases at high resolution.

3.7.1.4 *Phasing with bromine*

The nonamer d(GC[Br]GAAAGCT) was crystallized as a bromo-5-cytosine-2 derivative in the presence of cobalt hexamine. The structure was solved by the MAD method at the Br K-edge. Parameters of the bromine atom and phases were calculated with SHARP (La Fortelle and Bricogne 1997) and the enantiomorph was selected. After solvent flattening with SOLOMON (Abrahams and Leslie 1996), tracing of the phosphate backbone was unclear at some points. At this stage, a cobalt hexamine was identified. After including this cobalt and its weak anomalous scattering into the second round of heavy atom refinement with SHARP, the tracing of the entire DNA molecule became trivial. The very high quality map revealed an unexpected zipper-like motif in the middle of a standard B-DNA duplex. Four central adenines are intercalated and stacked on top of each other, without interstrand Watson-Crick base pairing (Shepard et al. 1998).

3.7.2 SAD

Recent works have demonstrated that the combination of high accuracy data collection, direct methods for solving the partial structure and statistical methods giving unbiased phases for solvent flattening can produce interpretable electron density maps from SAD data even with very small anomalous signals (La Fortelle, private communication). A striking example is the test SAD phasing of native tetragonal hen egg white lysozyme (HEWL) (Dauter et al. 1998). Accurate and highly redundant data were collected in four passes of 180° rotation with different exposure times, using synchrotron radiation ($\lambda = 1.54$ Å). The R-merge of the 17964 unique reflections (resolution 1.53 Å) is 0.043 and the overall signal-to-noise ratio is 73 (10 in the highest resolution shell). The positions of the native 10 S atoms (four Cys–Cys bridges and two Met) and seven chlorine ions in the solvent were obtained from 'half-baked' direct methods implemented in SHELXD. These atoms were treated as heavy atoms for phasing by ML methods, utilizing only their anomalous scattering signal ($f'' = 0.5$ and $0.7\,e^-$, respectively). Easily interpretable electron density maps were obtained after density modification.

3.7.3 SIRAS

3.7.3.1 Phasing with mercury

Mercury is a very heavy atom ($Z=80$), giving a strong isomorphous signal. The L_{III} absorption edge is at 1.0091 Å, a wavelength that is commonly accessible at synchrotron radiation facilities, and just below this edge the imaginary component for cysteinyl mercury is 9.991 e$^-$ (Tesmer *et al.* 1994). Reactive sulphydryl groups are not only common, but in addition they can be engineered into a protein by site-directed mutagenesis. A new cysteine can be introduced to create a potential site for mercury, or conversely a pre-existing cysteine can as well be removed if its reaction with mercury perturbs the crystal structure. SIRAS phasing with mercury optimized by the use of 1 Å wavelength is thus especially attractive. The structure of NDPK from *Dictyostelium discoideum* (Dumas *et al.* 1992) is an example of a protein structure solved by this method. As the protein was devoid of cysteines, the mutant with histidine 122 replaced by a cysteine was engineered and crystallized in space group P6$_3$22. The mercury derivative was highly isomorphous with the native structure and the chain tracing after solvent flattening was easy. The phases of the final refined model are 42.4° away on average from the initial SIRAS phases or 31.6° after solvent flattening.

3.7.3.2 Phasing with noble gases

For krypton, the maximum practical value of f'', just below the K edge at 0.86 Å, is 3.91 e$^-$. A test SIRAS experiment (Schiltz *et al.* 1997b) was done on a crystal of porcine pancreatic elastase (PPE), a protein with a molecular weight of 25.9 kDa. Data to a resolution of 1.9 Å were collected at 0.86 Å for a native PPE crystal and the same crystal pressurized under a krypton atmosphere of 56 bar.

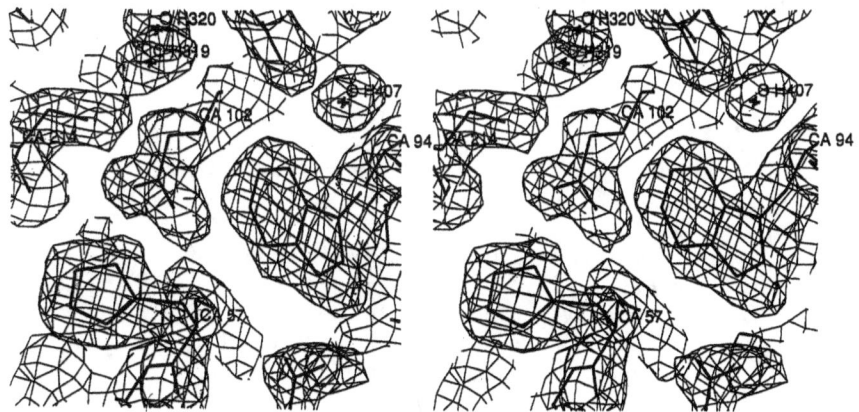

Fig. 3.13. Representative portion of the 1.87 Å solvent-flattened electron density map of porcine pancreatic elastase (stereo view). Initial phases were obtained by the SIRAS method with a krypton derivative. The refined native structure model (PDB entry: 3EST) is superimposed to this map. Most of the internal water molecules of this model show up as spherical density peaks in the experimental map. [Results communicated by M. Schiltz (LURE).]

After a careful data analysis and local scaling, a single Kr site was found on an anomalous difference Patterson map. Parameters of the partial structure were refined and SIRAS phases were calculated using ML (La Fortelle and Bricogne 1997). The solvent-flattened map at 1.87 Å resolution (see Fig. 3.13) has a correlation of 0.90 with the map calculated from the final refined model, which is remarkable, considering that the occupancy of the single krypton site is 0.5, so that the SIRAS phasing is based on an isomorphous difference $\Delta f = 18\,e^-$ and $f'' = 1.9\,e^-$. This result demonstrates the effectiveness of combining a highly isomorphous noble gas derivative, accurate and redundant data collection on a single sample at the optimal wavelength, local scaling of diffraction data, bias-free statistical treatment of the isomorphous and anomalous signals and solvent flattening.

The same approach would be more powerful with xenon derivatives, as xenon provides stronger isomorphous and anomalous signals than krypton and requires lower pressure for similar heavy atom site occupancies. The accessible range of wavelengths is limited on the long wavelength side by the strong absorption of compressed xenon, which makes the use of L edges difficult. A fair compromise is found in the wavelength range from 1Å ($f'' = 3.55\,e^-$) to 1.7 Å ($f'' = 8.61\,e^-$). Extrapolating the results of the PPE-Kr experiment, an optimal SIRAS experiment at 1.54 Å might allow the structure of a $\approx 90\,\mathrm{kDa}$ macromolecule with a single high occupancy Xe site to be solved.

Acknowledgements

We are grateful to the editors of *Progress in Biophysics and Molecular Biology* for the right to reproduce certain extracts from a review written by R.F., R.K. and W.S. (Fourme *et al.* 1996). We would also like to thank the referee for constructive comments.

References

Abrahams, JP and Leslie, AGW (1996). *Acta Crystallographica*, **D52**, 30.
Arndt, UW, Greenhough, TJ, Helliwell, JR, Howard, JAK, Rule, SA and Thompson, AW (1982). *Nature*, **298**, 835.
Bella, J and Rossmann, MG (1998). *Acta Crystallographica*, **D54**, 159.
Blow, DM and Crick, FHC (1959). *Acta Crystallographica*, **12**, 794.
Bricogne, G (1974). *Acta Crystallographica*, **A30**, 395.
Bricogne, G (1985). Unpublished lecture given at the Bischenberg Conference on the Crystallography of molecular biology (Bischenberg, France).
Bricogne, G (1988). *Acta Crystallographica*, **A44**, 517.
Bricogne, G (1991). In *Isomorphous replacement and anomalous scattering, Proceedings of CCP4 Study Week-end*, Wolf, W, Evans, PR and Leslie, AGW, eds, SERC, Daresbury Laboratory, Warrington, UK, p. 80.
Bricogne, G (1993). *Acta Crystallographica*, **D49**, 37.
Bricogne, G (1997). In *Methods in Enzymology: Macromolecular Crystallography*, **276**, Carter, CW and Sweet, RM, eds, Academic Press: New York, p. 361.

Brünger, AT, Temple-Burling, F, Grost, P, Flaherty, K and Weis, WI (1995). *American Crystallographic Association Meeting*, Montreal, Canada, Abstracts, p. 51.
Carter, CW, Crumley, KV, Coleman DE, Hage, F and Bricogne, G (1990). *Acta Crystallographica*, **A46**, 57.
Chiadmi, M, Kahn, R, La Fortelle E de and Fourme, R (1993). *Acta Crystallographica*, **D49**, 522.
Dauter, Z, La Fortelle, E de and Sheldrick (1998). *American Crystallographic Association Meeting*, Collected Abstracts.
Drenth, J (1994). *Principles of protein X-ray crystallography*, Springer-Verlag: New York.
Dumas, C, Lascu, I, Morera, S, Glaser, P, Fourme, R, Wallet, V, Lacombe, ML Véron, M and Janin, J (1992). *EMBO Journal*, **11**(9), 3203.
Dyda, F, Hickman, AB, Jenkins, TM, Engelman, A, Craigie, R and Davies, DR (1994). *Science*, **266**, 1981.
Einstein, RJ (1977). *Transactions of the American Crystallographic Association*, **A33**, 75.
Evans, G, Pettifer, RF and Wilson, KF (1996). *Review of Scientific Instruments*, **67**(10), 3428.
Fan, HF (1965). *Acta Physica Sinica*, **21**, 1105.
Fan, HF and Gu, YX (1985). *Acta Crystallographica*, **A41**, 280.
Fanchon, E and Hendrickson, WA (1990). *Acta Crystallographica*, **A46**, 809.
Fourme, R, Bahri, A, Kahn, R and Bosshard, R (1991). *Proceedings of european workshop on X-ray detectors for synchrotron radiation sources*, Aussois (France), Walenta, AH, ed. Siegen University, Germany, p. 16.
Fourme, R and Hendrickson, WA (1990). In *Synchrotron radiation and biophysics*, Hasnain, SS, ed. Chichester: Ellis Horwood Limited, p. 156.
Fourme, R, Shepard, W, Kahn, R, L'Hermite, G and Li de La Sierra, I (1995). *Journal of Synchrotron Radiation*, **2**, 36.
Fourme, R, Shepard, W and Kahn, R (1996). *Progress in Biophysics and Molecular Biology*, **64**(2–3), 167.
Friedman, AM, Fischmann, TO and Shamoo, Y (1994). *American Crystallographic Association Meeting*, Collected Abstracts TRN07.
Friedman, AM, Fischmann, TO and Steitz, TA (1995). *Science*, **268**, 1721.
Furey, W and Swaminathan, S (1990). *American Crystallographic Association Meeting*, Collected Abstracts PA33, **18**, 73.
Giacovazzo, C (1983). *Acta Crystallographica*, **A39**, 585.
Glover, ID, Denny, RC, Nguti, ND, McSweeney, SM, Kinder, SH, Thompson, AW, Dodson, EJ, Wilkinson, AJ and Tame, JRH (1995). *Acta Crystallographica*, **D51**, 39.
Green, EA (1979). *Acta Crystallographica*, **A35**, 351.
Green, DW, Ingram, VM and Perutz, MF (1954). *Proceedings of the Royal Society*, **A225**, 287.
Hao, Q and Woolfson, MM (1989). *Acta Crystallographica*, **A45**, 794.
Hauptman, H (1982). *Acta Crystallographica*, **A38**, 632.
Hendrickson, WA (1985). *Transactions of the American Crystallographic Association*, **21**, 11.
Hendrickson, WA (1991). *Science*, **254**, 51.
Hendrickson, WA (1994). In *Resonant anomalous X-ray scattering, theory and applications*, Materlik, G, Fischer, K and Sparks, CJ, eds, Amsterdam, Elsevier, p. 159.
Hendrickson, WA and Lattman, EA (1970). *Acta Crystallographica*, **B26**, 136.
Hendrickson, WA and Teeter, MM (1981). *Nature*, **290**, 107.
Hendrickson, WA and Ogata, CM (1997). In *Methods in enzymology: macromolecular crystallography*, **276**, Carter, CW and Sweet, RM, eds, Academic Press: New York, p. 494.
Hendrickson, WA, Smith, JL, Phizackerley, RP and Merritt, EA (1988). *Proteins*, **4**, 77.

Hendrickson, WA, Pähler, A, Smith, JL, Satow, Y, Merritt, EA and Phizackerley, RP (1989). *Proceedings of the National Academy of Sciences USA*, **86**, 2190.
Hendrickson, WA, Horton, JR and LeMaster, DM (1990). *EMBO Journal*, **9**, 1665.
James, RW (1958). *The optical principles of the diffraction of X-rays*, Bell & Sons, London, p. 135.
Kabsch, W (1988). *Journal of Applied Crystallography*, **21**, 916.
Kahn, R and Fourme, R (1996). In *Methods in Enzymology: Macromolecular Crystallography*, **276**, Carter, CW and Sweet, RM, eds, Academic Press, New York, p. 268.
Kahn, R, Fourme, R, Bosshard, R, Chiadmi, R, Risler, JL, Dideberg, O and Wery, JP (1985). *FEBS Letters*, **179**, 133.
Kahn, R, Fourme, R, Bosshard, R, Chiadmi, M, Risler, JL, Brunie S, Wery, JP, Dideberg, O and Janin, J (1986a). In *Structural biological applications of X-ray absorption, scattering and diffraction*, Bartunik, HD and Chance, B, eds, Academic Press, New York, p. 297.
Kahn, R, Fourme, R, Bosshard, R and Saintagne, V (1986b). *Nuclear Instruments and Methods*, **A246**, 596.
Karle, J (1980). *International Quantum Chemistry Symposium*, **7**, 357.
Karle, J (1984). *Acta Crystallographica*, **A40**, 1.
Karle, J (1989). *Acta Crystallographica*, **A45**, 303.
Klapper, MH (1977). *Biochemical and Biophysical Research Communications*, **78**, 1018.
Korszun, ZR (1987). *Journal of Molecular Biology*, **196**, 413.
La Fortelle, E de and Bricogne, G (1997). In *Methods in Enzymology: Macromolecular Crystallography*, **276**, Carter, CW and Sweet, RM, eds, Academic Press, New York, p. 472.
Lee, PL and Ogata, CM (1995). *Journal of Applied Crystallography*, **28**, 661.
Lee, JO, Rieu, P, Arnaout, MA and Liddington, R (1995). *Cell*, **80**, 631.
Lehmann, MS, Müller, HH and Stuhrmann, HB (1993). *Acta Crystallographica*, **D49**, 308.
Leslie, A (1994). *Mosflm User Guide, Mosflm version 5.20*, MRC Laboratory of Molecular Biology, Cambridge, UK
Liu, Y, Ogata, CM and Hendrickson, WA (1995). *American Crystallographic Association Meeting*, Collected. Abstracts p. 52.
Millau, J, Beuville, E, Cork, C, Earnest, T, Nygren, D, Padmore, H, Xuong, NH and Datte, P (1995). *5th International Conference on Biophysics and Synchrotron Radiation (BSR95)*, Grenoble, France. Collected Abstracts 06/06.
Miller, R, Gallo, SM, Khalak, HG, and Weeks, CM (1994). *Journal of Applied Crystallography*, **27**, 613.
Müller, CW, Rey, FA, Sodeoka, M, Verdine, GL and Harrison, SC (1995). *Nature*, **373**, 311.
Nagar, B, Jones, RG, Diefenbach, RJ, Isenman, DE and Rini, JM (1998). *Science*, **280**, 1277.
Narayan, R and Ramaseshan, S (1981). *Acta Crystallographica*, **A37**, 636.
Oed, A (1988). *Nuclear Instruments and Methods*, **A263**, 351.
Ogata, CM, Hendrickson, WA, Gao, X, and Patel, DJ (1989). *Abstracts of the American Crystallographic Association Meeting*, Series 2, **17**, p. 53.
Okaya, Y and Pepinsky, R (1956). *Physical Review*, **103**, 1645.
Okaya, Y, Saito, Y and Pepinsky, R (1955). *Physical Review*, **98**, 1857.
Otwinowski, Z (1991). In *Isomorphous replacement and anomalous scattering. Proceedings of CCP4 Study Weekend*, Wolf, W, Evans, PR and Leslie, AGW, eds, SERC, Daresbury Laboratory, Warrington, UK, p. 80.
Otwinowski, Z (1993). In *Data collection and processing*, Sawyer, L, Isaacs, N and Bailey, S, eds, SERC Daresbury Laboratory, Warrington, UK, p. 56.

Pähler, A, Smith, JL and Hendrickson, WA (1990). *Acta Crystallographica*, **A46**, 537.
Peterson, MR, Harrop, SJ, McSweeney, SM, Leonard, GA, Thompson, AW, Hunter, WN and Helliwell, JR (1996). *Journal of Synchrotron Radiation*, **3**(1), 24.
Phillips, JC and Hodgson, KO (1980). *Acta Crystallographica*, **A36**, 856.
Prangé, T, Schiltz, M, Pernot, L, Colloc'h, N, Longhi, S, Bourguet, W and Fourme, R (1998). *Proteins*, **30**, 61.
Raiz, VSH and Andreeva, NS (1970). *Kristallographya*, **15**, 246.
Ralph, AC and Woolfson, MM (1991). *Acta Crystallographica*, **A47**, 553.
Ramachandran, JN and Raman, S (1956). *Current Science*, **25**, 348.
Ramin, M (1999). Doctoral Thesis, University of Paris-Sud, Orsay, France.
Ramin, M, Shepard, W, Kahn, R and Fourme, R (1999). *Acta Crystallographica* **D55**(1), 157.
Ramakrishnan, V and Biou, V (1997). In *Methods in Enzymology: Macromolecular Crystallography*, **76**, Carter, CW and Sweet, RM, eds, Academic Press, New York, p. 538.
Ramakrishnan, V, Finch, JT, Graziano, V, Lee, PL and Sweet, RM (1993). *Nature*, **362**, 219.
Rossmann, MG (1961). *Acta Crystallographica*, **14**, 383.
Sasaki, S (1984). *Anomalous scattering factors for synchrotron radiation users calculated using the Cromer and Liberman method*, National Laboratory for High Energy Physics, KEK, Tsukuba, Japan.
Schiltz, M, Prangé, T and Fourme, R (1994). *Journal of Applied Crystallography*, **27**, 950.
Schiltz, M, Kvick, A, Svensson, OS, Shepard, W, La Fortelle, E de, Prangé, T, Kahn, R, Bricogne, G and Fourme, R (1997a). *Journal of Synchrotron Radiation*, **4**, 287.
Schiltz, M, Shepard, W, Fourme, R, Prangé, T, La Fortelle, E de and Bricogne, G (1997b). *Acta Crystallographica*, **D53**, 78.
Schoenborn, BP, Watson, HC and Kendrew, JC (1965). *Nature*, **207**, 28.
Shapiro, L, Fannon, AM, Kwong, PD, Thompson, A, Lehmann, MS, Grübel, G, Legrand, JF, Als-Nielsen, J, Colman, DR and Hendrickson, WA (1995). *Nature*, **374**, 327.
Shepard, W, Cruse, WBT, Fourme, R, La Fortelle, E de and Prangé, T (1998). *Structure*, **6**(7), 849.
Smith, JL (1991). *Current Opinion in Structural Biology*, **1**, 1002.
Smith, GD, Nagar, B, Rini, JM, Hauptman, HA and Blessing, RH (1998). *Acta Crystallographica*, **D54**, 799.
Stuhrmann, H, Goerigk, G and Munk, B (1991). In *Handbook on synchrotron radiation*, Vol. 4, Ebashi, S, Koch, M and Rubenstein, E, eds, North Holland, Amsterdam. p. 555.
Temple-Burling, F, Weis, WI, Flaherty, KM and Brünger, AT (1995). *Science*, **271**, 72.
Templeton, LK and Templeton, DH (1982). *Acta Crystallographica*, **A44**, 1045.
Terwilliger, TC (1994a). *Acta Crystallographica*, **D50**, 11.
Terwilliger, TC (1994b). *Acta Crystallographica*, **D50**, 17.
Terwilliger, TC and Eisenberg, D. (1983). *Acta Crystallographica*, **A39**, 813.
Tesmer, JJG, Stemmler, TL, Penner-Hahn, JE, Davisson, J and Smith, JL (1994). *Proteins*, **18**, 394.
Turner, MA, Yuan, C-S, Borchardt, RT, Hersfield, MS, Smith, GD and Howell, PL (1998). *Nature Structural Biology*, **5**(5), 369.
Vitali, J, Robbins, AH, Almo, SC and Tilton, RF (1991). *Journal of Applied Crystallography*, **24**, 931.
Wang, BC (1995). *American Crystallographic Association Meeting*, Montreal, Canada, Abstracts p. 52.

Weis, WI, Kahn, R, Fourme, R, Drickamer, K and Hendrickson, WH (1991). *Science*, **254**, 1608.

Weston, SA, Camble, R, Colls, J, Rosenbrock, G, Taylor, I, Egerton, M, Tucker, AD, Tunnicliffe, A, Mistry, A, Mancia, F, de la Fortelle, E, Irwin, J, Bricogne, G and Pauptit, RA (1998), *Nature Structural Biology*, **5**(3), 213.

Woolfson, MM (1987). *Acta Crystallographica*, **A43**, 593.

Woolfson, MM and Yao J-X (1994). *Acta Crystallographica*, **D50**, 7.

Yamamoto, M, Kumasaka, T, Fujisawa, T and Ueki, T (1998). *Journal of Synchrotron Radiation*, **5**(3), 222.

Yang, W, Hendrickson, WA, Crouch, RJ and Satow, Y (1990). *Science*, **249**, 1398.

4
Electron density—calculation, modification and interpretation

Morten Kjeldgaard

4.1 Electron density maps

4.1.1 The electron density equation

For every diffracted beam in a diffraction experiment, the intensity I_{hkl} of reflection hkl is proportional to the square of the structure factor F_{hkl}. The structure factor can be expressed as a summation of waves scattered from every infinitesimal volume element dV by the expression

$$F(S) = \int_{\text{cell}} \rho(r) \exp(2\pi i r \cdot S)\, dV, \qquad (4.1)$$

where $\rho(r)$ is the electron density at position $r = (x, y, z)$, and S is the scattering vector (h, k, l). If we introduce the atomic scattering factor f_j, the integral reduces to a sum over the atoms j in the unit cell, and the structure factor can be expressed as

$$F(S) = \sum_j f_j(S) \exp(2\pi i r_j \cdot S), \qquad (4.2)$$

where r_j describes the coordinates of the jth atom. As seen from this equation, the structure factor F_{hkl} is a complex number, composed of an amplitude and a phase. Each structure factor can be thought of as representing a wave, which is the sum of the individual waves scattered from every atom in the unit cell. This 'sum-wave' has an associated phase shift, which is lost in the diffraction experiment, and must be determined by other means, normally using heavy atom derivatives. The wavelength of the component waves is determined by the wavelength of the incoming radiation in the diffraction experiment.

From eqn (4.1) it can be seen that $F(S)$ is the Fourier transform of $\rho(r)$, so using the inverse Fourier transform and the fact that the structure factors are discrete, the electron density can be expressed in terms of a triple sum over all reflections,

$$\rho(r) = \frac{1}{V} \sum_h \sum_k \sum_l F(S) \exp(-2\pi i r \cdot S).$$

Using Euler's notation for the structure factor $F(S) = |F_{hkl}| \exp(i\alpha)$, the electron density function can be rewritten in the familiar form

$$\rho(x, y, z) = \frac{1}{V} \sum_h \sum_k \sum_l |F_{hkl}| \exp(-2\pi i(hx + ky + lz) + i\alpha_{hkl}). \quad (4.3)$$

From this equation it can be seen that the electron density function is composed of a sum of *density waves* in the unit cell, each generated by a single reflection. The wavelength of each density wave is determined by the scattering vector $S = (h, k, l)$. In Fig. 4.1, examples of such density waves are shown, and it is seen that errors in the amplitudes result in electron density peaks of the wrong magnitude, whereas errors in the phases result in peaks at wrong positions. Therefore, the interpretability of the electron density is much more sensitive to

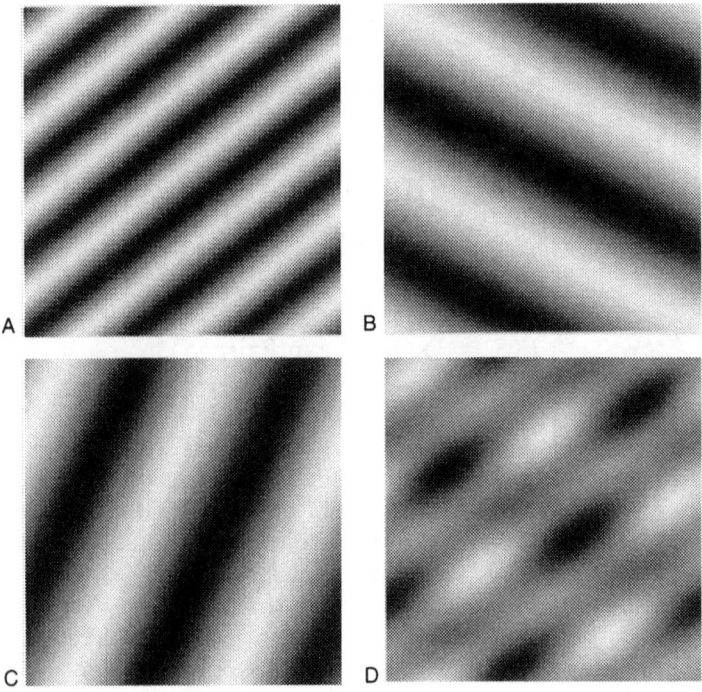

Fig. 4.1. Density waves in two dimensions, x-direction is down, y-direction across. A: Density waves from the reflection [3 4 0] with the phase 0°. The density wave intersects the x-direction four and the y-direction three times. B: Reflection [−1 2 0], phase 180°. The phase shift affects the position of the wave relative to the origin. C: Reflection [2 1 0], phase 90°. D: The summation of the density waves from the three reflections A–C. Peaks appear where the component waves reinforce each other; here all reflections have the same amplitude so they contribute equally to the sum. Waves with high *hkl* numbers have short wavelengths and contribute to the details of the image, whereas the low order terms contribute to the gross features of the density. The figure was produced using the FOURDEM program (Welberry and Owen 1992).

errors in the phases, as these tend to result in discontinuous density. However, errors in the measurement of amplitudes are also reflected in the phases; it is thus crucial to have high quality experimental data!

From eqn (4.3), it is seen that the electron density at every position in the unit cell is a sum over all reflections, so each structure factor contributes at every position in the cell. Therefore, it is possible to calculate the electron density from any subset of structure factors. Conversely, it is seen from eqn (4.2) that every atom contributes to every structure factor, so that the structure factor amplitude and phase can be calculated with the knowledge of the atomic positions.

Calculating the electron density using only a subset of the structure factors results in a number of effects. Thus, using only the low resolution reflections results in a considerable loss of detail in the electron density map, and a 'blurred' image of the molecule is obtained. In practice, the resolution of data from a particular crystal will be limited, due to inherent disorder in the crystal, and the magnitude of the structure factors will approach zero near the resolution limit. In other words, there is a 'natural termination' of the Fourier series. In general, truncating the Fourier series by leaving out significant terms will result in ripples in the electron density. This effect can be minimized by using all available data.

In a diffraction experiment, there will be errors in the measured intensities, which again result in errors in the phases. As long as these errors are random, they will contribute only to noise.

A lot of introductory material to help the understanding of the Fourier transform in crystallography, as well as diffraction theory in general, can be found on the Internet. The reader is encouraged to look at Kevin Cowtan's *Book of Fourier* (Cowtan 1996), and the *Teaching guide for X-ray and neutron diffraction* (Neder and Proffen 1996). Finally, an interactive tutorial called *The Interactive Structure Factor Tutorial* allows the reader to learn about structure factors, phases, symmetry and the relationship between the structure factors and the electron density map. The tutorial uses an interactive Java applet and runs on any computer system equipped with a recent Internet browser (Cowtan 1998).

4.1.2 *Calculation of maps*

The final experimental result of a crystallographic diffraction experiment is in the form of an *electron density map*. This is the calculation of the three-dimensional electron density function on a regular three-dimensional grid. Because the Fourier transform consists of a sum over a large number of terms involving sines and cosines, the calculation is normally quite slow. Therefore, the electron density map is usually calculated using the fast Fourier transform (FFT) (Drenth 1994). The grid-spacing used for the calculation is by rule-of-thumb about one-third of the maximum resolution, but should never be more than approximately 1 Å.

When two or more heavy atom derivatives are available, the phase angles can be determined unambiguously from the experiment (multiple isomorphous replacement, MIR), whereas a single heavy atom derivative results in two equally probable protein phase angles (single isomorphous replacement, SIR).

Further addition of independent derivatives permits the choice between these two solutions. However, the SIR method can occasionally provide electron density maps sufficiently accurate for an initial interpretation, provided that the arrangement of heavy atoms is not centrosymmetric. Phases determined by these methods are referred to as *observed* phases.

If the wavelength of the incident X-ray beam is close to an absorption edge of an atom, the scattering process can cause changes in the quantum state of the electrons within that atom. This leads to a small, but significant, phase shift of the scattered radiation that does not depend very much on scattering angle. The anomalous scattering (AS) from the elements carbon, nitrogen, oxygen and hydrogen that constitute macromolecules is negligible at the wavelengths used for X-ray diffraction. However, the AS component from heavy elements is significant and can be used for the determination of phases, if the location of the anomalous scatterer can be found. Intensity differences due to AS are most often used in combination with the isomorphous differences in the SIRAS or MIRAS phasing methods.

When a wavelength tunable synchrotron X-ray beam is available, AS at several wavelengths can be used for phase determination, as is done in the multi-wavelength anomalous diffraction (MAD) method (see Chapter 3 by R. Fourme *et al.*). Most recently, with the statistical phasing program SHARP (de la Fortelle and Bricogne 1997), it is feasible to determine accurate phases using only the AS at a single wavelength [the single anomalous diffraction (SAD) method].

4.1.3 Phase combination

If other sources of phase information are available, this information can be included by employing a *phase combination* procedure. The phase information available from an isomorphous replacement experiment is usually described in terms of the *phase probability curve* for each reflection. Hendrickson and Lattman (1970) realized that the phase probability distribution could be described by its four Fourier coefficients A, B, C and D,

$$P(\alpha) = N \exp[K + A \cos \alpha + B \sin \alpha + C \cos 2\alpha + D \sin 2\alpha],$$

where $P(\alpha)$ is the probability of the phase angle α, and N and K are constants. Phase combination can then be carried out by simply adding the Hendrickson–Lattman coefficients A, B, C and D of the probability distributions for each reflection.

4.1.4 Difference maps

The initial density map calculated from the MIR map is used for the initial interpretation of the structure. But once an atomic model exists, it is possible to include the phase information resident in the atomic model. This is commonly done using difference maps with coefficients of the form $mF_o - nF_c$, where m and n are integers, F_o are the observed structure factor amplitudes, and F_c those calculated from the model. The phases used in the Fourier synthesis are most commonly α_c.

A difference map computed with coefficients $F_o - F_c$ shows positive peaks in the electron density where atoms should be added to the model, and negative peaks indicates atoms that should be removed from the model. Thus, an atom at a slightly incorrect position will give rise to a positive peak at the correct position with a negative peak next to it.

Another use of the $F_o - F_c$ difference Fourier map is the determination of the temperature factors of the atoms. When calculating structure factors from the model, a term for the correction of thermal vibration of the form $\exp(-B_j \sin^2 \theta / \lambda^2)$ is applied. If the parameter B_j is overestimated, the calculated electron density peak for the atom will be broader and lower than it should be. In such a case, the atom will give rise to a positive peak surrounded by a valley. Similarly, if B_j is underestimated, there will be a positive peak with a depression in the middle.

It can be shown that the heights of the peaks in an $F_o - F_c$ map are only half of what they would be in a normal Fourier map (Drenth 1994). The $F_o - F_c$ maps are commonly used to identify errors or not-yet built features in an existing model, or they can be used to identify ligands or changes due to single point mutations.

Maps calculated with coefficients $2F_o - F_c$ are normally used to guide the construction of the model. A $2F_o - F_c$ map not only shows density for the errors in the model, but also density for the regions that are correct, which makes building and interpretation of the map much easier. Whenever the model has been modified during model building, new features added, or errors corrected, a new and better $2F_o - F_c$ map can be calculated that will contain less noise than the previous one, because the phases have improved. This process can be repeated iteratively as the model building progresses. Many crystallographers prefer to use the $3F_o - 2F_c$ map for model building. If one rewrites the coefficients in the form $F_o + 2(F_o - F_c)$, it is seen that the $3F_o - 2F_c$ map can be thought of as an observed map plus two times a difference map, so that features in the difference map are being added to the observed map on the correct scale. Thus, errors and not-yet built features are more clearly seen in a $3F_o - 2F_c$ map. As the model becomes better and better, the features in the difference map tend to disappear, and the $3F_o - 2F_c$ and $2F_o - F_c$ maps come to look similar. Figure 4.2 shows a piece of a protein model with a $2F_o - F_c$ density map.

4.1.4.1 *The σ_A weighted maps*

One problem with normal difference maps is that the Fourier synthesis is carried out using the phases α_c calculated from the model. As seen in Section 4.1.1, the electron density map is extremely sensitive to errors in the phases.

Incomplete and unrefined models of macromolecules generally contain large coordinate errors, or may be partially incorrect, and electron density maps calculated using partial structure phases are *biased* towards the incorrect model. This means that errors in the model will tend to be reproduced in the $2F_o - F_c$ electron density map.

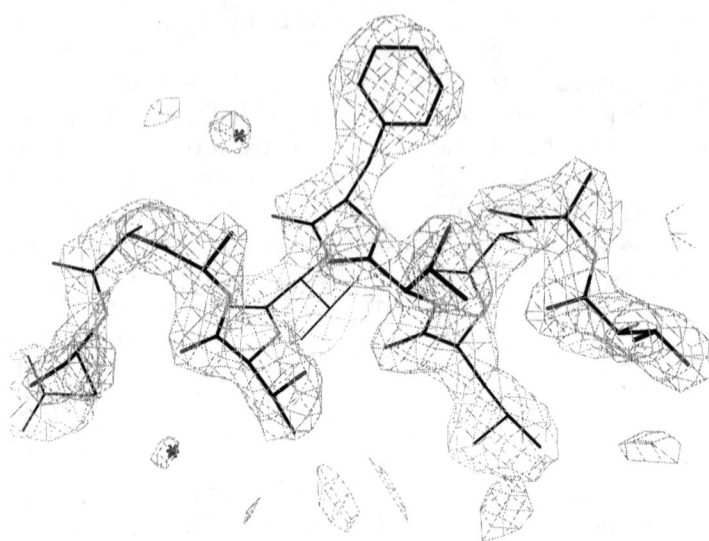

Fig. 4.2. A 2.5 Å $2F_o - F_c$ electron density map from a bit of helix A from elongation factor EF-Tu:GDPNP from *Thermus aquaticus* (Kjeldgaard *et al.* 1993) (see Plate Section).

The parameter σ_A is a combined measure of the completeness and accuracy of the partial structure, and Read (1986) devised a method for estimating the mean coordinate error of the model from the variation of σ_A with resolution. This leads to a new expression for the Fourier coefficients $2mF_o - DF_c$ that takes the coordinate errors into account and will suppress the model bias.

It is now common practice to compute σ_A weighted electron density maps by using $2mF_o - DF_c$ rather than the ordinary $2F_o - F_c$ Fourier coefficients.

4.1.5 *Phase improvement by density modification*

When an observed set of phases has been obtained, either from MIR, SIR or MAD, an initial electron density map can be calculated. Due to errors in the phases, this experimental electron density map is ordinarily quite noisy. This can be expressed symbolically as

Experimental map = Protein map + Noise map.

The noise map gives rise to positive density features in the solvent region, differences between non-crystallographically related molecules in the asymmetric unit, poor connectivity of electron density, negative density peaks and other spurious features. An unambiguous tracing of the molecule might be severely hampered by an insufficient quality of the electron density map, and it is therefore desirable to attempt to improve the map, through a process of *phase refinement*. This can be done by adding information external to the crystallographic experiment, by a *density modification* procedure. In practice, the criterion for quality of the electron density map is the crystallographer's perception of its interpretability, meaning the connectivity of density perceived as main chain, and how clearly the side-chains are imaged in the density.

Density modification takes place in real space, by imposing constraints on the values of the electron density function. An inverse Fourier transform generates a set of structure factor amplitudes and phases from the modified electron density map, and to maintain the experimental content, these are combined with the observed amplitudes and phases. The phase refinement takes place iteratively in the density modification cycle, shown schematically in Fig. 4.3.

4.1.5.1 Solvent flattening

The simplest form of density modification is solvent flattening, which takes advantage of the fact that there should be no significant features in the bulk solvent regions of the unit cell. In the protein molecule, the atoms are bonded with

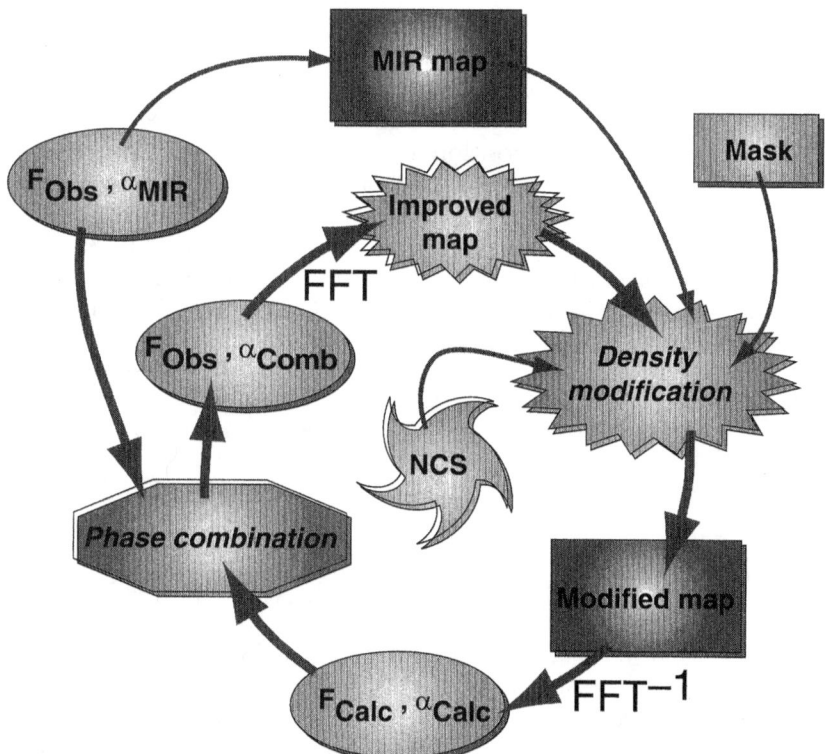

Fig. 4.3. The density modification cycle. Density modification is first carried out on the MIR map, and phases are computed from the modified map. These are combined with the original MIR phases, and a new, improved map is calculated. This procedure is iterated 5–8 times, while watching the statistical indicators $\sum |F_o - F_c|/\sum |F_o|$ and $\alpha_{\text{MIR}} - \alpha_{\text{Comb}}$ (see Plate Section).

a distance of approximately 1.5 Å. The solvent region primarily consists of hydrogen-bonded water molecules with a distance of approximately 2.7 Å. Furthermore, the solvent is in the liquid phase and owing to its dynamic character, the time-averaged electron density has a low constant value. The electron density is thus higher in the protein region. Most crystals of macromolecules have solvent content in the range 30–60%.

The phasing power of the solvent flattening procedure can be demonstrated mathematically. If $\mathcal{U}$ is the volume of the molecule, and V is the volume of the unit cell, the structure factor can be written as

$$F_k = \frac{\mathcal{U}}{V} \sum_h F_h G_{hk}, \qquad (4.4)$$

where

$$G_{hk} = \int_\mathcal{U} \exp(2\pi i (h-k) r) \, dV$$

as shown by Arnold and Rossman (1986) (see also Vellieux and Read 1997). This equation expresses the value of each structure factor F_k as a weighted sum over the whole of reciprocal space. The weight factor G_{hk} given by the volume integral term depends on the position and shape of the molecular envelope $\mathcal{U}$. G_{hk} has its largest value when $h \approx k$, and falls off steeply. Therefore reflections h with indices close to k will contribute most to a determination of F_k. The smaller the molecular envelope $\mathcal{U}$, the more structure factors will be related by eqn (4.4), or in other words, the more solvent there is to flatten, the more phasing power is obtained from the method. In Fig. 4.4, an example of the improvements that can be obtained by a solvent flattening procedure is shown.

Truncation of the density From the structure factor expression (4.2), it is readily realized that

$$F_{000} = \frac{1}{V} \sum_j Z_j. \qquad (4.5)$$

Because F_{000} coincides with the incident beam, it is not a measurable data point. When the electron density is calculated in the absence of F_{000}, its average value is ideally equal to 0. If the chemical composition of the macromolecule is known, F_{000} can be calculated according to eqn (4.5), and when the constant F_{000}/V is added to the electron density, the features should be positive everywhere.

When carrying out the solvent flattening procedure, each grid point in the solvent region should be set to the average value of the electron density outside the mask. Inside the mask, the effective F_{000}/V should be added to all grid points before truncation of negative densities. Figure 4.5 shows schematically how the electron density is modified by the solvent flattening procedure.

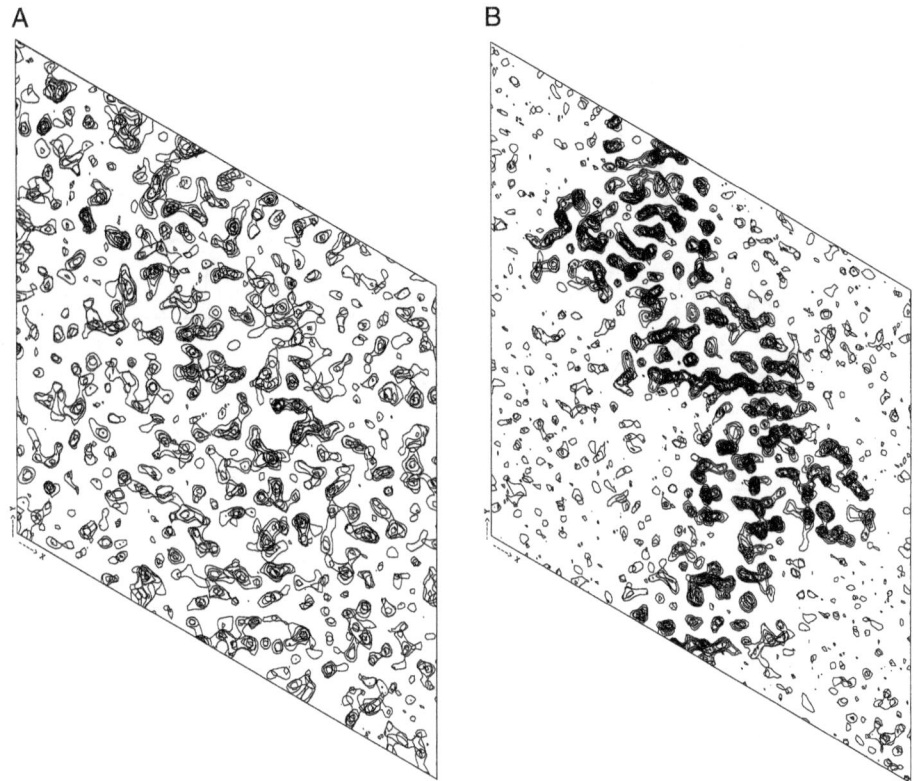

Fig. 4.4. Improvement achieved by solvent flattening of the electron density map from the receptor binding domain of bovine α_2-macroglobulin (Jenner *et al.* 1998). A: Selected sections from the experimental MIR map using phases to 2.8 Å. B: The same sections after density modification and phase extension to 1.9 Å has taken place using the SOLOMON program.

Solvent flipping From eqn (4.4) it is seen that the major contributor to the phase of reflection k is reflection k itself. Recently, Abrahams (1997) has introduced the γ correction, which uncouples this bias by subtracting the contribution of a reflection to itself. He demonstrated that the γ correction could be carried out in real space, by introducing a 'solvent flip' outside the mask,

$$\rho' = \rho + \frac{1}{(1-\gamma)}(\rho_{\text{avg}} - \rho).$$

When the factor $1/(1-\gamma)$ is 2, the features of the solvent are exactly inverted about ρ_{avg}, so peaks will be made into holes, and vice versa. Conventional solvent flattening corresponds to a zero value of γ. Electron density maps resulting from the 'solvent flipping' procedure are generally superior to those obtained from conventional solvent flattening. The γ correction is correlated with the solvent content and the degree of non-crystallographic symmetry (NCS) in the crystal. The procedure is implemented in the program SOLOMON (Abrahams 1997).

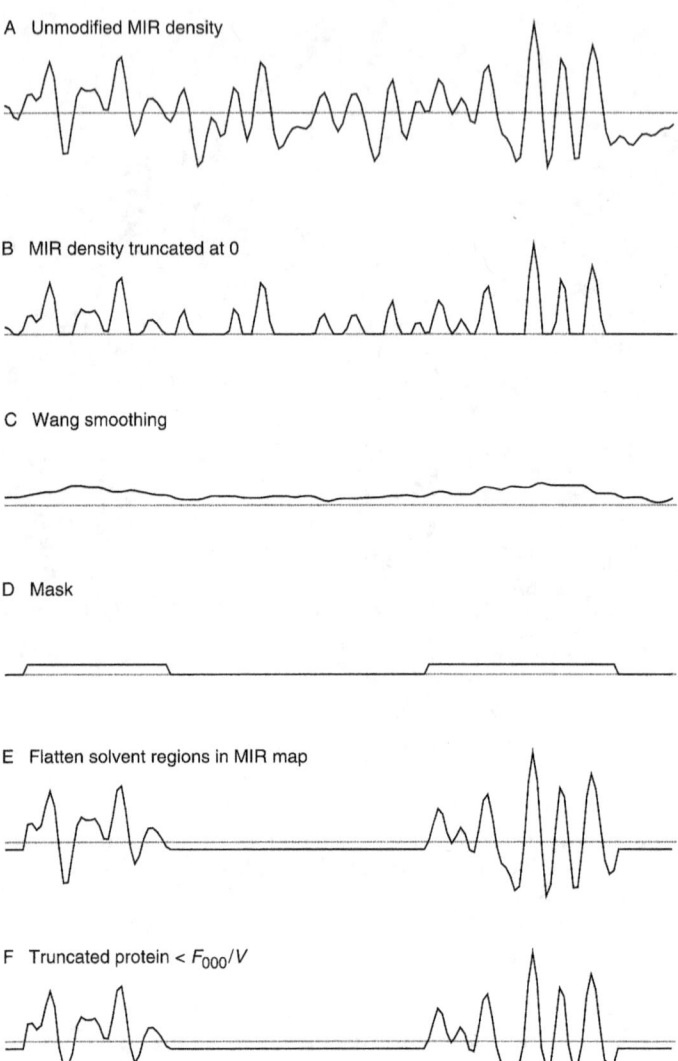

Fig. 4.5. Schematic illustration of the first step of a Wang type solvent flattening, performed in one dimension on the data from Fig. 4.4. A line of density along the entire z-axis of the unit cell is shown. A: The original unmodified density. B: Negative densities have been truncated. C: Wang smoothing within a sphere of radius 10 Å has been performed. D: Mask region has been assigned, corresponding to an estimated solvent content of 51%. E: Solvent regions have been set to the average solvent density value. F: In the protein region, negative peaks have been truncated at $-F_{000}/V$.

Other density modification methods   The program DM (Cowtan 1994) implements a number of other density modification procedures, such as histogram matching, Sayre's equation, skeletonization, and averaging (Zhang *et al.* 1997). The histogram matching technique attempts to bring the distribution of electron density values to match that of an ideal map. Ideal histograms for protein structures depend on the resolution and overall temperature factor, but they do not depend on the particular structure. Therefore, standard histograms can be obtained from existing structures, or they can be determined theoretically.

Sayre's equation describes the electron density in a structure consisting of equal and completely resolved atoms, using the additional information that atoms are spherical. For this reason, it is not strictly valid for macromolecular structures. In practice, however, Sayre's equation can still be applied in a modified form, using a sort of shape function for a spherically averaged polypeptide group, which can be derived empirically from the electron density map.

Obtaining the molecular envelope In order to perform solvent flattening, a molecular envelope must be defined. The envelope is usually represented by a *mask* function having the value 1 inside the molecular envelope and 0 outside.

An initial mask can be calculated automatically using the procedure of Wang (1985). The first step in this procedure is to calculate a smoothed MIR map by replacing the density at each grid point by a weighted average of the electron density at all grid points within a sphere of radius 5–10 Å. The problem with the method is that the calculation is extremely slow. Figure 4.5 illustrates the Wang mask generation procedure.

Leslie (1987) realized that Wang's smoothing procedure is actually a convolution in real space, which is equivalent to the multiplication of the structure factors with the Fourier transform of a sphere. The Wang calculation can thus be made very much more efficient in reciprocal space based on the FFT method. This procedure is implemented in the BNDRY program from the PHASES-95 package (Furey and Swaminathan 1997), in the DEMON package (Vellieux *et al.* 1995) and in DM (Cowtan 1994).

One problem with the Wang–Leslie type mask is that it is inaccurate, as it may cut ill-defined surface loops and other features; however, this is probably more critical when doing NCS averaging (see Section 4.1.5.2). It may therefore be preferable to construct an initial mask interactively, for example using the O program, which allows the user to view and edit the mask (Jones *et al.* 1991; Jones and Kjeldgaard 1997). Alternatively, the mask can be defined manually by digitalization of the electron density map using the program MAPVIEW from the PHASES-95 package, which allows the user to trace the molecular boundary section by section (Furey and Swaminathan 1997).

Regardless of the method used to generate the mask, it will need to be improved. In the MAMA program, the mask can be smoothed, islands and interior holes removed, and the mask can be trimmed for symmetry overlaps (Kleywegt and Jones 1994). A smooth mask decreases the risk of getting high order ripples in the Fourier transform of the mask; this is important when back transforming modified maps. Figure 4.6 shows a molecular mask that has been improved in this way.

4.1.5.2 *Non-crystallographic averaging*

A significant number of macromolecules form crystals that contain two or more identical copies in the asymmetric unit. The symmetry operators relating these

Fig. 4.6. The surface of a molecular mask, around one molecule of elongation factor EF-Tu: GDPNP (see Plate Section).

molecules are not reflected in the crystallographic symmetry, but are local to the unit cell, which is why they are called NCS operators. Because the NCS related molecules are assumed to be identical, any deviation from equality is treated as coming from errors in the phases. Therefore, the electron density map can be modified so each copy is replaced by the average density, and in addition solvent flattening is carried out. Molecular averaging of NCS related molecules is probably the most powerful density modification technique, especially if the number of copies is large. The theoretical basis for the phase refinement in the NCS averaging technique can be developed in a formalism equivalent to that given in expression (4.4), with the addition of the constraints imposed by the NCS (Arnold and Rossmann 1986; Kleywegt and Read 1997; Vellieux and Read 1997). In fact, the solvent flattening procedure discussed in the previous section can be regarded as a special case of one-fold averaging. Figure 4.7 schematically shows the molecular averaging operation.

The non-crystallographic averaging technique was first developed in a reciprocal space formulation as an *ab initio* phasing technique for virus structures in the classic paper by Rossmann and Blow (1962). Bricogne (1974) formally proved that NCS averaging could also be carried out in real space, and he provided a set of computer programs to carry out the computations, using an elegant double sorting technique to eliminate excessive computer memory requirements (Bricogne 1976). With the capabilities of modern computers, it is feasible to store the entire electron density maps in the random access memory of the computer, and several new program packages have emerged, capable of doing averaging and density modification. Several of the key references are given in this section, but an exhaustive account of the program packages for density modification can be found in Kleywegt and Read (1997).

Determining the NCS operators Determining the NCS operators is not always straightforward, but can be carried out in a number of ways (Kleywegt and Read 1997). The orientation of an NCS axis can be determined from an analysis

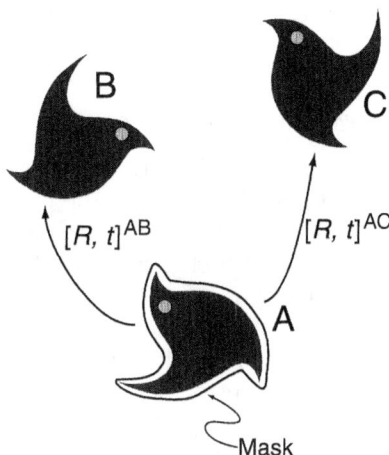

Fig. 4.7. Schematic illustration of real-space molecular averaging. For each grid point inside the mask around molecule A, the corresponding density values in molecules B and C are fetched and added to molecule A. When the average density has been constructed in the place of molecule A, it is copied to molecules B and C.

of the self-rotation function, or it may be obtained from the heavy atom positions of isomorphous derivatives. Probably the most powerful technique is by manual or automated inspection of the initial map that may reveal similar shaped pieces of density. By inserting guide atoms, it is then possible to derive the NCS operators. Once approximate values for the NCS operators $[R, t]_{ij}$ from molecule i to molecule j have been obtained, they can be refined by maximizing the correlation coefficient between the map at positions x and $R_{ij}x + t_{ij}$ with respect to the operator $[R, t]_{ij}$. This operation can be carried out by the RAVE package (Kleywegt and Jones 1994).

If accurate NCS operators are available, a molecular envelope can be determined by means of a local correlation coefficient map (Vellieux *et al.* 1995). This method specifically requires the presence of NCS.

Phase extension Unfortunately, the heavy atom derivative data are usually of lower resolution than the native data set. This means that phases for reflections in the outer resolution shells cannot be determined experimentally. From eqn (4.4), it is seen that density modification imposes restraints on neighbouring structure factor amplitude and phases. For this reason, the density modification technique can be used to determine the phases of the reflections that have not been phased by the experimental method. When the density modification has been performed, the map is inverted, and structure factors and phases are obtained for all reflections within a given resolution cutoff. The calculated phases can now be assigned to measured reflections that did not have an experimentally assigned phase. This process is called *phase extension* and works reasonably well, as long as the unphased data are included in very thin shells—in principle, a shell of a thickness corresponding to one reciprocal lattice unit can be

4.1.6 wARP—automated refinement protocol

As an additional step after density modification methods have been employed, the wARP procedure can substantially improve the phases (Lamzin and Wilson 1997; Perrakis et al. 1997). The density modification phase set is used to generate a number of dummy atom models, which are then subjected to refinement and iterative model updating in ARP, an automated refinement procedure (Lamzin and Wilson 1993). The ARP procedure works by adding free oxygen atoms to patches of spherical electron density. The free-atom model is then subjected to refinement, after which atoms are removed again if the density around them disappears. In wARP, the phase sets obtained from the ensemble of output models are averaged and weighted by their similarity to an average F vector. A weight, w_{wARP} is assigned to each structure factor. If there are n dummy models

$$w_{wARP} = \frac{F_o^2}{F_o^2 + (1/n) \sum_i |F_{aver} - F_i|^2},$$

where F_o is the observed structure factor amplitude, F_{aver} the vector average structure factor and F_i the vectors used to construct the average.

The wARP procedure results in a substantial improvement of the initial phases, but requires that the native data set extend to a resolution beyond ~ 2.4 Å, depending on the solvent content of the crystal.

4.2 Interpreting the electron density

4.2.1 Strategy for working with the density

In the early days of macromolecular crystallography an ingenious device called a 'Richards Box' was used to build the molecular model. It consisted of a semi-transparent mirror behind which the electron density was plotted on transparent sheets. There was a light behind the sheets. In front of the mirror, a physical model was constructed using the so-called 'Kendrew' brass models of peptides, carbon atoms, OH groups, etc. When looking at the reflection of the model in the half-silvered mirror, it would appear to be superimposed on the electron density sheets behind the mirror. It was thus possible to build the entire model by adding residue by residue. In the early seventies, this process was computerized, and FRODO was the first successful program to be used by crystallographers worldwide (Jones 1978; Hall 1995). The molecule was represented by lines connecting the atoms, and the electron density was contoured at a single level in the 'chicken wire' representation. FRODO in reality implemented an electronic Richards Box, where one would create a new residue, orient it in space to fit the density, and connect it to the previous residue.

However, having to deal with residues, side-chains, and atoms is an added complexity in the early stages of map interpretation. It removes attention from

- Determine molecular boundary
- Generate a main-chain trace
- Determine where the sequence matches the density
- Assign Cα positions
- Generate remaining atoms and side-chains
- Optimize the fit of the model to the density
- Evaluate the model

Fig. 4.8. Key steps in building a first model.

the primary goal: tracing the polypeptide chain. There are several examples in the literature of serious errors introduced in this first stage of interpretation.

With the **O** program (Jones *et al.* 1991; Jones and Kjeldgaard 1994, 1997), a completely different approach to map interpretation and model building is adopted. First, an approximation to the entire Cα trace is constructed, and much effort is spent making this preliminary Cα model as correct as possible. As it turns out, the remaining atoms of the protein chain, including the side-chains, can be modelled with surprising accuracy once the positions of the Cα atoms are known.

The starting point of the map interpretation is the *skeleton*, which is a very simple representation of the electron density, basically consisting of ridge lines along the maximum values of the map. One of the problems with the normal contoured electron density representation is that it is very difficult to get an overview of the whole structure. This problem is overcome with the skeleton, because it contains only the most important information. In addition, the skeleton can be modified and coloured to preference as more and more is discovered about the new structure. Figure 4.8 shows the principal steps in constructing a new protein model.

4.2.2 Editing the skeleton

Electron density interpretation is a complicated mental process, and during this process, the scientist needs to create in the mind the three-dimensional folding and features of the structure. The job of the computer is to simplify and assist in this respect. The electron density contains an extremely large amount of information, and it is desirable to represent this information in a simplified manner. The skeleton is such a simple representation. An initial skeleton can be calculated from an electron density map using the BONES program (provided with **O**), MAPMAN from the RAVE suite (Kleywegt and Jones 1994), or other programs, using an algorithm developed by Greer in 1974 (Greer 1974, 1975, 1976a, b).

In **O**, a skeleton is represented by connected points, called 'bone atoms' or just 'bones'. What is the difference between 'bones' and 'skeletons'? The term *skeleton*

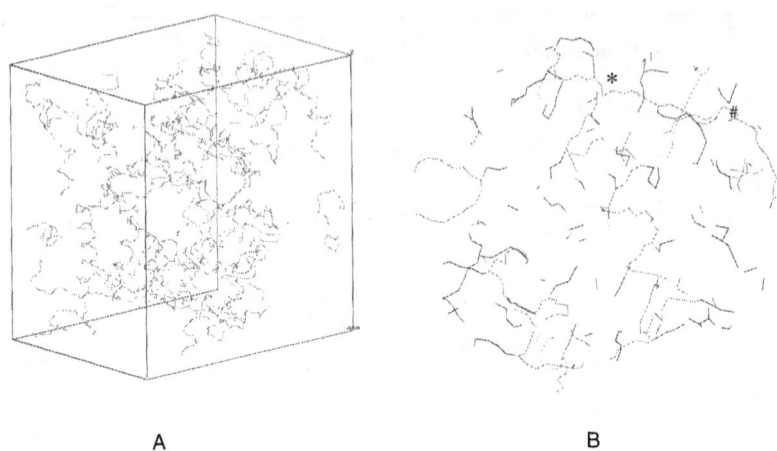

Fig. 4.9. Skeletons made from 1.9 Å MAD data on human psoriasin, an all-helical dimer of 200 residues (Brodersen et al. 1998). A: The overview display, showing only bone atoms of level main-chain within a sphere of 50 Å. The outline of the unit cell is also displayed. In this view, the boundary between the individual molecules can easily be seen. B: Detailed display, showing both main- and side-chain level bone atoms, within a 15 Å sphere. A helix can be identified with the axis pointing from the lower right to the upper left. The connection between the spots marked '*' and '#' is really side-chain erroneously assigned to main-chain, and the connection should be removed in the editing process. At '#', the main-chain actually continues up and to the right through the V-shaped side-chain bone (see Plate Section).

is just a more straightforward synonym for the term *skeletonized density*. *Bones* is the word describing the graphical representation of a skeleton. The distinction is the same as between 'density' and 'contours'. Each bone atom in a skeleton has a *position*, *level* and *connectivity*. On the computer graphics display, the level is represented by an associated colour. The positions, levels and connectivity of the skeleton can be changed interactively. This process is called editing. When generating the skeleton, the BONES program attempts to decide what skeleton segments are main-chain and what are side-chain. Main-chain skeleton is indicated by level 3, and side-chain skeletons by level 2.

When starting out with a new structure, one typically creates a number of different displays of the skeleton. First, an overall display that shows only bone atoms of 'main-chain' level, in a large radius, say 50 Å. Another object shows both main- and side-chain levels in a smaller radius, say 20 Å (see Fig. 4.9). While interpreting the map, one edits the skeleton, improving the appearance of the skeleton to make it look more and more like a $C\alpha$ trace. In this process, the levels of the skeleton can be manipulated to display the current trace in a distinct colour.

4.2.3 Determination of the molecular boundary

The unit cell contains crystallographically related molecules, but may also contain molecules related by local symmetry or NCS. The goal is to isolate a single molecule in the electron density map. The skeleton is well suited for this purpose, if one edits the skeleton levels to mark regions belonging to a single molecule. To check whether mistakes have been made, i.e. density from a symmetry mate has been included, it is possible to apply the crystallographic symmetry operators to the current trace. Wrong assignments will then show up as overlapping areas.

After a single molecule has been isolated, it is possible to construct a molecular envelope ('mask') using the bone atoms. The mask is useful for further improvement of the map by solvent flattening or averaging (Section 4.1.5).

4.2.4 *Strategy for working with the skeleton*

The next step in the interpretation is to recognize stretches of secondary structure and correct local errors in the skeleton by breaking and remaking the connectivity. Errors in the phases from the crystallographic experiment lead to spurious features in the electron density map. In particular, systematic errors, such as incompleteness of the data set or missing low resolution reflections can result in long continuous features in the density that look very much like polypeptide chains. It is obvious that extreme care should be taken not to build such features into the model. However, the program that calculates the skeleton knows nothing about this, and happily skeletonizes these bogus features. Other examples of errors that make the density appear continuous are interacting side-chains or hydrogen bonds between strands.

When editing the skeleton, one must pay attention to such false connections, and break the connectivity of the skeleton. Also, stretches of skeleton in side-chain density may erroneously have been assigned to main-chain level, and must be corrected. Frequently, one will observe breaks in the density. When in doubt, it is reasonable to carry on along the direction of the current secondary structural element. Secondary structural elements can be traced in both directions until the density fades and the connectivity stops.

In each segment of secondary structure, the local chain direction can be determined by examining the characteristic patterns of the different secondary structural elements. The α-helix tends to resemble a Christmas tree because the direction of the $C\alpha-C\beta$ bond makes the side-chains point towards the N-terminal end of the helix (see the helix depicted in Fig. 4.2). In β-strands, the side-chains display the characteristic up–down–up–down pattern, and the direction of the chain can be determined because the distance between the side-chain branch and the carbonyl group of the same residue is smaller than the distance to the next side-chain.

4.2.5 *Sequence placement*

Occasionally, all or most of the polypeptide chain can be built into the electron density in one continuous stretch. In most situations, however, this is not the case, since loop regions in proteins are often poorly defined in the electron density, but it may be possible to trace a number of unconnected segments of secondary structure. Having traced the secondary structural regions, one needs to connect these. At this point, the amino acid sequence of the protein comes into play, and it is necessary to associate the side-chain densities with the amino acids given by the primary sequence. It is often possible to associate sequence positions to the segments of secondary structure, and thus obtain information on their connectivity. Some amino acid side-chains such as Trp, Tyr, Phe, Pro and Gly display characteristic patterns in the density, and can often easily be recognized. Unfortunately, the appearance of side-chain density depends on where it is localized in the structure, and large side-chains on the surface of the molecule may not be visible in the density due to thermal mobility. Placing the sequence in the density is thus a critical and crucial step in building a model, and mistakes made in sequence placement propagate into other regions of the protein, resulting in further errors being made. Subsequent interpretation becomes more and more difficult, and such errors are difficult to undo.

Often, the connectivity of the density is poor, and the identity of side-chain densities is uncertain. In such a case, one typically builds polyalanine or polyserine into the density. **O** offers tools to assist in defining the placement of the sequence, by allowing the crystallographer to enter a guess of the sequence, typically 8–12 residues, from the look of the density. The guess is compared with the real sequence using a scoring matrix to decide where in the sequence the guess fits best. The scoring matrix has been established on the basis of experience with how the different side-chains typically appear in the density.

Gradually, sequence information can be included to tie stretches of skeleton to pieces of the sequence. At this stage of the model building process it is advantageous to employ knowledge about the structure of proteins. For example, in globular proteins hydrophobic side-chains are predominantly found in the interior of the protein whereas polar side-chains are mainly located on the surface. Figure 4.10 shows a list of desirable and undesirable features of a protein model.

One should also start employing the known information about the particular protein in question. It may be possible to identify the active site, special cofactors and metal ions. In addition, the positions of heavy atoms are known from Patterson maps and can give clues to positions of the side-chains they bind to, such as mercury compounds, which typically bind to cysteine.

Also, one can compare with homologous sequences. It is reasonable to assume that such a protein would have a similar structure, so a homologous sequence should fit into the structure and still make sense. Insertions and deletions in related sequences should occur in loop regions. It should also be checked that active site residues and disulphide pairings are the ones expected from biochemical evidence.

4.2.6 Improving the skeleton

The next goal is to make the skeleton look more and more like a Cα trace. Once a trace has been established, a branch point (a main-chain bone atom with a side-chain branch) should be placed at every residue position found in the electron density. Branch points should be moved in order to place them near

Good
- Long stretches of secondary structure
- Hydrogen bonds in the interior of protein
- Hydrophobic clusters in the interior. Propeller-shaped clusters of aromatics
- Pro and Gly are expected on the surface, and mostly in loop and turn structure

Bad
- Little secondary structure
- Many unsatisfied hydrogen bond donors and acceptors
- Beta turns in the interior of the protein
- Polar side-chains in the interior not involved in salt bridge interactions
- Polar against non-polar side-chain—interaction of non-local side-chains must make chemical sense
- Many phi–psi angles in disallowed regions of Ramachandran plot

Fig. 4.10. Good and bad features of a protein model.

expected Cα positions (positions where the side-chain density branches from the main-chain density). The secondary structure can be used to decide where the side-chains are, and in which direction they are pointing. With a branch point at every Cα position, the autobuilding tools in **O** work very efficiently.

4.2.7 *Building a rough first model*

At some stage in the model building process, it will be necessary to build at least part of a protein molecule. This may not necessarily be part of the unknown structure, but could just as well be a polyalanine or polyserine structure that would be available for experimenting, because the sequence placement may be uncertain.

The first step when generating the first rough model is placing the Cα coordinates from the positions found when editing the skeleton trace. In the **O** program, tools are available to ensure the right direction from one Cα to the next and the right distance of approximately 3.8 Å. Knowing the positions of the Cα atoms, the remaining atoms in the main-chain are easily defined. Five residue fragments from a database of known well-refined structures are fitted to the Cα positions, and the main-chain atoms from the best fitting fragment are copied into the central three positions. This process is repeated along the entire polypeptide chain.

Having placed the main-chain atoms, side-chain atoms can be added in their most common rotamer conformation. An analysis of structures solved to very high resolution shows that the side-chain rotamer values cluster tightly around the staggered values predicted from stereochemistry (Morris *et al.* 1992). Using this scheme, $\sim 75\%$ of the residues will be close to the right conformation.

4.2.8 Improving the model

At this point, a rough molecular model of the protein exists. An additional pass through the structure is required to remove obvious errors introduced by the autobuilding of main- and side-chain atoms. The crystallographer must examine each residue in turn, and evaluate the fit of the model to the density. **O** provides a number of tools that can be employed at this stage. Real-space refinement can be employed to optimize the fit of the model to the density, and structure regularization can be used to optimize the geometry of the model. Also, the database of structures can be searched to find peptide segments of similar structure.

Several indicators of structural quality can be calculated by **O**, and should be used as an aid in defining trouble spots of the structure. The program allows colouring of the molecule as a function of any such property, making it easy to perform these checks.

The 'pepflip' value measures the root mean square (rms) distance from the carbonyl oxygen atom in a peptide group to an ensemble of the 20 best-fitting peptides from the fragment database. The pepflip value lies between 0 and ~ 3 Å, in which case the peptide plane of the probe peptide is flipped compared to all or most of the fragments in the database. Peptides with high pepflip values should be examined carefully. Figure 4.11A shows plots of the pepflip score as a function of the residue number.

The *rotamer side-chain* score RSC measures the rms distance from side-chain atoms to the closest rotamer, and such a plot can be used to discover the problematic side-chain conformations (Fig. 4.11B).

O also allows the user to calculate the RSfit, which is a measure of how well the structure fits the density. The RSfit measure can be calculated either as an R-factor, or as a correlation coefficient:

On identical grids $\mathcal{G}_1, \mathcal{G}_2$ and $\mathcal{G}_3$ calculate ρ_{obs}, ρ_{calc} and μ_{mask}, where ρ_{obs} is the experimental electron density, ρ_{calc} the calculated electron density and μ_{mask} is a binary mask from the model, containing the fragment of interest. For all non-zero μ_{mask}, calculate the *real-space R-factor*

$$\text{RSRF} = \frac{\sum |\rho_{obs} - \rho_{calc}|}{\sum |\rho_{obs} + \rho_{calc}|}$$

or the *real-space correlation coefficient*

$$\text{RSCC} = \frac{n \sum \rho_{obs} \times \rho_{calc}}{\sqrt{n \sum \rho_{obs}^2 - (\sum \rho_{obs})^2} \times \sqrt{n \sum \rho_{calc}^2 - (\sum \rho_{calc})^2}}.$$

The RSfit measure is useful to display where the electron density does not fit well with the model that has been built. This can indicate an error in the model or flexibility in the region (Fig. 4.11C).

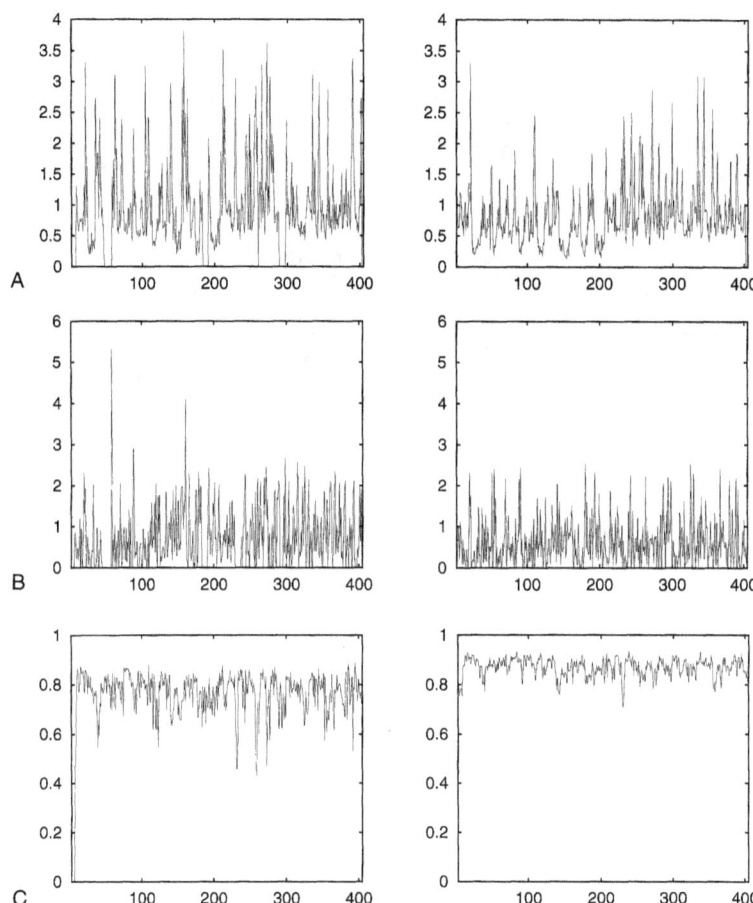

Fig. 4.11. Structural indicators used in the refinement of EF-Tu:GDPNP. A: The pepflip scores of model M20 and M37 (final). There are still spikes in the diagram, but these are dictated by the density. Outliers are generally found in loops. B: The RSC scores of model M20 and M37, respectively. Many outliers have been removed during the refinement. C: The RSfit score, from models M28 and M37. The curve to the right is less spiky, and the average correlation is higher.

In addition to the properties that can be calculated within **O**, structural quality indicators computed by external programs can be imported and used for colouring. The purpose of these methods is to avoid the common pitfalls and errors that can be made when building a model. A listing of the common errors in protein models is shown in Fig. 4.12.

The model building is often combined with the early stages of refinement, in which the coordinates of the model are changed to minimize the differences between the observed and calculated structure factor amplitudes while maintaining the geometry. However, it is important to distinguish between the use of refinement programs for *phase improvement* (that is, to make a more interpretable map) and *structure refinement*. Structure refinement is outside the scope of this

Totally wrong fold	The molecule could be traced backwards
Locally wrong fold	The folding of one domain could be incorrect. The connectivity between a few secondary structural elements could be wrong
Locally wrong structure	Directionality of the chain could be wrong locally. Using side-chain density for main-chain
Out of register errors	The placement of sequence may be out of register. An additional error corrects the problem
Wrong main-chain conformation	Wrong orientation of the peptide plane
Wrong side-chain conformation	Incorrect rotamer used for side-chains

Fig. 4.12. Common errors in protein models, listed in decreasing degree of seriousness.

chapter, but the reader is referred to the numerous articles on the subject (Brünger and Rice 1997; Kleywegt and Jones 1997; Sheldrick and Schneider 1997; Tronrud 1997).

The final result of any crystallographic experiment is the electron density map calculated using the observed phases. The building of a model into this density is the human interpretation of that electron density. As with any other human enterprise, this is error-prone, but with the tools available to the crystallographer in the **O** program, it is feasible to construct a molecular model that is correct to a high degree of confidence.

References

Abrahams, JP (1997). *Acta Crystallographica*, **D53**, 371–376.
Arnold, E and Rossmann, M (1986). *Proceedings of the National Academy of Sciences USA*, **83**, 5489–5493.
Bricogne, G (1974). *Acta Crystallographica*, **A30**, 395–405.
Bricogne, G (1976). *Acta Crystallographica*, **A32**, 832–847.
Brodersen, DE, Etzerodt, M, Madsen, P, Celis, JE, Thøgersen, HC, Nyborg, J and Kjeldgaard M (1998). *Structure*, **6**, 477–489.
Brünger, AT and Rice, LM (1997). *Methods in Enzymology*, **277**, 243–269.
Collaborative Computation Project, No. 4 (1994). *Acta Crystallographica*, **D50**, 760–763.
Cowtan, K (1994). *CCP4/ESF-EACBM Newletter on Protein Crystallography*, **31**, 34–38.
Cowtan, K (1996). http://www.yorvic.york.ac.uk/~cowtan/fourier/fourier.html
Cowtan, K (1998). http://www.yorvic.york.ac.uk/~cowtan/sfapplet/sfintro.html
Drenth, J (1994). *Principles of X-ray Crystallography*, Springer Verlag, New York.
de la Fortelle, E and Bricogne, G (1997). *Methods in Enzymology*, **276**, 472–494.
Furey, W and Swaminathan, S (1997). *Methods in Enzymology*, **277**, 590–620.
Greer, J (1974). *Journal of Molecular Biology*, **82**, 279–301.
Greer, J (1975). *Journal of Molecular Biology*, **98**, 649–653.
Greer, J (1976a). *Journal of Molecular Biology*, **104**, 371–386.
Greer, J (1976b). *Journal of Molecular Biology*, **100**, 427–458.

Hall, SS (1995). *Science*, **267**, 620–624.
Hendrickson, WA and Lattman, EE (1970). *Acta Crystallographica*, **B26**, 136–143.
Jenner, L, Husted, L, Thirup, S, Sottrup-Jensen, L and Nyborg, J (1998). *Structure*, **6**, 595–604.
Jones, TA (1978). *Journal of Applied Crystallography*, **11**, 268–272.
Jones, TA and Kjeldgaard, M (1994). *From first map to final model*, Bailey, S, Hubbard, R and Waller, DA (eds), SERC Daresbury Laboratory, Daresbury, U.K., pp. 1–13.
Jones, TA and Kjeldgaard, M (1997). *Methods in Enzymology*, **277**, 173–208.
Jones, TA, Cowan, S, Zou, J-Y and Kjeldgaard, M (1991). *Acta Crystallographica*, **A47**, 110–119.
Kjeldgaard, M, Nissen, P, Thirup, S and Nyborg, J (1993). *Structure*, **1**, 35–50.
Kleywegt, G and Jones, TA (1994). *From first map to final model*, Bailey, S, Hubbard, R and Waller, DA(eds), SERC Daresbury Laboratory, Daresbury, U.K., pp. 59–66.
Kleywegt, GJ and Jones, TA (1997). *Methods in Enzymology*, **277**, 208–230.
Kleywegt, GJ and Read, RJ (1997). *Structure*, **5**, 1557–1569.
Lamzin, VS and Wilson, KS (1993). *Acta Crystallographica*, **D49**, 129–149.
Lamzin, VS and Wilson, KS (1997). *Methods in Enzymology*, **277**, 269–305.
Leslie, AGW (1987). *Acta Crystallographica*, **A43**, 134–136.
Morris, AL, MacArthur, MW, Hutchinson, EG and Thornton, JM (1992). *Proteins: Structure Function and Genetics*, **12**, 345–364.
Neder, RB and Proffen, Th (1996). http://www.kri.physik.uni-muenchen.de/crystal/teaching
Perrakis, A, Sixma, TK, Wilson, KS and Lamzin, VS (1997). *Acta Crystallographica*, **D53**, 448–455.
Read, RJ (1986). *Acta Crystallographica*, **A42**, 140–149.
Rossman, M and Blow, D (1962). *Acta Crystallographica*, **15**, 24–31.
Sheldrick, GM and Schneider, TR (1997). *Methods in Enzymology*, **277**, 319–343.
Tronrud, DE (1997). *Methods in Enzymology*, **277**, 306–319.
Vellieux, FMD and Read, RJ (1997). *Methods in Enzymology*, **277**, 18–53.
Vellieux, FMDAP, Hunt, JF, Roy, S and Read, RJ (1995). *Journal of Applied Crystallography*, **28**, 347–351.
Wang, BC (1985). *Methods in Enzymology*, **115**, 90–112.
Welberry, TR and Owen, K (1992). *Journal of Applied Crystallography*, **25**, 443–447.
Zhang, KYJ, Cowtan, K and Main, P (1997). *Methods in Enzymology*, **277**, 53–64.

5
Neutron crystallography of biological molecules

Eva Pebay-Peyroula and Dean Myles

5.1 Introduction

In the past decade, the number of protein structures solved at atomic resolution by X-ray crystallography has increased dramatically. Developments in molecular biology, crystallization techniques, synchrotron radiation facilities and the associated progress in two-dimensional detectors now make possible the solution of very complex structures with a large number of atoms, at atomic or near atomic resolution, even from very small crystals. However, at the medium resolution limits (>1.5 Å) typical of most X-ray protein structure determinations, these analyses are invariably incomplete: hydrogen atoms—which account for 50% of the contents of the cell—often cannot be seen at all. Whilst the position of many hydrogen atoms can be inferred reliably from the chemical groups to which they are bound, the positions of other more labile—and more interesting—atoms cannot, and must be determined by other techniques. Hydrogen and/or deuterium (H/D) atoms can more readily be located in a corresponding neutron analysis because the scattering amplitudes of H/D for neutrons are closely similar to those of other biological atoms (Table 5.1). Moreover, as neutrons scattered from hydrogen and deuterium have different magnitude and phase, neutrons are uniquely able to distinguish between hydrogen and deuterium exchanged positions in the crystal. The location and bonding interactions of hydrogen atoms are of fundamental importance in biology and their precise location by high resolution neutron crystallography is described in Section 5.2.

Proteins and more general macromolecular assemblies of biological molecules are not static and even in a crystalline arrangement, part of the complex can be disordered. In fact, for many biological systems, it is likely that these less ordered domains play important functional roles, precisely because of their flexibility and dynamical properties. The large difference in neutron scattering lengths between hydrogen and deuterium isotopes has been exploited to powerful effect in contrast variation experiments that are able to match or enhance the scattering from specific components of such systems. We focus in the second part of this chapter, Section 5.3–5.5, on how low resolution neutron crystallography in combination with the contrast variation method can better localize these disordered regions. Some examples are developed [for a review, see Timmins *et al.* (1994a)].

Table 5.1. Selected scattering lengths for neutrons. For comparison, the number of electrons represents the interaction of these atoms with X-rays at zero scattering angle. These values point out the capability of neutrons to locate hydrogen atoms. This potential can even be enhanced by substituting hydrogen (H) by its isotope deuterium (D)

Atoms	H	D	C	N	O	P	S
Neutron scattering length 10^{-15} m	−3.7	6.6	6.6	9.4	5.8	5.3	3.1
Number of electrons	1	1	6	7	8	15	16

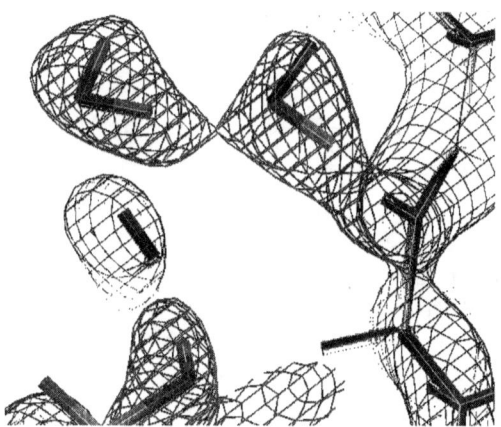

Fig. 5.1. Coordinated D_2O molecules in the neutron structure determination of triclinic lysozyme at 1.7 Å resolution. Positive scattering length contoured at 1.5σ (Bon 1998).

5.2 Neutron protein crystallography near atomic resolution

Neutron diffraction provides a unique and powerful complement to X-ray protein crystallography at medium resolution (1.5 Å) by enabling the hydrogen structure of biological molecules to be revealed (Fig. 5.1). Hydrogen bonding interactions mediate much of biological structure and function and this complementary information is likely to have a profound impact upon the key structural issues such as enzyme reaction mechanisms, protein dynamics and protein–water interactions. The major disadvantage, however, is that the inherent brilliance of neutron beams is orders of magnitude less than for X-ray beams and data collection times are proportionately longer. For this reason, biological neutron diffraction is a specialized technique that focuses specifically on those problems that cannot be resolved by X-ray structure analysis alone.

5.2.1 Methodology

The methods and concepts of neutron protein crystallography are directly analogous to those of X-ray diffraction [see Chapter 3 by Fourme et al. in this book, or Blundell and Johnson (1976) for a review]. In a diffraction experiment, the measured intensities, I_{hkl}, reflected from Bragg planes, hkl, are proportional to $|\mathbf{F}_{hkl}|^2$, where $\mathbf{F}_{hkl}$, the structure factor, is the Fourier transform of the scattering density $\rho(x, y, z)$ at position x, y, z in the unit cell of the crystal:

$$\mathbf{F}_{hkl} = \iiint \rho(x, y, z) \exp(-2\pi i(hx + ky + lz)) \, dV. \tag{5.1}$$

As the scattering density distribution $\rho(x, y, z)$ is due to N atoms positioned at x_j, y_j, z_j, in the crystal, (5.1) can be written as a summation over the scattering contribution of all atoms,

$$\mathbf{F}_{hkl} = \sum_{j=1}^{N} b_j \exp(-2\pi i(hx_j + ky_j + lz_j)) T_j(\sin \vartheta / \lambda), \tag{5.2}$$

where b_j is the scattering length of atom j, T_j is the temperature factor that accounts for thermal motion and disorder, λ is the wavelength, and 2ϑ is the angle of the diffracted beam. Knowledge of $\mathbf{F}_{hkl}$ therefore enables the scattering density and atomic distribution in the crystal to be determined. However, $\mathbf{F}_{hkl}$ is complex and only the amplitude $|\mathbf{F}_{hkl}|$ is measured in a diffraction experiment. The phase information ϕ_{hkl} must be recovered by other means. In neutron protein crystallography, the initial estimates of the phases are typically derived from the non-hydrogen atoms in an existing X-ray structure.

The crucial distinction between neutron and X-ray diffraction is due to the fundamental difference in their scattering interactions with matter, and how this is manifested in atomic scattering lengths, b. X-rays interact with electrons and b_e increases monotonically with atomic number. In contrast, neutrons interact with nuclei in a complex manner and b_n shows no simple progression between elements, nor between their isotopes, and is closely similar for all atoms. For example, the neutron scattering lengths of hydrogen, $b_n = -0.374 \times 10^{-12}$ cm, and deuterium, $b_n = +0.667 \times 10^{-12}$ cm, differ in magnitude and phase, but are comparable to those of other biological atoms (Table 5.1). As the diffracted intensity, I_{hkl}, is proportional to $|\mathbf{F}_{hkl}|^2$, the relative contribution of hydrogen and deuterium to a Fourier synthesis is then much more significant for neutrons $[b_n^2(D) \approx b_n^2(C)]$ than for X-rays $[b_e^2(H \text{ or } D) \approx 0.03 b_e^2(C)]$. This enhanced ability to visualize and discriminate individual hydrogens has been used to powerful effect. For example, Kossiakoff and Spencer (1980) used neutron data collected to 2.2 Å resolution to help elucidate the catalytic mechanism of the serine proteinase, trypsin, by showing that the imidazole of His57 is protonated and the carboxyl of Asp102 remains deprotonated during catalysis. The significant difference in b_n for oxygen and nitrogen also means that the orientation of glutamine and histidine residues can be determined properly.

The difference in magnitude and phase of b_n for hydrogen and deuterium is of specific advantage in H/D exchange techniques because neutron diffraction is uniquely able to distinguish between hydrogen (negative scattering length) and deuterium (positive scattering length) exchanged positions in the crystal. An analysis of the pattern and extent of H/D substitution can help reveal specific information on the solvent accessiblity of different groups, on the mobility and flexibility of interesting domains and on the exchange dynamics themselves (Kossiakoff 1985). For example, labile hydrogen atoms belonging to amide or carboxyl groups may not fully exchange if they are protected by their local environment. Hydrogen atoms buried in the core of the protein may be sterically protected or inaccessible to solvent D_2O, or may be involved in stabilizing a secondary structure, such as has been observed in beta sheets (Kossiakoff 1982). Conversely, the observation of H/D exchange at sites that seem inaccessible to solvent in a static X-ray crystallographic model can indicate dynamic motions of the protein that may have functional significance. Even the magnitude of the H/D signal may provide additional dynamic information as this may be correlated with the average time that protons occupy a particular site.

Despite these advantages, the large sample sizes ($> 5\,\text{mm}^3$) and extended data collection periods of weeks or months (necessary to compensate for the relatively weak neutron beams) have limited neutron crystallography to only a few small biological systems [reviewed in Lehmann (1994)].

5.2.2 Quasi-Laue neutron diffraction

Conventional neutron single crystal diffractometers measure individual or small groups of reflections sequentially using a monochromatic beam ($\delta\lambda/\lambda \sim 3\%$). Laue diffraction provides a more rapid and efficient survey of reciprocal space. In the Laue method, a stationary crystal is illuminated with a broad spectrum of wavelengths ($\delta\lambda/\lambda > 20\%$), stimulating large numbers of reflections ($\propto \delta\lambda/\lambda$) at all possible wavelengths ($\lambda_{min} < \lambda < \lambda_{max}$) and at many different angles. For maximum efficiency, a large area detector is required that can subtend this large solid angle of diffraction and record all stimulated diffraction at once [for review, see Helliwell and Wilkinson (1994)]. A Laue diffractometer (LADI), designed specifically to exploit this advantage for macromolecular crystallography, has been constructed at ILL/EMBL—Grenoble. The detector comprises a large Gd_2O_3-doped neutron sensitive image plate (400×800 mm) mounted on a cylindrical camera (318 mm diameter) that completely encircles the sample (Cipriani et al. 1995). Diffracted neutrons are absorbed in the plate to form a latent image that is read out after exposure under photo-stimulation with visible laser radiation (Fig. 5.2).

The instrument is optimized for single crystal studies of small protein systems at medium or high resolution (> 1.5 Å), sufficient to locate individual hydrogen atoms of special interest, water structures or other small molecules that can be marked with deuterium to make them particularly visible. The size of the unit cell that can be studied is clearly dependent on the size of the crystal, and beyond around $300\,000$ Å^3 crystals of >1 mm are required. A basic problem, in

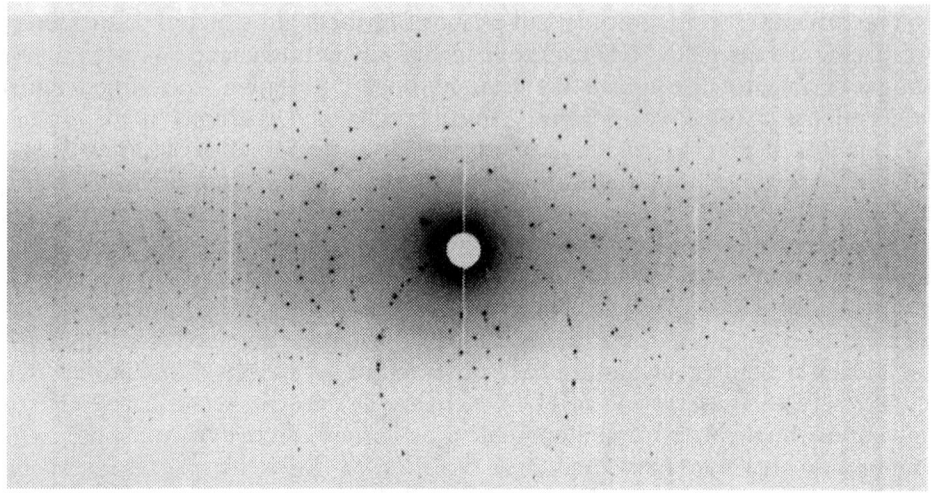

Fig. 5.2. A neutron quasi-Laue diffraction image recorded from triclinic lysozyme on the LADI image plate detector. Exposure time was 12 h using 3.0–4.0 Å neutrons on cold guide T17 at ILL.

addition to the inherent weakness of the diffraction, is that because protein crystals typically contain about 50% hydrogen, diffracted spots sit upon a substantial incoherent background that comes largely from the sample itself. This can be compensated for, in part, by replacing exchangeable hydrogen atoms and H_2O with deuterium by soaking in D_2O. Limiting the beam to a narrower bandpass optimizes the density of the peaks on the detector surface and further enhances the peak-to-background ratio. A distinct advantage of this narrow 'quasi-Laue' bandpass (3.5 Å, $\delta\lambda/\lambda = 20\%$) is that, as the proportion of energy overlapped (harmonic) reflections is dependent on the ratio $\delta\lambda/\lambda$ and is independent of the size of unit cell (Cruickshank et al. 1987) then for a 'typical' protein with $d_{min} = 1.8$ Å almost all reflections ($> 99\%$) recorded on the LADI detector are singlets and thus amenable to standard crystallographic analysis. A complexity is that the data must first be wavelength normalized, to account for the spectral distribution of the source, using an internal spectral function derived using programs from the LAUEGEN suite (Campbell 1995). Tests show that data collection rates using LADI are at least one or two orders of magnitude faster than by conventional diffractometry.

In a pioneering experiment, Niimura et al. (1997) used the LADI detector at ILL to determine the 2.0 Å resolution neutron structure of hen egg white lysozyme in its tetragonal crystal form at pH 7.0, using a narrow 25% quasi-Laue bandpass of neutrons in the range 3.0–4.0 Å. Of particular interest was the protonation state of the catalytic pairing Asp52–Glu35 at pH 7.0. Earlier neutron studies conducted on D19 at ILL showed that Glu35 is protonated at pH 5.0 (Mason et al. 1984). It is this proton that is transferred to bring about breakage of the glycosidic bond as part of the catalytic mechanism. Significantly, however, neither Glu35 nor Asp52 appear protonated in the Niimura study, a result consistent with the reduced enzymatic behaviour of lysozyme at pH 7.0. Concanavalin A, a well-studied plant lectin, has an intriguing Asp28–Glu8 pair reminiscent of the catalytic grouping in lysozyme. Using neutron data collected to Bragg resolution of 2.75 Å on LADI, Habash *et al.* (1997) were able to show that the proton on Asp28 had not exchanged for deuterium under the deuteration conditions used. This suggests that, despite the intriguing juxtaposition, the Asp28–Glu8 pairing does not mimic lysozyme in conferring enzymatic activity on the protein.

The speed and efficiency with which data can be collected using this new detector should make possible the investigation of larger biological systems and/or smaller crystals than has previously been possible.

5.3 Principles of low resolution crystallography

The disorder contained in crystals of biological macromolecules or assemblies of such molecules can sometimes be related to the macromolecule itself, for instance nucleic acids in viruses, or the crystalline assembly can contain less ordered regions. In both cases, high resolution X-ray diffraction experiments lead to structural information on the highly ordered parts only. The low resolution information is obtained from diffraction at very small Bragg angles. The major advantage of neutrons for low resolution crystallography is the difference in scattering lengths between protons and deuterons which allows contrast variation experiments in the crystals and thus considerably enhances specific components of the crystals which otherwise would remain difficult to detect. This is particularly the case of detergent or lipid molecules for which the contrast to water is very weak and thus undetectable for low resolution X-ray diffraction. Additionally, long wavelength neutrons facilitate the low resolution data collection.

5.3.1 The contrast variation method

In low resolution crystallography, the scattering from a small part of the cell located in u, is not sensitive to the scattering length of each atom located in this part, but only to the mean scattering length within this part. The mean scattering length $r(x)$ is represented in Fig. 5.3 for several molecules as a function of x, the D$_2$O concentration of the solvent. When exchangeable hydrogen atoms are present in the molecule (involved in O–H or N–H bonds) then $r(x)$ is not constant. The contrast $\rho(x)$ is defined as the difference between the scattering length density of the molecule and of the solvent,

$$\rho(x) = r_{\text{protein}}(x) - r_{\text{solvent}}(x). \tag{5.3}$$

It varies linearly as a function of x, and can even become zero (Fig. 5.3),

$$\rho(x) = \rho(0) + x\,\Delta\rho. \tag{5.4}$$

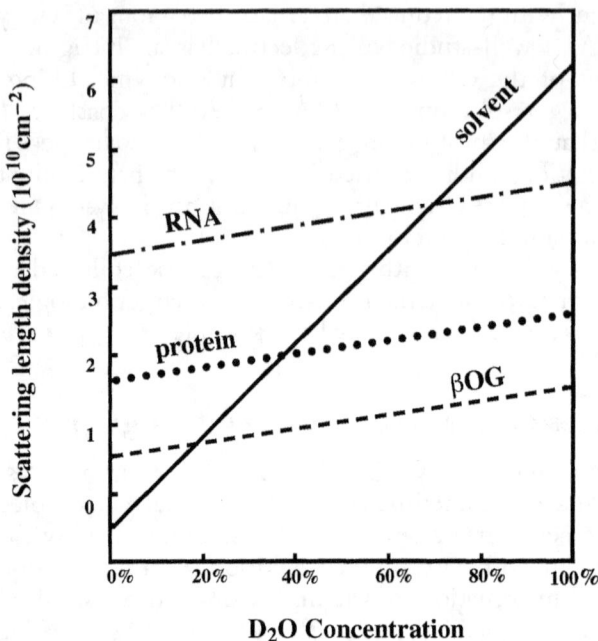

Fig. 5.3. Scattering length densities vs D$_2$O concentration. The scattering length densities are plotted for water, RNA, a typical protein and β-octyl-glucoside (βOG) a detergent commonly used in membrane protein crystallization. The matching points, where the contrast is zero, are around 70% for RNA, 40% for proteins and 20% for βOG.

The zero-contrast value is called the matching point; for a typical protein, RNA, hydrophobic tails of detergent molecules or lipids the matching points occur at 40%, 70% and 20% D$_2$O, respectively. By exchanging the solvent with the partially deuterated solvent, it is possible to enhance the 'visibility' of one type of molecule in the crystal (for instance, the detergent) and, on the contrary, to make invisible some other parts of the crystal (for instance, the proteins) or the reverse. These experiments are called contrast variation experiments and can be performed for small angle scattering studies of a solution as well as for low resolution crystallographic diffraction experiments. The contrast can also be varied by specifically deuterating some components or part of them. In a crystal, the contrast can be defined for each elementary volume of the cell located in u:

$$\rho(x, u) = \rho(0, u) + x \Delta\rho(u). \tag{5.5}$$

The structure factor, which reflects the amplitude of the wave scattered by the molecule(s), is related to the contrast $\rho(x, u)$.

5.3.2 The phase problem

The structure factor, $F(h)$, is defined by an amplitude (directly derived from the measured intensities) and a phase. The inverse Fourier reconstruction leading to the density requires knowledge of both pieces of information, as for X-ray crystallography. However, contrast variation places strong constraints on the possible phases. The Fourier transform of (5.5) gives the relation of the complex structure factor as a function of x,

$$F_x(h, k, l) = F_0(h, k, l) + x\, F_{HD}(h, k, l). \tag{5.6}$$

The diffracted intensity obtained from

$$\begin{aligned}I_x(h, k, l) &= F_x(h, k, l) \cdot F_x(h, k, l)^* \\ &= (F_0(h, k, l) + x\, F_{HD}(h, k, l)) \cdot (F_0(h, k, l) + x\, F_{HD}(h, k, l))^* \end{aligned} \tag{5.7}$$

is thus a quadratic function of x,

$$I_x(h, k, l) = |F_0(h, k, l)|^2 + 2x \cos\phi |F_0||F_{HD}| + x^2 |F_{HD}|^2, \tag{5.8}$$

where ϕ is the phase difference between F_0 and F_{HD}.

In the special case of centric reflections, (5.6) simplifies to

$$F_x(h, k, l) = F_0(h, k, l) + x\, F_{HD}(h, k, l). \tag{5.9}$$

Equation (5.6) can be represented by a vector diagram (Fig. 5.4). Determination of the parabolic constants in (5.8), $|F_0|$, $|F_{HD}|$, and $\cos\phi$, requires the collection

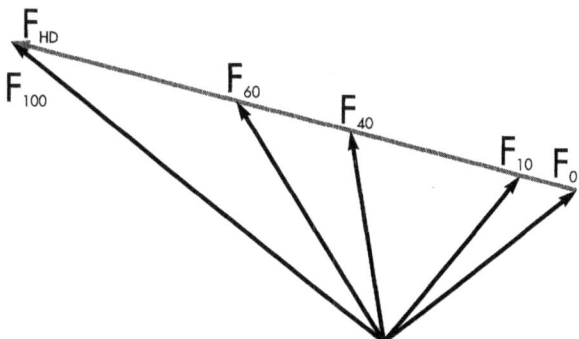

Fig. 5.4. Argand diagram representing the variation of the complex structure factors with the D$_2$O concentration for one reflection. This diagram gives the relative variation, and thus the relative phases for each contrast. The absolute phases are determined by knowing part of the structure from high resolution X-ray crystallography.

of data sets for at least four different contrasts. Scaling together of the different data sets has been described by Roth et al. (1984). The solvent exchange with deuterated or partially deuterated solvent is performed by soaking the crystals for 2–3 weeks, a time that takes into account the exchange of all accessible hydrogen atoms involved in O−H or N−H bonds. The centric reflections are first used to scale together the four experimental data sets obtained for different D_2O concentrations. After this the parabolic constants for each general reflection can then be determined.

5.3.3 Absolute phase determination

Figure 5.4 shows that the determination of the parabola parameters yields the relative locations of the $F(h, k, l)$, i.e. the relative phases associated with the structure factors in the various contrasts. As a consequence, the structure factors (amplitudes and relative phases) from a crystal containing any D_2O concentration can be interpolated from this figure. In the absence of any structural information, however, the orientation of the figure as a whole is undefined.

Knowing the structure of one component in the crystal (i.e. the protein in a membrane protein crystal or the protein in a virus crystal), the structure factors can be calculated at the D_2O concentration for which the second component (detergent molecule or RNA) is matched out. This additional information will therefore fix the phase of this vector in the diagram of Fig. 5.4 and allow the absolute phases to be determined. However, fixing one vector in the diagram still leaves a twofold ambiguity for the absolute phases of the other vectors, since taking the symmetry with respect to the first vector gives another solution compatible with the data. This problem is similar to the phase determination from one heavy atom derivative: if no supplementary information is present, the centroid phases are calculated. A first determination of the complete structure allows the ambiguity to be lifted and phases can be estimated for any D_2O concentration. Best phases are calculated (Roth 1987) in a manner similar to MIR phasing in X-ray crystallography (Blundell and Johnson 1976).

A typical application of this method is the determination of the structure of a disordered part in a crystal where the highly ordered part is already known to atomic resolution from X-ray diffraction experiments. For example, in the case of a membrane protein, the neutron structure factors at about 20% D_2O (matching point of ordinary detergents, Fig. 5.3) can be calculated directly from the coordinates of the protein atoms, previously determined by X-ray diffraction. Performing a contrast variation experiment as described above will give access to structure factors at any contrast. From the density maps, calculated at the appropriate D_2O concentration (i.e. 40% D_2O in order to visualize detergent) an envelope for the detergent can be modelled as shown in Fig. 5.5. Combining the protein coordinates with the detergent envelope improves the absolute phase determination and the procedure from phasing to envelope modelling can be iterated until convergence is reached.

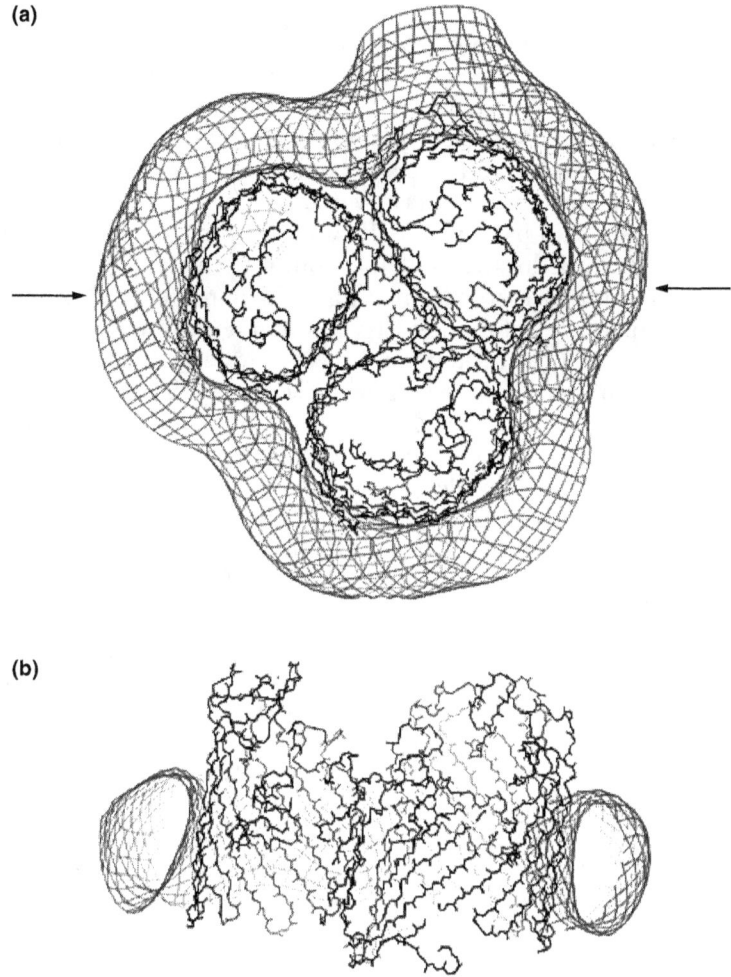

Fig. 5.5. The grey density represents the detergent ring surrounding the porin trimer as seen from the neutron diffraction. The main Cα chain of the protein, determined from high resolution X-ray crystallography is shown in black. (a) View from the top looking through the pores, (b) a cut through the trimer along the arrows shown in Fig. 5.5(a).

5.4 Some examples

5.4.1 The detergent in membrane protein crystals

Several membrane proteins have been crystallized, solved by X-ray crystallography and studied by neutron diffraction: two photosynthetic reaction centres (PRCs) from *Rhodopseudomonas viridis* (Deisenhofer et al. 1985) and from *Rhodobacter sphaeroides* (Arnoux et al. 1989), as well as the Ompf porin from *E. coli* (Cowan et al. 1992). Because of the insolubility of membrane proteins in water they have to be extracted from their natural environment using detergent. The detergent is still present in the crystals and its choice is of capital importance for the crystallization. Nevertheless, in all of these X-ray studies, no detergent (or at most one molecule) has been seen in the density maps. This is due to the disorder of the detergent molecules (therefore not contributing to the high resolution data) and to the low contrast between detergent and water for X-ray diffraction at low resolution. Contrast variation experiments give access to density at any contrast, in particular at 40% D_2O where the protein is invisible and the detergent can be located. The location of the detergent with respect to the protein is an approach to the investigation of protein–membrane interactions. In particular, it was found in PRCs and in porins that two rings of aromatic residues surround the protein, and these rings are assumed to anchor the protein in the membrane. The detergent locations in the crystal showed that these rings are at the limit between head groups and hydrophobic tails. It could also be seen from the organization of the detergent within the crystal that the length of the detergent molecule should not prevent protein–protein interactions and is crucial for the crystal cohesion. Neutron diffraction has been performed to 12 or 16 Å resolution, for PRC from *Rhodopseudomonas* (Roth *et al.* 1989) and *Rhodobacter sphaeroides* (Roth *et al.* 1991), and Ompf porin from *E. coli* (Pebay-Peyroula *et al.* 1995). Figure 5.5 shows the detergent belt modelled from the neutron data surrounding the porin trimer as seen in the tetragonal Ompf porin crystal. In all these cases, except the last, the cohesion of the crystal is due to protein–protein interactions. In the case of Ompf porin the protein crystallizes in two intertwined lattices of trimer chains with no protein connections between them. It could be shown that the connections are made by detergent–detergent head group contacts. In order to have a better view of the detergent organization in the crystal, it is possible to distinguish between head groups and hydrophobic tails. This can be achieved by specific deuteration of the head groups and the tails separately, which differentiates their contrasts.

5.4.2 Viruses

The folding of the polypeptide chain of the coat protein of the tomato bushy stunt virus (TBSV) has been determined by X-ray crystallography to 2.9 Å resolution (Harrison *et al.* 1978). Nevertheless, neither the RNA nor a portion of the protein was visible in these high resolution maps. Neutron diffraction studies at 16 Å resolution were made at ILL (Timmins *et al.* 1994b). From the density maps in 70% D_2O, which is the matching point of RNA, two shells of protein are visible: the outer shell which was determined from X-ray studies and used for the initial phasing of the neutron data, and an inner shell which is not connected to the outer one and which was invisible in X-ray density maps. The 40% D_2O map, where the protein is matched out, shows the RNA density sandwiched between the two protein shells. A more detailed analysis of several contrast maps showed a more complex organization with additional minor layers of protein and RNA. The maps showed structural differences in the interactions between the various subunits of the outer protein shell with RNA, which could play an important role in the assembly and disassembly of the virions.

5.5 Experimental set-up

Low resolution neutron diffraction experiments are carried out at ILL, Grenoble on a specially designed diffractometer, DB21, built jointly by ILL and EMBL and installed on a cold neutron source. A 7.5 Å wavelength is selected by a potassium intercalated pyrolytic graphite monochromator. Crystals are mounted on a standard four-circle device. Diffraction intensities can be measured on a two-dimensional detector operating in an argon atmosphere to minimize the background scattering from air. The experimental set-up allows measurement of diffraction data from crystals with unit cells up to 1000 Å to a resolution from infinity to about 12 Å.

As neutron Bragg reflections from these crystals are relatively weak, it is of primordial importance to work on a powerful neutron cold source. Crystal sizes are typically around 0.5 mm in each dimension. Data collection for one data set to 12 Å resolution lasts about one week. However, it has been possible to collect usable diffraction data from smaller crystals such as ribosomal particle crystals. These crystals were thin plates of $0.5 \times 0.5 \times 0.05$ mm^3, and data sets to 30 Å resolution could be collected within 3–5 weeks. These long data collection periods are possible because of the absence of radiation damage for the crystals in neutron experiments and the stability of the instrumental set-up. For contrast variation experiments it is important to collect a very high fraction of the data, particularly at the lowest resolution.

Data treatment presents some difficulties due to the low resolution of the data, in particular, the small number of strong spots makes the use of automatic crystal orientation programs difficult and even impossible. A first orientation matrix is calculated by an interactive graphics program working on Silicon Graphics [Maindex program, Penel and Legrand (1997)] which displays the spots as well as the crystallographic lattice in reciprocal space. The lattice is then rotated until it superimposes with the spots. This program is a direct way of visualizing the reflections in reciprocal space, independent of any lattice information and is therefore very helpful in checking the quality of the crystals.

5.6 Conclusion

The power to localize and discriminate protonation states at individual sites in large and complex protein structures, even at medium resolutions, is an established advantage of neutron diffraction. The great potential of the LADI instrument is the speed and efficiency with which good quality neutron Laue data can be obtained in order to exploit this advantage. The order of magnitude gains in data collection rates already achieved, relative to monochromatic neutron diffractometry, will make feasible studies of larger biological molecules and/or smaller crystals than previously possible. These combined benefits promise a new phase in high resolution neutron protein crystallography. In addition, developments in molecular biology make it possible to obtain perdeuterated or partially deuterated proteins or lipids which can be used to highlight specific regions with a better signal-to-noise ratio due to the larger contrast and a lower level of incoherent scattering from protons. This possibility should be especially helpful for high resolution studies as deuterated proteins should give reasonable intensities from smaller crystals (Gamble et al. 1994).

Low resolution crystallography remains of interest for the location of weakly ordered components in large macromolecular complexes. In particular, one of the key problems in protein crystallography is the crystallization of membrane proteins. A systematic study of the location of detergent molecules in membrane protein crystals combined with solution studies using small angle neutron scattering should provide valuable insight into the crystallization processes of membrane proteins.

Acknowledgements

We thank M.S. Lehmann and P.A. Timmins for useful discussions and for critical reading of the manuscript. We thank C. Bon for kind provision of Fig. 5.1.

References

Arnoux, B, Ducruix, A, Reiss-Husson, F, Lutz, M, Norris, J, Schiffer, M and Chang, CH (1989). *FEBS Letters*, **258**, 47–50.
Bailey, D and Cooper, JB (1994). *Protein Science*, **3**, 2129–2143.
Blundell, T and Johnson, L (1976). *Protein crystallography*, London, Academic Press.
Bon, C (1998). PhD thesis, Université Joseph Fourier, Grenoble.
Campbell, JW (1995). *Journal of Applied Crystallography*, **28**, 228–236.
Cipriani, F, Castagna, J-C, Lehmann, MS and Wilkinson, C (1995). *Physica B*, **213**, 975–977.
Cowan, SW, Schirmer, T, Rummel, G, Steiert, M, Ghosh, R, Pauptit, RA, Jansonius, JN and Rosenbusch, JP (1992). *Nature*, **358**, 727–733.
Cruickshank, DWJ, Helliwell, JR and Moffat, KR (1987). *Acta Crystallographica*, **A43**, 656–674.
Deisenhofer, J, Epp, O, Miki, K, Huber, R and Michel, H (1985). *Nature*, **318**, 618–624.
Gamble, TR, Clauser, KR and Kossikoff, AA (1994). *Biophysical Chemistry*, **53**, 15–25.
Habash, AJ, Rafferty, J, Weisgerber, S, Cassetta, A, Lehmann, MS, Hoghoj, P, Wilkinson, C, Campbell, JW and Helliwell, JR (1997). *Journal of the Chemical Society Faraday Transactions*, **93**(24), 4313–4317.
Harrison, SC, Olson, A, Schutt, CE, Winkler, FK and Bricogne, G (1978). *Nature*, **276**, 368–373.
Helliwell, JR and Wilkinson, C (1994). *Neutron and synchrotron radiation for condensed matter studies Vol. III. Applications to soft condensed matter and biology*, Baruchel, J, Hodeau, JL, Lehmann, MS, Regnard, JR and Schlenker, C (eds), Les Editions de physique and Springer Verlag.
Kossiakoff, AA (1982). *Nature*, **296**, 713.
Kossiakoff, AA (1985). *Annual Review of Biochemistry*, **54**, 1195.
Kossiakoff, AA and Spencer, SA (1980). *Nature*, **288**, 414–416.

Lehmann, MS (1994). *Neutron and synchrotron radiation for condensed matter studies Vol. III. Applications to soft condensed matter and biology*, Baruchel, J, Hodeau, JL, Lehmann, MS, Regnard, JR, and Schlenker, C (eds), Les Editions de physique and Springer Verlag.

Mason, SA, Bentley, GA and McIntyre, GJ (1984). *Neutrons in biology*, Schoenborn, BP (ed.), New York, Plenum Press, p. 323.

Niimura, N, Minezaki, Y, Nonaka, T, Castagna, JC, Cipriani, F, Høghøj, P, Lehmann, MS and Wilkinson, C (1997). *Nature Structural Biology*, **4**, 909–917.

Pebay-Peyroula, E, Garavito, Rosenbusch, JP, Zulauf, M and Timmins, PA (1995). *Structure*, **3**, 1051–1059.

Penel, S and Legrand P (1997). *Journal of Applied Crystallography*, **30**, 206.

Roth, M (1987). *Acta Crystallographica*, **A43**, 780–787.

Roth, M, Lewit-Bentley, A and Bentley GA (1984). *Journal of Applied Crystallography*, **17**, 77–84.

Roth, M, Lewit-Bentley, A, Michel, H, Deisenhofer, J, Huber, R and Oesterhelt, D (1989). *Nature*, **340**, 659–662.

Roth, M, Arnoux, B, Ducruix, A and Reiss-Husson, F (1991). *Biochemistry*, **30**, 9403–9413.

Timmins, PA, Pebay-Peyroula, E and Welte, W (1994a). *Biophysical Chemistry*, **53**, 27–36.

Timmins, PA, Wild, D and Witz, J (1994b). *Structure*, **2**, 1191–1201.

6

Reactions in crystals and time-resolved crystallography

Ilme Schlichting

In the general excitement of a time when new protein structures are published every week, it is often forgotten that a structure in itself does not tell one how an enzyme works. For that, one needs to know a great deal about mechanism, intermediates,[1] and dynamics; a structure is only a beginning, although an important one.

Up to now, mechanism and dynamics have formed the sphere of chemists and spectroscopists rather than crystallographers. This is because intermediate states are usually short-lived and thus not amenable to conventional crystallographic approaches. In the last few years, however, crystallographic determination of structures of intermediates that are normally short-lived has become feasible either via Laue crystallography on fast time scales or via slowing the reaction through use of temperature, pH, or chemical modification of the macromolecule [e.g. mutation, see Bolduc et al. (1995); Murphy et al. (1997)]. The first approach is usually called 'time-resolved crystallography' (Hajdu et al. 1988; Hajdu and Johnson 1990; Hajdu and Andersson 1993; Helliwell 1992; Johnson 1992; Moffat 1989; Pai 1992). However, the second approach, which we term 'kinetic crystallography', can often address the same scientific questions with better, more reliable data and fewer technical problems (Hajdu and Andersson 1993). Kinetic and time-resolved crystallography have the unique ability of providing direct and global information at the atomic level about the mechanism and structural dynamics of biomolecules.

The crystallographic determination of structures of reaction intermediates is usually not straightforward due to the large gap between protein motion and enzymatic reaction rates on the one hand (typically milliseconds to seconds) and the long crystallization and data acquisition times on the other hand [see Cruickshank et al. (1992)]. Due to the long time required to obtain large single crystals (typically days to weeks) it is in general not possible to co-crystallize enzyme–substrate complexes; the substrate would be converted to product long before the crystals are large enough for data collection. A way out of this dilemma is to co-crystallize enzyme and substrate in an inactive form that can be transformed into an active form shortly before collection of the diffraction

---

[1] We use 'intermediate' for an unstable complex that can be appreciably populated under certain conditions. This definition includes relatively stable enzyme–substrate complexes as well as very short-lived high-energy intermediates, such as may be encountered in covalent bond rearrangements.

data. Therefore, a major task in time-resolved and kinetic crystallographic studies is to find ways to initiate reactions in crystals rapidly, gently, and uniformly. In addition, physicochemical analysis methods need to be established to follow the reaction as it proceeds in the crystal.

6.1 Crystals

A prerequisite for the structural characterization of a reaction by time-resolved or kinetic crystallography is of course that the reaction takes place in the crystal. This is indeed the case for many crystalline enzymes (Makinen and Fink 1977; Rossi 1992; Rossi et al. 1992) on which we will focus. Protein crystals contain extended networks of solvent filled channels that make up 40–80% of the volume of a crystal. Due to this high solvent content and due to the limited number of—usually weak—interactions that hold the crystal together, the environment of the molecules in the crystal resembles that of a solution, albeit very concentrated (see below). Thus, the environment of the molecules in a crystal is much less artificial than may appear at first sight.

Nevertheless, one has to be aware that in the conditions of crystallization (e.g. ionic strength, pH, organic solvents, etc.), a reduced mobility of the molecules in the crystal lattice, or restricted access to active sites can cause changes in the binding constants and/or reaction rates determined in crystals relative to those measured in solution. If the rate constants in solution and the crystal differ, one may collect diffraction data at a time inferred from solution kinetics when the intermediate has not yet been formed (Stoddard et al. 1991) or has already decayed in the crystal. Therefore, it is important to determine the kinetics of the reaction in the crystal.

Generally, the kinetics of the reaction taking place in a crystal can differ from the one in solution due to (i) a changed affinity of the substrate, (ii) a changed (apparent) k_{cat}, and (iii) the reduced rates due to slow diffusion of the substrate in the crystal (or in case of reactions under steady-state conditions, of the product out of the crystal). These three effects can be analysed (Stoddard and Farber 1996) by studying the kinetics in solution under near crystallization conditions (provided no aggregation occurs, which can be verified, e.g. by dynamic light scattering). In that case, diffusion and steric hindrance are not an issue, and the effect of the crystallization conditions (e.g. ionic strength, pH) on affinity and (apparent) k_{cat} can be determined. The effect of steric hindrance can be analysed by using a microcrystalline slurry, the dependence of the (apparent) k_{cat} on the crystal size indicates the limiting effect of diffusion.

The characterization of the detailed kinetics of the reaction in the crystal is also of relevance for knowing *when* to start data collection after initiating the reaction. As radiation damage may limit the number of usable X-ray exposures, the intermediate should be captured when it has accumulated to maximum occupancy in the crystal. Ideally, the kinetics in the crystal should be analysed non-invasively (i.e. spectroscopically) *in situ* (Fülöp et al. 1994; Hadfield and Hajdu 1993; Williams et al. 1997). This adds the advantage that the reaction can be followed during collection of the X-ray data. Thus, unpleasant surprises, such as faster-than-expected reactions caused by heating by the X-ray beam (Schlichting *et al.* 1990), other-than-expected reactions due to production of electrons by the X-ray beam, and incomplete photolysis due to misalignment of the apparatus (Duke *et al.* 1994) can be detected and corrected during the experiment.

6.2 Kinetic considerations

The kinetics of the reaction to be studied dictate the proceedings of the experiment. Thus, of vital importance are not only the general time scale of events to be followed in relation to the rapidity of the trigger and data collection method used, but also the ratio of the time constants of build-up and decay of the intermediate in question. The latter factor determines whether the intermediate can be observed crystallographically: a necessary but not sufficient requirement is that the apparent rate constant for the generation of the intermediate should be larger than the rate constant for its disappearance. It is only in this case that the intermediate may be occupied sufficiently (detection limit *ca.* 30% occupancy).

If an intermediate accumulates to high occupancy under steady-state conditions of the reaction, the structure determination of the intermediate is straightforward experimentally: the reaction is allowed to proceed under steady-state conditions during X-ray data collection [see e.g. Malashkevich *et al.* (1993)]. In that case, the duration of the data acquisition is of no principal significance. The situation is completely different when the reaction is studied under single turnover conditions. In this case short-lived intermediates can be observed only if they live much longer than it takes to generate them and to collect their diffraction data. This is made feasible by either slowing the observed reaction (e.g. by cooling, or/and using poor substrates, suboptimal pH, and slow mutants) or by decreasing the data acquisition time (by using the high-intensity X-rays provided by synchrotrons, where a further reduction in data collection time can be achieved with fast data collection strategies employing the Weissenberg or Laue geometries (Hajdu and Andersson 1993). Both approaches require that the intermediate of interest can be generated fast, gently, and effectively. This translates into finding a way of synchronizing the initiation of the reaction of the $10^{13}-10^{15}$ molecules in the crystal.

6.3 Reaction initiation

Reactions can be initiated or 'triggered' by changes in thermodynamic parameters such as the temperature or pressure, by changes in the concentration of substrates, cofactors, protons or electrons, by light or other radiation (Schlichting and Goody 1997). The choice of the trigger depends largely on the physicochemical properties of the system and the reaction studied. In the ideal case the system has a built-in trigger, as is the case for the light–sensitive carbon monoxide complexes of heme proteins (Hartmann *et al.* 1996; Schlichting *et al.* 1994; Šrajer *et al.* 1996; Teng *et al.* 1994), photosynthetic reaction centres (Stowell *et al.* 1997), or photosensors such as the photoactive yellow protein (Genick *et al.* 1997, 1998; Perman *et al.* 1998).

6.3.1 Photolysis

A more general access to the reaction initiation with light is provided, for example, by the modification of substrates (Duke *et al.* 1994; Scheidig *et al.* 1994; Schlichting *et al.* 1990), cofactors, or catalytically important groups of the enzyme (Stoddard *et al.* 1990a,b, 1991) with biochemically inactivating groups that can be removed photolytically (Gurney and Lester 1987). Commonly used 'cage groups' (Corrie *et al.* 1992; Corrie and Trentham 1993; Kaplan *et al.* 1978; McCray and Trentham 1989) are substituted 2-nitrobenzyls such as 2-nitrophenylethyl which can be cleaved with light around 350 nm wavelength with concomitant production of a nitroso ketone. Apart from potential inactivating side reactions of the liberated cage group with the macromolecule (e.g. Duke *et al.* 1994; Hajdu and Johnson 1990), difficulties using caged compounds in crystallographic studies are rooted in the high concentration[2] (tens of millimolar) of enzymes in crystals. Since one wants to study at least a 1:1 complex between enzyme and photolysed caged compound, equally high concentrations of the caged compound are required. Even in the favourable case of high affinity between the protein and the (caged) compound, so that no excess of caged substance over enzyme needs to be used, the required concentration is at least equimolar, i.e. tens of millimolar. Thus, rapid and complete photolysis may be difficult to achieve. Despite the relatively high photolysis rate of caged compounds ($10^1-10^5\,\text{s}^{-1}$), photolysis of sufficient amounts of a caged compound for crystallographic purposes may take minutes since several light flashes may be required; the intensity and frequency of the photolysis flashes using a laser or xenon flash lamp (Rapp and Güth 1988; Rapp and Goody 1991) must be balanced against heating of the crystal [see also Ng *et al.* (1995)] and may be limited by the light source used. However, it is important to note that these limitations are not inherent to the technique itself; in the ideal case of a caged compound with a high quantum yield and a low extinction coefficient at the exciting wavelength, reaction initiation by a single flash of a powerful laser should be feasible.

Crystals containing caged compounds can be obtained by co-crystallization with the macromolecule [under no or subdued lighting (Schlichting *et al.* 1989) or by diffusion of the caged compound into the crystal (Stoddard *et al.* 1991), see also next section]. If a choice on the caged compound can be made, it is advantageous to use a compound that binds to the active site. This minimizes the potentially long diffusion time of the deprotected compound to the active site, thus making the most of the fast time scale of photolysis.

---

[2] The concentration of the enzyme in the crystal is calculated from the ratio of the number of molecules in the unit cell (given by: [number of molecules/asymmetric unit] · [number of asymmetric units/unit cell]) divided by the volume of the unit cell to Avogadro's number.

6.3.2 Diffusion

Reaction initiation by concentration jumps can be achieved not only photolytically but also by diffusion. The big advantage of this method is that it is experimentally straightforward. The disadvantage is that it is rather slow since diffusion inside crystals is limited to the solvent channels, which may be small or sterically blocked. Typical diffusion times for small molecules across a 200 μm thick crystal are one to many minutes. Because of the intrinsic generation of temporal and spatial concentration gradients and the competing effects of diffusion and enzymatic reaction (Makinen and Fink 1977), reaction initiation by diffusion is suitable only for slow enzymatic processes. It may pay to pre-diffuse a large, hence a slowly diffusing, ligand into the crystal under non-reactive conditions and then to start the reaction rapidly on completion of diffusion, e.g. by a diffusive change in pH, a light flash, or by a temperature jump.

Changes in pH constitute a powerful tool as long as they are tolerated by the crystal lattice, since they can be performed quite rapidly and efficiently by change of buffer. pH changes can be used in a dynamic way to initiate a reaction whose time course will be followed (Singer *et al.* 1992, 1993) or in a static way by trapping intermediates of the reaction at a pH where they are stable (Verschueren *et al.* 1993). This is very powerful if only one protonation stage of an amino acid, e.g. a histidine, is catalytically competent.

6.4 Data collection

6.4.1 Mono- and polychromatic diffraction

An intuitive way towards X-ray data collection strategies is provided by Bragg's law. Bragg explains X-ray diffraction by reflection of X-rays from equally spaced parallel planes cutting equivalent points in the crystal lattice. Rays reflected from successive lattice planes with spacing d are reinforced only if they interfere constructively, that is if the path difference between them is a multiple of the wavelength λ: $2 \cdot d \cdot \sin\theta = n \cdot \lambda$ where θ is the angle between incident or reflected ray and the reflecting plane, and n is an integer.

The Bragg equation contains a constant, the plane spacing d, and two experimentally accessible variables, the wavelength λ and the angle θ between the direction of the incident X-ray beam and the reflecting plane. Correspondingly, data collection (sampling of reciprocal space) can be done by changing either θ (the orientation of the crystal in the X-ray beam) or λ. The first approach translates to the monochromatic rotation method and the latter to the Laue method in which a stationary crystal is exposed to polychromatic X-rays.

Although X-ray diffraction studies started in 1912 with the Laue method, the monochromatic approach superseded it in the 1930s, and it was only with the advent of the inherently polychromatic synchrotron radiation that the Laue method was revived (Hajdu *et al.* 1987). The Laue method has disadvantages (e.g. sensitivity to imperfections of the crystal lattice, need for wavelength normalization, harmonic and spatial overlaps, increased radiation damage) but also advantages over monochromatic approaches (Moffat 1997). The biggest asset is that it allows very rapid data acquisition: in the Laue geometry all reciprocal lattice points lying between the two limiting Ewald spheres $1/\lambda_{min}$ and $1/\lambda_{max}$ are in diffraction condition simultaneously. When making use of the time structure of synchrotron radiation, exposure times down to the duration of the X-ray pulse (ca. 150 ps) are feasible (Bourgeois et al. 1996).

6.4.2 Experimental set-up

The experimental set-up for collection of diffraction data contains a 'mounting' unit that provides for the stability of the crystal and allows its orientation to be changed in the X-ray beam. Time-resolved and kinetic crystallographic experiments require, in addition, a geometry that allows the reaction to be started in the crystal. Routinely, crystals of macromolecules are mounted in glass or quartz capillaries whose ends are filled with mother liquor to protect against drying. This geometry allows reaction initiation by photolysis but not by diffusion. For the latter purpose the crystals are mounted in capillaries filled with mother liquor, and immobilized by pipe cleaner fibres or sephadex beads to prevent the crystal from slippage during data collection. Tubing glued to the ends of the capillary permits the flow of different solutions across the crystal (Petsko 1985). Such 'flow cells' can also be used at cryogenic temperatures if the solvent does not freeze (e.g. 70% methanol) (Douzou 1979; Douzou and Petsko 1984; Fink and Petsko 1981; Ding et al. 1994; Rasmussen et al. 1992).

In general, data collection at cryogenic temperatures has to be performed in such a way that no ice crystals are formed that would destroy the protein crystal lattice. Therefore, crystals are flash cooled so that the water in the solvent channels freezes amorphously. Often, this requires the addition of cryoprotectants such as glycerol, sugars, ethylene glycol, etc. To maximize the heat transfer during cooling, the crystals are suspended in a film that is formed by surface tension in a small loop (Teng 1990). This mounting method can be used to freeze-trap intermediates after they have formed at a higher temperature (Lee et al. 1997; Williams et al. 1997) or to photo-initiate a reaction at cryogenic temperatures (Genick et al. 1998; Hartmann et al. 1996; Schlichting et al. 1994; Stowell et al. 1997; Teng et al. 1994).

6.4.3 Strategy

The first question to answer is whether the experiment should be performed using the time-resolved or the kinetic approach. Many factors come into play, but some may make one approach more cumbersome than the other. Mosaic crystals, for example, render the Laue method unfavourable. Following a reaction on a fast time scale is much more difficult if a reaction is not reversible. This is a fundamental difference between systems that have been rendered light-sensitive and inherently light-sensitive systems. In the latter case the reaction is cycled during data collection (Šrajer et al. 1996), whereas several crystals are needed if the reaction is not reversible. In that case data are collected at different time points after reaction initiation from one crystal, and the data set is completed by collecting data at the same time points from other, differently oriented crystals. Since the 'kinetic' quality of the merged data depends on how similar crystals are at the various data collection time points, it is worth considering alternative experimental approaches, e.g. trapping by low temperature.

References

Bolduc, JM, Dyer, DH, Scott, WG, Singer, P, Sweet, RM, Koshland Jr, DE and Stoddard, BL (1995). *Science*, **268**, 1312–1318.
Bourgeois, D, Ursby, T, Wulff, M, Pradervand, C, Legrand, A, Schildkamp, W and Laboure, S (1996). *Journal of Synchrotron Radiation*, **3**, 65–74.
Corrie, JET and Trentham, DR (1993). *Bioorganic photochemistry*, Vol. 2 (ed. Morrison, H), Wiley, pp. 203–305, New York.
Corrie, JET, Katayama, Y, Reid, GP, Anson, M and Trentham, DR (1992). *Philosophical Transactions of the Royal Society of London*, **A340**, 233–244.
Cruickshank, DWJ, Helliwell, JR and Johnson, LN (eds) (1992). *Time-resolved Macromolecular Crystallography*, **A340**, Cambridge University Press, Cambridge.
Ding, X, Rasmussen, BF, Petsko, GA and Ringe, D (1994). *Biochemistry*, **33**, 9285–9293.
Douzou, P (1979). *Quarterly Reviews of Biophysics*, **12**, 521–569.
Douzou, P and Petsko, GA (1984). *Advances in Protein Chemistry*, **36**, 245–361.
Duke, EMH, Wakatsuki, S, Hadfield, A and Johnson, LN (1994). *Protein Science*, **3**, 1178–1196.
Fink, AL and Petsko, GA (1981). *Advances in Enzymology and Related Areas of Molecular Biology*, **52**, 177–246.
Fülöp, V, Phizackerley, RP, Soltis, SM, Clifton, IJ, Wakatsuki, S, Erman, J, Hajdu, J and Edwards, SL (1994). *Structure*, **2**, 201–208.
Genick, UK, Borgstahl, GEO, Ng, K, Ren, Z, Pradervand, C, Burke, PM, Šrajer, V, Teng, TY, Schildkamp, W, McRee, DE, Moffat, K and Getzoff, E (1997). *Science*, **275**, 1471–1475.
Genick, UK, Soltis, SM, Kuhn, P, Canestrelli, IL and Getzoff, ED (1998). *Nature*, **392**, 206–209.
Gurney, AM and Lester, HA (1987). *Physiological Reviews*, **67**, 583–617.
Hadfield, AT and Hajdu, J (1993). *Journal of Applied Crystallography*, **26**, 839–842.
Hajdu, J and Andersson, I (1993). *Annual Reviews in Biophysics and Biomolecular Structure*, **22**, 467–498.
Hajdu, J and Johnson, LN (1990). *Biochemistry*, **29**, 1669–1678.
Hajdu, J, Machin, PA, Campbell, JW, Greenhough, TJ, Clifton, IJ, Zurek, S, Gover, S, Johnson, LN and Elder, M (1987). *Nature*, **329**, 178–181.
Hajdu, J, Acharya, KR, Stuart, DI, Barford, D and Johnson, LN (1988). *Trends in Biochemical Science*, **13**, 104–109.
Hartmann, H, Zinser, S, Komninos, P, Schneider, RT, Nienhaus, GU and Parak, F (1996). *Proceedings of the National Academy of Sciences, USA*, **93**, 7013–7016.
Helliwell, JR (1992). *Macromolecular crystallography with synchrotron radiation*, Cambridge University Press, Cambridge.
Johnson, LN (1992). *Protein Science*, **1**, 1237–1243.
Kaplan, JH, Forbush III, B and Hoffman, JF (1978). *Biochemistry*, **17**, 1929–1935.
Lee, Y-H, Olson, TW, Ogata, CM, Levitt, DG, Banaszak, LJ and Lange, AJ (1997). *Nature Structural Biology* **4**, 615–618.
Makinen, M and Fink, AL (1977). *Annual Reviews in Biophysics and Bioengineering*, **6**, 301–342.

Malashkevich, VN, Toney, MD and Jansonius, JN (1993). *Biochemistry*, **32**, 13 451–13 462.
McCray, JA and Trentham, DR (1989). *Annual Reviews in Biophysical Chemistry*, **18**, 239–270.
Moffat, K (1989). *Annual Reviews in Biophysics and Biophysical Chemistry*, **18**, 309–332.
Moffat, K (1997). *Methods in Enzymology*, **277**, 433–447.
Murphy, JE, Boguslaw, S, Ma, L and Kantrowitz, ER (1997). *Nature Structural Biology*, **4**, 618–621.
Ng, K, Getzoff, ED and Moffat, K (1995). *Biochemistry*, **34**, 879–890.
Pai, EF (1992). *Current Opinion in Structural Biology*, **2**, 821–827.
Perman, B, Šrajer, V, Ren Z, Teng, TY, Pradervand, C, Ursby, T, Bourgeois, D, Schotte, F, Wulff, M, Kort, R, Hellingwerf, K and Moffat, K (1998). *Science*, **279**, 1946–1950.
Petsko, GA (1985). *Methods in Enzymology*, **114**, 141–146.
Rapp, G and Güth, K (1988). *European Journal of Physiology*, **411**, 200–203.
Rapp, G and Goody, RS (1991). *Journal of Applied Crystallography*, **24**, 857–865.
Rasmussen, BF, Stock, AM, Ringe, D and Petsko, GA (1992). *Nature*, **357**, 423–424.
Rossi, GL (1992). *Current Opinion in Structural Biology*, **2**, 816–820.
Rossi, GL, Mozzarelli, A, Peracchi, A and Rivetti, C (1992). *Philosophical Transactions of the Royal Society*, **A340**, 191–207.
Scheidig, AJ, Sanchez-Llorente, A, Lautwein, A, Pai, EF, Corrie, JET, Reid, G, Wittinghofer, A and Goody, RS (1994). *Acta Crystallographica*, **D50**, 512–520.
Schlichting, I, John, J, Rapp, G, Wittinghofer, A, Pai, EF and Goody, RS (1989). *Proceedings of the National Academy of Sciences USA*, **86**, 7687–7690.
Schlichting, I, Almo, SC, Rapp, G, Wilson, K, Petratos, K, Lenfter, A, Wittinghofer, A, Kabsch, W, Pai, EF, Petsko, GA and Goody, RS (1990). *Nature*, **345**, 309–315.
Schlichting, I, Berendzen, J, Phillips Jr, GN and Sweet, RM (1994). *Nature*, **371**, 808–812.
Schlichting, I and Goody, RS (1997). *Methods in Enzymology*, **277**, 467–490.
Singer, PT, Carty, RP, Berman, LE, Schlichting, I, Stock, A, Smalås, A, Cai, Z, Mangel, WF, Jones, KW and Sweet, RM (1992). *Philosophical Transactions of the Royal Society of London*, **A340**, 285–300.
Singer, PT, Smalås, A, Carty, RP, Mangel, WF and Sweet, RM (1993). *Science*, **259**, 669–673.
Šrajer, V, Teng, TY, Ursby, T, Pradervand, C, Ren, Z, Adachi, S, Schildkamp, W, Bourgeois, D, Wulff, M and Moffat, K (1996). *Science*, **274**, 1726–1729.
Stoddard, BL and Farber, GK (1996). *Structure*, **3**, 991–996.
Stoddard, BL, Bruhnke, J, Porter, N, Ringe, D and Petsko, GA (1990a). *Biochemistry*, **29**, 4871–4879.
Stoddard, BL, Koenigs, P, Porter, N, Ringe, D and Petsko, GA (1990b). *Biochemistry*, **29**, 8042–8051.
Stoddard, BL, Koenigs, P, Porter, N, Petratos, K, Petsko, GA and Ringe, D (1991). *Proceedings of the National Academy of Sciences USA*, **88**, 503–507.
Stowell, MHB, McPhillips, TM, Rees, DC, Soltis, SM, Abresch, E and Feher, G (1997). *Sciences*, **276**, 812–816.
Teng, T-Y (1990). *Journal of Applied Crystallography*, **23**, 387–391.
Teng, T-Y, Šrajer, V and Moffat, K (1994). *Nature Structural Biology*, **1**, 702–705.
Verschueren, KH, Seljee, F, Rozeboom, HJ, Kalk, KH and Diikstra, BW (1993). *Nature*, **363**, 693–698.
Williams, PA, Fülöp, V, Garman, EF, Saunders, NFW, Ferguson, SJ and Hajdu, J (1997). *Nature*, **389**, 406–412.

7
Local structure of metalloproteins at atomic resolution by XAFS

S. S. Hasnain

7.1 Introduction

7.1.1 XAFS

Soon after the first X-ray experiment using X-rays from a synchrotron for muscle diffraction (Rosenbaum et al. 1971), the Kronig and Kossel structures of the physics text books were transformed into a ubiquitous technique called XAFS (X-ray absorption fine structure) (Sayers et al. 1971; Kincaid and Eisenberger 1975), which found immediate applications in biology, chemistry, geology and materials science [see, e.g. Goulon et al. (1997)]. The growth of the subject has been very rapid and can be judged by the size of the proceedings for first international XAFS conference (140 pp.) held in March 1981 at Daresbury (Garner and Hasnain 1981) and the tenth conference in August 1998 in Chicago (nearly 900 pp. of the *Journal of Synchrotron Radiation* (1999), **6**, Part 3). Worldwide, the number of XAFS instruments continues to increase, some 25% of all SR instruments are either devoted or partially used for doing XAFS.

XAFS arises from the scattering of ejected photoelectrons and is a final state effect. It can thus be pictured as an electron diffraction experiment where the source of the electron is the photo-excited metal centre and the diffracted electron's detector is also the same metal centre. From this, surface scientists will recall LEED (low energy electron diffraction), a technique routinely used for probing the structure of surfaces (Pendry 1974). Thus, it is not surprising that the theory of XAFS is based on electron scattering (Ashley and Doniach 1975; Lee and Pendry 1975). The pathlength of electrons (mean free path) for most of the XAFS range (electron energy > 50 eV) is such that single scattering events dominate the extended range. Near the absorption edge, i.e. for electron energies < 50 eV, many multiple scattering paths exist and the data interpretation still requires a heroic effort (Binsted and Hasnain 1996). These two regions of XAFS are referred to as EXAFS (extended X-ray absorption fine structure) and XANES (X-ray absorption near-edge structure). In addition to the electron scattering effects, the near-edge part of the XANES spectrum contains electronic structure information, and thus even a qualitative comparison of this region is powerful in providing information about the redox state of the metal centre. Even though the EXAFS region is generally dominated by single scattering events, strong multiple scattering events can exist for much of the EXAFS region. This is the case for protein ligands such as histidine and tyrosines as well as for ligands/substrates/inhibitors such as CO, NO, CN, etc. (Blackburn et al. 1987; Strange et al. 1987). The multiple scattering is strongest when two atoms are collinear with the metal atom and sharply drops off as the angle reduces from 180°, with little contribution at angles less than 110°. These multiple scattering effects can easily be treated in data analysis if proper care is taken, and offer the opportunity for extracting angular information between the scattering centres.

XAFS is equally applicable both to the aqueous and the crystalline state (Bianconi et al. 1985), even though the majority of applications have been to the aqueous proteins. In a number of cases, data have been obtained on crystalline slurry (Ascone et al. 1997). One of the primary reasons for the lack of XAFS data on single crystals is the lack of suitable instruments [some multiwavelength anomalous diffraction (MAD) instruments are capable of this] where high quality fluorescence data from single crystals (< 100 µm) can be obtained.

XAFS provides very high (atomic) resolution information about a specific centre such as the metal atom in a metalloprotein. In fact, the resolution of XAFS is essentially the same for small molecules and large metalloproteins. Despite the high resolution of the technique and its equal applicability to crystalline and aqueous state, its full power is achieved only when it is combined with the three-dimensional high resolution structure, generally available only from X-ray crystallography. This three-dimensional structure may be of the protein under study itself or of a closely related system. The purpose of this review is to provide a working knowledge of the XAFS technique and its application to some metalloproteins. The usefulness of this combined approach for structure–function studies of metalloproteins will be demonstrated through some recent examples.

7.1.2 *Metalloproteins*

Metalloproteins form a large fraction (between one-fourth to one-third) of all known proteins. These contain metal ions either as single atom or as part of a cluster and play a variety of life-sustaining roles in the bacterial, plant and animal kingdoms (Harrison 1985). In the list of 25 elements that have been recognized as essential and indispensable to life, 15 are metals. Some of the fundamental biological processes in which metalloproteins participate include electron storage and transfer, dioxygen binding, storage and activation, and substrate transport, catalysis and activation. In addition to their abundance and importance in the biosphere, many of the metalloproteins are coloured due to metal–ligand charge transfer bands. The colour of these biologically important molecules not only makes them aesthetically attractive but also offers much practical advantage in biochemical manipulation. Thus, it is not surprising that these proteins have attracted much attention from biophysicists including some of the pioneers (Dorothy Hodgkin, Perutz, Kendrew, Phillips) of protein crystallography [e.g. see Kendrew et al. (1958); Perutz (1970)].

Metalloproteins utilize the chemistry of metals to their advantage to perform varied biological functions with specification and control. The redox and ligand chemistry of metals is used to perform a wide variety of chemical reactions in the biosphere. Quite often, these chemical reactions are accompanied by only a small structural change around the metal atom. Ideally therefore, atomic resolution structures of these proteins is required in different reaction states but a knowledge of the very high resolution structure of the metal centre in itself is of major benefit for understanding the chemistry/reaction mechanisms of metalloenzymes. It is perhaps fair to claim that nowhere in the determination of molecular structure is precision more at a premium than in the case of metalloproteins. It is thus understandable why high resolution (< 2 Å) structure determination of metalloproteins has attracted so much attention (see, e.g. Dauter *et al.* 1997). In many cases, it has not been possible to obtain atomic resolution (< 1.2 Å, where individual atoms become visible) crystallographic data owing to inherent diffraction limits of protein crystals arising from static disorder. Thus, the combination of very high resolution crystallographic studies of metalloproteins in the resting state and atomic resolution XAFS of resting and reactive intermediate states is a very powerful approach for studying structure/function relationships in these important (and colourful) proteins.

7.2 Theoretical background

7.2.1 *Single scattering*

An approximate formulation of the EXAFS theory was worked out during the mid-1970s by Stern (1974) and developed extensively by Lee and Pendry (1975) and Ashley and Doniach (1975). During the next 20 years many other contributions to the development of EXAFS theory occurred (Gurman *et al.* 1984, 1986; Foulis *et al.* 1990; Mustre de Leon *et al.* 1990; Pettifer *et al.* 1986; Rehr *et al.* 1994). A detailed theoretical discussion is not within the scope of this review. However, the basic concepts and their relationship to structural parameters are discussed.

It has been shown that, except for the energies very close to the absorption threshold, a single scattering formalism is sufficient to describe the observed data in most cases (Lee and Pendry 1975; Ashley and Doniach 1975). When the energy of the photoelectron is sufficiently high, the curvature of the electron wave can be neglected and thus the theory can be greatly simplified into what became known as the 'plane wave approximation'. In this high energy approximation the oscillatory EXAFS function $\chi(k) = (\mu - \mu_0)/\mu_0$ may be written, for the K-absorption edge,

$$\chi(k) = \sum_j -\frac{N_j}{kR_j^2} |f_j(k, \pi)| \sin(2kR_j + 2\delta_1 + \psi_j) e^{-2R_j/\lambda} e^{-2\sigma_j^2 k^2}, \qquad (7.1)$$

where μ is the absorption of the sample, μ_0 the atomic absorption and k is the wave vector of the photoelectron.

Equation (7.1) shows the structural basis of EXAFS in that $\chi(k)$ is dependent on the number of scattering atoms N_j, the location of the scattering atoms R_j, and on the type of scattering atom through the characteristic energy dependence of the backscattering amplitude $|f_j(\pi)|$. In eqn (7.1) $2\delta_1$ is a phase shift due to the potential of the emitting atom and ψ_j is the phase of the backscattering factor. The mean square variation in the interatomic distance between the metal ion and the scattering atom is related by the Debye–Waller factor, σ_j^2, which assumes a harmonic distribution. λ is the elastic mean free path of the photoelectron and it is this damping term $[\exp(-2R_j/\lambda)]$ that limits the backscattering contribution to approximately 6 Å from the metal ions in a metalloprotein.

The simplified expression for EXAFS given in eqn (7.1) is valid at high photoelectron energies. At lower electron energies, the plane wave approximation breaks down and leads to errors in the calculated phase, which in turn can result in incorrect determination of the interatomic distances. Furthermore, it is this low energy part of the EXAFS spectrum that contains information from the more distant shells and the multiple scattering events. This is of further importance in the case of metalloproteins where the range of the EXAFS data is limited in view of the weaker scattering from the low Z atoms which are typically the metal ligands (e.g. nitrogen and oxygen).

The low energy part of the EXAFS spectrum can be analysed by using the exact theory given by Lee and Pendry (1975), which takes account of the curvature of the electron wave and thus has been named the 'curved wave method'. The exact theory has not been used in a majority of metalloprotein studies due to its mathematical complexity and requirement for large computational time. However, the theory has been simplified by performing an average over the angular positions of the scattering atoms relative to the X-ray beam direction (Gurman et al. 1984). This simplification does not compromise the exact nature of the theory for polycrystalline or amorphous samples and is well suited for studies of metalloproteins in solution. This simplification has been called 'fast curved wave' theory.

7.2.2 Multiple scattering (EXAFS and XANES)

So far we have considered only the single scattering formalism for EXAFS. Initially, multiple scattering contributions were not considered to be important in the EXAFS region, except for a collinear arrangement of scattering atoms. However, it was noted by several workers [see, e.g. Blackburn et al. (1984)] in the case of imidazole ligands that strong multiple scattering contributions are present in the EXAFS data up to 300 eV above the absorption edge. Perutz et al. (1982) had pointed out that if such contributions could be analysed, then stereochemical details such as the position of the iron atom with respect to the haem plane could be determined. These multiple scattering contributions arise mainly due to backscattering from the distal (outer shell) atoms of imidazole/pyrole groups, where the photoelectron is forward scattered by the intervening N atoms before being backscattered by the outer shell atom. Forward scattering is generally strong at lower electron energies ($E < 200$ eV) over a significant angular range (0–70°). At higher electron energies the strong forward scattering is increasingly confined to a narrower cone and the multiple scattering contributions become unimportant except when an approximately collinear geometry exists.

The higher order scattering terms become even more important in the X-ray absorption near-edge region. Multiple scattering of the excited electron provides the possibility of obtaining information about bond angles and relative orientations of metal ligands, thus giving a fuller stereochemical picture of the metal site in a protein. Recent theoretical developments have made this region of the spectrum tractable to theoretical interpretation and offer a unique method of obtaining higher correlation functions in aqueous protein samples [see e.g. Durham *et al.* (1981); Bianconi *et al.* (1985); Joly (1997)].

Obviously, it is highly desirable to be able to utilize the whole XAFS region in structural studies and much effort has been made towards a unified analysis approach for the two regions. A comparison of matrix inversion and finite path sum methods shows that the latter method is more promising for fitting the edge region. Recent developments have made the calculations of both the scattering and atomic components practicable and thus have made the fitting of an entire X-ray spectrum rather than its components, EXAFS and XANES, possible (Binsted and Hasnain 1996). However, a number of improvements are still required before this approach can be used for 'routine' structure determination.

7.2.3 Data analysis

Much of the early work was based on the plane wave approximation. More recently, much effort has been put towards the use of *exact* methods. The curved wave approach based on the Lee and Pendry formalism has long been used for structural studies of metalloproteins (Bordas *et al.* 1980; Strange *et al.* 1987; Blackburn *et al.* 1987; Murphy *et al.* 1993, 1997; Shiro *et al.* 1997; Baugh *et al.* 1997). These exact methods have been combined with restrained and constrained refinement approaches in order to utilize the known stereochemical information (bond distances and angles within a group) and maintain a reasonable ratio of parameters to observations (Binsted *et al.* 1992).

7.3 Experimental requirements

This section is intended to provide a basic understanding of the requirements of XAFS for metalloproteins. The first requirement is that of a continuous and stable X-ray source; to date, there has been no EXAFS study of a metalloprotein using a laboratory X-ray source. The EXAFS signal is small compared to the atomic absorption resulting from the excitation of the core electron of the atom of interest. This fact necessitates an intense and stable X-ray source. For biological systems, the situation is even worse as the atom of interest around which the environment needs to be determined is normally present at $< 0.1\%$ level. There is also a solubility limit of a metalloprotein which ultimately defines the maximum concentration; most experiments are performed at a metal (absorbing atom) concentration of ~ 0.5–5 mM. For these reasons, fluorescence detection is necessary. In fact, it is the application of XAFS to proteins that has driven the development of multi-element detection systems based on Ge/Si(Li) detector technology (Cramer *et al.* 1988) with parallel development in purpose-built front end electronics (Derbyshire *et al.* 1992). In addition, a complete tunability of wavelength with high resolution ($\Delta\lambda/\lambda \sim 10^{-4}$) is required to record the near edge and extended fine structure, thus necessitating the use of a double crystal monochromator. Again, it is the application to dilute systems including proteins that has prompted the rapid scanning of these monochromators without compromising the wavelength stability or spectral purity (Murphy *et al.* 1995). Thus, it is now practical to sweep the whole of the XAFS spectrum within a matter of seconds. The recent emergence of several third generation X-ray sources is likely to further catalyse these developments so that structural biologists can approach the ultimate dream of obtaining a structural movie at atomic resolution.

The source and monochromator requirements of single crystal XAFS is very similar to that of a MAD beam line. Indeed, the alignment requirements of the sample are also the same. Single crystals of metalloproteins are generally no more than 100 µm in at least two of the directions and thus a brilliant source such as a third generation undulator is ideal. The rapid tunability requires either the use of two slightly different undulators or an undulator with some taper. On a lower energy low emittance SR source, a multipole wiggler is most suitable. The fluorescence detector for single crystals needs to be somewhat special given the small volume and limited access to the sample. Recent progress for a low profile nine element detector (C-TRAIN) for a small volume sample in a reaction cell is likely to prove of interest in this respect (Derbyshire *et al.* 1999).

7.4 Applications

XAFS has been applied to a large number of metalloproteins over the last 20 or so years. It is worth remarking that several of the pioneering studies have stood the test of time. Table 7.1 summarizes a selection of these early works.

7.4.1 XANES

Two types of information can be obtained from this region. As has been said before, this region is sensitive to the electronic configuration of the metal site, and thus changes in the redox state of the metal or in its stereochemistry cause changes in the spectrum in this region.

Figure 7.1 shows the XANES spectra of oxidized and reduced forms of superoxide dismutase for the Cu and Zn sites of the enzyme. The changes in the Zn spectra are subtle, confirming that very subtle changes in the ligand geometry take place upon reduction at the Zn site. In contrast, the Cu data show a major change. In particular, the appearance of a distinct peak at 8983 eV in the reduced enzyme is a characteristic feature observed for a three-coordinate Cu(I) site with trigonal geometry (Kau *et al.* 1987; Blackburn *et al.* 1989), thus confirming that upon reduction, the bridging imidazole His-61 is protonated and that Cu moves to form a trigonal geometry with the remaining three His ligands. The features, marked α and β, are absent in the reduced Cu site and are consistent with the Cu^+ redox state.

Table 7.1. Some early XAFS (1975–80)

Year	Authors	Main results
1975	R.G. Schulman, P. Eisenberger, W.E. Blumberg and N.A. Stombaugh	EXAFS of rubredoxin showed that the four Fe–S distances were equal in length, in contrast to the then available crystal structure results.
1978b	S.P. Cramer, K.O. Hodgson, W.O. Gillum and L.E. Mortenson	The first application of EXAFS to determine *de novo* a metal site in a metalloprotein was the remarkable discovery of an unprecedented polynuclear Mo–Fe–S cluster in the nitrogenase enzyme from the EXAFS data analysis.
1978a	S.P. Cramer, J.H. Dawson, K.O. Hodgson and L.P. Hager	Applications to cytochrome P-450 quantified the presence of an axial sulphur ligand.
1978	T.D. Tullius, P. Frank and K.O. Hodgson	The 'blue copper protein' azurin was found to have an unusually short Cu–S ligand which defined the electronic structure of the active site.
1979, 1980	J. Bordas, R.C. Bray, C.D. Garner, S. Gutteridge and S.S. Hasnain	First difference EXAFS to show that functional form of xanthine oxidase contained a cyanosable sulphur (essential for catalysis) at the Mo site.

7.4.2 EXAFS

Before we discuss any specific examples, we need to ask ourselves why use XAFS when protein crystallography has become a 'routine' technique. The major strengths of EXAFS, namely that of very high resolution and its applicability to solution and solid state samples have already been mentioned. In this section three examples from our own work are discussed to illustrate how these advantages manifest themselves in real systems and as such each represents a different case for using EXAFS when X-ray crystallography is available.

7.4.2.1 *Cupredoxins*

A family of functionally related metalloproteins, known as 'blue' copper proteins (or cupredoxins) mediate electron transfer—a fundamental biological process. These proteins have attracted much attention as model electron transfer proteins because of their relative simplicity, their richness in electronic spectra (colour—in fact, the colour could be blue to green) and their accessibility by a variety of physical methods (Adman and Jensen 1981; Reinhammar 1979). The crystallographic structure of several of these have been determined at a variety of resolutions (Adman and Jensen 1981; Nar *et al.* 1991; Guss and Freeman 1983;

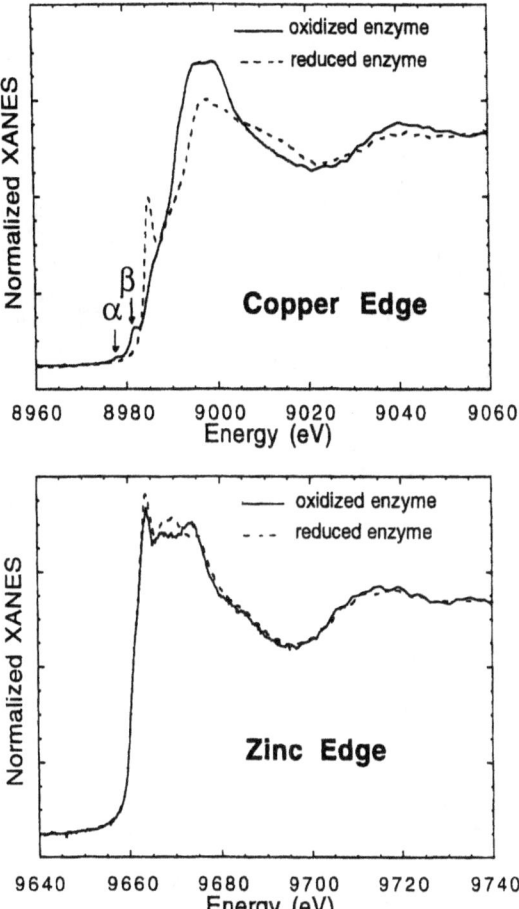

Fig. 7.1. K edge XANES spectra of oxidized and reduced forms of superoxide dismutase for the Cu and Zn sites of the enzyme. CuK edge shows a distinct peak at 8983 eV in the reduced enzyme; a characteristic feature for a three-coordinate Cu(I) site with trigonal geometry. Changes in Zn edge XANES are subtle and can be accounted for by the rotation of His-61 observed in the crystallographic structure (see text).

Guss et al. 1992; Baker 1988; Norris et al. 1983; Dodd et al. 1995). The coordination sphere around copper consists of two nitrogens (residues His-117 and His-46 in the case of azurins) and two sulphur donors (residues Cys-112 and Met-121).

Figure 7.2 shows a comparison of EXAFS data for the oxidized and reduced forms of azurin. It is clear that a change in Cu–ligand distances occurs upon reduction. A detailed analysis shows that the extent of the change is very significant from the EXAFS point of view and is what will be expected from the removal/addition of a single electron. An increase of 0.1 Å in the Cu–S_{cys} bond and a slight asymmetry in the Cu–N_{His} distances is observed in the EXAFS of reduced protein. These changes are compared with the crystallographic values

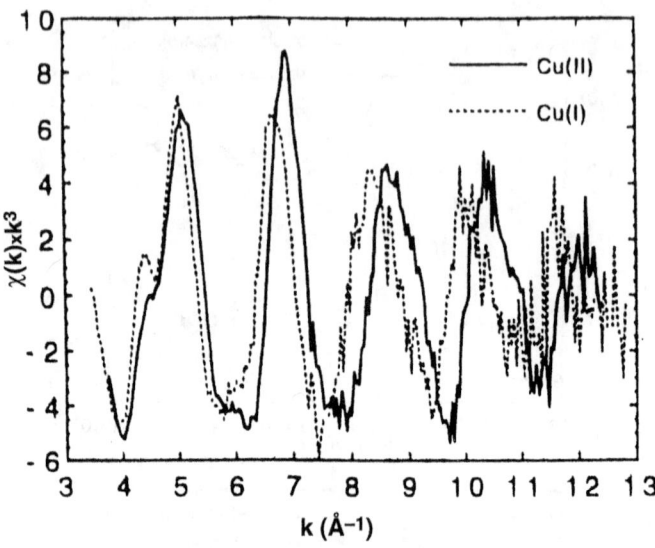

Fig. 7.2. K edge EXAFS data for the oxidized and reduced forms of azurin demonstrating that a change in Cu–ligand distances occur upon reduction (see Table 7.2).

Table 7.2. A comparison of EXAFS and crystallographic structural parameters for the oxidized and reduced azurin II from *Alcaligenese xylosoxidans*. The EXAFS values for Sδ are uncertain and arrived at when included in the model at the crystallographic positions before refinement

Cu ligand	Crystallographic (1.75 Å)*		EXAFS†	
	Oxidized	Reduced	Oxidized	Reduced
46 Nδ	2.04	2.03	1.86	1.90
117 Nδ	1.99	2.02	1.94	2.01
112 Sγ	2.11	2.17	2.12	2.17
45 O	2.74	2.75	2.82	2.98
121 Sδ	3.28	3.28	3.40	3.35

*Dodd and Hasnain, unpublished results.
†Strange, Murphy and Hasnain, unpublished results.

for the oxidized and reduced form of azurin (Dodd and Hasnain, unpublished results), Table 7.2. This comparison reveals a general agreement between the two techniques, with the agreement for the Cu–S_{cys} distance being excellent. We note, however, that the Cu–His distance for the native oxidized protein is systematically longer in the crystal structure by about 0.1 Å; thus little meaning can be attached to the very small variation observed for these ligands in the crystallographic structure of the reduced protein. This is an example where the resolution of crystallographic studies is at the margins of the expected changes that are likely to take place upon reduction.

7.4.2.2 *Cu–Zn superoxide dismutase*

Superoxide dismutases are a ubiquitous family of functionally related enzymes responsible for the removal of toxic radical superoxides in biological systems. The widely accepted mechanism of superoxide dismutation [eqns (7.2) and (7.3)] involves a cyclic reduction and oxidation of Cu(II):

$$O_2^{\bullet-} + \text{Enz-Cu}^{2+} \rightarrow O_2 + \text{Enz-Cu}^+ \tag{7.2}$$

$$\text{Enz-Cu}^+ + O_2^{\bullet-} + 2H^+ \rightarrow \text{Enz-Cu}^{2+} + H_2O_2 \tag{7.3}$$

A 2 Å resolution crystal structure of the oxidized bovine Cu–Zn SOD was published by Tainer *et al.* (1982), which showed that the active site consisted of a binuclear site where the Cu(II) and Zn(II) ions were bridged by an imidazolate ring—His-61. The copper ion was found to be coordinated to four histidine residues arranged in a distorted square pyramid with a water molecule located at ~3 Å. The zinc ion was found to be coordinated by three histidine residues and an aspartate in a distorted tetrahedron. Based on this structure, Tainer *et al.* proposed a detailed mechanism of dismutation in 1983, which involved the dissociation of the Cu–His-61 bond upon reduction. The presence of three-coordinate copper in the reduced form had been suspected from many spectroscopic studies and its first structural evidence was provided by EXAFS in 1984 (Blackburn *et al.* 1984). It took another ten years before the first crystallographic structure of reduced SOD became available, but surprisingly this showed that the coordination of both Cu and Zn remains unchanged (Rypniewski *et al.* 1995). As we have seen above under the XANES section, clear changes take place upon reduction where the Cu site adopts a trigonal geometry. This reduction in coordination is also clearly evident in the EXAFS region as has been shown by Blackburn *et al.* (1984) and Murphy *et al.* (1997). We have obtained bovine Cu–ZnSOD in two crystal forms $P2_12_12_1$ (1.65 Å resolution) and $C2_12_12_1$ (2.3 Å resolution); in the 2.3 Å structure both monomers are in the oxidized state while in the 1.65 Å structure monomer A is in the reduced state and the Cu is tri-coordinate having lost the coordinated water molecule and His-61 from ligation (Hough and Hasnain, 1999). In the reduced site, a clear movement of Cu in the plane of coordinating nitrogens is observed together with a rotation of His-61 away from the Cu. The Cu–N(His-61) distance in the reduced site is 3.1 Å (Fig. 7.3). Interestingly, the Cu–Zn distance in the reduced form is increased by about 0.6–0.7 Å. The mixed state (one site reduced and one oxidized) is confirmed by the XANES data obtained from a crystalline slurry (Hough and Hasnain, 1999). The evidence for a tri-coordinate Cu site has also been provided in yeast SOD (Ogihara *et al.* 1996).

This is an example that illustrates the importance of ensuring the integrity of a reaction state (native or an intermediate stable functional state) in the crystal. It is important to remember that it is not sufficient to carry out controls on

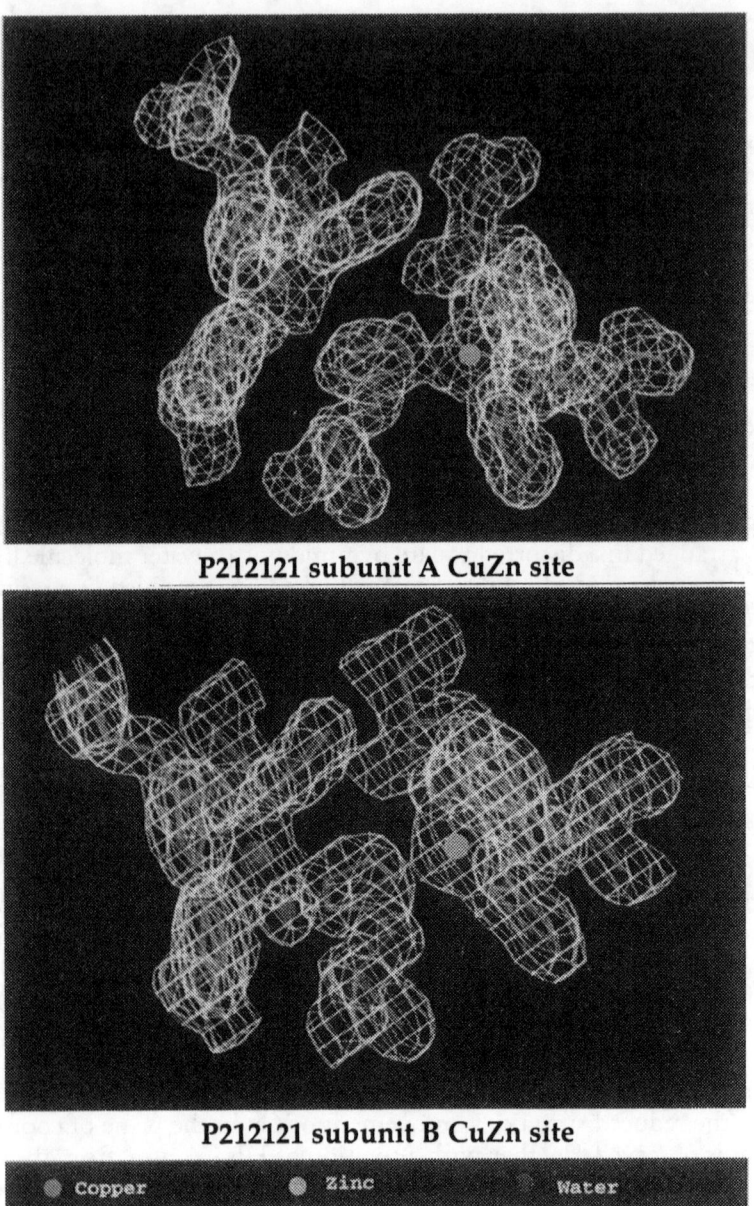

Fig. 7.3. Cu–Zn site of bovine SOD as revealed by 1.65 Å resolution crystallographic structure (see Plate Section).

a crystal of the same batch. This does, indeed, argue for the need of online monitoring. For metalloproteins, this is ideally provided by the online recording of XANES data, as these data can be obtained for any metal in any oxidation state. Optical or EPR data are useful supplementary information but are limited only to electronically active states; thus Cu^+ or Zn, for example, have no optical or EPR signals.

7.4.2.3 *Domain closure in transferrins*

Recent X-ray crystallographic and solution X-ray scattering studies have shown that transferrins (serum transferrin, lactoferrin and ovotransferrin) undergo a major conformational change when iron is incorporated into the molecule. Apo-proteins show a structure with open interdomain clefts that close when iron is bound. This closed conformation has been suggested as an important step in the receptor recognition. XAFS and X-ray solution scattering of a mutant of the half molecule, where one of the ligands of Fe, tyrosine 95, is mutated to histidine (Y95H), has revealed that the large scale domain movement is a two step process (Grossmann et al. 1998). This mutant is still not crystallized and thus represents an example where useful structure/function questions can be addressed by XAFS in the absence of a three-dimensional structure. It is likely that several of these states may not be amenable to crystallographic studies as it is not necessary that the protein may crystallize in these substates.

7.5 Conclusions

Very high resolution structural information can be obtained by XAFS and XANES in both crystalline and aqueous states for the metal centres in a protein. The resolution at which this information is achieved is the same in the two states and in fact is the same as that routinely achieved in 'small molecule' crystallography. The agreement between XAFS and crystallography is excellent for small molecules. However, this agreement is rather limited in the case of proteins mainly because of the lower resolution of the protein structures (generally > 1.5 Å). The resolution of protein X-ray structures can be expected to improve with improvements in SR sources and detectors but for a foreseeable future XAFS will continue to be an important structural tool for structure–function studies of metalloproteins. It has an additional important role to play, namely that of acting as a control for ensuring the integrity of the reaction state of the protein. If our dream of a 'structural movie' is to be realized, then online monitoring by XAFS should be an integral part of the experiment for such studies on metalloproteins. It is only through the rigour of the experiment that we shall be able to provide structural data of quality to underpin the chemistry of this class of proteins.

Acknowledgements

I would like to thank my colleagues, collaborators and the current and past members of the group. I am particularly pleased to acknowledge the contributions of Norman Binsted, Gareth Derbyshire, Fraser Dodd, Gunter Grossmann, Mike Hough, Lorrie Murphy and Richard Strange.

References

Adman, ET and Jensen, LH (1981). *Israeli Journal of Chemistry*, **21**, 8.
Ascone, I and Castanor, R (1997). *Biochemical and Biophysical Acta*, **241**, 119–121.
Ashley, CA and Doniach, S (1975). *Physical Review B*, **11**, 1279.
Baker, EN (1988). *Journal of Molecular Biology*, **203**, 1071.
Baugh, PE, Garner, CD, Charnock, JM, Collison, D, Davies, ES, McAlpine, AS, Bailey, S, Lane, I, Hanson, G and McEvan, AG (1997). *JBIC*, **2**, 634.
Bianconi, A, Congiu-Castellano, A, Durham, PJ, Hasnain, SS and Phillips, S (1985). *Nature*, **318**, 685.
Binsted, N, Strange, RW and Hasnain, SS (1992). *Biochemistry*, **31**, 12 117.
Binsted, N, and Hasnain, SS (1996). *Journal of Synchrotron Radiation*, **3**, 185.
Blackburn, NJ, Hasnain, SS, Binsted, N, Diakun, GP, Garner, CD and Knowles, PF (1984). *Biochemical Journal*, **219**, 985.
Blackburn, NJ, Strange, RW, McFadden, LM and Hasnain, SS (1987). *Journal of the American Chemical Society*, **109**, 7162.
Blackburn, NJ, Strange, RW, Reedijk, J, Volbeda, A, Farooq, A, Karlin, KD and Zubieta (1989). *Journal of Inorganic Chemistry*, **28**, 1349.
Bordas, J, Bray, RC and Garner, CD (1979). *Journal of Inorganic Biochemistry*, **11**, 181.
Bordas, J, Bray, RC, Garner, CD, Gutteridge, S and Hasnain, SS (1980). *Biochemical Journal*, **191**, 499.
Cramer, SP, Dawson, JH, Hodgson, KO and Hager, LP (1978a). *Journal of the American Chemical Society*, **100**, 7282.
Cramer, SP, Hodgson, KO, Gillum, WO and Mortensen, LE (1978b). *Journal of the American Chemical Society*, **100**, 3398.
Cramer, SP, Tench, O, Yocum, M and George, GN (1988). *Nuclear Instruments and Methods*, **A266**, 586.
Dauter, Z, Lamzin, VS and Wilson, KO (1997). *Current Opinion in Structural Biology*, **7**, 681.
Derbyshire, GE, Dent, AJ, Dobson, BR, Farrow, RC, Felton, A, Greaves, GN, Morrell, C and Wells, MP (1992). *Review of Scientific Instruments*, **63**, 814–815.
Derbyshire, GE, Cheung, KC, Sangsingkeow, P and Hasnain, SS (1999). *Journal of Synchrotron Radiation*, **6**, 62.
Dodd, FE, Hasnain, SS, Abraham, ZHL, Eady, RR and Smith, BE (1995). *Acta Crystallographica*, **D51**, 1052.
Durham, PJ, Pendry, JB and Hodges, CH (1981). *Solid State Communications*, **38**, 159.
Foulis, DL, Pettifer, RF, Natoli, CR and Benfatto, M (1990). *Physical Review A*, **41**, 6922.
Garner, CD and Hasnain, SS (1981). *Proceedings of the First EXAFS conference*, DL/SCI/R17.
Goulon, J, Goulon-Ginet, C and Brookes, N (1997). *Proceedings of the 9th XAFS conference, Journal de Physique (Paris)*, **7**.
Grossmann, JG, Crawley, W, Strange, RW, Patel, KJ, Murphy, LM, Neu, M, Evans, RW and Hasnain, SS (1998). *Journal of Molecular Biology*, **279**, 461–472.
Gurman, SJ, Binsted, N and Ross, I (1984). *Journal of Physics C*, **C17**, 143.
Gurman, SJ, Binsted, N and Ross, I (1986). *Journal of Physics C*, **C19**, 1845.
Guss, JM and Freeman, HC (1983). *Journal of Molecular Biology*, **169**, 521.
Guss, JM, Bartunik, HD and Freeman, HC (1992). *Acta Crystallographica*, **B48**, 790.
Gutteridge, S and Hasnain, SS (1980). *Biochemical Journal*, **191**, 499.
Harrison, PM (1985). *Metalloproteins (Parts I and II)*, Academic Press, New York.
Hough, M and Hasnain, SS (1999). *Journal of Molecular Biology*, **287**, 579.

Joly, Y (1997). *Journal de Physique C2 (Paris)*, **7**, 111.
Kau, L, Spira-Solomon, DJ, Penner-Hahn, JE, Hodgson, KO and Solomon, EI (1987). *Journal of the American Chemical Society*, **109**, 6433.
Kendrew, JC, Bodo, G, Dintzis, H, Parrish, RG, Wyckoff, H and Phillips, DC (1958). *Nature*, **181**, 662.
Kincaid, B and Eisenberger, P (1975). *Physical Review Letters*, **34**, 1361.
Lee, PA and Pendry, JB (1975). *Physical Review B*, **11**, 2795.
Murphy, LM, Strange, RW, Karlsson, G, Lundberg, L, Pascher, T, Reinhammar, B and Hasnain, SS (1993). *Biochemistry*, **32**, 1965.
Murphy, LM, Dobson, BR, Neu, M, Ramsdale, CA, Strange, RW, and Hasnain, SS (1995). *Journal of Synchrotron Radiation*, **2**, 64.
Murphy, LM, Strange, RW and Hasnain, SS (1997). *Structure*, **5**, 371.
Mustre de Leon, J, Yacoby, Y, Stern, EA and Rehr, JJ (1990). *Physical Review B*, **42**, 10 843.
Nar, H, Messerschmidt, A, Huber, R, van de Kamp, M and Canters, GW (1991). *Journal of Molecular Biology*, **221**, 765.
Norris, GE, Anderson, BF and Baker, EN (1983). *Journal of Molecular Biology*, **165**, 501.
Ogihara, N, Parge, HE, Hart, PJ, Weiss, MS, Goto, JJ, Crane, BR, Tsang, J, Slater, K, Roe, JA, Valentine, JS, Eisenberg, D and Tanier, JA (1996). *Biochemistry*, **35**, 2316.
Pendry, JB (1974). *Low energy electron diffraction*, Academic Press, New York.
Perutz, MF (1970). *Nature*, **228**, 726.
Perutz, MF, Hasnain, SS, Duke, PJ, Sessler, JL, and Hahn, JE (1982). *Nature*, **295**, 535.
Pettifer, RF, Foulis, DL and Hermes, C (1986). *Journal de Physique C (Paris)*, **8**, 545.
Rehr, JJ, Booth, CH, Bridges, F and Zabinsky, SI (1994). *Physical Review B*, **49**, 12 347.
Reinhammar, B (1979) *Advances in inorganic biochemistry*, eds. Eichhorn, GL *et al.*, **92**, Elsevier, North Holland.
Rosenbaum, G, Holmes, KC and Wiltz, J (1971). *Nature*, **230**, 434.
Rypniewski, WR, Mangani, S, Bruni, B, Orioli, PL, Casati, M and Wilson, KS (1995). *Journal of Molecular Biology*, **251**, 282.
Sayers, DE, Stern, E and Lytle, F (1971). *Physical Review Letters*, **27**, 1204.
Schulman, RG, Eisenberger, P, Blumberg, WE and Stombaugh, NA (1975). *Proceedings of the National Academy of Sciences USA*, **72**, 4003.
Shiro, Y, Obayashi, E, Adachi, SI, Iizuka, T, Nomura, M and Shoun, H (1997). *Journal de Physique (Paris)*, **7**, 587.
Stern, EA (1974). *Physical Review B*, **10**, 3027.
Strange, RW, Blackburn, NJ, Knowles, PF and Hasnain, SS (1987). *Journal of the American Chemical Society*, **109**, 7157.
Tainer, JA, Getzoff, ED, Beem, KM, Richardson, JS and Richardson, D C. (1982). *Journal of Molecular Biology*, **160**, 181.
Tainer, JA, Getzoff, ED, Richardson, JS and Richardson, DC (1983). *Nature*, **306**, 284.
Tullius, TD, Frank, P and Hodgson, KO (1978). *Proceedings of the National Academy of Sciences USA*, **75**, 4069.

Dynamics of proteins

8
Molecular dynamics

Martin J. Field

8.1 Introduction

Experiment and theory have always been partners in the scientific endeavour. Experiments probe the behaviour of the material world while theory provides a framework for the interpretation of the results of experiments and for the design of new ones. In recent years, however, the explosive growth in the power of computers has led to the emergence of a field intermediate between those of theory and experiment—that of computer simulation. Simulations allow the investigation of models which are both more complicated and more realistic than those that can be handled by traditional theoretical approaches.

The aim of this chapter is to provide a brief introduction to techniques for the computer simulation of biomacromolecular systems. There are a huge variety of simulation methods and so only a brief overview of some of the more important techniques can be attempted here. Fuller discussions can be found in the monographs by Allen and Tildesley (1987), by Field (1999) and by Leach (1996). Greater detail about simulation methods for biomacromolecular systems is given in the books by McCammon and Harvey (1987) and by Brooks *et al.* (1988) and in the collections of articles edited by van Gunsteren *et al.* (1989, 1993, 1997).

The organization of this chapter is as follows. Section 8.2 describes methods for calculating the potential energy of a system. Ways in which this energy can be used in a range of molecular simulation techniques are discussed in Sections 8.3–8.5 and Section 8.6 concludes.

8.2 Calculating the potential energy of a system

One of the most important prerequisites of any molecular simulation method is a way to determine the energy of a system. This is because it is the energy, via the formulae of statistical thermodynamics, that allows the connection to be made between the microscopic world that is being simulated and the macroscopic world that is observed experimentally. The energy is normally made up of two components—the kinetic energy which is the energy arising from the movement of the particles in the system and the potential energy which comes from the interparticle interactions. Although the kinetic energy is normally quite easy to handle, the determination of the potential energy requires special methods which will be the subject of this section.

8.2.1 The Born–Oppenheimer approximation

In principle, it is known that a complete knowledge of the behaviour of a non-relativistic atomic system can be obtained by solving its *time-dependent Schrödinger equation* (Davydov 1976). This equation has the form

$$\hat{H}\Psi = i\hbar \frac{\partial \Psi}{\partial t}, \qquad (8.1)$$

where $\hat{H}$ is the *Hamiltonian operator* for the system, Ψ is its *wave function*, $\hbar$ is Planck's constant h divided by 2π, and t is the time. The Hamiltonian contains terms governing the kinetic and potential energies of the system while the square of the wave function is the particles' probability density distribution function. For systems in *stationary states*, whose nature does not alter with time, eqn (8.1) can be simplified to a *time-independent* form

$$\hat{H}\Psi = E\Psi, \qquad (8.2)$$

where E is the energy of the state.

As they stand eqns (8.1) and (8.2) are too difficult to solve for all but the smallest molecular systems and so simplifications must be sought. An approximation that is almost systematically invoked is the *Born–Oppenheimer approximation* which says that the dynamics of the electrons and nuclei can be treated separately as their masses are so different. In practice, this means that the *electronic problem* is tackled first by solving the Schrödinger equation for the electronic wave function and energy at different, fixed values of the nuclear coordinates. The nuclear problem is then dealt with afterwards. The electronic equation is

$$\hat{H}_{el}\Psi_{el}(r_\alpha, r_i) = E_{el}(r_i)\Psi_{el}(r_\alpha, r_i), \qquad (8.3)$$

where r is a vector of Cartesian coordinates for a particle, the subscript 'el' refers to electronic quantities and the Greek and Roman subscripts correspond to electrons and nuclei, respectively. The electronic Hamiltonian $\hat{H}_{el}$ differs from that in eqn (8.2) by the removal of the operator for the nuclear kinetic energy.

The electronic energy E_{el} is the effective interaction potential between the nuclei in the system. Because it is a parametric function of the nuclear coordinates r_i, it is a multidimensional function which forms the *potential energy surface* for the system (Sutcliffe 1995).

8.2.2 Empirical force fields

The potential energy surface is a quantity which is crucial in determining the behaviour of a system and so much effort has gone into developing methods which can be used to calculate it. In principle, the most precise are the *ab initio quantum mechanical* methods which attempt to solve eqn (8.3) directly with as few approximations as possible. Methods which fall into this category are

Hartree–Fock molecular orbital (Szabo and Ostlund 1989) and *valence bond* algorithms as well as approaches based upon *density functional theory* (Parr and Yang 1989). The problem with all the *ab initio* techniques is that they are expensive and cannot be applied to systems comprising more than a few hundred atoms at most.

Semi-empirical quantum mechanical methods try and circumvent these limitations by employing the same basic strategy as the *ab initio* methods but by approximating and simplifying the most time-consuming parts of the computation. These methods can be applied to much larger systems but, due to their formulation, they must be parametrized vs the experimental data or the results of higher level quantum mechanical calculations to ensure that they give results of the desired precision (Pople and Beveridge 1970).

Even with further simplification, it proves difficult to employ any quantum mechanical method, either semi-empirical or *ab initio*, for the simulation of macromolecular systems and so alternative methods are needed. The usual approach is to forget all about quantum mechanics and to employ *empirical potential energy functions* or *force fields*. These functions are designed so that they mimic as closely as possible the behaviour of the function E_{el} in the regions of the potential energy surface that are of interest. There is an enormous variety of force fields which are adapted to studying a wide range of different systems and phenomena (Burkett and Allinger 1982). However, a typical force field for the simulation of macromolecular systems at an atomic level of detail will consist of the sum of two types of terms—*bonding terms* which describe the energy due to the covalent structure of the molecule and *non-bonding terms* which give the interaction between atoms which are not directly bound together.

If the total potential energy of the system is denoted as V, then

$$V = V_{\text{bonding}} + V_{\text{nb}}, \tag{8.4}$$

where V_{bonding} and V_{nb} are the energies due to the bonding and non-bonding interactions, respectively. It is normal to subdivide the bonding and non-bonding energies further. Thus, the bonding energy will often consist of a sum of bond, angle, dihedral and out-of-plane terms:

$$V_{\text{bonding}} = V_{\text{bond}} + V_{\text{angle}} + V_{\text{dihedral}} + V_{\text{out-of-plane}}. \tag{8.5}$$

A common form for the bond energy is a sum of harmonic terms, one for each bond in the system,

$$V_{\text{bond}} = \sum_{\text{bonds}} \frac{1}{2} k_b (b - b_0)^2. \tag{8.6}$$

In this equation, b is the actual distance between the two atoms involved in the bond, b_0 is an equilibrium distance characteristic of the bond and k_b is the force constant for the bond which determines the steepness of the potential well and, hence, the bond's frequency of oscillation.

The angle energy is usually similar except that energy terms are functions of the angle θ subtended by three atoms:

$$V_{\text{angle}} = \sum_{\text{angles}} \frac{1}{2} k_\theta (\theta - \theta_0)^2. \tag{8.7}$$

For the dihedral or torsional energy, a harmonic form is not appropriate as the energy must be a periodic function of the torsion angle about the bond. A suitable form is an expansion in terms of trigonometric functions, i.e.

$$V_{\text{dihedral}} = \sum_{\text{dihedrals}} \frac{1}{2} V_n (1 + \cos(n\phi - \delta)). \tag{8.8}$$

Here, V_n is the height of the torsional barrier, n is the periodicity of the term and δ is its phase.

The out-of-plane or improper torsional energy $V_{\text{out-of-plane}}$ is used to keep atoms, such as those which are sp^2 hybridized, planar. In some force fields, a harmonic form for this energy term is employed while in others, terms reminiscent of the dihedral energy are preferred.

The non-bonding energies are the sum of a electrostatic term and a Lennard-Jones term

$$V_{\text{nb}} = V_{\text{elect}} + V_{\text{LJ}}. \tag{8.9}$$

The charge distribution of a molecule is represented in many force fields by partial charges centred on the nuclei. In these cases, the electrostatic energy, V_{elect}, is a simple Coulomb-type expression

$$V_{\text{elect}} = \sum_{ij \text{ pairs}} \frac{q_i q_j}{\epsilon r_{ij}}, \tag{8.10}$$

where q_i and q_j are the partial charges on atoms i and j and r_{ij} is the distance between them. ϵ is the dielectric constant for the interaction which will be unity for two atoms in vacuum.

The Lennard-Jones energy mimics the quantum mechanical *exchange-repulsion* interaction arising when two charge clouds overlap at short-range and the attractive *van der Waals* inverse sixth power interaction at longer range. It has the form

$$V_{\text{LJ}} = \sum_{ij \text{ pairs}} \frac{A_{ij}}{r_{ij}^{12}} - \frac{B_{ij}}{r_{ij}^6}, \tag{8.11}$$

where A_{ij} and B_{ij} are constants whose values depend upon the nature of the atoms i and j. The Lennard-Jones energy between two atoms will be non-zero even when they have no net charge.

The non-bonding interaction energies of eqns (8.10) and (8.11) are normally calculated for all pairs of atoms in the system. However, it is usual to *exclude* pairs of atoms from the sum which are separated by only one or two covalent bonds so as to avoid the overcounting that would result if both bonding and non-bonding terms were calculated for these atoms.

The calculation of the non-bonding interactions, and particularly the longer range electrostatic terms, is almost always the most expensive part of an energy calculation. This is because the number of interactions scales as the square of the number of particles, N, whereas the number of internal, covalent terms scales roughly linearly with the size of the system. There are a number of ways in which the cost of the non-bonding energy calculation can be reduced. One of these is to employ a *truncation* technique which either neglects or tapers the interactions to zero beyond a certain *cutoff distance*. Truncation techniques reduce the cost of calculating the non-bonding terms but suffer from the fact that they have little basis in physical reality. More accurate and faster *linear scaling methods* which attempt to calculate the full non-bonding energy of a system to a certain precision are still the subject of active research.

The terms listed for the 'typical' force field above are by no means the only ones in use. Thus, for example, in more complicated functions there will be bonding cross-terms which couple various internal coordinate deformations as well as non-bonding polarization terms which describe the interactions due to the changes in the charge distribution of a molecule in different environments.

An advantage of empirical force fields of the type discussed in this section is that they can be used for the simulation of systems comprising many thousands of atoms. In addition, their analytic form is such that it is straightforward to calculate the derivatives of the energy with respect to various atomic quantities. The most important derivative in practical applications is the first derivative of the potential energy with respect to the atomic coordinates, g_i, which is equal to the negative of the force on the atom, f_i, i.e.

$$g_i = \frac{\partial \mathcal{V}}{\partial r_i} = -f_i. \qquad (8.12)$$

A disadvantage of empirical force fields is that they contain many parameters (b_0, k_θ, V_n, q_i, etc.) which must come from somewhere. In some applications it may not be necessary to have very precise values but, in most cases, it will be necessary to parametrize the force field against data from experiments or from high quality quantum mechanical calculations. Such parametrizations are often laborious and can demand great effort if force fields of reasonable precision are to be obtained. Another disadvantage of force fields of the type discussed above is that they will be inappropriate for the investigation of certain processes. Thus, for example, it will not be possible to study reactions in which bonds are broken and formed because the harmonic bond term of eqn (8.6) does not allow dissociation. For these problems, other formulae for the potential will have to be used and parametrized accordingly (Warshel 1991).

8.2.3 Condensed phase systems

The vast majority of molecular biological experiments are done in the condensed phase in which macromolecules are immersed in solvent and have strong interactions with other molecules. Such an environment can have a profound influence on the properties of the system and so it is important that it is properly accounted for in any simulation methodology. The major problem when simulating a condensed phase is that an effectively infinite system has to be represented by a finite one. There are essentially two ways to tackle this problem. Either some order can be imposed upon the infinite system so that the interactions within the system can be handled analytically or the bulk of the system can be neglected and replaced by some simplified representation.

The first approach is perhaps the most accurate and it forms the basis of the method of *periodic boundary conditions*. In this approximation, the infinite system is constructed as a periodically repeated array or 'crystal' of a smaller, finite system which contains the molecules that are to be investigated. The fact that the system is periodic means that the calculation of the non-bonding interactions becomes manageable. These can either be evaluated by applying one of the truncation methods mentioned in the last section or by using *Ewald summation* which evaluates the full non-bonding energy for the periodic system [see e.g. Williams (1971)]. The Ewald method as originally formulated was not especially efficient but recently faster methods, such as the *particle mesh Ewald method* (Essman et al. 1995), have been developed which make practicable the treatment of periodic systems in a routine fashion.

In contrast to the periodic boundary methods which treat the environment explicitly, there are several methods which treat the solvent *implicitly*. The simplest methods are empirically based and neglect the environment completely. An example of one of these is the *distance-dependent dielectric method* which uses a dielectric permittivity proportional to the distance between the atoms (i.e. $\epsilon \propto r_{ij}$) in eqn (8.10). Other more rigorous approximations treat the solvent as a dielectric medium with the macromolecule immersed within it. The electrostatic energy of interaction between the macromolecule and the medium can be computed most accurately by solving the Poisson–Boltzmann equation while the non-polar repulsive and van der Waals interactions must be calculated with other models. For reviews of implicit solvation methods readers are referred to the articles by Gilson (1995), Honig and Nicholls (1995) and Sharp (1994).

8.3 Local exploration of potential energy surfaces

The potential energy surface is a multidimensional function of the coordinates of the atoms in the system and can display an extremely complicated topology. While in many cases it is most valuable to employ techniques, such as the molecular dynamics method, that explore large areas of the potential energy surface, useful information can also be gained by more limited search techniques.

Fortunately, it is unnecessary to investigate all parts of a potential energy surface. This is because it is often the regions around *stationary points* on the surface which are of most interest. A stationary point is characterized by being a point at which the gradient vectors for all the atoms in the system are zero, i.e.

$$\mathbf{g}_i = \mathbf{0} \quad \forall i = 1, N. \tag{8.13}$$

The type or the *order* of the stationary point can be further classified by determining the *eigenvalues* of the $3N \times 3N$ matrix of second derivatives, $\mathbf{H}$, of the energy with respect to the atomic coordinates. This matrix has the form

$$\mathbf{H} = \begin{pmatrix} \frac{\partial^2 V}{\partial x_1^2} & \frac{\partial^2 V}{\partial x_1 \partial y_1} & \cdots & \frac{\partial^2 V}{\partial x_1 \partial z_N} \\ \frac{\partial^2 V}{\partial y_1 \partial x_1} & \frac{\partial^2 V}{\partial y_1^2} & \cdots & \frac{\partial^2 V}{\partial y_1 \partial z_N} \\ \vdots & \vdots & \vdots & \vdots \\ \frac{\partial^2 V}{\partial z_N \partial x_1} & \frac{\partial^2 V}{\partial z_N \partial y_1} & \cdots & \frac{\partial^2 V}{\partial z_N^2} \end{pmatrix}. \tag{8.14}$$

The eigenvalues λ are determined by solving the *secular equation*

$$|\mathbf{H} - \lambda \mathbf{I}| = 0, \tag{8.15}$$

where $\mathbf{I}$ is the $3N \times 3N$ identity matrix and the straight lines denote the determinant.

A stationary point is a *minimum* if all the eigenvalues of the matrix are positive while it is an *nth order saddle point* if there are n negative eigenvalues. The two most important types of stationary points are minima and first-order saddle points (or simply saddle points). A minimum corresponds to a stable state of a system because the fact that all its eigenvalues are positive means that any small displacement away from it will result in an increase in energy. For saddle points with one negative eigenvalue, there will be a single direction along which a displacement will produce an energy lowering while displacements in all other directions will result in an energy increase. The identification of saddle points is useful when studying transitions, or reactions, between two stable states. This is because a saddle point will be the configuration of highest energy on the lowest energy *reaction path* joining the two minima. A simple two-dimensional potential energy surface with minima and saddle points is shown in Fig. 8.1.

The location of minima, saddle points and reaction paths is an important aspect of many simulation studies and a wide variety of algorithms has been developed for this task. The problem of finding a minimum of a multidimensional function is mathematically well defined and readers are referred to the book by Press *et al.* (1993) for more details. *Conjugate gradient* algorithms which use the energy and its first derivatives are probably the most appropriate for macromolecular systems. The determination of saddle points is more difficult although efficient methods exist for systems comprising a small number of

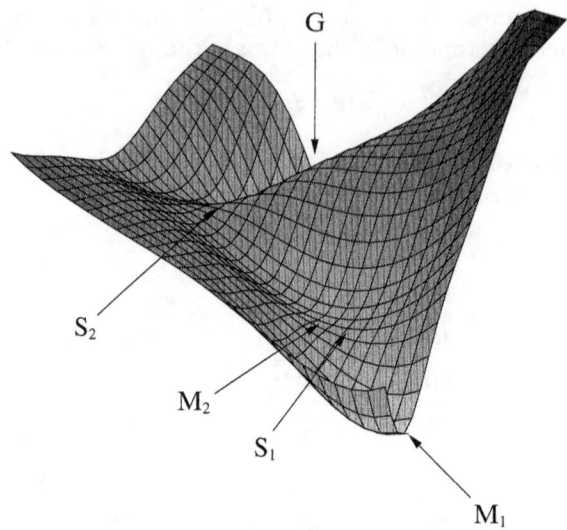

Fig. 8.1. A two-dimensional model potential energy surface showing a global minimum (G), two additional minima (M_1 and M_2) and two saddle points (S_1 and S_2). The lowest energy path connecting the minima goes through the saddle points.

atoms. Likewise, reasonable reaction path location algorithms are available for small systems but the determination of paths in large molecules requires special methods (Field 1999). A companion technique to those discussed here is *normal mode analysis* which is useful for giving information about the dynamics of a system in the neighbourhood of a stationary point.

8.4 Molecular dynamics simulations

Whereas searching locally the potential energy surface may reveal useful insights into a system's behaviour, a proper analysis requires techniques, such as the method of *molecular dynamics*, which explore the surface more thoroughly. In principle, it is possible to study the dynamics of the nuclei in a system quantum mechanically by solving a time-dependent Schrödinger equation similar to that of eqn (8.1) using the electronic energy E_{el} or the potential energy $\mathcal{V}$ as the potential of interaction. This is only feasible for very small systems comprising a few quantum particles and so it is normal to resort to a classical description of the dynamics. Classical descriptions have been found to be reasonably accurate but they can neglect important effects such as *tunnelling* and *zero-point motion*.

The total energy, $\mathcal{H}$, of a classical system will consist of a sum of the kinetic energy of the particles, $\mathcal{K}$, and the potential energy $\mathcal{V}$:

$$\mathcal{H} = \mathcal{K} + \mathcal{V}. \tag{8.16}$$

The kinetic energy has the form

$$\mathcal{K} = \frac{1}{2}\sum_{i=1}^{N} m_i v_i^2, \quad (8.17)$$

where m_i is the mass of the particle i and $\boldsymbol{v}_i$ is its velocity vector which is equivalent to the first time derivative of its position vector, $d\boldsymbol{r}_i/dt$. The equations of motion appropriate for atoms in a system with an energy given by eqn (8.16) are Newton's equations

$$\boldsymbol{f}_i = m_i \boldsymbol{a}_i = m_i \frac{d^2 \boldsymbol{r}_i}{dt^2}, \quad (8.18)$$

where $\boldsymbol{a}_i$ is the acceleration of particle i.

There is a well-known result from statistical mechanics which relates the thermodynamic temperature of a system to its average kinetic energy. The formula is

$$T = \frac{2}{N_{df} k_B} \langle \mathcal{K} \rangle \quad (8.19)$$

where k_B is Boltzmann's constant, N_{df} is the number of the system's *degrees of freedom*, which is usually of the order of $3N$ and the angled brackets denote an *ensemble average*.

The addition of a kinetic energy or a temperature can be viewed as 'lifting' the system above the potential energy surface (see Fig. 8.2 for a schematic). The ratio of the magnitude of the kinetic energy and the depths of the wells in the surface has important consequences for the behaviour of the system. At low temperatures or on a surface with deep wells the system is trapped in a few well-defined configurations and so its properties will be closely related to the properties of these wells. At high temperatures or on surfaces with many shallow minima, the system can adopt a wide range of configurations each of which may have widely differing properties. Such surfaces are called *rugged* and are characteristic of proteins (Frauenfelder and Wolynes 1994; Parisi 1993).

8.4.1 *The simulation methodology*

In a classical molecular dynamics simulation, Newton's equations of motion [eqn (8.18)], which are second-order ordinary differential equations, are solved for each atom as a function of time. It is usual to treat the solution as an *initial value problem* in that values are assigned to the positions and the velocities of the atoms at an initial time and the integration of the equations is continued for as long as required.

No single integration algorithm will be appropriate in all circumstances but, of the many algorithms that have been proposed, the most common is probably one based upon an algorithm originally applied to molecules by Verlet (1967). The algorithm is easy to derive. The position of an atom at a future time $t + \Delta$

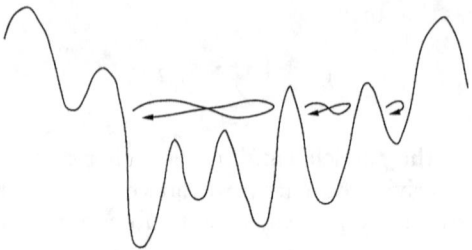

Low temperature or deep wells

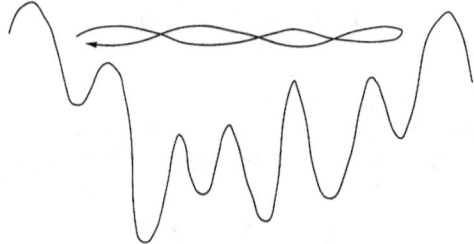

High temperature or shallow wells

Fig. 8.2. A schematic of the dynamics of a system. In the top figure, the wells in the potential energy surface are deep relative to the available kinetic energy and so only relatively few distinct configurations are sampled. In the bottom figure, the ratio of kinetic energy to well depth is larger and so much more of the surface is accessible.

can be expanded as a Taylor series of the position at the current time t, i.e.

$$r_i(t + \Delta) = r_i(t) + \Delta \dot{r}_i(t) + \frac{\Delta^2}{2} \ddot{r}_i(t) + O(\Delta^3), \tag{8.20}$$

where the notation $O(\Delta^3)$ denotes terms of cubic order or higher in the *time step*, Δ. Similarly, the position at a time $t - \Delta$ is obtained from the expansion

$$r_i(t - \Delta) = r_i(t) - \Delta \dot{r}_i(t) + \frac{\Delta^2}{2} \ddot{r}_i(t) - O(\Delta^3). \tag{8.21}$$

Adding these equations and rearranging gives an expression for the position of an atom at $t + \Delta$ in terms of the positions and forces on the atom at earlier times:

$$r_i(t + \Delta) = 2r_i(t) - r_i(t - \Delta) + \Delta^2 \ddot{r}_i(t) + O(\Delta^4)$$
$$\sim 2r_i(t) - r_i(t - \Delta) + \Delta^2 \frac{f_i(t)}{m_i}, \tag{8.22}$$

where eqn (8.18) has been used on going from the first to the second equation. Subtracting eqns (8.20) and (8.21) gives an equation for the velocity of the atom,

v_i, at the current time t:

$$v_i(t) = \dot{r}_i(t)$$
$$\sim \frac{1}{2\Delta}(r_i(t+\Delta) - r_i(t-\Delta)). \tag{8.23}$$

Equations (8.22) and (8.23) are sufficient to integrate the equations of motion but they are slightly inconvenient. This is because the velocities at time t are only available once the positions at time $t + \Delta$ have been calculated which means that at the start of the simulation, when $t = 0$, it is necessary to use another formula. A slight modification of these equations produces an algorithm called the *velocity Verlet method* which avoids these problems and can be shown to produce results that are equivalent to the standard Verlet method (Swope et al. 1982). The equations are

$$r_i(t+\Delta) = r_i(t) + \Delta v_i(t) + \frac{\Delta^2}{2}\frac{f_i(t)}{m_i}, \tag{8.24}$$

$$v_i(t+\Delta) = v_i(t) + \frac{\Delta}{2m_i}(f_i(t) + f_i(t+\Delta)). \tag{8.25}$$

Performing a simulation with the velocity Verlet integration algorithm mentioned above is straightforward. A typical scheme would be as follows.

1. Define the system, including the number and type of particles and the interaction potential $\mathcal{V}$.
2. Choose a set of starting coordinates $r_i(0)$, and velocities $v_i(0)$ for the atoms. Calculate the forces on the atoms at the initial configuration, $f_i(0)$. Set $t = 0$.
3. Calculate the positions of the atoms at the next time step $t + \Delta$ using eqn (8.24).
4. Calculate the forces on the atoms using the new atomic positions.
5. Calculate the velocities on the atoms at the next time step using eqn (8.25).
6. Periodically save the new coordinates and velocities for later analysis if desired. The sequence of coordinate and velocity sets produced in this way is known as a *trajectory*.
7. Set $t = t + \Delta$ and go back to step 3 for as many steps as required.

There are a number of points which are raised by this scheme. The first is what value should the time step Δ take? Ideally, it would be best to have a value that is large so that as much of the configuration space available to the system can be explored as possible. In practice, the largest time step that can be used has to be less than the oscillation period of the highest frequency motions in the system because otherwise the integration algorithm becomes unstable. The highest frequency motions in organic molecular systems are typically carbon–hydrogen bond stretching motions with frequencies of the order of $3000\,\mathrm{cm}^{-1}$. The periods of such motions are of the order of a few femtoseconds ($1\,\mathrm{fs} \equiv 10^{-15}\,\mathrm{s}$) meaning that the time step has to be less than this. Values of the time step of up to 1 fs are commonly used.

One way of increasing the time step is to remove or 'freeze' the high frequency motions out of the dynamics. An example algorithm of this type is SHAKE (Ryckaert 1985). Such methods may allow the time step to be increased by a factor of two or three but they are often expensive to apply and so any savings resulting from being able to use a large time step is counteracted by the extra time that is needed to implement the modified dynamics algorithm. Another disadvantage of such approaches is that freezing various degrees of freedom in the system can have significant effects on its dynamics (van Gunsteren and Karplus 1982).

Because the maximum size of the integration time step is about 1 fs, there is an upper limit on the length of time for which a system can be simulated due to limitations in computer resources. Consider a 1 ns (1 ns $\equiv 10^{-9}$ s) simulation of a small globular protein in water with a total size of 20 000 atoms. On a reasonably fast work station or personal computer, such a simulation could take several months. Use of a powerful parallel computer could reduce the cost of this calculation to a week or two. It is clear that no matter what machine is used, it will not be possible to simulate directly processes with time scales of more than a few nanoseconds. In these cases other, indirect simulation schemes have to be employed.

A second point is how to choose the initial values of the coordinates and velocities for the atoms. In a macromolecular study the coordinates of a protein or nucleic acid are usually those obtained experimentally from NMR spectroscopic or X-ray crystallographic experiments. The structures from such sources are normally not suitable for immediate use in a simulation study and so they must be refined first using energy minimization techniques and algorithms that determine the positions of the hydrogen atoms, which are often unavailable in X-ray structures, and solvent molecules.

To choose the velocities it is usual to make use of another result from statistical mechanics which says that the distribution of velocities of the atoms in a classical system will follow the Maxwell–Boltzmann law. If the temperature of the system is T, the probability of each component of the velocity of the ith atom having a value between v and $v + dv$ is

$$f(v)\,dv = \sqrt{\frac{m_i}{2\pi k_B T}} \exp\left(-\frac{m_i}{2k_B T} v^2\right) dv. \tag{8.26}$$

To select velocities from this distribution, each velocity component is treated as an independent *Gaussian random variable* with a mean value of zero and a standard deviation of $\sqrt{k_B T/m_i}$. After the velocities have been assigned it is common to scale them uniformly so that the *instantaneous temperature* of the system will have a particular value (e.g. 300 K). This is defined by an expression similar to eqn (8.19) except that no averaging of the kinetic energy is performed.

An important aspect of any molecular dynamics study is how to estimate the precision of the simulation. Useful measures are what are known as *conservation conditions*. For the algorithm outlined above, it is not difficult to show that the total energy of the system, $\mathcal{H}$, should be a constant throughout the simulation. In a real simulation, of course, the total energy will fluctuate from its initial value due to inaccuracies in the integration algorithm but the size of the fluctuations about the average value or the drift in the total energy from its initial value can give an idea of how accurate the integration has been. Other important conserved quantities can include the system's momentum and angular momentum.

8.4.2 Simulations in other ensembles

In the molecular dynamics algorithm of the last section, the total number of particles N and the volume of the system V (for a condensed phase system) were kept fixed and the total energy was a conserved quantity. Such a simulation samples the *microcanonical* or *NVE ensemble* (McQuarrie 1976). In reality, most experiments are done in conditions better approximated by other ensembles. Examples include the *canonical* or *NVT*, the *isothermal–isobaric* or *NPT* and the *grand canonical* or *μVT* ensembles. In these ensembles, the temperature T, the pressure P and the chemical potential μ are the thermodynamic variables which can be fixed.

A wide variety of integration algorithms has been developed to allow simulations in all these ensembles, but it is only those that have been designed to maintain a constant temperature and, to a lesser extent, a constant pressure that are routinely used for macromolecular simulations. The principles behind algorithms for performing constant pressure and temperature simulations are similar. It is usual to introduce extra dynamical degrees of freedom which represent an *external bath* to which the system being simulated is coupled. In a constant temperature simulation, the bath acts as a *thermostat* and energy transfer occurs between it and the system. In a constant pressure simulation, the bath is a *barostat* and it is the volume of the system that is allowed to change. For further details, the reader should consult the book by Allen and Tildesley (1987).

8.4.3 Trajectory analysis

The point of doing a simulation is, of course, to calculate quantities that can be compared with those measured experimentally. The way that this is done is to analyse the trajectories generated during the molecular dynamics calculation using the formulae of statistical mechanics. In classical statistical mechanics an average of a property, $\mathcal{X}$, in a particular ensemble is computed as

$$\langle \mathcal{X} \rangle = \int d\Gamma\, \mathcal{X}(\Gamma) \rho(\Gamma), \tag{8.27}$$

where Γ are the ensemble variables and ρ is the probability density distribution function for the ensemble. In the canonical ensemble, for example, the ensemble variables will be the $3N$ coordinates and the $3N$ momenta of the atoms and the probability distribution will give the probability of a particular configuration as a function of these $6N$ variables. The average in eqn (8.27) is then the integral, over all possible configurations, of the value of the property at the configuration weighted by its probability of occurrence.

For accurate averages to be obtained from a molecular dynamics simulation, the trajectories calculated must contain configurations which are representative of the ensemble. The average of a property calculated from a trajectory is

$$\langle \mathcal{X} \rangle = \frac{1}{n_f} \sum_{n=1}^{n_f} \mathcal{X}_n, \tag{8.28}$$

where n_f is the total number of *frames* in the trajectory and $\mathcal{X}_n$ is the value of the property at each frame. The equivalence of an ensemble average and a time average calculated from a trajectory is the basis of the *ergodic principle* (Chandler 1987).

There are two other useful statistical quantities that can be calculated from simulation data which will be mentioned here. These are *fluctuations* and *time correlation functions*. The fluctuation is the average of the square of the deviation of the property from its average. If the deviation is denoted by $\delta\mathcal{X} = \mathcal{X} - \langle \mathcal{X} \rangle$, the fluctuation is

$$\langle (\delta\mathcal{X})^2 \rangle = \langle \mathcal{X}^2 \rangle - \langle \mathcal{X} \rangle^2$$
$$= \frac{1}{n_f} \sum_{n=1}^{n_f} \mathcal{X}_n^2 - \langle \mathcal{X} \rangle^2. \tag{8.29}$$

The *rms deviation* of the property, denoted $\sigma(\mathcal{X})$, is the square root of the fluctuation, i.e. $\sigma^2(\mathcal{X}) = \langle (\delta\mathcal{X})^2 \rangle$.

The calculation of correlation functions is more complicated as they are functions of time. The *autocorrelation* function for the property $\mathcal{X}$ is denoted by $\mathcal{C}_{\mathcal{XX}}(t)$ and has the expression

$$\mathcal{C}_{\mathcal{XX}}(t) = \langle \delta\mathcal{X}(t)\delta\mathcal{X}(0) \rangle$$
$$= \langle \mathcal{X}(t)\mathcal{X}(0) \rangle - \langle \mathcal{X} \rangle^2. \tag{8.30}$$

The function can be normalized by dividing by the fluctuation of the property. The normalized function $\hat{\mathcal{C}}_{\mathcal{XX}}(t)$ is

$$\hat{\mathcal{C}}_{\mathcal{XX}}(t) = \mathcal{C}_{\mathcal{XX}}(t)/\sigma^2(\mathcal{X}). \tag{8.31}$$

A *cross-correlation function* can be defined for two different properties, $\mathcal{X}$ and $\mathcal{Y}$, as

$$\mathcal{C}_{\mathcal{XY}}(t) = \langle \delta\mathcal{X}(t)\delta\mathcal{Y}(0) \rangle$$
$$= \langle \mathcal{X}(t)\mathcal{Y}(0) \rangle - \langle \mathcal{X} \rangle\langle \mathcal{Y} \rangle. \tag{8.32}$$

It can be normalized in the same way as the autocorrelation function by dividing by the product of the rms deviations of properties $\mathcal{X}$ and $\mathcal{Y}$, i.e. by $\sigma(\mathcal{X})\sigma(\mathcal{Y})$.

The importance of the three quantities discussed above—averages, fluctuations and time correlation functions—is that they enter into many of the statistical mechanical formulae that enable experimentally observable quantities to be calculated from simulation data. Equation (8.19) is one example of a relation enabling a thermodynamic quantity, the temperature, to be related to the average of the kinetic energy of a system. Another example of a relation involving an average is one for the pressure:

$$PV = Nk_{\text{B}}T + \left\langle \frac{1}{3}\sum_{i=1}^{N} r_i \cdot f_i \right\rangle, \qquad (8.33)$$

where the last term on the right-hand side of the equation is an average over the system's *internal virial*. Examples of expressions involving fluctuations are those that relate the specific heats (either at constant volume or constant pressure) to fluctuations in the potential and kinetic energies. Time correlation functions are important because they are fundamental to the derivation of formulae that permit *transport coefficients* for a system to be calculated. Examples include the *diffusion coefficient*, the *bulk* and *shear viscosities* and the *thermal conductivity*.

As a final point, it is to be noted that an important topic in simulation studies is how to estimate the statistical precision of the averages and other properties obtained from a molecular dynamics trajectory. Useful introductions to these topics can be found in Pastor (1994) and in Zwanzig and Ailawadi (1969).

8.4.4 Calculation of free energies

A knowledge of the free energy of the configuration of a system or of the difference in free energy between two configurations is crucial in helping to identify a configuration's stability and reactivity. The computation of free energies, however, is a very challenging application of simulation methods and requires special techniques. It is impossible in most instances to determine the absolute free energy of a system because the length and number of simulations required to obtain results of high enough precision would be beyond the current computational capabilities. What is done in practice, therefore, is to calculate *differences* in free energy between systems whose configurations or compositions differ by only a small amount. There are a large range of methods for the calculation of free energy differences but probably the most widely used techniques are the *thermodynamic integration*, *thermodynamic perturbation* and *umbrella sampling* methods.

The difference in free energy in the canonical ensemble between two systems I and J can be written as

$$\begin{aligned} \Delta A_{I \to J} &= A_J - A_I \\ &= -k_{\text{B}} T \ln \frac{Z_J}{Z_I}, \end{aligned} \qquad (8.34)$$

where Z_I is the classical partition function for system I. With a little manipulation and expanding the expression for the partition function, it is possible to arrive at the equation that forms the basis of the thermodynamic perturbation method,

$$\Delta A_{I \to J} = -k_B T \ln \langle \exp(-(\mathcal{H}_J - \mathcal{H}_I)/k_B T) \rangle_I, \quad (8.35)$$

where $\mathcal{H}_I$ is the total energy of system I, and the subscript I on the angled brackets implies that the average is done using the probability density distribution for the system, I. Practically, this means that a molecular dynamics trajectory is generated for the system I, and then the configurations in the trajectory are used to calculate the average in eqn (8.35). For this to be possible, though, the systems I and J have to be defined in such a way that calculating an energy for system J using a configuration of system I is meaningful.

The average in eqn (8.35) will not converge very rapidly unless the energy difference is small (of the order of $k_B T$). It is therefore usual to break up the problem into even smaller pieces by introducing an intermediate total energy function or Hamiltonian, $\mathcal{H}(\lambda)$, which is a function of a coupling parameter λ, whose value can vary between 0 and 1. The Hamiltonian is formulated such that $\mathcal{H}(0) = \mathcal{H}_I$ and $\mathcal{H}(1) = \mathcal{H}_J$. With this Hamiltonian it is possible to do simulations at various values of λ and then calculate the free energy change on going from one value of λ to another. The total energy change for the transition is the sum of the energies for each of the small transitions.

Thermodynamic integration methods also employ a coupling parameter but rely upon another identity for the free energy difference. It is

$$\Delta A_{I \to J} = \int_0^1 d\lambda \, \frac{\partial A}{\partial \lambda}. \quad (8.36)$$

The derivatives of the free energy A can be determined straightforwardly giving an expression in terms of the derivative of the intermediate Hamiltonian with respect to the coupling parameter:

$$\Delta A_{I \to J} = \int_0^1 d\lambda \left\langle \frac{\partial \mathcal{H}}{\partial \lambda} \right\rangle_\lambda. \quad (8.37)$$

This integral is determined by performing simulations to calculate the ensemble average in the integrand at various values of λ and then applying a standard numerical integration technique to the values that result.

A nice feature of the thermodynamic integration and perturbation methods is that they can be used to compute free energy differences for physically impossible, *alchemical perturbations* as well as physically realizable processes because of the introduction of the coupling parameter λ. An example of the latter would be the calculation of the free energy difference between two systems with the same composition but which differed in the value of certain geometrical parameters (such as the distance between two atoms). Each value of the coupling parameter would then correspond to a structure with a well-defined geometry.

An example of an alchemical change would be the calculation of the free energy difference between two different solutes in a solvent or of two different ligands bound to the same protein molecule. The intermediate states in this case would be superpositions of the two solute or ligand molecules.

Umbrella sampling is a general technique for enhancing the sampling in a molecular dynamics simulation in a particular region of configuration space and is not restricted to determining the free energy differences. It is, however, widely employed for the computation of a particular type of free energy, the *potential of mean force* (PMF), which is an important quantity in many statistical mechanical theories of molecular processes. The PMF is normally determined as a function of one or a small number of geometrical parameters. For the same reasons as given above, it is not possible to calculate the entire PMF with a single simulation, so a series of umbrella sampling simulations is performed each of which concentrates the dynamics of the system around particular values of the geometrical parameters. The PMF for the entire range of values is then pieced together from the individual simulation data using a special statistical procedure.

For more details of the application of free energy calculations to biomacromolecular systems, see the reviews by Beveridge and DiCapua (1989), by Kollman (1993) and by Straatsma and McCammon (1992).

8.4.5 *Simulated annealing*

In Section 8.3, *local* algorithms for exploring the potential energy surface of a system were discussed. In many instances, however, it is desirable to have a knowledge of the *global minimum*, which is the configuration with the lowest energy on the potential energy surface. Thus, it is, for example, generally believed that the folded, native state of a protein corresponds to the global minimum state.

A number of global optimization strategies have been developed. The most widely used method in molecular simulations is that of *simulated annealing* (Kirkpatrick *et al.* 1983) although other techniques, such as those involving *genetic algorithms* (Forrest 1993), are also becoming popular. The essential idea behind simulated annealing and the fact that distinguishes it from local optimization algorithms is in the consideration of the temperature of the system. As illustrated in Fig. 8.2, the presence of a temperature lifts the system off its potential energy surface. The higher the temperature the larger the number of configurations that are accessible because the system can surmount larger barriers.

In a simulated annealing calculation, the system is given a high temperature to start off with and is slowly cooled until its temperature is zero and it reaches the potential energy surface. The way in which the cooling is performed helps to determine the effectiveness of the simulated annealing method. In general, the cooling is done slowly so that the system can thoroughly explore the potential energy surface and avoid becoming trapped in regions of high potential energy. There is no guarantee that this will not happen, but it is known from statistical mechanics that the probability that a particular configuration will be favoured is proportional to its Boltzmann factor, $\exp(-\mathcal{V}/k_\text{B}T)$. Thus, the lower the potential energy of the configuration the more probable it is. The optimal cooling schedule cannot be found in most problems of interest and so the choice of schedule is a matter of experimentation.

Molecular dynamics algorithms, because they employ a temperature, can be used for simulated annealing calculations. In fact, the use of molecular dynamics simulations in this way for the refinement of structures obtained from NMR spectroscopic and X-ray crystallographic experiments is probably its most important and successful application to date (Brünger et al. 1987a,b).

8.5 Monte Carlo techniques

Another simulation technique for the investigation of molecular systems is the *Monte Carlo* method. Like the molecular dynamics method it can simulate systems in different thermodynamic ensembles and be used to determine various equilibrium thermodynamic quantities, such as free energies. Unlike the molecular dynamics method, it cannot be used to study time-dependent processes.

The Monte Carlo method has been widely used for systems comprising small molecules and for macromolecules with simplified representations of the energy function and internal geometry. It has, however, found less application to macromolecular systems described with the types of energy functions discussed in Section 8.2.2, for which the molecular dynamics method has been preferred. The reason for this is primarily because it is difficult to devise a Monte Carlo algorithm that handles internal changes in the geometry of a macromolecule in an efficient manner. Having said this, some innovative Monte Carlo calculations of macromolecular systems with energy functions of atomic detail have been performed [see, e.g. Essex et al. (1997)].

8.6 Perspectives

Simulations are increasingly important for studying condensed phase systems, either as tools for such tasks as structure refinement or as theoretical probes for the understanding of their structure and function. Without doubt the field will continue to expand rapidly as the power of computer hardware increases and its price falls.

The increased availability of computational power will, of course, allow the application of existing simulation techniques to systems of larger size and mean that the time scale of processes accessible in molecular dynamics simulations will be increased. Equally, it will permit more accurate methods, such as improved potential functions, to be used to study systems which have already been investigated with more approximate techniques. Either way, a lot more will be learned about how molecular systems behave.

Acknowledgements

The author would like to thank Patricia Amara for comments on the manuscript and the Institut de Biologie Structurale Jean-Pierre Ebel, the Commissariat à l'Energie Atomique and the Centre National de la Recherche Scientifique for support.

References

Allen, MP and Tildesley, DJ (1987). *Computer simulations of liquids*, Oxford University Press, Oxford.
Beveridge, DL and DiCapua, M (1989). *Annual Review of Biophysics and Biophysical Chemistry*, **18**, 431–492.
Brooks III, CL, Karplus, M and Pettitt, BM (1988). *Proteins: A theoretical perspective of dynamics, structure and thermodynamics*, Advances in Chemical Physics, Vol. LXXI, J. Wiley & Sons, New York.
Brünger, AT, Campbell, RL, Clore, GM, Gronenborn, AM, Karplus, M, Petsko, GA and Teeter, MM (1987a). *Science*, **235**, 1049–1053.
Brünger, AT, Kuriyan, J and Karplus, M (1987b). *Science*, **235**, 458–460.
Burkert, U and Allinger, NL (1982). *Molecular mechanics*, ACS Monograph, Vol. 177, American Chemical Society, Washington, DC.
Chandler, D (1987). *Introduction to modern statistical mechanics*, Oxford University Press, Oxford.
Davydov, AS (1976). *Quantum mechanics*, Second Edition, Pergamon, Oxford.
Essex, JW, Severance, DL, Tirado-Rives, J and Jorgensen, WL (1997). *Journal of Physical Chemistry*, **B101**, 9663–9669.
Essmann, U, Perara, L, Berkowitz, ML, Darden, T, Lee, H and Pedersen, LG (1995). *Journal of Chemical Physics*, **103**, 8577–8593.
Field, MJ (1999). *A practical introduction to the simulation of molecular systems*, Cambridge University Press, Cambridge.
Forrest, S (1993). *Science*, **261**, 872–878.
Frauenfelder, H and Wolynes, PG (1994). *Physics Today*, February, 58–64.
Gilson, MK (1995). *Current Opinions in Structural Biology*, **5**, 216–223.
Honig, B and Nicholls, A (1995). *Science*, **268**, 1144–1149.
Kirkpatrick, S, Gelatt, Jr, CD and Vecchi, MP (1983). *Science*, **220**, 671–680.
Kollman, PA (1993). *Chemical Reviews*, **93**, 2395–2417.
Leach, AR (1996). *Molecular modelling*, Addison-Wesley Longman, London.
McCammon, JA and Harvey, S (1987). *Dynamics of proteins and nucleic acids*, Cambridge University Press, Cambridge.
McQuarrie, DA (1976). *Statistical mechanics*, Harper Collins, New York.
Parisi, G (1993). *Physics World*, September, 42–47.
Parr, RG and Yang, W (1989). *Density-functional theory of atoms and molecules*, Oxford University Press, Oxford.
Pastor, RW (1994). Techniques and Applications of Langevin Dynamics Simulations. In *The molecular dynamics of liquid crystals* (eds Luckhurst, GR and Veracini, CA), pp. 85–138, Kluwer Academic Publishers, The Netherlands.
Pople, JA and Beveridge, DL (1970). *Approximate molecular orbital theory*, McGraw-Hill, New York.
Press, WH, Teukolsky, SA, Vetterling, WT and Flannery, BP (1993). *Numerical recipes in* FORTRAN 77: *the art of scientific computing*, Second edition, Cambridge University Press, Cambridge.
Ryckaert, JP (1985). *Molecular Physics*, **55**, 549–556.
Sharp, KA (1994). *Current Opinions in Structural Biology*, **4**, 234–239.

Straatsma, TP and McCammon, JA (1992). *Annual Review of Physical Chemistry*, **43**, 407–435.
Sutcliffe, BT (1995). *Journal of Molecular Structure (Theochem)*, **341**, 217–235.
Swope, WC, Andersen, HC, Berens, PH and Wilson, KR (1982). *Journal of Chemical Physics*, **76**, 637–649.
Szabo, A and Ostlund, NS (1989). *Modern quantum chemistry: introduction to advanced electronic structure theory*, First revised edition, McGraw-Hill, New York.
van Gunsteren, WF and Karplus, M (1982). *Macromolecules*, **15**, 1528–1544.
van Gunsteren, W, Weiner, P and Wilkinson, A (1989, 1993, 1997). *Computer simulation of biomolecular systems: theoretical and experimental applications*, Vols. 1, 2 and 3, ESCOM, Leiden.
Verlet, L (1967). *Physical Review*, **159**, 98–103.
Warshel, A (1991). *Computer modeling of chemical reactions enzymes and solutions*, John Wiley & Sons, New York.
Williams, DE (1971). *Acta Crystallographica*, **A27**, 452–455.
Zwanzig, R and Ailawadi, NK (1969). *Physical Review*, **182**, 193–196.

9

Inelastic and quasielastic neutron scattering: complementarity with biomolecular simulation

Jeremy C. Smith

9.1 Introduction

Inelastic and quasielastic neutron scattering can be used to examine the dynamics of biological macromolecules. Our emphasis here is on the complementarity between these experiments and computer simulation, which can provide a stepping stone between experiment and simplified descriptions of the physical behaviour of complex systems.

The spectroscopic techniques that have been most frequently used to investigate biomolecular dynamics are those that are commonly available in laboratories, e.g. nuclear magnetic resonance (NMR), fluorescence and Mössbauer. However, these methods involve motions on timescales that are not well sampled by molecular dynamics simulation using present computer power. Moreover, the establishment of relations linking NMR and fluorescence with atomic motion is fraught with theoretical difficulties. Neutron inelastic and quasielastic scattering spectroscopies probe dynamics on subnanosecond timescales that can be sampled with present-day molecular dynamics simulations. Moreover, underlying relations between dynamics and measurement are relatively easy to express formally for these techniques.

In what follows, we first present the theoretical background for quasielastic and inelastic neutron scattering with special emphasis on the relation with molecular simulation. Examples of applications to systems of small molecules of biological interest are given, and finally work on proteins is discussed.

9.2 Neutron scattering and its relation to simulation

9.2.1 *Dynamic structure factor*

We examine the relation between particle dynamics and the scattering of neutron radiation in the case where both the energy and momentum transferred between the sample and the incident radiation are measured. Linear response theory allows dynamic structure factors to be written in terms of the equilibrium fluctuations of the sample. For neutron scattering from a system of identical particles (van Hove 1954, 1958; Lovesey 1984):

$$S_{\text{coh}}(\mathbf{Q},\omega) = \frac{1}{2\pi} \iint dt\, d^3 r\, e^{i(\mathbf{Q}\cdot\mathbf{r}-\omega t)} G(\mathbf{r},t), \tag{9.1}$$

$$S_{\text{inc}}(\mathbf{Q},\omega) = \frac{1}{2\pi} \iint dt\, d^3 r\, e^{i(\mathbf{Q}\cdot\mathbf{r}-\omega t)} G_s(\mathbf{r},t), \tag{9.2}$$

where $\mathbf{Q}$ is the scattering wave vector, ω the energy transfer and the subscripts coh and inc refer to coherent and incoherent scattering, discussed later. $G_s(\mathbf{r},t)$ and $G(\mathbf{r},t)$ are van Hove correlation functions which, for a system of N particles undergoing classical dynamics, are defined as follows:

$$G(\mathbf{r},t) = \frac{1}{N} \sum_{i,j} \langle \delta(\mathbf{r} - \mathbf{R}_i(t) + \mathbf{R}_j(0)) \rangle, \tag{9.3}$$

$$G_s(\mathbf{r},t) = \frac{1}{N} \sum_{i} \langle \delta(\mathbf{r} - \mathbf{R}_i(t) + \mathbf{R}_i(0)) \rangle, \tag{9.4}$$

where $\mathbf{R}_i(t)$ is the position vector of the ith scattering nucleus and $\langle \cdots \rangle$ indicates an ensemble average; $G(\mathbf{r},t)$ is the probability that, given a particle at the origin at time $t=0$, any particle (including the original particle) is at $\mathbf{r}$ at time t; $G_s(\mathbf{r},t)$ is the probability that, given a particle at the origin at time $t=0$, the same particle is at $\mathbf{r}$ at time t.

Equation (9.1) has an equivalent form in X-ray scattering, where the scattered intensity is given as follows:

$$|F(\mathbf{Q},\omega)|^2 = \frac{1}{2\pi} \iint d^3 r\, dt\, P(\mathbf{r},t) e^{i(\mathbf{Q}\cdot\mathbf{r}-\omega t)}, \tag{9.5}$$

where $P(\mathbf{r},t)$ is the spatiotemporal Patterson function given by

$$P(\mathbf{r},t) = \iint d\mathbf{R}\, dt\, \rho(\mathbf{R},t)\rho(\mathbf{r}+\mathbf{R},t+\tau) \tag{9.6}$$

and $\rho(\mathbf{r},t)$ is the time-dependent electron density. Unfortunately, X-ray photons with wavelengths corresponding to atomic distances have energies much higher than those associated with thermal fluctuations. For example, an X-ray photon of 1.8 Å wavelength has an energy of 6.9 keV corresponding to a temperature of 8×10^7 K. X-ray detectors have not been sufficiently sensitive to measure the minute fractional energy changes associated with molecular fluctuations, and so the practical exploitation of eqn (9.5) has been difficult. Therefore, X-ray scattering has mostly been used only in cases where inelastic and elastic scattering are indistinguishable (although recent work with third generation synchrotron sources have produced some interesting and useful inelastic X-ray scattering results). In contrast to X-rays, the mass of the neutron is such that the energy exchanged in exciting or deexciting ps timescale thermal motions is a large fraction of the incident energy and can be measured precisely. A thermal neutron of 1.8 Å wavelength has an energy of 25 meV corresponding to $k_B T$ at 300 K. To further examine the neutron scattering case, we perform space Fourier transformation of the van Hove correlation functions:

$$S_{coh}(\mathbf{Q}, \omega) = \frac{1}{2\pi} \int_{-\infty}^{+\infty} dt\, e^{-i\omega t} I_{coh}(\mathbf{Q}, t), \quad (9.7)$$

$$I_{coh}(\mathbf{Q}, t) = \frac{1}{N} \sum_{i,j} b_{coh}^2 \langle e^{-i\mathbf{Q}\cdot\mathbf{R}_i(0)} e^{i\mathbf{Q}\cdot\mathbf{R}_j(t)} \rangle, \quad (9.8)$$

$$S_{inc}(\mathbf{Q}, \omega) = \frac{1}{2\pi} \int_{-\infty}^{+\infty} dt\, e^{-i\omega t} I_{inc}(\mathbf{Q}, t), \quad (9.9)$$

$$I_{inc}(\mathbf{Q}, t) = \frac{1}{N} \sum_{i} b_{inc}^2 \langle e^{-i\mathbf{Q}\cdot\mathbf{R}_i(0)} e^{i\mathbf{Q}\cdot\mathbf{R}_i(t)} \rangle. \quad (9.10)$$

Neutrons are scattered by the nuclei of the sample. Due to the random distribution of nuclear spins in the sample, the scattered intensity contains a *coherent* part arising from the average neutron–nucleus potential and an *incoherent* part arising from fluctuations from the average. The coherent scattering arises from self- and cross-correlations of atomic motions and the incoherent scattering from single atom motions. Each isotope has a coherent scattering length $b_{i,coh}$ and an incoherent scattering length $b_{i,inc}$ which defines the strength of the interaction between the nucleus of the atom and the neutron. We see from eqns (9.7) and (9.9) that the coherent and incoherent dynamic structure factors are time Fourier transforms of the coherent and incoherent *intermediate scattering functions*, $I_{coh}(\mathbf{Q}, t)$ and $I_{inc}(\mathbf{Q}, t)$. Further, $S_{inc}(\mathbf{Q}, \omega)$ and $S_{coh}(\mathbf{Q}, \omega)$ may contain elastic ($\omega = 0$) and inelastic ($\omega \neq 0$) parts. Elastic scattering probes the correlations of atomic positions at long times whereas the inelastic scattering process probes position correlations as a function of time.

9.2.2 Incoherent neutron scattering

Neutron scattering from organic molecules is often dominated by incoherent scattering from the hydrogen atoms. This is largely because the incoherent scattering cross-section ($4\pi b_{inc}^2$) of hydrogen is $\simeq 15$ times greater than the total scattering cross-section of carbon, nitrogen or oxygen. In the systems examined here incoherent scattering thus essentially gives information on self-correlations of hydrogen atom motions. A program for calculating neutron scattering properties from molecular dynamics simulations has been published (Kneller et al. 1995).

In practice, the measured incoherent scattering energy spectrum is divided up into elastic, quasielastic and inelastic scattering. Inelastic scattering arises from vibrations. Quasielastic scattering is typically Lorentzian or a sum of Lorentzians centred on $\omega = 0$ and arises from diffusive motions in the sample. Elastic scattering gives information on the self-probability distributions of the hydrogen atoms in the sample. We now examine these forms of scattering in more detail.

9.2.2.1 Quasielastic incoherent scattering

It is useful for the subsequent analysis to review here the procedure commonly used to extract the dynamical data directly from experimental incoherent quasielastic neutron scattering profiles (Bee 1988). It is assumed that the atomic position vectors can be decomposed into two contributions, one due to diffusive motion, $r_{i,d}(t)$, and the other from vibrations, $u_{i,v}(t)$, i.e.

$$\mathbf{R}_i(t) = \mathbf{r}_{i,d}(t) + \mathbf{u}_{i,v}(t). \tag{9.11}$$

Combining eqn (9.11) with eqn (9.10) and assuming that $r_{i,d}(t)$ and $u_{i,v}(t)$ are uncorrelated (this requires that their characteristic timescales be separated) one obtains

$$I_{\text{inc}}(\mathbf{Q}, t) = I_d(\mathbf{Q}, t) I_v(\mathbf{Q}, t), \tag{9.12}$$

where $I_d(\mathbf{Q}, t)$ and $I_v(\mathbf{Q}, t)$ are obtained by substituting $\mathbf{R}_i(t)$ in eqn (9.10) with $r_{i,d}(t)$ and $u_{i,v}(t)$, respectively.

The Fourier transform of eqn (9.12) gives

$$S(\mathbf{Q}, \omega) = S_d(\mathbf{Q}, \omega) \otimes S_v(\mathbf{Q}, \omega), \tag{9.13}$$

where $S_d(\mathbf{Q}, \omega)$ and $S_v(\mathbf{Q}, \omega)$ are obtained by Fourier transformation of $I_d(\mathbf{Q}, t)$ and $I_v(\mathbf{Q}, t)$ and the symbol $\otimes$ denotes the convolution product.

The vibrational intermediate scattering function is given by (Lovesey 1984)

$$I_v(\mathbf{Q}, t) = \sum_i b_i^2 e^{-\langle (\mathbf{Q} \cdot \mathbf{u}_{i,v})^2 \rangle} e^{\langle [\mathbf{Q} \cdot \mathbf{u}_{i,v}(0)][\mathbf{Q} \cdot \mathbf{u}_{i,v}(t)] \rangle}. \tag{9.14}$$

To derive an analytically tractable form of $S_{\text{inc}}(\mathbf{Q}, \omega)$ in the quasielastic energy window (typically $-15 \text{ cm}^{-1} < \hbar\omega < 15 \text{ cm}^{-1}$) we align $\mathbf{Q}$ with the Cartesian axis x in the laboratory frame. Assuming that (i) $\langle u_{v,x}^2 \rangle$, the x-axis vibrational mean-square displacement, is the same for all the hydrogen atoms and (ii) $Q^2 \langle u_{v,x}^2 \rangle \ll 1$, $S_v(\mathbf{Q}, \omega)$ can be expressed as

$$S_v(\mathbf{Q}, \omega) = e^{-Q^2 \langle u_{v,x}^2 \rangle} [\delta(\omega) + S_v^{\text{inel}}(\mathbf{Q}, \omega)], \tag{9.15}$$

where $e^{-Q^2 \langle u_{v,x}^2 \rangle} S_v^{\text{inel}}(\mathbf{Q}, \omega)$ is the vibrational inelastic dynamic structure factor. Combining eqns (9.13) and (9.15) one obtains,

$$S_{\text{inc}}(\mathbf{Q}, \omega) = e^{-Q^2 \langle u_{v,x}^2 \rangle} [S_d(\mathbf{Q}, \omega) + S_d(\mathbf{Q}, \omega) \otimes S_v^{\text{inel}}(\mathbf{Q}, \omega)]. \tag{9.16}$$

$S_v^{inel}(\mathbf{Q},\omega)$ contains high-frequency inelastic peaks due to intramolecular vibrations that fall outside the quasielastic energy window and may also contain intensity within the energy window. Further considerations, given in Souaille et al. (1996a), allow us to represent $S_d(\mathbf{Q},\omega) \otimes S_v^{inel}(\mathbf{Q},\omega)$ as an energy-independent background, $B(\mathbf{Q})$, leading to the equation

$$S_{inc}(\mathbf{Q},\omega) = e^{-Q^2 \langle u_{v,x}^2 \rangle} S_d(\mathbf{Q},\omega) + B(\mathbf{Q}). \tag{9.17}$$

In directionally averaged versions of eqn (9.17) the above mean-square displacement is replaced by the corresponding two- or three-dimensional quantity divided by a factor of two in the two-dimensional case and six for a spherically averaged dynamic structure factor.

9.2.2.2 Elastic incoherent structure factor

$I_d(\mathbf{Q},t)$ can be separated into time-dependent and time-independent parts as follows:

$$I_d(\mathbf{Q},t) = A_0(\mathbf{Q}) + I_d'(\mathbf{Q},t). \tag{9.18}$$

The elastic incoherent structure factor (EISF), $A_0(\mathbf{Q})$, is defined as follows (Bee 1988):

$$A_0(\mathbf{Q}) = \lim_{t \to \infty} I_d(\mathbf{Q},t) = \int d^3 r\, e^{i\mathbf{q}\cdot\mathbf{r}} \lim_{t \to \infty} G_d(\mathbf{r},t), \tag{9.19}$$

where $G_d(\mathbf{r},t)$ is the contribution to the van Hove self-correlation function due to diffusive motion. $A_0(\mathbf{Q})$ is thus determined by the diffusive contribution to the space probability distribution of the hydrogen nuclei.

Taking the Fourier transform of eqn (9.18) and combining it with eqn (9.17) yields

$$S(\mathbf{Q},\omega) = e^{-Q^2 \langle u_v^2 \rangle} [A_0(\mathbf{Q})\delta(\omega) + S_d'(\mathbf{Q},\omega)] + B(\mathbf{Q}). \tag{9.20}$$

This equation contains three terms, representing the elastic $[e^{-Q^2 \langle u_v^2 \rangle} A_0(\mathbf{Q}) \times \delta(\omega)]$, quasielastic $[e^{-Q^2 \langle u_v^2 \rangle} S_d'(\mathbf{Q},\omega)]$ and inelastic $[B(\mathbf{Q})]$ scattering.

Experimentally, the scattering spectra have a finite energy resolution, given by a resolution function, $R(\omega)$. Incorporating this effect in eqn (9.20) the dynamic structure factor becomes

$$S(\mathbf{Q},\omega) = e^{-Q^2 \langle u_v^2 \rangle} [A_0(\mathbf{Q})R(\omega) + S_d'(\mathbf{Q},\omega) \otimes R(\omega)] + B(\mathbf{Q}). \tag{9.21}$$

$A_0(\mathbf{Q})$ and $S'(\mathbf{Q},\omega)$ may be extracted from experiment by fitting eqn (9.21) to the measured scattering profiles. For this, it is necessary to assume *a priori* parametric forms for $A_0(\mathbf{Q})$ and $S'(\mathbf{Q},\omega)$; these depend on the dynamical model that one wishes to fit. Several such models, such as continuous diffusion on a circle or sphere or jumps between sites, are described in Bee (1988). It turns out that $A_0(\mathbf{Q})$ and $S'(\mathbf{Q}, \omega)$ obtained from experiment may also depend on the instrumental resolution function. If slow motions occur in the system, the dynamic structure factor may contain quasielastic contributions with widths much narrower than that of $R(\omega)$. These contributions are then indistinguishable experimentally from the elastic scattering and the extracted experimental EISF are different from the EISF in the long-time limit.

Extraction of $A_0(\mathbf{Q})$ from a molecular dynamics simulation This assumes that we are able to determine the diffusive contribution to the atomic trajectories. In this case the EISF can be obtained in two ways: from the long-time limit of $I_d(\mathbf{Q}, t)$ using eqn (9.19) or, assuming that the position vector of any given atom is uncorrelated with itself at infinite time, the EISF can be written as follows [cf. eqn (9.10)]:

$$A_0(\mathbf{Q}) = \sum_i b_i^2 |\langle e^{i\mathbf{Q} \cdot \mathbf{r}_{i,d}} \rangle|^2. \tag{9.22}$$

If the full molecular dynamics trajectories are used, without separation into diffusive and non-vibrational components, a different EISF, which we call $A_{0,\text{tot}}(\mathbf{Q})$, that includes contributions from all types of motions, can be calculated:

$$A_{0,\text{tot}}(\mathbf{Q}) = \sum_i b_i^2 |\langle e^{i\mathbf{Q} \cdot \mathbf{R}_i} \rangle|^2. \tag{9.23}$$

Given the assumptions used in deriving eqn (9.20), we can write

$$A_{0,\text{tot}}(\mathbf{Q}) = e^{-Q^2 \langle u_v^2 \rangle} A_0(\mathbf{Q}). \tag{9.24}$$

9.2.2.3 Inelastic incoherent scattering

Scattering intensity For a system executing harmonic dynamics, the transform in eqn (9.2) can be performed analytically and the result expanded in a power series over the normal modes in the sample. The following expression is obtained

$$\begin{aligned} S_{\text{inc}}(\mathbf{Q}, \omega) = \sum_i b_{\text{inc}}^2 \exp[-2W_i(\mathbf{Q})] \\ \times \prod_\lambda \left[\sum_{n_\lambda} \exp(n_\lambda \hbar \omega_\lambda \beta/2) \mathcal{I}_{n_\lambda} \left(\frac{\hbar (\mathbf{Q} \cdot \mathbf{e}_{\lambda,i})^2}{2M\omega_\lambda \sinh(\hbar \omega_\lambda \beta/2)} \right) \right] \\ \times \delta\left(\omega - \sum_\lambda n_\lambda \omega_\lambda\right). \end{aligned} \tag{9.25}$$

In eqn. (9.25) M is the hydrogen mass, λ labels the mode, $e_{\lambda,i}$ is the atomic eigenvector for hydrogen i in mode λ, and ω_λ the mode angular frequency. n_λ is the number of quanta of energy $\hbar\omega_\lambda$ exchanged between the neutron and mode λ, and $\mathcal{I}_{n_\lambda}(x)$ represents the modified Bessel function of order n_λ. $W_i(\mathbf{Q})$ is the exponent of the Debye–Waller factor, $\exp[-2W_i(\mathbf{Q})]$, for hydrogen atom i and is given as follows:

$$2W_i(\mathbf{Q}) = \frac{1}{2NM}\sum_\lambda \frac{\hbar(\mathbf{Q}\cdot e_{\lambda,i})^2}{\omega_\lambda}[2n(\omega_\lambda)+1] = Q^2\langle u^2_{Q,i}\rangle. \qquad (9.26)$$

In eqn (9.26), N is the number of modes, $n(\omega_\lambda)$ is the Bose occupancy and $\langle u^2_{Q,i}\rangle$ is the mean-square displacement for the atom i in the direction of $\mathbf{Q}$.

Equation (9.25) is an exact quantum mechanical expression for the scattered intensity. A detailed interpretation of this equation is given in Smith et al. (1986). Inserting the calculated eigenvectors and eigenvalues in the equation allows the calculation of the incoherent scattering in the harmonic approximation for processes involving any desired number of quanta exchanged between the neutrons and the sample, e.g. one-phonon scattering involving the exchange of one quantum of energy $\hbar\omega_\lambda$, two-phonon scattering, and so on.

The label λ in eqn (9.25) runs over all the modes of the sample. In the cases examined here normal mode analyses have been performed for isolated molecules (proteins) and molecular crystals. In the case of an isolated molecule, λ runs over the $3N-6$ normal modes of the molecule, where N is the number of atoms. In the case of a crystal, λ runs over the phonon modes in the asymmetric unit of the first Brillouin zone.

Vibrational density of states The vibrational density of states, $G(\omega)$, is related to the classical dynamical structure factor by

$$G(\omega) = \lim_{Q\to 0}\frac{\omega^2}{Q^2}S_{cl}(Q,\omega)$$

$$= \frac{1}{N}\sum_i \frac{1}{2\pi}\int_{-\infty}^{\infty} dt\, e^{-i\omega t}\langle \mathbf{v}_i(0)\mathbf{v}_i(t)\rangle, \qquad (9.27)$$

where $\langle \mathbf{v}_i(0)\mathbf{v}_i(t)\rangle$ is the autocorrelation function of the velocities $\mathbf{v}_i(t)$ of atom i, and $G(\omega)$ is the kinetic energy of the hydrogen atoms in the system as a function of frequency. Equation (9.27) holds formally also in the quantum case. Experimentally, $G(\omega)$ can be obtained, in principle, by performing the extrapolation of $S(\mathbf{Q},\omega)$ to $Q=0$. From molecular dynamics simulations, $G(\omega)$ can be calculated as the Fourier transform of the velocity autocorrelation function. For harmonic analysis $G(\omega)$ is simply the cross-section-weighted frequency distribution, i.e.

$$G(\omega) = \sum_i \frac{b^2_{inc,i}e^2_{\lambda,i}}{m_i}. \qquad (9.28)$$

9.2.2.4 Coherent inelastic neutron scattering

The use of coherent neutron scattering with simultaneous energy and momentum resolution provides a probe of time-dependent pair correlations of atomic motions. Coherent inelastic neutron scattering is therefore particularly useful for examining lattice dynamics in molecular crystals and holds promise for the characterization of correlated motions in biological macromolecules. For accurate measurements the samples must be completely deuterated.

A property of lattice modes is that for particular wave vectors there are well-defined frequencies; the relations between these two quantities are the phonon dispersion relations. Neutron scattering is the only effective technique for determining phonon dispersion curves. The following momentum conservation law is obeyed:

$$\mathbf{k}_i - \mathbf{k}_f = \mathbf{Q} = \boldsymbol{\tau} + \mathbf{q}, \qquad (9.29)$$

where $\mathbf{k}_i$ and $\mathbf{k}_f$ are the initial and final neutron wave vectors. The vibrational excitations have a wave vector $\mathbf{q}$ which is measured from a Brillouin zone centre (Bragg peak) located at $\boldsymbol{\tau}$, a reciprocal lattice vector.

If the displacements of the atoms are given in terms of the harmonic normal modes of vibration for the crystal, the coherent one-phonon inelastic neutron scattering cross-section is given by

$$S_{\text{coh}}(\mathbf{Q}, \omega) = \sum_j S_j(\mathbf{Q}, \omega), \qquad (9.30)$$

where the summation is over all vibrational modes of the crystal. For one mode one has

$$S_j(\mathbf{Q}, \omega) = \frac{\langle n(\omega_j(\mathbf{q})) + 1 \rangle}{\omega_j(\mathbf{q})} \mid F_j(\mathbf{Q}, \mathbf{q}) \mid^2 \delta(\omega \pm \omega_j(\mathbf{q})) \delta(\mathbf{Q} \pm \mathbf{q} - \boldsymbol{\tau}). \qquad (9.31)$$

In this expression $\omega_j(\mathbf{q})$ is the frequency of the phonon with wave vector $\mathbf{q}$ belonging to phonon dispersion branch j; $\langle n(\omega_j(\mathbf{q})) + 1 \rangle$ is the Bose factor. $\mid F_j(\mathbf{Q}, \mathbf{q}) \mid^2$ is given by

$$\mid F_j(\mathbf{Q}, \mathbf{q}) \mid^2 = \sum_k m_k^{1/2} b_k (\mathbf{Q} \cdot \mathbf{e}_k^j(\mathbf{q})) \exp(i\mathbf{Q} \cdot \mathbf{R}_k) \exp(-2W_k(\mathbf{Q})), \qquad (9.32)$$

where m_k is the mass of the atom k, b_k its coherent scattering length and $\mathbf{R}_k$ its position. $W_k(\mathbf{Q})$ in eqn (9.32) is the exponent of the Debye–Waller factor and is given by eqn (9.26). $\mathbf{e}_k^j(\mathbf{q})$ is the eigenvector of the kth atom in the jth mode and describes the pattern of the displacements in one unit cell. $\mathbf{e}_k^j(\mathbf{q})$ has $3s$ components, where s is the number of atoms in the unit cell. For any direction of $\mathbf{q}$ in the Brillouin zone there are $3s$ dispersion curves.

9.3 Dynamics in small molecule condensed phases

In this section we give examples of the combination of molecular simulation using the CHARMM program (Brooks *et al.* 1983) with experiment for the determination of ps timescale dynamics of crystals of molecules of biological interest. The examples, taken from work involving the Saclay molecular simulation laboratory, are chosen to cover a wide range of dynamical phenomena: local and collective vibrations, lattice modes, and diffusive motions of molecules.

9.3.1 Vibrational dynamics

9.3.1.1 Anharmonic local vibrations in acetanilide

The crystalline state provides structurally well-characterized systems enabling detailed studies of environmental effects on molecular motions. Using a molecular mechanics force field it is possible, in principle, to obtain a complete description of the ground state nuclear dynamics of a molecular crystal, by working in the full, $3N$-dimensional configurational space (where N is the number of atoms in the crystal) using computer simulation methods. In this way an attempt can be made to describe the structural and dynamical features of the crystal in a unified fashion.

An optimized molecular mechanics potential function has been obtained for the acetanilide ($C_6H_5CONHCH_3$) crystal by performing energy minimizations and harmonic analyses of the crystal and adjusting the parameters of the function so as to reproduce low-temperature structural and spectroscopic data (Hayward *et al.* 1995). The resulting normal modes provide a description of the low-temperature intramolecular and lattice vibrations. With one exception all the fundamental frequencies of the intramolecular modes in the refined force field were within 3% of their values obtained at the centre of the Brillouin zone by optical spectroscopy. Most of the mode assignments were in agreement with previous assignment schemes. Moreover, the calculated crystal field splitting of the vibrational bands (into eight distinct components for acetanilide) was also found to be in quantitative agreement with experiment.

Using the results of the harmonic analysis, the incoherent inelastic neutron scattering intensities were calculated using eqn (9.25) assuming the presence of one- two- and three-phonon scattering processes. Results at 25 K are compared with experiment in Fig. 9.1. Because of the large hydrogen displacement in methyl torsion, this peak is by far the strongest feature of the experimental spectrum. The peak is narrow and well resolved at 145 cm^{-1}. The average intensity of the lattice mode peaks (< 100 cm^{-1}) is $\sim 20\%$ of the methyl torsional peak at 145 cm^{-1}.

Having refined the force field in the harmonic approximation, it was possible to use the full, anharmonic potential function in molecular dynamics simulations. Simulations of the acetanilide crystal were performed using periodic boundary conditions, at 80, 140 and 300 K, and the temperature dependence of the hydrogen-weighted vibrational density of states calculated using eqn (9.27). The results, which will not be discussed in detail here, show anharmonic

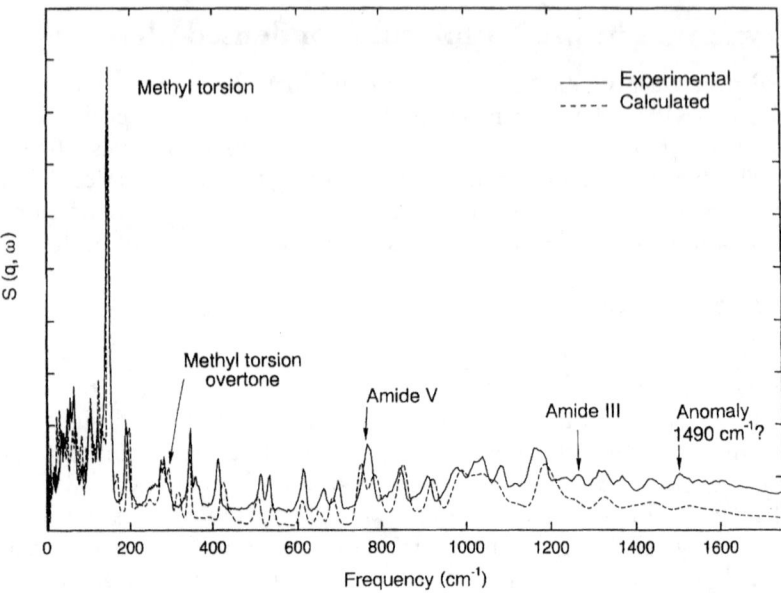

Fig. 9.1. The 25 K incoherent neutron scattering dynamic structure factor, $S_{inc}(\mathbf{Q}, \omega)$, measured using the spectrometer TFXA at the spallation source at the Rutherford–Appleton Laboratory, Oxford, and calculated using the results of a normal mode analysis of the crystal and assuming the presence of one- two- and three-phonon scattering, using eqn (9.25). [From Hayward et al. (1995).]

features of the methyl librational and H-bond vibrations that are in quantitative agreement with experiments.

9.3.1.2 Lattice vibrations in L-alanine

Zwitterionic L-alanine ($^{+}H_3N-C(CH_3)-CO_2^-$) is a dipolar molecule that forms large, well-ordered crystals in which the molecules form hydrogen-bonded columns. The strong interactions lead to the presence of well-defined intra- and intermolecular vibrations that can usefully be described using harmonic theory.

Coherent inelastic neutron scattering experiments have been combined with normal mode analyses with the molecular mechanics potential function to examine the collective vibrations in L-alanine (Micu et al. 1995). Ab initio quantum chemical calculations were performed to determine the ammonium:carboxylate hydrogen-bonding interaction energy curve and bond rotational potentials. The molecular mechanics potential function was parametrized to fit the *ab initio* results. Using the potential function, normal mode calculations were performed in the full configurational space of the crystal. Experiments were performed to obtain coherent inelastic neutron scattering intensities, $S_{coh}(\mathbf{Q}, \omega)$, and to trace the phonon dispersion relations along the three principal axes of the first Brillouin zone.

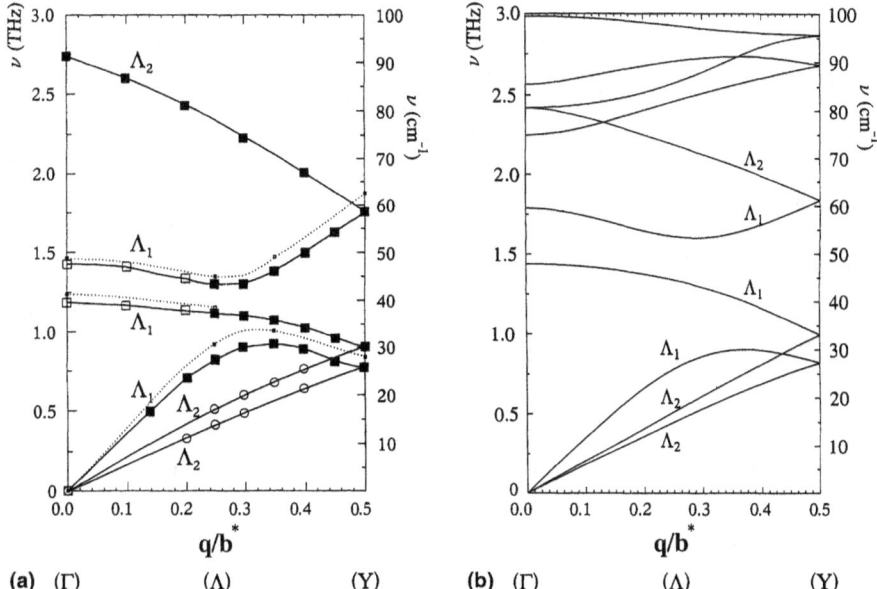

Fig. 9.2. (a) Dispersion curves for crystalline zwitterionic L-alanine at room temperature along the $\mathbf{b}^*$ crystallographic direction determined by coherent inelastic neutron scattering. The full circle and full square symbols are associated with phonon modes observed in predominantly transverse and purely longitudinal configurations, respectively, i.e. for vectors $\mathbf{Q}$ and $\mathbf{q}$ perpendicular and parallel to one another, respectively. They correspond to measurements performed around the strong Bragg reflections (200), (040) and (002). The empty-square symbols are neutron data points obtained around the (330), (103) and (202) reciprocal lattice point in a mixed configuration. Solid lines indicate the most probable connectivity of the dispersion curves and dashed lines correspond to the measurements performed at low temperature $T = 100$ K. (b) Theoretical dispersion curves for L-alanine determined from normal mode analysis. [From Micu et al. (1995).]

The experimental phonon frequencies $\nu_i(\mathbf{q})$ ($\nu = \omega/2\pi$) for several modes propagating along the crystallographic direction $\mathbf{b}^*$ are shown in Fig. 9.2. The solid lines represent the most probable paths for the dispersion curves $\nu_i(\mathbf{q})$. The theoretical dispersion curves are also given. The forms of the calculated branches are similar to those determined experimentally. At the border of the zone the frequencies are in good agreement, ranging from 46 to 62 cm^{-1} compared with 50 to 62 cm^{-1} experimentally. At the zone centre the frequencies of the optical modes are ~ 10 cm^{-1} higher than the experimental values and the gap between the frequencies of the second and the third optical modes is smaller in the calculations than in the experiment.

The calculated sound velocities, mean-square displacements, dispersion curves and coherent inelastic neutron scattering intensities were found to be mostly in quantitative agreement with experiment. An exception is the low-wave vector portion of the longitudinal acoustic branch in the $\mathbf{c}^*$ direction, for which the associated experimentally determined sound velocity is remarkably high—6.6 km/s, twice the theoretical value. The c^* direction is roughly parallel to the aligned zwitterionic dipoles. It is possible that polarization effects and/or long-range dipolar correlations, such as those discussed for water in the previous section, play a sensitive role in determining the high-sound velocity and were not correctly represented in the potential function.

A further example of the combination of molecular simulation with inelastic neutron scattering is given in Dianoux *et al.* (1993, 1994) in which the low-frequency density of states of polyacetylene and its modification on sodium doping are investigated.

9.3.2 Diffusive motions in molecular crystals

Molecules or parts of molecules in crystals can undergo diffusive motion on timescales accessible to molecular dynamics simulation. These motions can also be probed using incoherent quasielastic and elastic neutron scattering.

9.3.2.1 Alkane diffusion in urea inclusion compounds

Urea inclusion compounds are attractive systems for the characterization of the dynamics of n-alkane chains in a confined environment. In these systems the urea 'host' molecules form a hydrogen-bonded network containing parallel channels into which linear 'guest' molecules pack (see Fig. 9.3). Incoherent

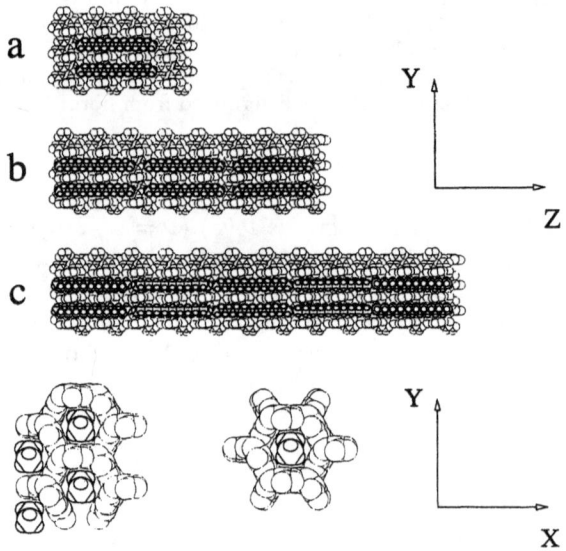

Fig. 9.3. Simulation primary boxes for the urea/n-nonadecane inclusion compound. *Upper:* in the YZ plane (a) model MU1, (b) model MU3 and (c) model MU5. The primary box of MU10 is that of MU5 but doubled in the Z-direction. *Lower:* in the XY plane. [From Souaille *et al.* (1996a).]

quasielastic neutron scattering experiments have been performed on n-nonadecane-urea at 180 K in which $S_{\text{inc}}(\mathbf{Q}, \omega)$ was determined with $\mathbf{Q}$ oriented parallel perpendicular to the channel axes. A detailed discussion of a simulation/experiment comparison for this system furnishes insight into the problems typically encountered and information that can be obtained.

Varying the simulation model The effect of varying the molecular dynamics simulation model system on the calculated quasielastic neutron scattering profiles has been examined (Souaille *et al.* 1996a,b). Simulations were performed with differing numbers of n-nonadecane molecules per channel and by varying the packing distance between the molecules. The effect of varying the alkane repeat distance along the channel axis on the calculated quasielastic scattering is shown in Figs 9.4(a) and (b) in the $\mathbf{Q}_\parallel$ and $\mathbf{Q}_\perp$ geometries together with the corresponding experimental data. Clearly, the calculated quasielastic profiles depend strongly on the intrachannel alkane−alkane interactions. In simulations MU1 and MU3 the alkane molecule centres of mass are separated by ∞ and ~ 29 Å respectively. In simulations MU5 and MU10 the repeat distance is 26.44 Å, the experimental value. In MU10 there are 10 molecules per channel in the primary box, whereas there are five in MU5. Figure 9.4(a) shows that simulations MU1 and MU3, in which the guest molecules are further apart than observed experimentally, are not in agreement with experiment. Their quasielastic spectra in the $\mathbf{Q}_\parallel$ direction are too broad, indicating the presence of very fast diffusive motion and the chains are not sufficiently confined. Figure 9.4(b) indicates that the guest−guest interactions also noticeably affect the dynamics in the orthogonal, $\mathbf{Q}_\perp$, directions. In both the $\mathbf{Q}_\parallel$ and $\mathbf{Q}_\perp$ geometries the best agreement with experiment is obtained with the experimentally determined n-nonadecane repeat distance. Other calculations indicated that fixing the urea molecules also leads to quantitative disagreement with the experimental quasielastic spectra. More details on the urea−alkane work, including an analysis of the geometries of the motions involved (the elastic incoherent structure factor) can be found in Souaille *et al.* (1996a).

9.4 Protein dynamics

Picosecond timescale and Å lengthscale dynamics in native proteins are of particular interest as they are accessible to molecular dynamics simulation. The ps timescale is also interesting physically as all the different types of motion discussed in the previous section on small molecules occur in proteins on this timescale at physiological temperatures, i.e. underdamped vibrations, overdamped vibrations, continuous and jump diffusion. Thus, ps timescale protein dynamics possesses considerable complexity and the combination of experiment and simulation is necessary to unravel the components of the atomic motions present.

The combination of simulation and neutron scattering in the analysis of internal motions in globular proteins was reviewed by Smith (1991). Here we briefly recall these results and complement them with some newer findings.

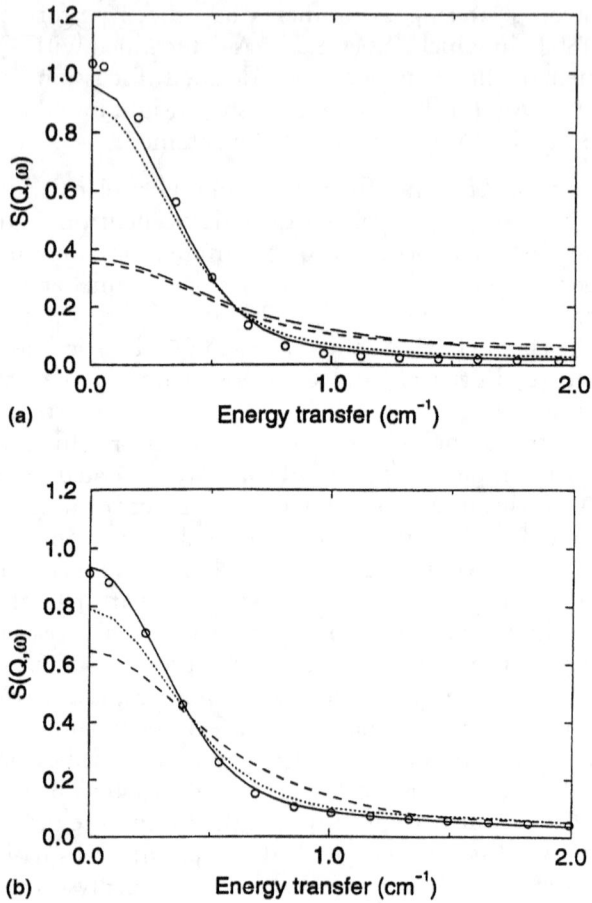

Fig. 9.4. (a) Dynamic structure factor for urea–alkane inclusion compound at 180 K in the $Q_\parallel$ geometry; $Q = 1.0$ Å^{-1}. Experiment (o) simulation MU10 (———) simulation MU5 (···) simulation MU3 (- - -) simulation MU1 (– –). (b) Dynamic structure factor in the $Q_\perp$ geometry; $Q = 1.0$ Å^{-1}. Experiment (o) simulation MU10 (———) simulation MU5 (···) simulation MU3 (- - -) simulation MU1 (– –). [From Souaille et al. (1996a).]

9.4.1 Vibrations

Vibrations in proteins can conveniently be examined using normal mode analysis of isolated molecules. The results of such analyses indicate the presence of a variety of vibrations, with frequencies upward of a few cm^{-1}. In most cases the very lowest-frequency modes dominate the calculated mean-square displacements. For example, in a normal mode analysis of lysozyme 80% of the mass-weighted mean-square fluctuation was found to originate from a small number of modes with frequencies < 30 cm^{-1} (Go 1990). The very low frequency modes are large-amplitude, delocalized, correlated vibrations.

9.4.1.1 *Incoherent inelastic neutron scattering—vibrational amplitudes and damping properties.*

Incoherent inelastic neutron scattering, combined with normal mode analysis, is well suited to examine low-frequency vibrations in proteins. This is primarily due to the fact that large-amplitude displacements scatter neutrons strongly. Experiments on bovine pancreatic trypsin inhibitor (BPTI), combined with normal mode analysis of the isolated protein, demonstrated that low-frequency underdamped vibrations do exist in the protein (Cusack *et al.* 1988). A comparison of absolute scattering cross-sections indicated that the average vibrational amplitudes were in quantitative agreement. Improved agreement with experiment was obtained by introducing a friction coefficient for each mode in a damped Langevin oscillator description (Smith *et al.* 1990a). The distribution of friction coefficients $p(\omega)$, obtained by fitting to experiment, follows a Gaussian form, i.e.

$$p(\omega) = A \exp\left[-\frac{\omega^2}{2\sigma^2}\right] \tag{9.33}$$

with $A = 30 \text{ cm}^{-1}$ and $\sigma^2 = 225 \text{ cm}^{-2}$. Therefore, modes with frequencies $< 16 \text{ cm}^{-1}$ are overdamped, i.e. they do not oscillate. Thus, the very lowest-frequency modes predicted by harmonic models, e.g. the lysozyme hinge bending mode, do not vibrate at the calculated frequencies. They are either absent or overdamped. If overdamped, however, they can still be a source of correlated motions in proteins.

Femtosecond spectroscopic experiments have provided evidence implicating low-frequency vibrations in the primary electron transfer in functional photosynthetic reaction centres (Vos *et al.* 1993). The lowest-frequency vibration identified had a frequency of 15 cm^{-1} and was underdamped, close to being critically damped. Although the form of this vibration and how it influences the electronic transition are not yet clear, it is interesting to note that 15 cm^{-1} corresponds to the frequencies of the lowest-frequency, underdamped collective vibrations detected in small proteins using neutrons. Moreover, the damping characteristics of the vibration inferred from the femtosecond spectroscopic studies are similar to what would be expected from the damping scheme introduced phenomenologically in eqn (9.33), and also with the damping properties of a 15 cm^{-1} rigid-helix vibration characterized in a molecular dynamics simulation of myoglobin (Furois-Corbin *et al.* 1993).

9.4.1.2 *Vibrational density of states*

The low-frequency portion of the vibrational density of states, $G(\omega)$, for BPTI has been determined experimentally using eqn (9.27). Subsequently, attempts were made to reproduce this frequency distribution using molecular simulation. In an initial study normal mode analyses were performed with different electrostatic truncation schemes (Smith *et al.* 1990b). The resulting $G(\omega)$s are compared

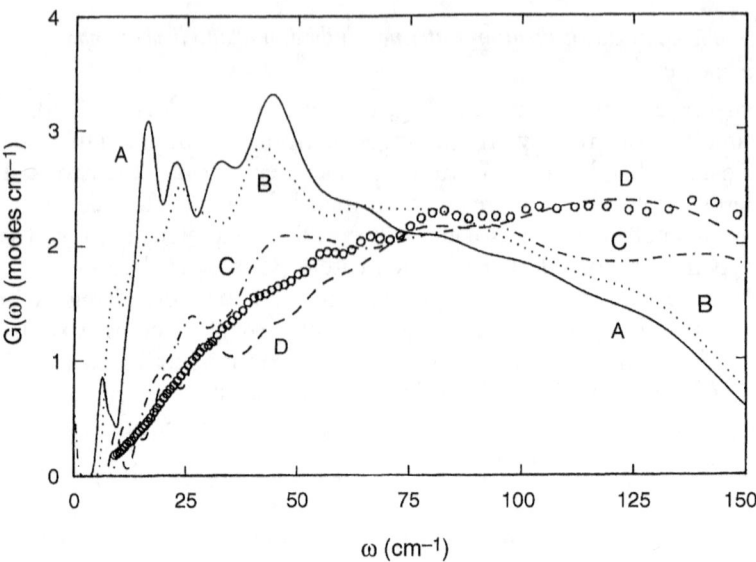

Fig. 9.5. Density of states, $G(\omega)$, for BPTI from experiment (circles) and from four normal mode analyses. The analysis corresponding to curve A was performed in the extended atom approximation with no electrostatic truncation. Curve B used the extended atom approximation and shift electrostatic truncation at 7.5 Å. Curve C used the extended atom method and switch electrostatic truncation (from 6.5 Å to 7.5 Å). Curve D included all the atoms explicitly and used switch truncation. [From Smith *et al.* (1990).]

with experiment in Fig. 9.5. $G(\omega)$ obtained using electrostatic truncation smoothed by a cubic switching function was found to be in better agreement with experiment than that obtained using no electrostatic truncation. One possible explanation for this in that effect of the switch function mimics the effect of the environment in the experimental powder sample. By analysis of two 100 ps simulations of BPTI, one in water and one in vacuum, a model of frictional damping was developed that describes the effect of water on $G(\omega)$ (Hayward *et al.* 1993). Of the two simulations, $G(\omega)$ calculated for the protein in water resembled more closely the form of the experimental function. It was shown that treating each vacuum principal mode as an independent damped Langevin oscillator with the natural frequency of each mode given by its vacuum effective frequency, and assigning to all modes a friction coefficient of 47 cm^{-1}, gives a $G(\omega)$ closely similar to that obtained in the solution simulation and in the experiment. This damping scheme is different from that proposed in eqn (9.33) but the frequency of critical damping (23.5 cm^{-1}) is similar.

9.4.2 High-frequency vibrations

Until recently, neutron sources and instruments have not allowed inelastic spectra to be measured with useful energy resolutions in the range $\sim$ 50–3500 cm^{-1}.

The construction of the time-focusing crystal analyser spectrometer (TFXA) at the ISIS pulsed neutron source near Oxford has changed this situation, allowing high-resolution spectra to be obtained over the whole frequency range of interest. In recent work experimental TFXA spectra were measured for the enzyme, Staphylococcal nuclease, and compared with results from a normal mode analysis of the enzyme (Lamy-Goupil et al. 1997). The calculated spectral intensities, which were obtained using a force field that was not refined to fit the neutron data, are in general agreement with experiment and were used to assign the peaks.

9.4.3 Diffusive motion

9.4.3.1 Incoherent quasielastic neutron scattering

Above ~ 200 K there is a non-vibrational component to protein atom dynamics that has been detected using several experimental techniques including neutron scattering (Smith 1991; Doster et al. 1989). The dynamical transition is also present in molecular dynamics simulations (Smith et al. 1990a,b). There is evidence that the non-vibrational dynamics might play a role in the functioning of some proteins, e.g. in ligand binding (Rasmussen et al. 1992) or proton transfer reactions (Ferrand et al. 1993). Inelastic neutron scattering measurements on bacteriorhodopsin have shown that the ability of the protein to functionally relax and complete its photocycle is strongly correlated with the onset of anharmonic dynamics in the membrane (Ferrand et al. 1993).

Various models for the non-vibrational atomic motions in proteins at 300 K have been proposed. Most are based on the idea of transitions between conformational substates and assume individual or collective stochastic jump dynamics of the atoms between minima on the potential energy surface of the folded protein. However, the observed neutron scattering profiles could originate instead from continuous diffusive motion, and/or from overdamped harmonic motion. To determine the nature of the non-vibrational component requires an 'ab initio' test of a given model hypothesis. This test can be made using molecular dynamics simulations, by determining to what extent the hypothetical atomic motion contributes to the simulated atomic trajectories. The contribution to the simulation-derived intensity from the simplified model dynamics can then be calculated and compared with experiment.

The dynamical transition of proteins is often discussed within the framework of the liquid–glass transition (Angell 1995). In this context one may ask whether a granularity of the high-temperature 'liquid' phase exists, i.e. whether it is possible to define subunits of proteins that can be treated analogously to molecules in a liquid. In a recent analysis the individual side-chains attached to the protein backbone were considered as rigid subunits and their contribution to the neutron scattering profiles of myoglobin at physiological temperatures was calculated (Kneller and Smith 1994).

To determine the contribution of the 'side-chain liquid' to the dynamics, rigid reference structures of each side-chain were fitted to the corresponding structures in each time frame of a molecular dynamics trajectory of myogobin. In this way a trajectory of the fitted atomic positions was built up and could be analysed in the same manner as the full trajectory, enabling a quantitative calculation of the rigid side-chain contribution to the neutron scattering. For the fitting procedure the C_α atoms on the protein backbone were included in the side-chain reference structures and constrained to coincide with the C_α positions in each time frame of the full molecular dynamics trajectory. In other words, the fitted rigid side-chains were pinned to the C_α atoms.

Figure 9.6 shows the EISF obtained from experiment and simulation. The experimental and simulation data match well. Moreover, clearly the rigid side-chain motions account completely for the full dynamics.

That the diffusive motion leading to the EISF arises from rigid side-chain motions may seem surprising as most side-chains contain rotatable dihedral degrees of freedom. Indeed, torsional jump models have been applied to describe quasielastic scattering from proteins. However, it is clear from Fig. 9.6 that conformational transitions of the side-chains, although present in the simulation, do not contribute significantly to the scattering.

The quasielastic scattering and average mean-square displacement contain a dominant component from rigid-body diffusive motions of the side-chains. The displacements are caused by collisions between atoms in different side-chains. These 'kicks' are transmitted through the side-chains via stiff covalent forces. The result is rigid-body displacement of all the atoms in a side-chain. Side-chain collisions are very frequent since the atomic packing density in a protein is comparable to that of a solid. After some time the form of a side-chain may change due to a torsional transition, giving rise to an error in the fitted

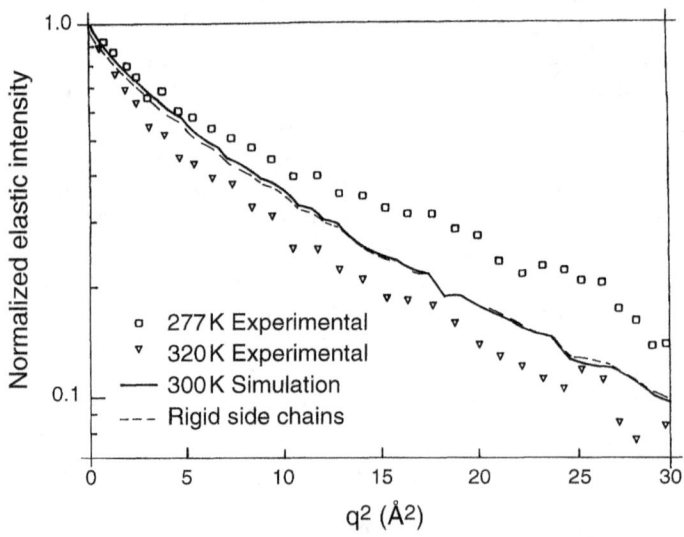

Fig. 9.6. EISF for myoglobin from experiment, full simulation trajectory and rigid side-chain fit. [From Kneller and Smith (1994).]

atomic positions. However, the continuous diffusive motion of the side-chains is still well described since the requirement for this is that consecutive side-chain conformations be similar. Although the side-chains are flexible, they behave as rigid bodies with respect to the diffusive, liquid-like motion detected in the neutron scattering experiments.

Quasielastic scattering from proteins has recently been extended to the examination of a strongly denatured state of phosphoglycerate kinase (Receveur et al. 1997). The results indicated a modification of the EISF on denaturation that could be interpreted in terms of an increase of the proportion of atoms in the protein that undergo diffusive motion on the accessible timescale. Experiments such as these are likely to provide useful information on the potential energy surfaces accessible to proteins at different stages along their folding funnels. Generally speaking, a large effort is indeed underway by several research groups trying to understand quasielastic incoherent neutron scattering in proteins.

A discussion of the relation between the dynamics probed by X-ray diffuse scattering using synchrotron radiation and quasielastic neutron scattering is given in Faure et al. (1994).

9.5 Conclusions

The combination of simulation with neutron scattering experiments allows a wide range of dynamical phenomena to be characterized in condensed phase biomolecular systems. The work described in the present chapter testifies to the versatility of empirical potential energy functions of the molecular mechanics type in describing motions on a range of timescales from 10^{-15} to 10^{-10} s, i.e. from fs, localized vibrations to ~ 100 ps activated processes.

Neutrons are not easy to get hold of. To produce them in controlled conditions requires a nuclear reactor or a spallation source, the cost of which does not fall into the budget of an average structural biology laboratory! Moreover, even from these sources neutron fluxes are very low, typically $\sim 10^7$ neutrons cm^{-2} s^{-1} compared, for instance, to X-ray fluxes at a synchrotron ($\sim 10^{12}$ photons cm^{-2} s^{-2}). Thus, neutron experiments suffer from a counting statistical problem that has limited the range of applications to those with large samples ($\sim 10^{-1}$ g) and long counting times ($\sim$ days). However, there is some hope that a future European neutron source with considerably higher flux will be built. This would open up a new range of inelastic experiments on biological macromolecules involving specific H/D labelling, and spin-echo measurements of coherent scattering that would provide information on ns timescale correlations. However, by the time such a source is built concurrent progress in computer power and simulation methodology is likely to have pushed the timescale of events accessible to atomic-detail computer simulation well beyond the ns regime. Nevertheless, the general strategy outlined here for combining simulation with experiment will still be applicable and simulations will be required for the reliable interpretation of experiments probing the long-time dynamics of complex biological systems.

References

Angell, CA (1995). *Proceedings of the National Academy of Sciences USA*, **92**, 6675.
Bee, M (1988). *Quasielastic neutron scattering: principles and applications in solid state chemistry, biology and materials science*, Adam Hilger, Bristol and Philadelphia.
Brooks, B, Bruccoleri, R, Olafson, B, States, D, Swaminathan, S and Karplus, M (1983). *Journal of Computational Chemistry*, **4**, 187.
Cusack, S, Smith, JC, Finney, JL, Tidor, B and Karplus, M (1988). *Journal of Molecular Biology*, **202**, 903.
Dianoux, AJ, Kneller, GR, Sauvajol, JL and Smith, JC (1993). *Journal of Chemical Physics*, **99**, 5586.
Dianoux, AJ, Kneller, GR, Sauvajol, JL and Smith, JC (1994). *Journal of Chemical Physics*, **101**, 634.
Doster, W, Cusack, S and Petry, W (1989). *Nature*, **337**, 754.
Faure, P, Micu, A, Doucet, J, Smith, JC and Benoît, J-P (1994). *Nature Structural Biology*, **2**, 124.
Ferrand, M, Dianoux, AJ, Petry, W and Zaccai, G (1993). *Proceedings of the National Academy of Sciences USA*, **90**, 9668.
Furois-Corbin, S, Smith, JC and Kneller, GR (1993). *Proteins: Structure Function and Genetics* **16**(2), 141.
Go, N (1990). *Biophysical Chemistry*, **35**, 105.
Hayward, S, Kitao, A, Hirata, F and Go, N (1993). *Journal of Molecular Biology*, **234**, 1207.
Hayward, RL, Middendorf, HD, Wanderlingh, U and Smith, JC (1995). *Journal of Chemical Physics*, **102**, 5525.
Kneller, GR and Smith, JC (1994). *Journal of Molecular Biology*, **242**, 181.
Kneller, G, Keiner, V, Kneller, M and Schiller, M (1995). *Computational Physics Communications*, **91**, 191.
Lamy-Goupil, AV, Smith, JC, Yunoki, J, Parker, SF and Kataoka, M (1997). *Journal of the American Chemical Society*, **119**, 9268.
Lovesey, S (1984). *Theory of thermal neutron scattering from condensed matter*, International Series of Monographs on Physics, Vol. 72, Oxford Science, Oxford.
Micu, A, Durand, D, Quilichini, M, Field, MJ and Smith, JC (1995). *Journal of Physical Chemistry*, **99**, 5645.
Rasmussen, BF, Stock, AM, Ringe, D and Petsko, GA (1992). *Nature*, **357**, 423.
Receveur, V, Calmettes, P, Smith, JC, Desmadril, M, Coddens, G and Durand, D (1997). *Proteins: Structure Function and Genetics*, **28**, 380–387.
Smith, JC (1991). *Quarterly Review of Biophysics*, **24**(3), 227.
Smith, JC, Cusack, S Brooks, B, Pezzeca, U and Karplus, M (1986). *Journal of Chemical Physics*, **85**, 3636.
Smith, JC, Cusack, S, Tidor, B and Karplus, M (1990a). *Journal of Chemical Physics*, **93**(5), 2974.
Smith, JC, Kuczera, K and Karplus, M (1990b). *Proceedings of the National Academy of Sciences USA*, **87**, 1601.
Souaille, M, Guillaume, F and Smith, JC (1996a). *Journal of Chemical Physics*, **104**(4), 1516.
Souaille, M, Guillaume, F and Smith, JC (1996b). *Journal of Chemical Physics*, **104**, 1529.
van Hove, L (1954). *Physical Review*, **95**, 249.
van Hove, L (1958). *Physica*, **24**, 404.
Vos, MH, Rappaport, F, Lambry, J-C, Breton, J and Martin, J-L (1993). *Nature*, **363**, 320.

10

Diffuse X-ray scattering from molecular crystals

D. S. Moss, G. W. Harris and A. Wostrack

10.1 Introduction

In addition to Bragg reflections, real crystals give rise to diffuse scattering that occurs both at and between the Bragg positions. This diffuse scattering arises from the temporary or permanent breakdown of space group symmetry in the crystal and in crystals of macromolecules often accounts for most of the diffracted energy except at low Bragg angles. Unlike Bragg reflections, diffuse scattering contains information about the correlation of the atomic displacements in the crystal and is thus particularly relevant in the study of the conformational flexibility of molecules.

The availability of highly monochromatic, well-collimated X-ray sources at synchrotron laboratories together with the use of electronic area detectors has caused a resurgence in the study of X-ray diffuse scattering. Traditional studies of crystal diffuse scattering have concentrated largely on simple ionic crystals, alloys and small organic molecules where the computational resources required for quantitative interpretation are modest. However, the diffraction patterns of proteins and nucleic acids often show pronounced diffuse scattering (see Fig. 10.1), and the availability of powerful workstations has made possible its analysis in terms of molecular parameters. We are now reaching a position where we shall be able to exploit the whole diffraction pattern in crystal structure analysis.

Diffuse blackening on an X-ray diffraction picture of a crystal may arise from many sources and the most notable are (a) thermal diffuse scattering, (b) static disorder scattering, (c) solvent disorder, (d) Compton scattering, (e) fluorescence, (f) scattering from mounting rods or tubes and (g) air scattering and background fog on film. In order to interpret diffuse scattering, all these sources may have to be considered but interesting structural information arises mostly from (a) and (b) and these will be the main subjects of this review.

All crystals exhibit thermal diffuse scattering (TDS). Thermal vibrations of the atoms cause the breakdown of translational symmetry in a crystal and this gives rise to TDS and also produces a concomitant reduction in the Bragg intensities. In hard crystals TDS occurs almost exclusively around the Bragg reflections. This is *acoustic scattering*, which is due to ultrasonic waves propagating in the crystal and can be interpreted in terms of the elastic constants of the crystal. In softer crystals much more extensive TDS may be observed which is also due to other low-frequency vibrations in the crystal.

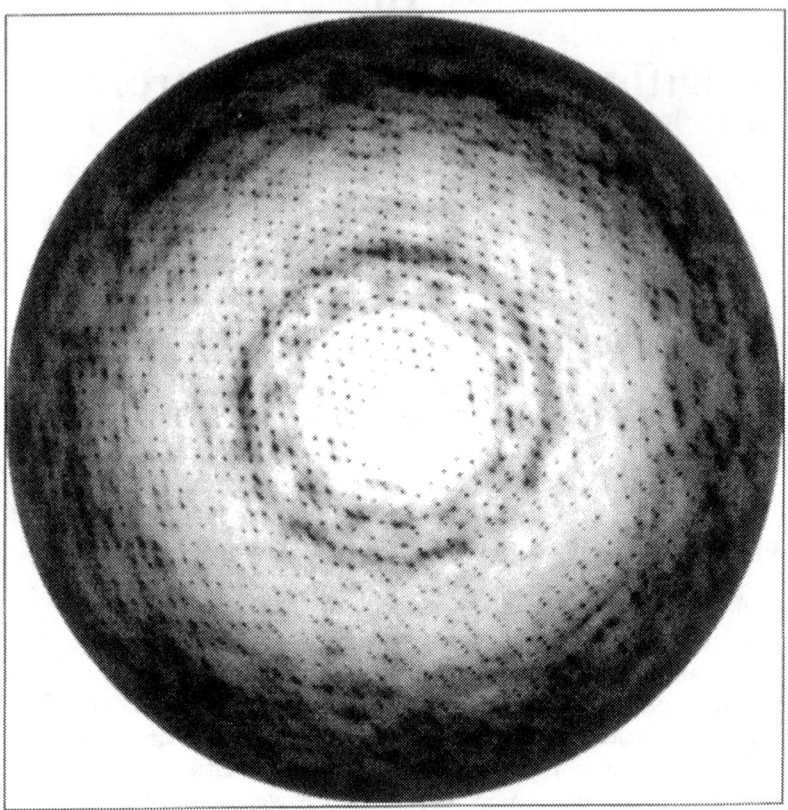

Fig. 10.1. One-degree oscillation X-ray image of a high-pH crystal of the avian eye lens protein δ-crystallin recorded with synchrotron radiation at Daresbury Laboratory using a Mar Research imaging plate. The resolution at the edge of the image is 3.5 Å. The 14-fold pseudo-symmetry of the innermost diffuse ring arises from the 7_3 pseudo-screw axes relating the seven molecules in the unit cell. These axes are approximately parallel to the X-ray beam. (Image by courtesy of A. Simpson and C. Slingsby.)

Crystal vibrations occur on a time scale ranging from about 10^{-6} to 10^{-14} s but motions in protein crystals such as reorientation of surface side-chains and local unfolding often have longer relaxation times. However, all these times are small compared with the time scale of an X-ray diffraction experiment, even using synchrotron radiation (10^{-4}–10^5 s). The implication of this is that the observed X-ray diffraction pattern is a time-average over all the crystal motions. These thermal perturbations are an example of *dynamic disorder*. By contrast, in *static disorder* unit cells exist with different arrangements of the time-averaged atomic positions. Interconversion of these arrangements is inhibited by energy barriers which are large compared with $k_B T$, where k_B is Boltzmann's constant and T is the absolute temperature. Both dynamic and static disorder may be characterized in terms of the type of probability distribution that describes the atomic positions. In *displacement disorder* the atomic positions can be described in terms of a continuous probability distribution. Thermal vibrations, for example, are usually described in terms of a Gaussian distribution.

Many forms of static disorder must be described in terms of discrete distributions. *Substitutional disorder* occurs in alloys and is due to different chemical species occupying corresponding sites in different unit cells. A related type of disorder occurs in protein crystals where ligands may bind statistically to the protein molecules in the crystal, giving rise to apparent partial site occupancy. *Orientational disorder* occurs in molecular crystals where molecules or flexible domains or side groups may take up different orientations, thus breaking down the translational symmetry. Both the latter types of disorder are characterized by discrete distributions (binary or multinomial). When thermal perturbations are also taken into account, the total time-averaged and lattice-averaged disorder may be described in terms of multimodal continuous distributions whose parameters are a function of temperature.

10.2 Thermal diffuse scattering

10.2.1 *Phonon scattering*

The thermal and acoustic waves in a crystal lattice are characterized by an angular frequency ω and a wave vector q, which is a vector in the direction of wave propagation, with a magnitude equal to $2\pi/\lambda$ where λ is its wavelength. A wave consists of atomic displacements and for a given wave there is associated with each atom in the unit cell a displacement vector e and an amplitude a which together describe the displacement undergone by the atom. The determination of e requires the solution of the equations of motion of the crystal and this has so far only been achieved for molecular crystals containing small molecules (Pawley 1969; Kroon and Vos 1979; Criado *et al.* 1985).

The waves propagated in crystals can be classified in terms of 'normal modes', which are the independent modes of crystal vibration. The contribution of a normal mode to the X-ray diffuse scattering depends on the magnitude of the resulting atomic displacements, which in turn depend on the energy of the mode. The energies associated with the normal modes, known as phonons, cannot easily be studied using X-rays and most information comes from inelastic neutron scattering. This technique enables the wave vectors q of the normal modes of a crystal to be plotted in reciprocal space and surfaces of constant frequency ω (proportional to energy) can be constructed. A one-dimensional schematic plot of q vs ω is shown in Fig. 10.2 for wave vectors perpendicular to the (100) planes in an orthorhombic crystal. The curves shown in the figure are known as 'branches' and for a crystal with N atoms per unit cell, there are $3N$ branches.

The curves passing through the origin are known as *acoustic branches*. A three-dimensional crystal possesses three such branches, although degeneracies may occur in crystals of high symmetry. In high-symmetry crystals, one of these branches is often associated with atomic displacements that are parallel to q

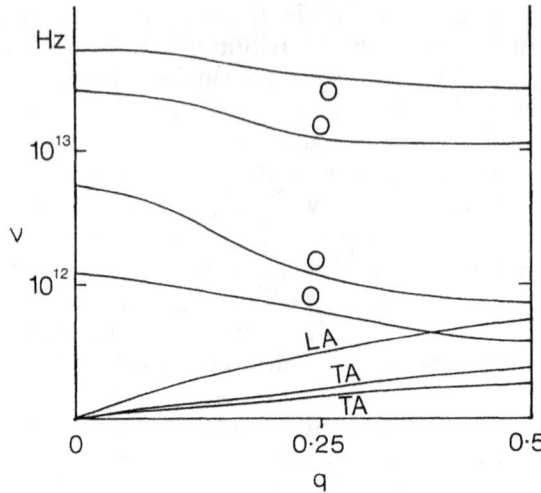

Fig. 10.2. Dispersion curves which are plots of frequency ν (vertical axis) against wavenumber (horizontal axis). The wavenumber is the reciprocal of the wavelength of the atomic displacements expressed in unit cell lengths. The curves which originate from the origin describe the acoustic modes. There are generally three such curves and two of them (labelled TA) represent approximately transverse vibrations where the direction of displacement is perpendicular to the direction of propagation. The third acoustic curve (labelled LA) represents an approximately longitudinal vibration where the displacement is approximately parallel to the direction of propagation. Such a vibration usually is of higher frequency than a transverse vibration of the same wavenumber. The dispersion curves labelled O represent optic modes, so called because they may be active in the infrared or Raman spectrum. Unlike the acoustic modes, which produce diffraction effects close to the reciprocal lattice points, the optic modes may produce effects throughout reciprocal space.

(longitudinal) and the other two are associated with displacements perpendicular to q (transverse). In general a more complex relationship will exist between e and q, and the branches can only be described approximately as longitudinal or transverse.

The low-frequency normal modes lying on acoustic branches are ultrasonic vibrations. The slope of these branches near the origin is $v_s = \omega/q$ where v_s is the phase velocity of sound in the crystals, which may be measured by ultrasonic experiments. In soft crystals such as proteins v_s is of the order of 2000 m s^{-1} (Edwards et al. 1990) but in hard crystals such as tungsten, the initial slope of the longitudinal acoustic branches may be of the order of 5000 m s^{-1}. Transverse branches in most crystals have initial slopes that are about three times lower than the corresponding longitudinal value and this is due to the shear stresses set up by transverse molecular displacements being smaller than the compressive stress, set up by correspondingly large longitudinal displacements. In macromolecular crystals containing bulk solvent, transverse velocities are likely to be even lower due to the inability of liquids to withstand shear stresses.

Due to the approximate equipartition of energy among the low-frequency modes, those of lowest frequency are associated with the largest atomic displacements and hence produce the most significant diffraction effects. Thus acoustic modes lying near to the origin (see Fig. 10.2) give some of the strongest diffuse scattering, and as will be seen below, this occurs close to Bragg reflections.

If there are N ordered atoms per primitive cell, there are $3N - 3$ non-acoustic branches, described as *optic branches* because their associated normal modes of vibration can often interact with electromagnetic radiation. In a molecular crystal the optic branches are often fairly flat due to the weak coupling of the molecular modes to form the crystal optic modes. In such a case the atomic displacements will be similar to those of the normal modes of the isolated molecules.

10.2.2 *Acoustic scattering*

An important component of thermal diffuse scattering that is present in all crystalline diffraction patterns is acoustic scattering, produced by acoustic modes of vibration of the crystal. It gives rise to diffuse X-ray scattering which peaks at reciprocal points. This is because frequency ω_i is approximately proportional to the magnitude of the wave vector q_i on an acoustic branch (see Fig. 10.2) and because low-frequency modes give rise to larger diffuse intensity occurring close to $\mathbf{Q}$, the reciprocal lattice vector. The acoustic peaks are more diffuse than the Bragg peaks but are roughly proportional to the latter in integrated intensity.

Diffraction methods have been used in the past to determine elastic constants of crystals but this has declined as more precise techniques such as ultrasonics, Brillouin scattering and inelastic neutron scattering have become available. The reverse approach was applied by Helmholdt and Vos (1977) to calculate the errors in Bragg diffraction intensities due to first-order TDS from the elastic constants of the crystal. This was done for dibenzoyl ($C_6H_5COCOC_6H_5$) and ammonium hydrogen oxalate hemihydrate ($NH_4HC_2O_2 \cdot \frac{1}{2}H_2O$).

Since TDS occurs at the Bragg positions, accurate measurement of the Bragg data requires a correction for the acoustic scattering (Cochran 1969). Neglect of a TDS correction causes under-estimation of the thermal parameters. Kroon and Vos (1979) studied the influence of TDS errors on structural parameters for naphthalene at 100 K. Errors in the coordinates due to TDS are very small, but considerable errors occur in both the scale factor and thermal parameters if no TDS corrections are made, and these errors depend on the reflection range used in the refinement.

Generally acoustic modes have smaller frequencies than optic modes and therefore give the main contribution to the diffuse intensities. The frequency of the acoustic modes increases rapidly with increasing q (see Fig. 10.2), hence the falloff of intensity with increasing distance from a Bragg peak (Sommer *et al.* 1981). The acoustic peak that surrounds a Bragg reflection of ribonuclease-A (Glover *et al.* 1991) is shown in Fig. 10.3.

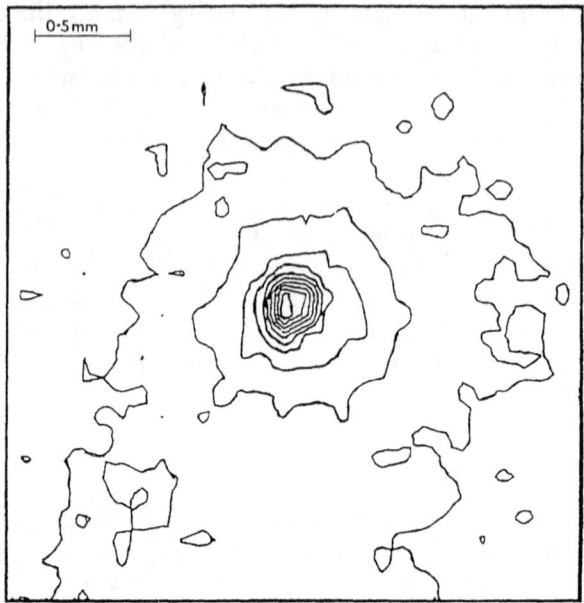

Fig. 10.3. The distribution of acoustic scattering around a 6 Å reflection of RNAse-A recorded on film using synchrotron radiation. A collimator of 0.2 mm diameter was used, along with slitting down of the beam premonochromator to reduce further the horizontal divergence. A 3 × 3 Å box was contoured on a Joyce–Loebl densitometer. [Taken from Glover et al. (1991).]

10.3 Disordered diffuse scattering of molecular crystals

10.3.1 General theory of diffuse X-ray scattering

The theory of X-ray diffuse scattering has been developed in different ways in the literature according to whether the focus of attention was on lattice vibrations, independent molecular vibrations or static disorder. In molecular crystals the internal modes of vibration of the molecules exhibit little dispersion, i.e. their frequencies are almost independent of the wavelength. Such vibrations give rise to diffuse scattering that does not peak on points or rows of the reciprocal lattice and the theory of such diffuse scattering can be developed without considering lattice waves. Static disorder often requires a statistical treatment, such as provided by the Ising model. The most difficult type of disorder to treat analytically is diffusional disorder that is exhibited by exposed side-chains of protein molecules and the diffraction effects of this type of disorder are better computed by molecular dynamics or Monte Carlo simulation. The following treatment is developed in terms of the dynamic disorder but the results apply equally to static disorder because of the time and lattice averaging inherent in a conventional diffraction experiment.

Diffuse scattering arises when translational symmetry temporarily or permanently breaks down in a crystal. The unit cells in the crystals then have different structure factors, written F_m for unit cell m. The instantaneous intensity is then given by

$$I(\mathbf{Q}) = \sum_m \sum_n F_m F_n^* \exp(i\mathbf{Q} \cdot (\mathbf{r}_m - \mathbf{r}_n)), \qquad (10.1)$$

where $\mathbf{r}_m$ and $\mathbf{r}_n$ are Bravais lattice vectors defining origins for primitive cells m and n. The coefficients $F_m F_n^*$ may be replaced by their time average to yield the time-averaged intensity or, in the case of static disorder by an ensemble average taken over all pairs of cells separated by the same vector $\mathbf{s}_{mn} = (\mathbf{r}_m - \mathbf{r}_n)$. Time averaging will be assumed in the following exposition unless static disorder is mentioned explicitly.

We may write the time-averaged intensity as

$$\langle I(\mathbf{Q}) \rangle = \sum_s \langle F_m F_n^* \rangle \exp(i\mathbf{Q} \cdot \mathbf{s}_{mn}), \qquad (10.2)$$

where $\mathbf{s}_{mn}$ is the vector between cells m and n. Assuming that the time-averaged structure factor of all cells is the same and equal to $\langle F \rangle$, we may write

$$\langle F_m F_n^* \rangle = |\langle F \rangle|^2 + \langle (F_m - \langle F \rangle)(F_n - \langle F \rangle)^* \rangle. \qquad (10.3)$$

Equation (10.2) may then be rewritten as

$$\langle I(\mathbf{Q}) \rangle = |\langle F \rangle|^2 \sum_s \exp(i\mathbf{Q} \cdot \mathbf{s}_{mn})$$
$$+ \sum_s \langle (F_m - \langle F \rangle)(F_n - \langle F \rangle)^* \rangle \exp(i\mathbf{Q} \cdot \mathbf{s}_{mn}). \qquad (10.4)$$

The first term on the right-hand side gives rise to the Bragg scattering and may be written as

$$I_B(\mathbf{Q}) = |\langle F \rangle|^2 \sum_s \exp(i\mathbf{Q} \cdot \mathbf{s}_{mn}) = |\langle F \rangle|^2 \Delta(\mathbf{Q}). \qquad (10.5)$$

$\Delta(\mathbf{Q})$ peaks sharply at reciprocal lattice points where $\mathbf{Q} = \mathbf{h}$, a reciprocal lattice vector.

The extent of the summation over s depends on the volume of the crystal in which coherent diffraction takes place, and governs the natural width of the Bragg peak.

The second term in eqn (10.4) is responsible for diffuse scattering:

$$I_D(\mathbf{Q}) = \sum_s \langle (F_m - \langle F \rangle)(F_n - \langle F \rangle)^* \rangle \exp(i\mathbf{Q} \cdot \mathbf{s}_{mn}). \qquad (10.6)$$

Thus the diffuse scattering is given by the discrete Fourier transform of the covariances of the unit cell structure factors. These covariances can be expressed in terms of the structure factor variance and a correlation coefficient α_{mn}:

$$\alpha_{mn} = \frac{\langle (F_m - \langle F \rangle)(F_n - \langle F \rangle)^* \rangle}{\langle |F - \langle F \rangle|^2 \rangle}, \qquad (10.7)$$

which describes the correlation between the structure factors of unit cells related by the vector s_{mn}. If α_{mn} is assumed to be independent of $\mathbf{Q}$, the diffuse scattering term of eqn (10.4) may be written as

$$I_D(\mathbf{Q}) = \langle |F - \langle F \rangle|^2 \rangle \sum_s \alpha(s_{mn}) \exp(i\mathbf{Q} \cdot s_{mn}) = \epsilon(\mathbf{q}) \langle |F - \langle F \rangle|^2 \rangle, \qquad (10.8)$$

where

$$\epsilon(\mathbf{q}) = \sum_s \alpha(s_{mn}) \exp(i\mathbf{q} \cdot s_{mn}) \qquad (10.9)$$

and

$$\mathbf{q} = \mathbf{h} - \mathbf{Q}. \qquad (10.10)$$

Equation (10.8) shows that $I_D(\mathbf{Q})$ is the product of two factors. The first, $\epsilon(\mathbf{q})$, is a function of the wave vector $\mathbf{q}$ and depends on the correlation of atomic displacements *between* unit cells. It determines how the diffuse scattering is associated with the reciprocal lattice, along rows, planes or in diffuse peaks. The second term, $|F - \langle F \rangle|^2$, is the structure factor variance and depends on the atomic displacements *within* a unit cell.

The correlation of a unit cell with itself implies that $\alpha(0) = 1$ and if displacements in different unit cells are uncorrelated, we have

$$\epsilon(\mathbf{q}) = \alpha(0) = 1, \qquad (10.11)$$

which leads to

$$I_D(\mathbf{Q}) = \langle |F - \langle F \rangle|^2 \rangle. \qquad (10.12)$$

Equation (10.12) describes quite generally the diffuse scattering from a crystal where the disorder in one unit cell is independent of that in any other over the time scale of the diffraction experiment. Although this equation has been the basis of a number of diffuse scattering studies, the assumption of independent unit cells is never strictly justified. Thermally induced ultrasonic waves will ensure that the correlation between cells $\alpha(s)$ will be positive and decreasing as s

increases. The Fourier transform $\epsilon(q)$ of $\alpha(s)$ is then a function that peaks at $q = 0$. In soft molecular crystals, where $\alpha(s)$ falls off more quickly as s increases due to weak intermolecular forces, $\epsilon(q)$ has a broad peak which produces broad acoustic peaks at the Bragg positions.

Static disorder may also significantly contribute to $\epsilon(q)$. If alternate unit cells have the same structure but neighbouring unit cells in the *a* direction are different, $\alpha(2na) = 1$ and $\alpha([2n+1]a) < 1$, where a is the cell dimension and n is any integer. In this case $\epsilon(q)$ peaks midway between the Bragg reflections and gives rise to diffuse peaks or planes of intensity in these positions.

10.3.2 *Diffuse scattering and the molecular Fourier transform*

The general variation of TDS as a function of Q may be plotted for the Einstein crystal (a crystal where all atomic displacements are independent of each other). Figure 10.4 shows plots of

$$I(Q)_{TDS} = (\exp(Q^2 U) - 1) \sum_j g_j^2 \qquad (10.13)$$

for a carbon atom for three values of U. It will be noted that the TDS peaks between 4 and 5 Å resolution, followed by a gradual fall as Q increases. Comparison of this trend with what is actually observed in a mercury derivative of γB-crystallin (Fig. 10.5) shows that the diffuse scattering does rise to a broad peak in the resolution range predicted. However, the diffuse scattering falls more quickly with increasing Q than the TDS predicted from the Einstein crystal.

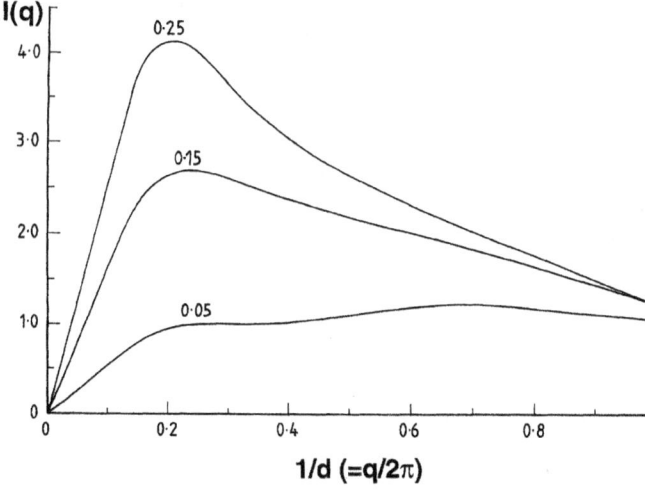

Fig. 10.4. Diffuse scattering curves due to uncorrelated isotropic overall mean-square atomic displacements of 0.05, 0.15 and 0.25 Å². The intensities $I(Q)$ are on an arbitrary scale and are plotted against the reciprocal of the resolution d. [Taken from Glover *et al.* (1991).]

190 *Diffuse X-ray scattering from molecular crystals*

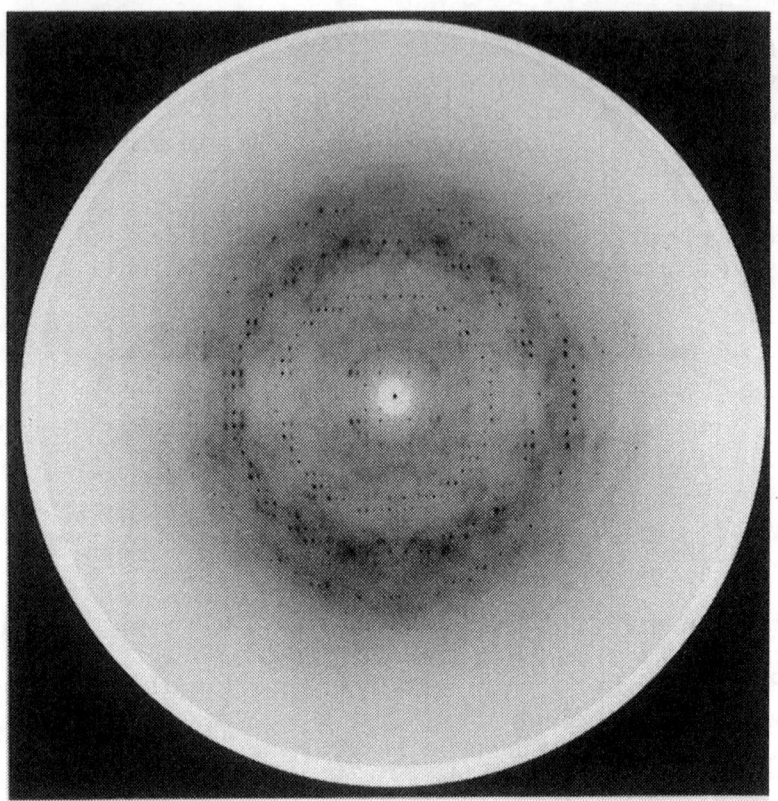

Fig. 10.5. Oscillation X-ray image of the ethyl-mercury-thiosalicylate (EMTS) derivative of γB-crystallin recorded on film with synchrotron radiation at Daresbury Laboratory. A prominent broad ring of diffuse scattering occurs between 4 and 6 Å resolution.

Clearly, a protein is a poor approximation to an Einstein crystal and correlated motion, in addition to static and diffusive disorder, is likely to play a significant role in determining the details of the diffuse scattering.

Hoppe (1960a,b) was able to derive a chemical model for $C_{20}H_{20}O_3$ (Biflorin) from X-ray diffuse scattering and projections of the weighted reciprocal lattice. With the use of X-ray diffuse scattering in conjunction with other methods, Hoppe and Rauch (1960) were able to derive a structural model for tetrachlor-1-keto-naphthalene-dihydride.

10.3.3 *Random orientational disorder*

In some molecular crystals, molecules can take up several discrete orientations. This type of orientational disorder gives rise to diffuse scattering. Interaction between molecules may give rise to short-range orientational correlations.

Moore and Lang (1973) have explained the diffuse X-ray scattering from the diamond-like adamantane molecules in terms of independent contributions from various types of molecular motion, such as the random translational vibrations, librations and random orientational disorder, the latter being the most important (Moore 1973, 1985). The molecules were considered to be rigid bodies and the contributions from finite librations could be described in terms of rigid body molecular motions. Computer generated contour maps of diffuse X-ray scattering intensity in reciprocal space were plotted.

Reynolds has focused attention on the high-temperature cubic phases (space groups $Fn3n$ and $F\bar{4}3n$) of adamantane (Reynolds 1978; Reynolds and Markey 1979). Unlike the Bragg X-ray data (Nordman and Schmitkons 1965), the X-ray diffuse scattering can be used to distinguish between the orientationally ordered ($F\bar{4}3n$) and the orientationally disordered ($Fn3n$) phases of adamantane (Reynolds and Markey 1979). The diffuse scattering of the ordered $F\bar{4}3n$ phase shows short-range correlation effects interpretable as pairs of neighbouring molecules trying to adopt local configurations characteristic of the low-temperature phase ($P\bar{4}2_1c$) of adamantane, i.e. a breakdown of symmetry (Reynolds 1978).

Flack (1970a–c) studied the X-ray diffuse scattering of anthrone ($C_{14}H_{10}O$) crystals, which possess long-range disorder, with a statical distribution of anthrone molecules among two centrosymmetrically related orientations. Measurement of the integral breadth of the diffuse scattering maxima gave an average short-range order domain length of nine molecules (36 Å) (Flack 1970b,c). In mixed crystals, up to 12 mol% of anthraquinone ($C_{14}H_8O_2$) can crystallize with anthrone without disturbing this short-range order (Flack 1970c). Similar disordered diffuse X-ray scattering effects due to short-range order have been observed for N-oxyphenazine (Glazer 1970), 9-bromo-10-methylanthracene (Prat 1961; Jones and Welberry 1980) and lead-n-butylxanthate (Hagihara et al. 1968).

10.3.4 Experimental considerations

Much attention has been focused on how to correct Bragg X-ray data for TDS not removed by normal background subtraction methods (see Section 10.2.1). However, when recording diffuse X-ray scattering data (at general points in reciprocal space), a difficult problem is encountered, as one is essentially measuring the background. It is not possible to make a normal background correction, as for Bragg data because one is measuring background diffuse scattering. Other sources of scattering, e.g. fluorescent, Compton (Walker 1956) and air scattering (Lonsdale 1942) also contribute to the general background intensity. There may also be extraneous sources of scattering, e.g. from the walls of the glass tube. For protein crystals, there will also be water scattering (see Section 10.4). These other sources of background scattering must be taken into account. Welberry (1983) has proposed an empirical background correction for scattering, which is a function of scattering angle only (Welberry and Glazer 1985).

Jagodzinski has pointed out many of the experimental problems associated with the measurement of diffuse scattering (Boysen et al. 1984). The advent of the synchrotron has provided an intense source of monochromatic radiation, which helps to eliminate many of the difficulties earlier researchers experienced.

The question also arises of the relative merits of film and diffractometer methods of data collection. For diffuse scattering, the differences between the two methods are more obvious than in the case of measuring only the Bragg scattering. With film techniques, a large part of reciprocal space is surveyed on one film, whereas with the diffractometer, measurements are made at those points in reciprocal space determined by the operator (Welberry and Glazer 1985).

The majority of X-ray diffuse scattering studies from molecular crystals have employed photographic methods. Annaka and Amorós (1960) made some counter diffractometer measurements on anthracene. There have been a few cases where a diffractometer has been used to measure whole reciprocal layer sections (Singh and Glazer 1981; Stadnicka and Glazer 1984). Welberry and Glazer (1985) have made a systematic study of diffuse X-ray scattering from a disordered molecular crystal (of 1,4-dibromo-2,5-diethyl-3,6-dimethylbenzene) measured by diffractometer and conventional Weissenberg film techniques. The data sets were compared. The interexperimental agreement factor between the two sets of intensities was 22%. The time required and the resolution obtained with the diffractometer were comparable to conventional film techniques and the routine measurement of diffuse scattering with the recording of the Bragg data on the diffractometer has been advocated.

10.4 Diffuse scattering studies of proteins

The size and complexity of protein molecules complicates the analysis of their diffraction pattern. As disorder may arise from many sources in protein crystals, e.g. multiple sites, side-chain mobility, disordered solvent, all possible sources of diffuse scattering must be taken into account. Diffuse scattering is potentially a valuable source of information regarding correlated motion within protein crystals and this could, for example, be related to the mode of function of an enzyme.

Glover et al. (1991) have commented on the diffuse scattering from crystals of avian pancreatic polypeptide (aPP), ribonuclease-A (RNAse), 6-phosphogluonate dehydrogenase, glutamine dehydrogenase, transferrin and t-RNA-(Met). Both aPP and RNAse give Bragg reflections to high resolution (< 1.2 Å) and hence show only weak diffuse scattering. The other proteins diffract to lower resolutions and show strong diffuse scattering, particularly in the resolution range 3–4 Å, often showing an anisotropic pattern. The t-RNA diffraction pattern shows very prominent streaking at higher resolution, indicating highly directional correlated displacements. These authors also showed how, in the case of γII-crystallin, there was a drastic increase of diffuse scattering when crystals were soaked in 10 mM thiomersal.

Temperature-dependent TDS has been observed for some protein crystals, e.g. actin-DNAse, porin, but not for others, e.g. prothrombin (Bartunik and Schubert 1982). Such studies could provide information on protein dynamics.

Boylan and Phillips (1986) have studied the X-ray diffuse scattering from tropomyosin crystals at 15 Å resolution. The observed scattering could be accounted for by simulating random displacements of the filaments from their average positions coupled with the displacements in neighbouring unit cells and calculating the diffraction pattern.

Doucet and Benoît (1987) have investigated the X-ray diffuse scattering from lysozyme crystals using synchrotron radiation. By considering correlated displacements of groups of sterically coupled molecules in planes, a simulated pattern was produced which reproduced the observed diffraction pattern. Doucet et al. (1989) applied similar techniques to account for the diffuse streaks in the oscillation photographs of a single crystal of an A-DNA octamer. These streaks were due to the presence of occluded B-DNA octamers within the channels of the A-DNA crystalline matrix. Doucet et al. (1992) have discussed the significance of earlier protein diffuse scattering studies.

More general approaches to the interpretation of diffuse scattering must start with less restrictive assumptions about the correlated displacements that may be present in the crystal. Such approaches are akin to the determination of temperature factors from Bragg data where only the harmonic approximation is usually assumed. Faure et al. (1994) calculated the diffuse scattering from eqn (10.12) using molecular dynamics simulation of a single molecule of lysozyme in order to simulate the diffuse pattern of the orthorhombic crystals. The averages were performed over the coordinate frames obtained from the simulation. Good qualitative agreement was obtained. These workers further used normal mode analysis and calculated the diffuse scattering independently for each mode. The latter method gave poorer agreement at higher Bragg angles, possibly due to the neglect of localized harmonic displacements in this approach.

Mizuguchi et al. (1994) have used normal mode refinement to calculate diffuse scattering simulations of X-ray films of human lysozyme from an image using the geometry of the Weissenberg camera. Their software implements a number of models of correlation. They were able to employ an isotropic V tensor, a generalization of the mean-square displacement tensor U which occurs in the crystallographic 'temperature factor', and could assume an exponential decay of the correlations as in the work of Caspar and co-workers. They concluded that higher order scattering is important at higher Bragg angles.

Caspar et al. (1988) have investigated the X-ray diffuse scattering from insulin crystals. They digitally separated the diffraction into Bragg reflections and associated haloes which they call variational scattering. By simulating the diffuse diffraction pattern they concluded that movements of atoms in the insulin molecules have rms amplitudes of about 0.4 Å and appear to be coupled over a range of about 6 Å.

This work was extended to tetragonal and triclinic lysozyme crystals (Caspar and Badger 1991; Clarage et al. 1992) where couplings also of about 6 Å were estimated from simulated diffraction patterns. It should be noted that although this equation does not take into account lattice correlations [see eqn (10.8)], these workers used their expressions to simulate the haloes around the Bragg reflections. However, further work on the diffuse scattering of tetragonal lysozyme was carried out by Pérez et al. (1996), who concluded that rigid-body motions were more important than the homogeneous disorder in accounting for the diffuse scattering. Kolatkar et al. (1992, 1994) have adapted Caspar's approach to the interpretation of the diffuse scattering of yeast-initiator t-RNA. Their model employed anisotropic lattice-coupled motions as well as short-range correlated disorder in the anti-codon arm to account for the observed pattern.

Significant diffuse X-ray scattering is observed in the oscillation photographs of crystals of glycogen phosphorylase *b* recorded using synchrotron radiation (Wilson *et al.* 1983). The diffuse scattering is thought to be due to correlated vibrations of the molecules in a plane perpendicular to a crystal lattice consistent with the packing of the molecules in the crystal lattice. The diffuse X-ray scattering observed from crystals of δ-crystallin (Fig. 10.1) also exhibits features due to molecular packing. The diffraction pattern shows an inner ring of peaks with 14-fold symmetry. This is due to the presence of supramolecular helices in the crystal that have seven-fold symmetry.

For myoglobin crystals, the overall value of the lattice disorder $\langle x^2 \rangle_{ld}$ was estimated as 0.0045 Å^2 by comparing the mean-square displacement $\langle x^2 \rangle$ values obtained for the iron atom by X-ray diffraction and by Mössbauer absorption (Hartmann *et al.* 1982). The latter separates the elastic scattering from the inelastic TDS (see Section 10.2). Subtraction of $\langle x^2 \rangle$ for the elastic scattering from $\langle x^2 \rangle$ for the total X-ray scattering measurement gives the $\langle x^2 \rangle_{ld}$ value for inelastic scattering due to static lattice disorder (Parak and Formanek 1971). The total diffracted X-ray intensity for myoglobin has also been calculated assuming uncorrelated motion (Kuriyan *et al.* 1986).

Mössbauer radiation can be used to distinguish between elastic and inelastic scattering. Nienhaus *et al.* (1989) have used it to study the molecular dynamics of a myoglobin crystal. The inelastic scattering due to molecular motions occurs near the Bragg reflections.

The background-corrected integrated Bragg (elastic) intensity is given by

$$I_{el}(Q) = |F_0(hkl)|^2 \exp(-Q^2(\langle x_{SC}^2 \rangle + \langle x_{LC}^2 \rangle)). \qquad (10.14)$$

The background-corrected integrated inelastic intensity is given by

$$I_{inel}(Q) = (1 - \exp(-Q^2 \langle x_{LC}^2 \rangle))|F_0(hkl)|^2 \exp(-Q^2 \langle x_{SC}^2 \rangle), \qquad (10.15)$$

where $\langle x_{SC}^2 \rangle$ is the mean-square displacement for short-range correlation and $\langle x_{LC}^2 \rangle$ is the displacement for long-range correlations.

At room temperature the root-mean-squared amplitude for intermolecular motions for myoglobin was found to be 0.13 Å for inelastic scattering. The above studies did not take into account the contribution to the X-ray diffuse scattering pattern of the disordered solvent (water). The X-ray diffraction pattern of water has been studied extensively (Morgan and Warren 1938). North and Smith (1985) have pointed out the importance of considering the solvent contribution to the diffraction pattern in accurate protein crystallographic structure determinations. The Bragg reflections are affected by the solvent contribution predominantly in the low-angle region. Various methods have been used to allow for the solvent contribution by modelling the solvent with a continuous electron density (Fraser *et al.* 1978; Phillips 1980; Langridge *et al.* 1960) and applied to

DNA (Langridge et al. 1960) and oxymyoglobin (Phillips 1980). Considering the solvent scattering is particularly important when studying diffuse scattering, as the disordered solvent contributes to the observed diffuse scattering pattern.

References

Annaka, S and Amorós, JL (1960). *Journal of the Physical Society of Japan*, **15**, 356.
Bartunik, HD and Schubert, P (1982). *Journal of Applied Crystallography*, **15**, 227.
Boylan, D and Phillips Jr, GN (1986). *Biophysical Journal*, **49**, 76.
Boysen, H, Frey, F and Jagodzinski, H (1984). *The Rigaku Journal*, **1**, 3.
Caspar, DLD and Badger, J (1991). *Current Opinion in Structural Biology*, **1**, 877.
Caspar, DLD, Clarage, J, Salunke, DM and Clarage, M (1988). *Nature*, **332**, 659.
Clarage, JB, Clarage, MS, Phillips, WC, Sweet, RM and Caspar, DLD (1992). *Proteins*, **12**, 145.
Cochran, W (1969). *Acta Crystallographica*, **A25**, 95.
Criado, A, Conde, A and Márquez, R (1985). *Acta Crystallographica*, **A41**, 158.
Doucet, J and Benoît, JP (1987). *Nature*, **325**, 643.
Doucet, J, Benoît, J-P, Cruse, WBT, Prange, T and Kennard, O (1989). *Nature*, **337**, 190.
Doucet, J, Benoît, J-P, Faure, P and Durand, D (1992). *Journal de Physique (Paris) I*, **2**, 981.
Edwards, C, Palmer, SB, Emsley, P, Helliwell, JR, Glover, ID, Harris, GW and Moss, DS (1990). *Acta Crystallographica*, **A46**, 315.
Faure, P, Micu, A, Pérahia, D, Doucet, J, Smith, JC and Benoît, JP (1994). *Structural Biology*, **1**, 124.
Flack, HD (1970a). *Philosophical Transactions of the Royal Society London*, **A266**, 561.
Flack, HD (1970b). *Philosophical Transactions of the Royal Society London*, **A266**, 575.
Flack, HD (1970c). *Philosophical Transactions of the Royal Society London*, **A266**, 583.
Fraser, RDB, MacRae, TP and Suzuki, E (1978). *Journal of Applied Crystallography*, **11**, 693.
Glazer, AM (1970). *Philosophical Transactions of the Royal Society London*, **A266**, 635.
Glover, ID, Harris, GW, Helliwell, JR and Moss, DS (1991). *Acta Crystallographica*, **B47**, 960.
Hagihara, H, Watanabe, Y and Yamashita, S (1968). *Acta Crystallographica*, **B24**, 960.
Hartmann, H, Parak, F, Steigemann, W, Petsko, GA, Ponzi, DR and Frauenfelder, H (1982). *Proceedings of the National Academy of Sciences USA*, **79**, 4967.
Helmholdt, RB and Vos, A (1977). *Acta Crystallographica*, **A33**, 38.
Hoppe, W (1960a). *Zeitschrift für Kristallographie*, **114**, 156.
Hoppe, W (1960b). *Zeitschrift für Kristallographie*, **114**, 393.
Hoppe, W and Rauch, R (1960). *Zeitschrift für Kristallographie*, **115**, 141.
Jones, RDG and Welberry, TR (1980). *Acta Crystallographica*, **B36**, 852.
Kolatkar, AR, Clarage, JB and Phillips Jr, GN (1992). *The Rigaku Journal*, **9**, 4.
Kolatkar, AR, Clarage, JB and Phillips Jr, GN (1994). *Acta Crystallographica*, **D50**, 210.
Kroon, PA and Vos, A (1979). *Acta Crystallographica*, **A35**, 675.
Kuriyan, J, Petsko, GA, Levy, RM and Karplus, M (1986). *Journal of Molecular Biology*, **190**, 227.
Langridge, R, Marvin, DA, Seeds, WE, Wilson, HR, Hooper, CW, Wilkins, MHF and Hamilton, LD (1960). *Journal of Molecular Biology*, **2**, 38.
Lonsdale, K (1942). *Reports on Progress in Physics*, **9**, 256.
Mizuguchi, K, Kidera, A and Go, N (1994). *Proteins*, **18**, 34.

Moore, M (1973). *Dissolution, defects and disorder in diamond and diamond-like substances*, Ph.D. thesis, University of Bristol.
Moore, M (1985). *Contributions to the diffuse intensity away from Bragg reflections*, BCA Meeting, Bristol.
Moore, M and Lang, AR (1973). *Diffuse X-ray scattering and molecular motion in adamantane*. 1st European Crystallographic Meeting, Bordeaux.
Morgan, J and Warren, BE (1938). *Journal of Chemical Physics*, **6**, 666.
Nienhaus, GU, Heinzl, J, Huenges, E and Parak, F (1989). *Nature*, **338**, 665.
Nordman, CE and Schmitkons, DL (1965). *Acta Crystallographica*, **18**, 764.
North, ACT and Smith, J (1985). *International Journal of Biological Macromolecules*, **7**, 223.
Parak, F and Formanek, H (1971). *Acta Crystallographica*, **A27**, 573.
Pawley, GS (1969). *Acta Crystallographica*, **A25**, 702.
Pérez, J, Fuare, P and Benoît, J-P (1996). *Acta Crystallographica*, **D52**, 722.
Phillips, SEV (1980). *Journal of Molecular Biology*, **142**, 531.
Prat, M-T (1961). *Acta Crystallographica*, **14**, 110.
Reynolds, PA (1978). *Acta Crystallographica*, **A34**, 242.
Reynolds, PA and Markey, BR (1979). *Acta Crystallographica*, **A35**, 627.
Singh, S and Glazer, AM (1981). *Acta Crystallographica*, **A37**, 804.
Sommer, R, Schulz, H and Kress, W (1981). *Acta Crystallographica*, **A37**, 219.
Stadnicka, K and Glazer, AM (1984). *Acta Crystallographica*, **B40**, 139.
Walker, CB (1956). *Physical Review*, **103**, 558.
Welberry, TR (1983). *Journal of Applied Crystallography*, **16**, 192.
Welberry, TR and Glazer, AM (1985). *Acta Crystallographica*, **A41**, 394.
Wilson, KS, Stura, EA, Wild, DL, Todd, RJ, Stuart, DI, Babu, YS, Jenkins, JA, Fourme, R, Kahn, R, Gadet, A, Bartels, KS and Bartunik, HD (1983). *Journal of Applied Crystallography*, **16**, 28.

Solutions and partially ordered systems

11

Small-angle X-ray scattering by solutions of biological macromolecules

Patrice Vachette and Dmitri Svergun

11.1 Introduction

This chapter on the application of Small-angle X-ray scattering (SAXS) to biological macromolecules can be read independently from other contributions in the series of volumes of *Hercules lectures* but it does not ignore them. The general considerations on the phenomenon of light scattering by matter are kept to a minimum and focused on the case of particles in solution. The reader is referred to the contributions by Geissler (1994) and Williams (1993) for further developments presented from a different standpoint. The section on data analysis and the formalism of spherical harmonics can be complemented by reading the contribution by Stuhrmann (1994). The question of interactions between particles is barely touched upon since it has been dealt with by Tardieu (1994). Finally, this chapter is obviously not the place for an exhaustive presentation of the theory of SAXS. Beyond the other contributions already mentioned, the interested reader is referred to the textbooks listed in the bibliography.

SAXS is a technique that provides us with structural information on inhomogeneities of electron density with characteristic dimensions between about ten and a few thousand angstrom. The applications cover all kinds of fields: from metal alloys to synthetic polymers (Geissler 1994), from colloids to micelles and micro-emulsions (Levelut 1994), from porous media (Williams 1993) to liquid crystals (Levelut 1994) and finally solutions of biological objects—proteins or viruses for instance.

We are interested in distances that are large compared to interatomic distances. Therefore, we will describe the scattering objects using a continuous function of electron density determined by the chemical composition of the object. If there are no strong variations of electron density, we will use the approximation of the mean (e.g. a protein in solution). In the case of more complex particles, several levels of electron density can be introduced (e.g. nucleoproteins or lipoproteins).

The scattering objects can be considered as isolated within the solution (ideal solution) or in interaction, thereby displaying spatial correlations that contribute to the scattering intensity. X-ray scattering patterns contain not only information on the size and shape of particles but also information about their interactions within the solution.

After a quick reminder of some basic notions on the phenomenon of X-ray scattering, the main body of the chapter will be focused on the case of particles in an ideal solution before briefly dealing with solutions of interacting particles.

11.2 X-ray scattering by matter

If we consider a sample placed in an X-ray beam, only the electrons will scatter X-rays because of their very small mass. In what follows, the electrons are considered to be free; this is a valid approximation as long as the energy of the photons is different from the binding energy of the electrons. We consider only elastic scattering, i.e. scattering without any change in the wavelength, or, in other words, without any exchange of energy between the photon and the electron.

Let $\rho(r)$ be the electron density of the sample at point r, and s_0 the characteristic vector of the incident beam ($|s_0| = 1/\lambda$, where λ is the wavelength of the X-rays). At distances much greater than the size of the sample, the expression for the amplitude of the scattered radiation is

$$F(s) = \int_{V_r} \rho(r) e^{-2i\pi rs}\, dV_r, \qquad (11.1a)$$

where $s = s_1 - s_0$ and V_r is the volume of the sample; s_1 is the characteristic vector of the scattered wave (see Fig. 11.1) and s is called the scattering vector. $F(s)$ is the *Fourier transform* of the electron density distribution $\rho(r)$:

$$s = |s| = \frac{2 \sin \theta}{\lambda} \cong \frac{2\theta}{\lambda}$$

is the modulus of the scattering vector.

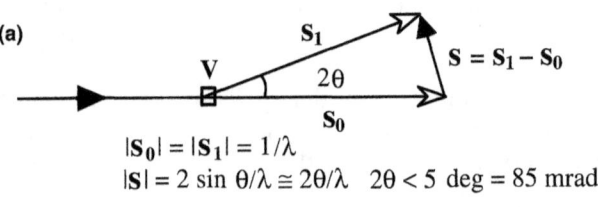

$|s_0| = |s_1| = 1/\lambda$
$|s| = 2 \sin \theta/\lambda \cong 2\theta/\lambda \quad 2\theta < 5 \text{ deg} = 85 \text{ mrad}$

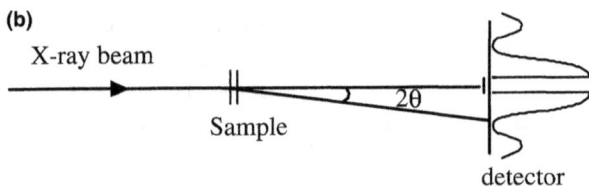

Fig. 11.1. (a) Vectorial representation of scattering by a volume V; (b) Schematic representation of a scattering experiment.

The last approximation holds true for small values of θ. In solution scattering experiments, 2θ is generally $< 5°$ (85 mrad) for $\lambda = 1.5$ Å.

Note that for $s = 0$,

$$F(0) = \int_{V_r} \rho(r) \, dV_r = m, \tag{11.1b}$$

where m is the number of electrons in the volume V_r.

The scattered intensity is

$$I(s) = F(s) \cdot F^*(s) = \int_{V_r} \int_{V_{r'}} \rho(r)\rho(r') e^{-2i\pi(r-r')s} \, dV_r \, dV_{r'}, \tag{11.2}$$

where $F^*(s)$ is the complex conjugate of $F(s)$.

Let $V_r \gamma(\mathbf{R})$ be the autocorrelation function of the electron density $\rho(r)$ of the sample (or the particle if this is the case) (Fig. 11.2(a)). Then

$$\gamma(\mathbf{R}) = \frac{1}{V_r} \int_{V_r} \rho(r)\rho(r + \mathbf{R}) \, dV_r. \tag{11.3}$$

From (11.2) and (11.3), it follows that

$$I(s) = \int_{V_R} V_R \gamma(\mathbf{R}) e^{-2i\pi \mathbf{R} \cdot s} \, dV_R \quad \text{with } \mathbf{R} = r - r'. \tag{11.4}$$

In the case of *particles in suspension in a solvent*, the scattering originates from the *contrast* of electron density between the particle and the homogeneous solvent, the latter having a constant electron density ρ_0:

$$\Delta\rho(r) = \rho(r) - \rho_0 \quad \text{and} \quad F(s) = \int_{V_r} \Delta\rho(r) e^{-2i\pi r \cdot s} \, dV_r.$$

Solutions of biological macromolecules, essentially composed of light atoms (H, C, N, O and P in the case of nucleic acids) in water, display little contrast of electron density and therefore weak scattering power. For instance, a 1 mg/ml solution of a 40 kDa protein scatters of the order of 1 photon per 100 000 incident photons. This immediately explains the advantage offered by an X-ray source with a very high flux, such as those provided by synchrotron radiation from storage rings.

Furthermore, in the case of scattering by *solutions*, the sample is *isotropic*: the particles in Brownian motion take all possible orientations with respect to the direction of the incident beam. Only the *spherical average of the intensity is*

(a) $\quad V\gamma(\mathbf{R}) = \int_{V_r} \rho(\mathbf{r})\, \rho(\mathbf{r}+\mathbf{R})\, dV_r \qquad -\langle\ \rangle$ spherical average

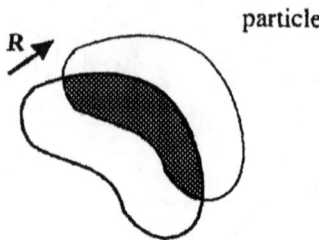
particle ∩ ghost

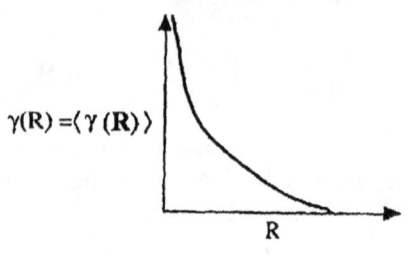
$\gamma(R) = \langle \gamma(\mathbf{R}) \rangle$

(b) $\quad p(R) = R^2 V\gamma(R)$

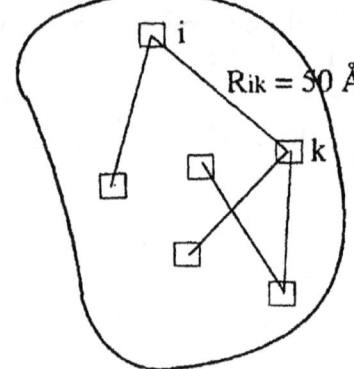

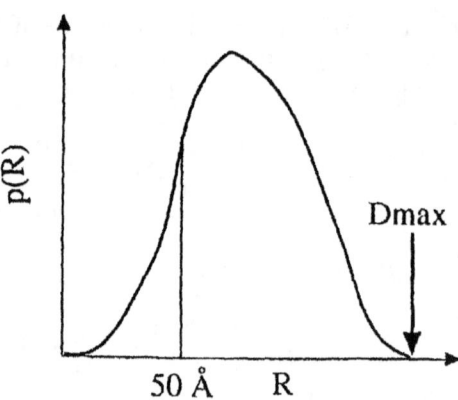

Fig. 11.2. (a) Autocorrelation function and its spherical average; (b) distance distribution function.

experimentally accessible:

$$I(s) = \langle I(\mathbf{s}) \rangle_\Omega,$$

where $\langle\ \rangle_\Omega$ signifies a spherical average. Ω is the solid angle in reciprocal space. After averaging over all orientations with respect to the beam, expressions (11.3) and (11.4) become

$$I(s) = 4\pi \int_0^\infty p(R) \frac{\sin 2\pi Rs}{2\pi Rs}\, dR \qquad (11.5a)$$

and

$$p(R) = R^2 V\gamma(R) = \frac{1}{\pi} \int_0^\infty Rs I(s) \sin 2\pi Rs\, ds \qquad (11.5b)$$

with

$$\gamma(R) = \langle \gamma(\mathbf{R}) \rangle_\Omega. \tag{11.5c}$$

Here $p(\mathbf{R})$ is the *pair distribution function* or *distance distribution function*. For homogeneous particles ($\rho(\mathbf{R}) = $ const), $p(\mathbf{R})$ is the histogram of distances between all pairs of points (volume elements) of the sample (Fig. 11.2(b)). $\gamma(\mathbf{R})$, called the *characteristic function* of the particle, represents the probability of finding a point within the particle at a distance R from a given point. Integrating this probability over the surface of the sphere of radius R and over the volume V yields the distance distribution function, as expressed by the eqn (11.5b).

11.3 Scattering by an ideal monodisperse solution

Definitions

Ideality is a thermodynamic notion that characterizes a solution of particles without interactions, in which case thermodynamic quantities show a linear (ideal) dependence on concentration. It is approximated experimentally by using as dilute a solution as possible (i.e. compatible with the measurement of a statistically meaningful scattered intensity).

In a *monodisperse* solution, all particles are identical. This holds true for particles with a defined shape, such as a native protein or a virus. In the case of a solution of a given stretch of DNA or of a denatured protein, the solution is chemically monodisperse (one molecular species), but polydisperse in shape (the molecules adopt many different conformations; see Section 11.3.3). Polydispersity in size can be observed between particles having the same shape, as in the case of a micelle suspension, while polydispersity both in size and shape are present in the case of aggregation.

In what follows, the solution under investigation is assumed to be ideal and monodisperse.

The solution is monodisperse. Therefore, there is only one kind of particle in solution with an associated electron density distribution $\rho(r)$ and a scattered intensity $i_1(s)$. The solution is ideal. All particles scatter independently and the resulting intensity scattered by the sample is simply the sum of all the contributions from individual molecules. The experimental scattering intensity $I(s)$ is related to $i_1(s)$ by the relationship

$$I(s) = Ni_1(s), \tag{11.6}$$

where N is the number of particles in the sample.

11.3.1 Global structural parameters

Intensity at the origin

The scattered intensity at $s = 0$ is given by

$$i_1(0) = \int\int_V \Delta\rho(r)\Delta\rho(r')\,dV_r\,dV_{r'} = \Delta m^2 = m^2 - m_0^2, \tag{11.7a}$$

where m is the number of electrons of the particle and m_0 that of the volume of water displaced by the molecule. Then from eqn (11.6),

$$I(0) = Ni_1(0) = Nm^2(1 - \rho_0\psi)^2, \qquad (11.7b)$$

where ψ is the electronic partial specific volume of the particle. Generally one uses the concentration in mass per volume (g/l) $c = N\mu m/N_A$, where N_A is Avogadro's number and μ the ratio M/m of the molecular weight to the number of electrons, which depends on the chemical composition of the particle (for proteins a good approximation is $M/m = 1.87$). It follows that

$$I(0)/c = \frac{N_A M}{\mu^2}(1 - \rho_0\psi)^2. \qquad (11.7c)$$

The intensity at zero angle scaled by the particle concentration is proportional to the molecular weight M. If ψ, ρ_0 and c are known, and if the intensity of the incident beam is known on an absolute scale (number of incident photons during the measurement), then the intensity at the origin provides a determination of the molecular weight.

Radius of gyration: the Guinier relation

The scattered intensity can be expanded in powers of s^2. Close to the origin, the expansion can be restricted to the first terms

$$I(s) = I(0)\left[1 - \frac{4\pi^2}{3}R_g^2 s^2 + ks^4 + \cdots\right], \qquad (11.8a)$$

$$I(s) \cong I(0)\exp\left(-\frac{4\pi^2}{3}R_g^2 s^2\right). \qquad (11.8b)$$

In the vicinity of the origin, the scattering pattern of any isolated particle can thus be approximated by a Gaussian, the width of which is proportional to the square of the radius of gyration of the particle: this is known as the Guinier relation (Guinier and Fournet 1955). The radius of gyration is defined by

$$R_g^2 = \frac{\int_{V_r} \Delta\rho(r) r^2 \, dV_r}{\int_{V_r} \Delta\rho(r) \, dV_r}, \qquad (11.9)$$

where the origin of the vectors is taken to be the centre of mass of the particle. The radius of gyration of an object measures its degree of (non)sphericity. For a given volume, the smallest radius of gyration is that of a sphere,

$$R_g = \sqrt{\frac{3}{5}}R, \qquad (11.10)$$

where R is the radius of the sphere. In practice, a linearized representation will be used, by plotting $\ln[I(s)]$ vs s^2 (Guinier plot). Simple linear regression yields the radius of gyration from the slope, together with the intensity at the origin. In the case of spherical particles, the Guinier approximation is valid up to $sR_g = 0.1$. With real proteins, large variations in the domain of validity can be observed, according to the value of the s^4 term in the expansion (11.8a), which depends on the precise shape of the protein, as illustrated in Fig. 11.3. The slopes (hence the radii of gyration) of the two curves labelled ATCase (*E. coli* aspartate transcarbamylase, see Section 11.7.1) and S1(+LC2) (subfragment S1 of the myosin head with the attached light chain 2), respectively, are the same, while the former is linear over a much wider range than the latter.[1] It must be kept in mind that, beyond recording data in the appropriate angular range, the Guinier approximation relies on the assumption of an ideal monodisperse solution.

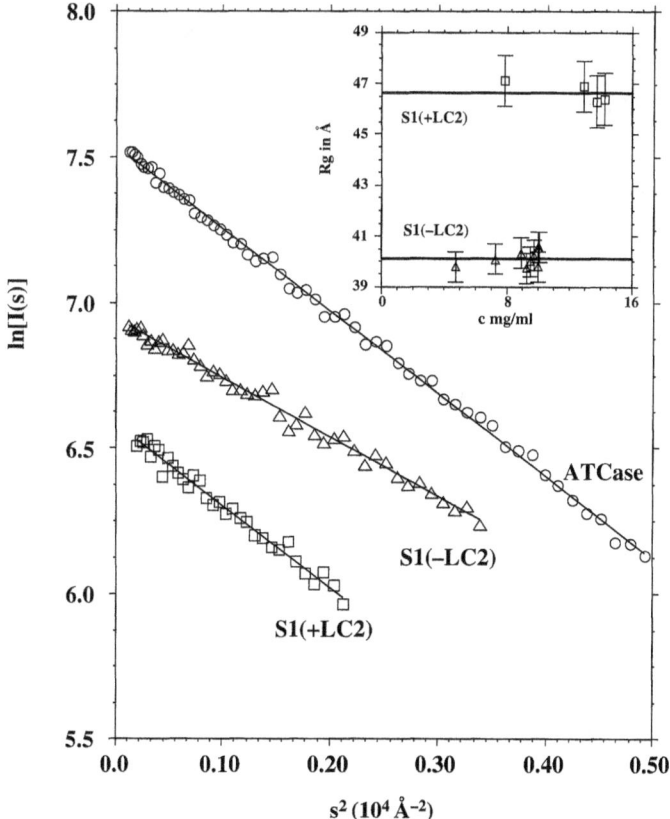

Fig. 11.3. Guinier plots. Uncorrected experimental data after background subtraction. Inset: radii of gyration vs protein concentration. [Reprinted from Garrigos *et al.* (1992); © Biophysical Society.]

---

[1] For a systematic discussion of the range of validity of the Guinier law, see Feigin and Svergun (1987).

If several species are present in solution (e.g. oligomers due to protein aggregation), their scattering adds up, and in most cases, the value of R_g of the monomer cannot be unambiguously retrieved. Besides, the determination of the radius of gyration will be erroneous if the interactions between proteins are not negligible. To keep them to a minimum, experiments will be performed at several concentrations, which will be as low as possible, before extrapolating to zero concentration (see inset in Fig. 11.3).

Rods and platelets

In the case of very elongated particles, the radius of gyration of the cross-section R_c can be derived, using a similar representation, plotting this time $sI(s)$ vs s^2,

$$sI(s) \propto \exp(-2\pi^2 R_c^2 s^2). \tag{11.11a}$$

Finally, in the case of a platelet, a thickness parameter is derived from a plot of $s^2 I(s)$ vs s^2:

$$s^2 I(s) \propto \exp(-4\pi^2 R_t^2 s^2), \quad R_t = T/\sqrt{12} \text{ (thickness } T\text{)}. \tag{11.11b}$$

For more details the reader is referred to Chapter 3 of Feigin and Svergun (1987).

Porod invariant

The autocorrelation function is given by eqn (11.5b). Its value at $R = 0$ is

$$V\gamma(0) = 2 \int_0^\infty s^2 I(s)\, ds = Q \tag{11.12a}$$

while, from the definition of $\gamma(R)$ [eqn (11.4)],

$$\gamma(0) = \frac{1}{V} \int_{V_r} \Delta\rho(r)\Delta\rho(r)\, dV_r = \overline{\Delta\rho^2}. \tag{11.12b}$$

Hence

$$Q = V\overline{\Delta\rho^2}. \tag{11.13}$$

The integral (11.12a) is an invariant, called the Porod invariant (Porod 1952), since its value does not depend on the structure of the particle (in fact, the result is also true for non-particulate systems) and is equal to the mean square of the electron density contrast of the sample.

Hydrated volume

The intensity $i_1(0)$ scattered at the origin by a particle can be written as

$$i_1(0) = (V\overline{\Delta\rho})^2, \tag{11.14}$$

where V is the volume of the hydrated particle. If the particle has a homogeneous density, then $(\overline{\Delta\rho})^2 = \overline{\Delta\rho^2}$ and therefore

$$V = \sqrt{\frac{i_1(0)}{\gamma(0)}} = \frac{i_1(0)}{Q}. \tag{11.15}$$

The asymptotic regime

We assume that the particle has a uniform electron density distribution and a sharp interface with the solvent. Porod (1951) showed that in this case the asymptotic behaviour of the scattered intensity is given by

$$8\pi^3 \lim_{s \to \infty}[s^4 i_1(s)] = S\overline{\Delta\rho^2} + Bs^4, \tag{11.16}$$

where S is the area of the interface between the solute and the solvent, while B is a correction term that takes into account the existence of short distance density fluctuations as well as experimental uncertainties in the value of $i(s)$ at wide angles. The latter can be quite large, considering the smallness of the signal in this angular range. This fact often limits quite severely the usefulness of this relation in the study of proteins.

11.3.2 Data analysis

11.3.2.1 Data processing

An ideal single particle scattering intensity $I(s)$ cannot be measured directly in a solution scattering experiment. Instead, a discrete data set of intensities $I_{\exp}(s_i)$, $i = 1, \ldots, N$, is determined in a restricted angular range $s_{\min} < s < s_{\max}$. This set contains statistical errors and, generally speaking, is smeared because of the beam divergence, polychromaticity, detector resolution, etc. (Feigin and Svergun 1987, Section 9.1). The main task of the data processing is to restore the ideal intensity $I(s)$ from the experimental set $I_{\exp}(s_i)$. For monodisperse systems, $I(s)$ is related to the pair distribution function of the particle $p(r)$ by the Fourier transform

$$I(s) = 4\pi \int_0^{D_{\max}} p(r) \frac{\sin 2\pi sr}{2\pi sr} dr, \tag{11.17}$$

where $D_{\max}$ is the maximum diameter of the particle. This is eqn (11.5a), but the upper bound of the integral is now $D_{\max}$. The function $p(r)$ contains the same information as $I(s)$, and the data processing can be done 'indirectly' by restoring the $p(r)$. This 'indirect transform' approach, first introduced by Glatter (1977), is usually superior to other data processing techniques as it introduces an important constraint, namely a boundedness of the characteristic function. The latter is parametrized on $[0, D_{max}]$ by a linear combination of orthogonal functions $\varphi_k(r)$,

$$p(r) = \sum_{k=1}^{K} c_k \varphi_k(r). \tag{11.18}$$

Substituting (11.18) into (11.17), one obtains

$$I_{exp}(s) = \sum_{k=1}^{K} c_k \psi_k(s), \tag{11.19}$$

where $\psi_k(s)$ are the Fourier transformed and smeared functions $\varphi_k(r)$. The coefficients c_k are determined by minimizing the functional

$$\Phi_\alpha = \sum_{i=1}^{N} \left[I_{exp}(s_i) - \sum_{k=1}^{K} c_k \psi_k(s_i) \right]^2 + \alpha \int_0^{D_{max}} [p'(r)]^2 \, dr, \tag{11.20}$$

where the regularizing multiplier $\alpha \geq 0$ allows compensation between the quality of fit to the data (first sum) and the smoothness of the $p(r)$ function (second integrand). For a given α, eqn (11.20) is reduced to a system of linear equations with respect to the set $\{c_k\}$. The solution of the system provides the $p(r)$ and the function $I(s)$ is automatically extrapolated beyond the measured range of scattering vectors back to the origin using (11.17). In particular, the radius of gyration is derived through the relationship

$$R_g^2 = \frac{\int r^2 p(r) \, dr}{2 \int p(r) \, dr}. \tag{11.21}$$

This expression, making use of the whole scattering curve, is much less sensitive to factors such as the presence of residual interparticle interactions or even of a small amount of aggregates than the computationally more straightforward Guinier relation. Indeed, like the Guinier relation, the effects of interparticle interactions are mostly restricted to the innermost part of the scattering pattern (Tardieu 1994).

The main problem in using an indirect transform technique is to select properly the regularizing multiplier α. Too small values of α yield solutions unstable to the experimental errors; increasing α too much leads to systematic deviations from the experimental data. Theoretical methods of estimating the value of α usually work properly only for high quality data sets. Most of the indirect transform packages (e.g. the original ITP program of Glatter) often leave the choice to the user based on a visual comparison of the solutions obtained for different α values.

The program GNOM (Svergun et al. 1988; Svergun 1992) performs the 'visual' search automatically. The solution is characterized by a set of perceptual

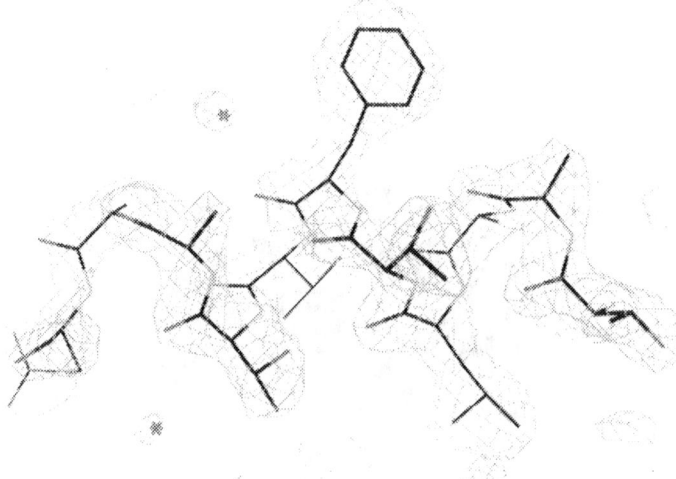

Fig. 4.2. A 2.5 Å $2F_o - F_c$ electron density map from a bit of helix A from elongation factor EF-Tu:GDPNP from *Thermus aquaticus* (Kjeldgaard et al. 1993) (see p. 84).

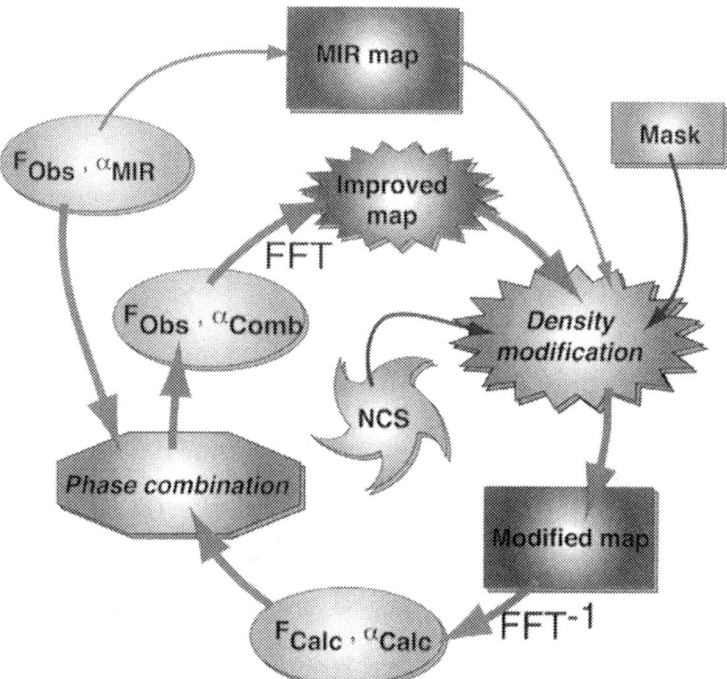

Fig. 4.3. The density modification cycle. Density modification is first carried out on the MIR map, and phases are computed from the modified map. These are combined with the original MIR phases, and a new, improved map is calculated. This procedure is iterated 5–8 times, while watching the statistical indicators $\Sigma |F_o - F_c| / \Sigma |F_o|$ and $\alpha_{MIR} - \alpha_{Comb}$ (see p. 85).

Fig. 4.6. The surface of a molecular mask, around one molecule of elongation factor EF-Tu: GDPNP (see p. 90).

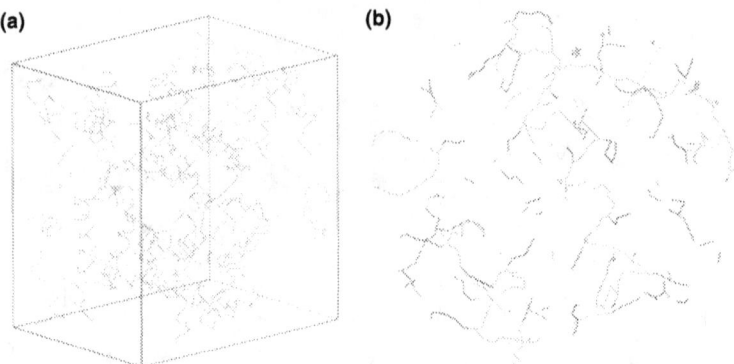

Fig. 4.9. Skeletons made from 1.9 Å MAD data on human psoriasin, an all-helical dimer of 200 residues (Brodersen *et al.* 1998). (a) The overview display, showing only bone atoms of level main-chain within a sphere of 50 Å. The outline of the unit cell is also displayed. In this view, the boundary between the individual molecules can easily be seen. (b) Detailed display, showing both main- and side-chain level bone atoms, within a 15 Å sphere. A helix can be identified with the axis pointing from the lower right to the upper left. The connection between the spots marked '*' and '#' is really side-chain erroneously assigned to main-chain, and the connection should be removed in the editing process. At '#', the main-chain actually continues up and to the right through the V-shaped side-chain bone (see p. 94).

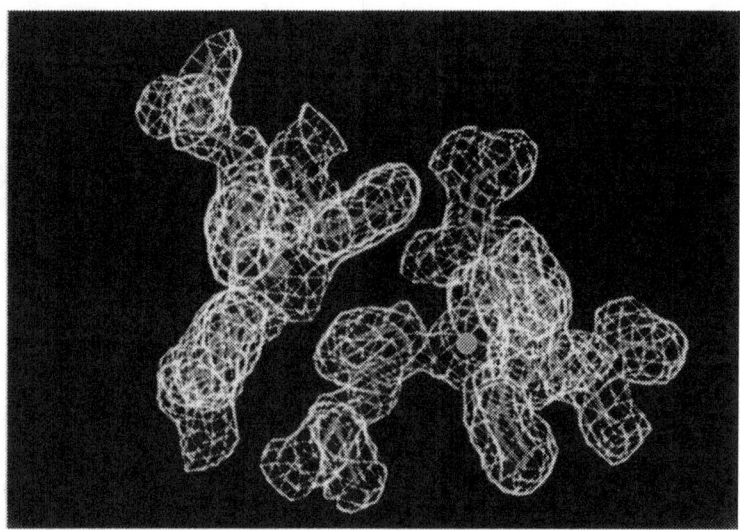

P212121 subunit A CuZn site

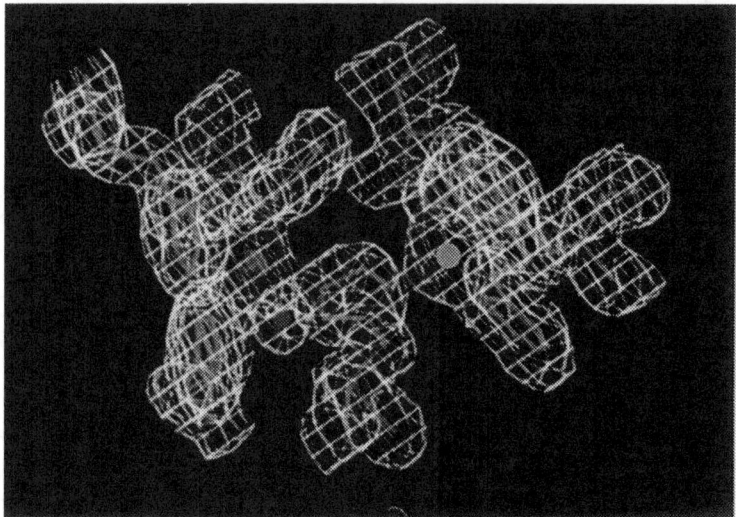

P212121 subunit B CuZn site

Fig. 7.3. Cu–Zn site of bovine SOD as revealed by 1.65 Å resolution crystallographic structure (see p. 134).

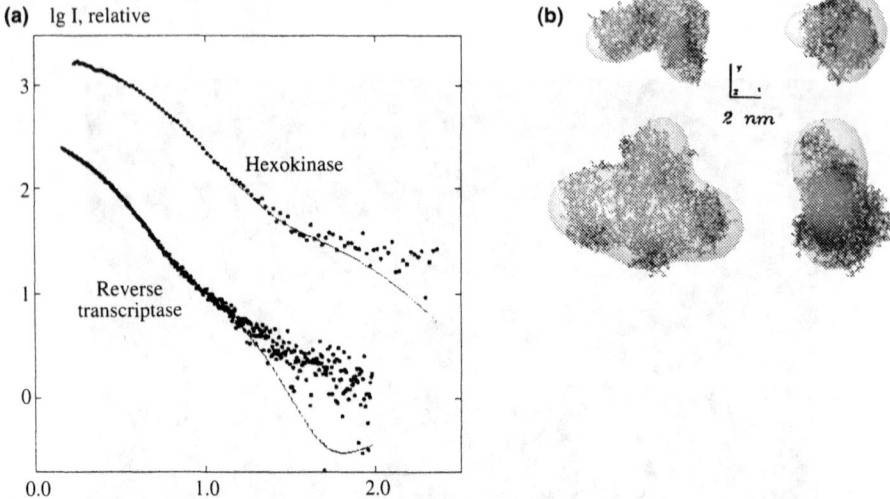

Fig. 11.5. Shape determination of the hexokinase and HIV-1 reverse transcriptase. (a) Experimental X-ray scattering data (dots) and the curves calculated from the restored envelopes (solid lines); (b) comparison between the envelopes (transparent solids) and the atomic structures in the crystal (dots). Top panel: hexokinase, bottom panel: reverse transcriptase. Right pictures are rotated 90° counterclockwise around Y (see p. 211).

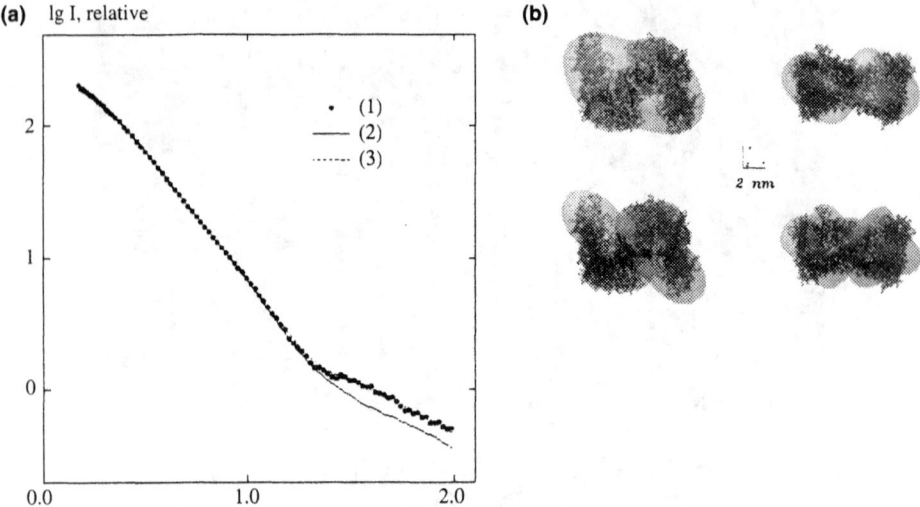

Fig. 11.6. Shape determination of the ribonucleotide reductase R1 protein. (a) Experimental X-ray scattering data (1), scattering from the single-envelope model (2) and from the two-envelope model (3). (b) Restored single-envelope model (top row) and two-envelope model (bottom row) merged with the atomic structure in the crystal (dots). Left orientation: twofold axis coincides with Z; right orientation is rotated 90° clockwise around X (see p. 212).

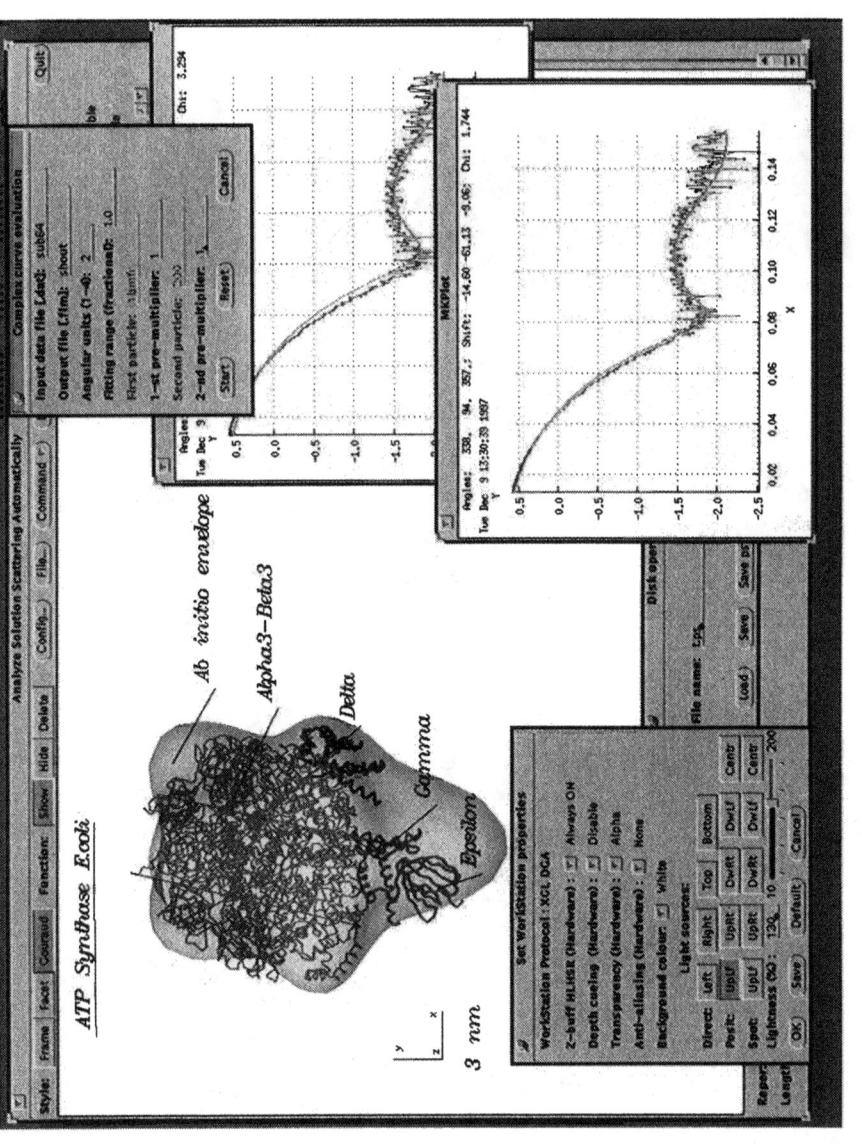

Fig. 11.7. Structure modelling of the ECF_1 ATP synthase. Semi-transparent envelope: low resolution model determined *ab initio* from the scattering data (threefold axis coincides with Y). The atomic structures of the domains are displayed as C_α traces (red: $\alpha_3\beta_3$ complex, magenta: γ subunit, green: ε subunit, blue: δ subunit). The domains are positioned interactively using ASSA and the two-dimensional plot in the bottom right corner shows the fit (blue curve) to the experimental data (red curve; the X-axis corresponds to $0.2\pi s$ in nm^{-1}) (see p. 216).

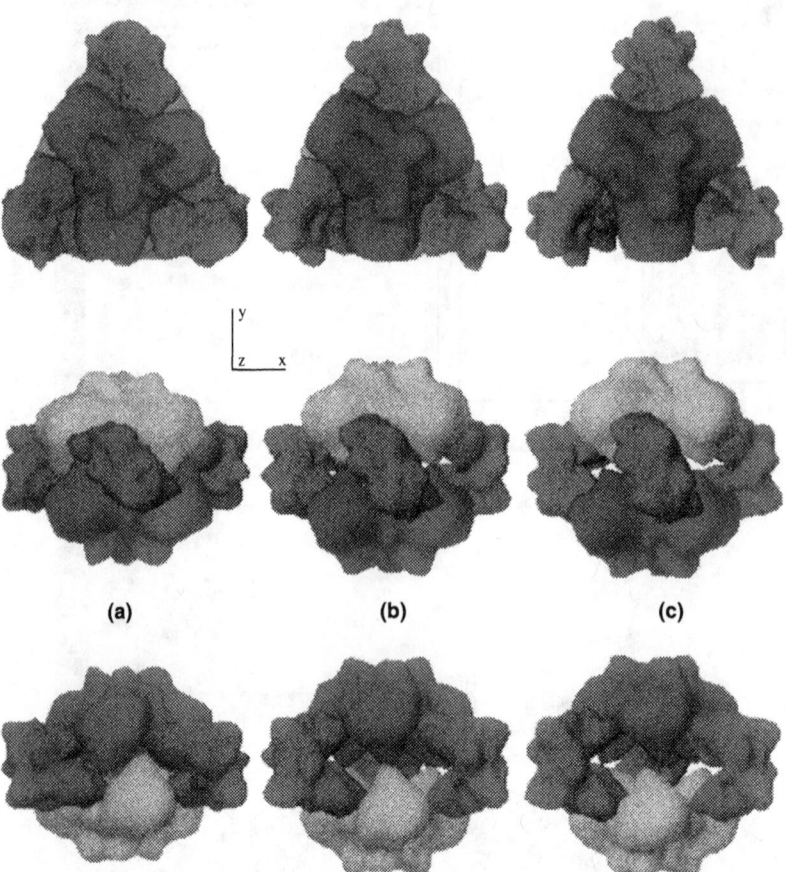

Fig. 11.14. Models of ATCase quaternary structures 'as seen' by X-rays at low resolution, since the envelope functions of the domains are represented using the spherical harmonics description for the scattering patterns. The two C trimers are coloured in yellow and cyan, the three R dimers in magenta. Top: View along the threefold axis (Z) from the side of the lower C trimer. Middle and bottom: View along one of the two fold axes (top view rotated along X-axis by 90° counterclockwise and clockwise, respectively). In each row, A and B are the crystal structures of the T and R states, respectively. C is the model of the R state in solution (see p. 228). [Reprinted with permission from Svergun *et al.* (1997a); © Wiley-Liss Inc.]

Fig. 13.11. Crystal packing of the two bR layers projected on the a–b plane (see p. 268).

Fig. 16.4. A zone plate made by electron beam lithography (see p. 343).

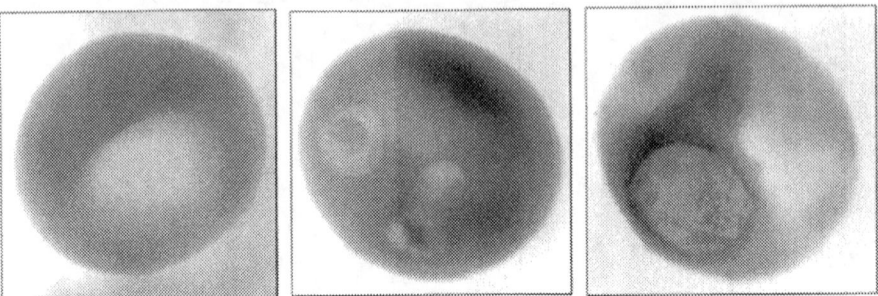

Fig. 16.5. Life cycle of malaria in human red blood cells. The images show from left to right an uninfected blood cell, a newly infected cell, and a 12-hour-old parasite. Image sizes are $7 \times 7\ \mu m$ (see p. 343).

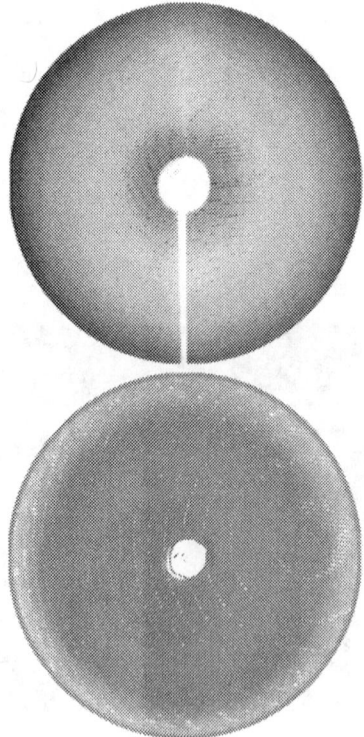

Fig. 18.1. *Top*: a 1° rotation diffraction pattern obtained from a crystal of treated small subunit from *Thermus thermophilus* (T30S), obtained at ID2/ESRF. *Bottom*: a 0.5° rotation diffraction pattern of a fresh crystal of H50S soaked in solution with 0.5 mM of W30, obtained in 20 s at the microfocus beamline (ID13) at ESRF (see p. 369).

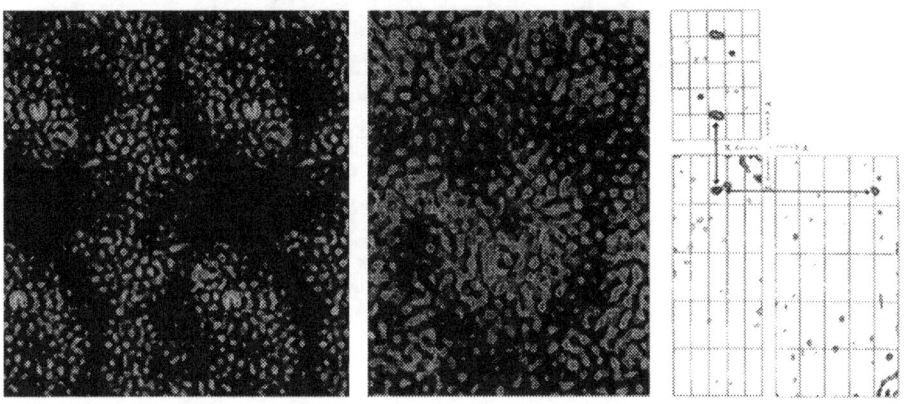

Fig. 18.2. *Left*: a part of the current 10–12 Å MIR map of H50S, showing the compact packing regions (around $z = 1/4$ and $3/4$) as well as the isolated contact area along the z-axis. For clarity, two unit cells are shown along the y-direction (horizontal). The dense areas represent the position of the most occupied heavy atom site (at the interface between two subunits). *Middle*: the Ta_6Br_{14} difference Patterson map of H50S, including the data of the 7.5–12.5 Å resolution shell. The corresponding Harker peaks are shown by arrows. *Right*: a part of the map oriented to show the entrance to the main ribosomal internal tunnel (Yonath et al. 1987). More than 11 000 reflections were measured, and a total of 15 heavy atom sites of the three derivatives (Ta_6Br_{14}, W12 and W17, Table 18.2) were included. The positioning of the heavy atom sites was performed by a combination of difference Patterson and Fourier methods, based on the major position of Ta_6Br_{14}, found to be stable and consistent in all resolution ranges until 7.5 Å. Each heavy atom position was cross-verified and refined by MLPHARE with maximum likelihood. Since the contribution of the two W clusters was negligible beyond 10 Å, their scattering curve could be approximated by spherical averages of their corresponding radii (W18: 10 Å and W12: 8–9 Å). The Ta_6Br_{14}, however, was treated as in Knäblein et al. (1997) owing to its potential contribution to the higher resolution shells. Mean figure of merit: 0.32 (0.57 for centric); R_{cullis}: 0.76–0.97; phasing power: 0.98–1.15. The map was solvent flattened: one cycle, assuming 54% solvent (see p. 372).

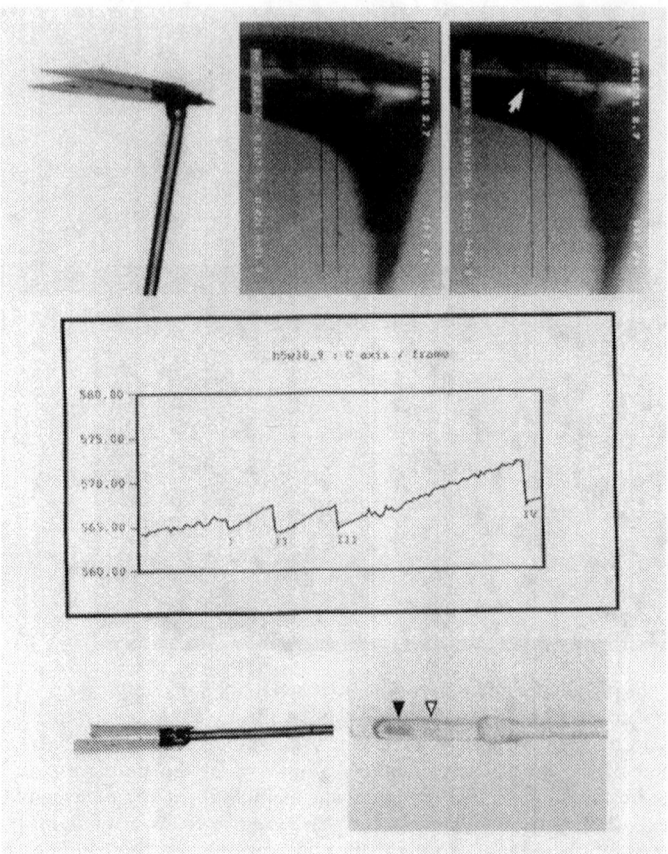

Fig. 18.3. *Top row* (left): a perpendicular double spatula used for mounting H50S crystals; (*Middle and Right*) two photographs of the irradiated crystal, at the beginning of the experiment (the square shows the area being exposed) and after the decay of this position. Note that the irradiated region became dark (indicated by the arrow). *Middle row*: the 'fluctuating' c-axis. Data were collected sequentially from a crystal of H50S, mounted as shown on the right. The crystal size was 400 × 380 × 8 μm and the beam cross-section 100 × 100 μm. The initial resolution was higher than 3.2 Å (at the edge of the MAR detector). The loss of resolution was monitored by visual inspection, and when it reached 6–7 Å (points I, II, III), the crystal was translated to a new position. At the position marked IV the resolution limit is 9.5 Å. The region exposed last suffered from the decay of its neighbour even before its own exposure. While evaluating the data it was found that the crystal decay was accompanied by an increase in the c-dimension, from 564 to 572 Å. *Bottom row* (left): a flat spatula, used for mounting T30S crystals; (right) a crystal of 300 × 50 × 30 μm, placed in a spatula similar to that shown on the left, was irradiated by a beam with a cross-section of 65 μm at ID19/APS. The first position was at the far end of the spatula (black arrow) and translated once decayed (total 4°, 20 rotations 0.2° each) to the middle of the spatula (white arrow). There it was exposed for 15 s. The intact crystal was transparent. Note that the intensities of the 'burns' of the crystals are proportional to the exposure time (see p. 373).

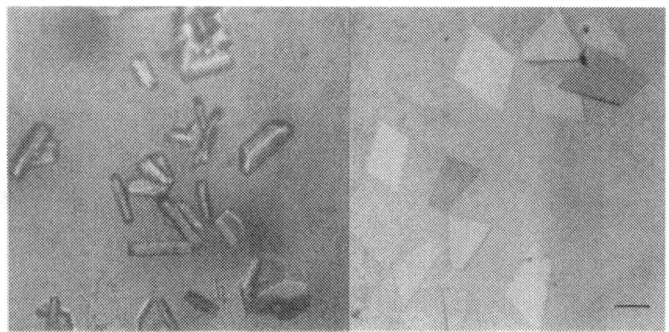

Fig. 18.4. *Left*: Crystals of large ribosomal subunits from strain H2 grown under conditions somewhat milder than those used for H50S (1.45 instead of 1.6 M KCl). *Right*: those of H50S. Bars: 0.2 mm (see p. 380).

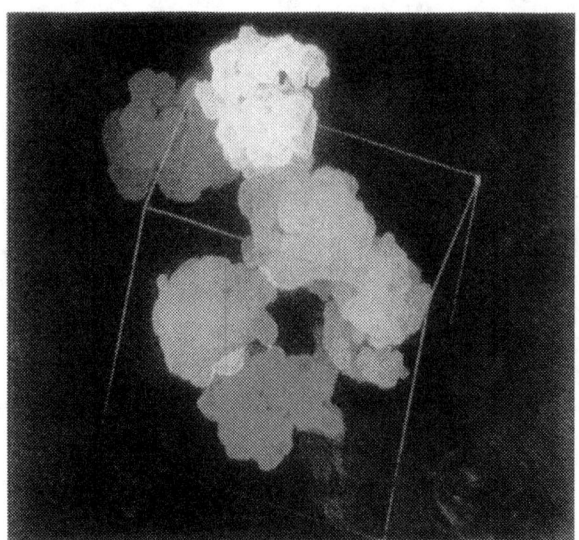

Fig. 18.5. The packing diagram of the crystallographic unit cell of whole ribosome (T70S), assembled by positioning the 26 Å electron microscope model in the crystallographic unit cell according to the most prominent result of the molecular replacement search, and applying the eight symmetry operations. Data were collected at BW6/DESY to 17 Å resolution. Note the similarity to (bottom) the model obtained by image reconstructions from negatively stained two dimensional sheets at 40 Å resolution (Arad *et al.* 1987) (see p. 381).

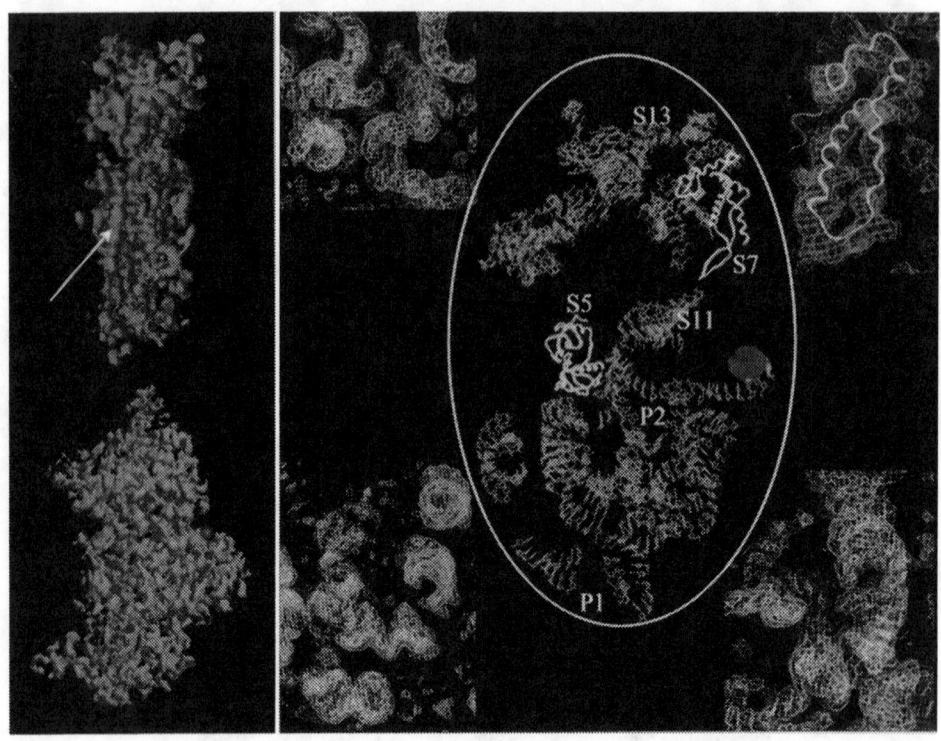

Fig. 18.6. *Left*: Two orthogonal views of the overall structure of the small subunit, as extracted from the 4.5 Å map. The arrow points to an exceptionally long dense region, suitable for hosting a double-helical RNA chain that may be interpreted as helix 44 in the model of the 16S RNA (Müller and Brimacombe 1997). For orientation, the location of the tail of the cDNA complementary to the 3' end of the 16S RNA is shown in red on both sides (here and on the *Right* side). *Right*: Part of the 16S RNA chain so far traced in the 4.5 Å map of T30S is shown within the white ellipse. The position of the centre of mass of the TAMM molecules that were bound to the mRNA analog is shown as a red sphere (artificially enlarged). The locations of proteins TS5, TS7 are represented by their backbone structure, as determined crystallographically. Two tentative locations for protein TS15 are marked P1 and P2. The positions of the exposed cysteines of proteins TS11 and TS13 are marked by their numbers (see p. 382).

criteria (discrepancy and systematic deviations in reciprocal space, smoothness, compactness and positiveness of the $p(r)$, stability to small changes in α). Their weighted average yields an estimate of the quality of the solution, and the value of α is found by maximizing the estimate. The 'perceptual' approach proved to be quite successful in practice: the program either finds the optimal solution automatically or indicates that the user's assumptions about the system (e.g. the value of D_{max}) are incorrect.

In most cases the maximum diameter is not known *a priori* with sufficient accuracy so that trial-and-error computations have to be made for different D_{max}. Svergun (1993) developed a 'direct indirect' method that does not require this value to be specified in advance. The $p(r)$ function is represented according to (11.18) by Hermite polynomials orthogonal on $[0, \infty]$. Starting from $K=1$, the number of polynomials is increased until the improvement of the fit to the experimental data is no longer statistically significant. Reduced coordinates in real and reciprocal space are used to ensure that the system of linear equations corresponding to problem (11.20) with $\alpha = 0$ is well posed, so that no regularization is required. The method implemented in the program ORTOGNOM yields fair estimates of the values R_g, D_{max}, and of the number of parameters K required to describe adequately the given data set. The function $p(r)$ can be further refined with GNOM [see example in Fig. 11.4(a) and (b)]. Moreover, both GNOM and ORTOGNOM can also be used for polydisperse systems to determine the size distribution functions.

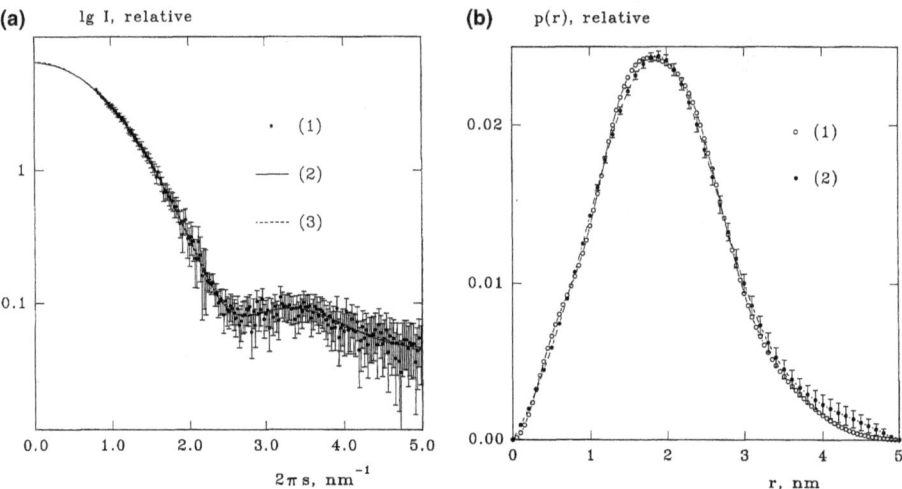

Fig. 11.4. Data processing using indirect transformation methods. (a) Experimental X-ray scattering data from the solution of lysozyme (1) and the curves restored by ORTOGNOM (2) and GNOM (3); (b) characteristic functions computed by ORTOGNOM using seven Hermite polynomials (1) and by GNOM at $D_{max} = 5$ nm (2). The radii of gyration are 1.48 nm and 1.52 ± 0.03 for the two methods, respectively. Note that the experimental data start at $2\pi s_{min} = 0.08$ nm^{-1}, i.e. beyond the range of validity for the Guinier approximation.

11.3.2.2 Shape determination

The information content in the scattering data is drastically reduced by random orientation of particles in solution. The Shannon sampling theorem (Moore 1980; Shannon and Weaver 1949; Taupin and Luzzati 1982) states that the number of parameters (Shannon channels) required to represent $I(s)$ in an interval $[s_{min}, s_{max}]$ is equal to $N_s = 2D_{max}(s_{max} - s_{min})$. Usually, this number does not exceed 10–15.

In keeping with the low resolution of the solution scattering studies, the data interpretation is usually performed in terms of homogeneous bodies. The standard trial-and-error approach involves the evaluation of the scattering patterns from different models and their comparison with the experimental data. The models are represented by hundreds or thousands of spheres and the number of parameters in the model N_p significantly exceeds N_s. An *ab initio* method presented below employs spherical harmonics to describe economically the particle shape, which makes possible its direct restoration from the scattering data.

The particle envelope is represented at low resolution by a two-dimensional angular function $r = F(\omega)$ describing its boundary in spherical coordinates (r, ω), which is conveniently parametrized as (Stuhrmann 1970b)

$$F(\omega) \approx F_L(\omega) = \sum_{l=0}^{L} \sum_{m=-l}^{l} f_{lm} Y_{lm}(\omega), \qquad (11.22)$$

where $Y_{lm}(\omega)$ are spherical harmonics, and the multipole coefficients f_{lm} are complex numbers. The truncation value L defines the number of parameters [for the general case, $N_p = (L+1)^2 - 6$] and the spatial resolution $\delta r \approx \sqrt{5\pi} R_g / [\sqrt{3}(L+1)]$. The density distribution in the homogeneous approximation can be written as

$$\rho(r) = \begin{cases} 1, & 0 \leq r < F(\omega) - \Delta, \\ [F(\omega) - r]/\Delta, & F(\omega) - \Delta < r \leq F(\omega), \\ 0, & r > F(\omega), \end{cases} \qquad (11.23)$$

where Δ is the width of the particle–solvent interface, which for dissolved macromolecules can be taken as $\Delta = 0.3$ nm to account for the first hydration shell. The scattering intensity is (Stuhrmann 1970a)

$$I(s) = 2\pi^2 \sum_{l=0}^{\infty} \sum_{m=-l}^{l} |A_{lm}(s)|^2, \qquad (11.24)$$

where the partial amplitudes $A_{lm}(s)$ are rapidly computed from the shape coefficients f_{lm} as described by Svergun and Stuhrmann (1991) and Svergun (1997). The coefficients f_{lm} are determined by a non-linear optimization procedure that minimizes the R factor between the calculated and the experimental curves:

$$R_I^2 = \frac{\sum_{j=1}^{N}\{W(s_j)[I(s_j) - I_{\exp}(s_j)]\}^2}{\sum_{j=1}^{N}[W(s_j)I_{\exp}(s_j)]^2}, \quad (11.25)$$

where the weighting function $W(s_j) = s_j^2/[\sigma(s_j)/I_{\exp}(s_j)]$, and $\sigma(s_j)$ denotes the standard deviation in the jth point. The minimization algorithm starts from a spherical initial approximation; additional penalties to keep the particle surface smooth and the envelope function positive-definite are described by Svergun et al. (1996, 1997b).

The *ab initio* shape determination program SASHA runs on IBM-PC and on major UNIX platforms (Svergun et al. 1997b). Its implementation on SUN Sparc-20ZX workstations is coupled with a three-dimensional rendering program ASSA allowing the user to monitor the refinement process (Kozin et al. 1997). Computer simulations on model bodies (Svergun et al. 1996) indicated that the low resolution shape determination for error-free data is unique, even when very limited ranges are used in the simulated curves. Unicity means that in the presence of errors, the shape restoration was found to be practically independent of the initial approximation and stable with respect to the random errors if $N_p \leq 1.5 N_s$.

The program was tested on X-ray scattering data from several proteins with known atomic structures in the crystal. Figure 11.5 presents the shape

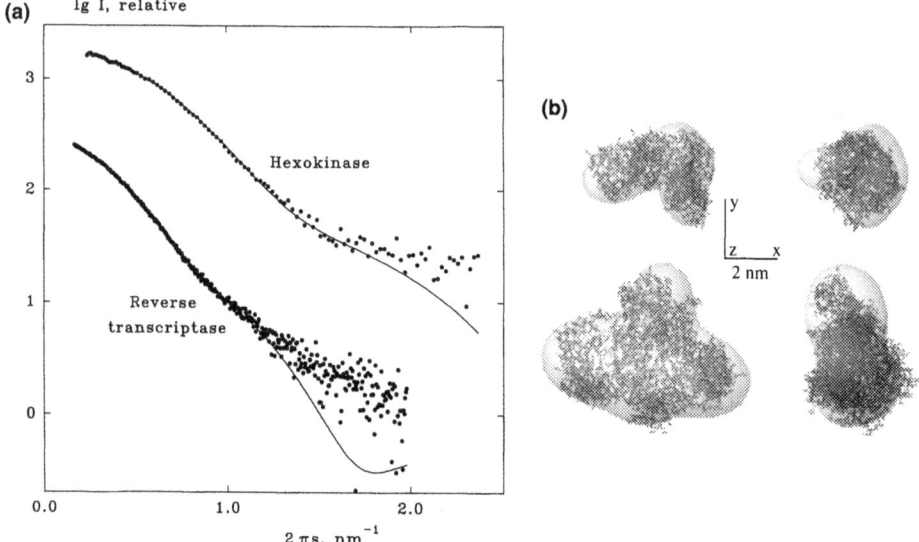

Fig. 11.5. Shape determination of the hexokinase and HIV-1 reverse transcriptase. (a) Experimental X-ray scattering data (dots) and the curves calculated from the restored envelopes (solid lines); (b) comparison between the envelopes (transparent solids) and the atomic structures in the crystal (dots). Top panel: hexokinase, bottom panel: reverse transcriptase. Right pictures are rotated 90° counterclockwise around Y (see Plate Section).

determination of two proteins, monomeric hexokinase and HIV-1 reverse transcriptase (molecular masses 52 and 105 kDa, respectively). In both cases, the value of $L=4$ was used (19 free parameters) yielding a spatial resolution $\delta r = 2.0$ and 3.0 nm, respectively. The restored envelopes are displayed along with the atomic structures of the hexokinase (Bennett and Steitz 1980), and of the reverse transcriptase (Wang et al. 1994) deposited in the Protein Data Bank (PDB) (Bernstein et al. 1977).

The reliability and/or resolution of the shape restoration can be enhanced if information about the particle symmetry is available. Consider, for example, an envelope function of a homodimer with a twofold symmetry axis. Assuming that the twofold axis coincides with Z, all f_{lm} coefficients with odd m vanish in the series (11.22), thus reducing the number of free parameters for a given L. Alternatively, following Svergun et al. (1997b), $F(\omega)$ can represent a shape of the monomer; the scattering amplitude from the symmetry related monomer is then generated automatically and the intensity from the dimer is calculated. This enhances the resolution by parametrizing the shape of a monomer rather than of a dimer at the expense of an additional free parameter (the distance between the centre of monomer and the twofold axis).

Figure 11.6 illustrates the shape determination of the R1 protein of ribonucleotide reductase, a homodimer of molecular mass 171 kDa. Symmetry restrictions were imposed as described above to restore the shapes of the entire dimer and of its monomeric part from the X-ray scattering data at $L=4$. As seen from the comparison with the structure of R1 in the crystal

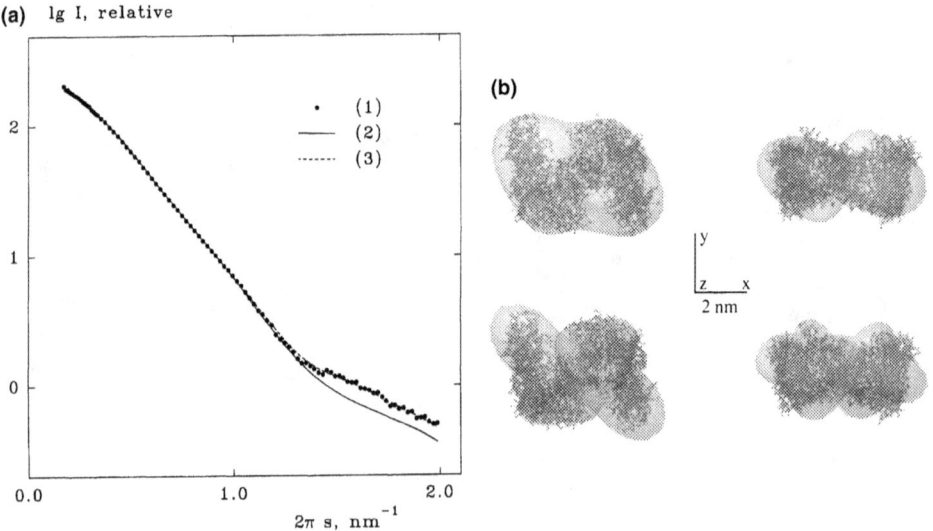

Fig. 11.6. Shape determination of the ribonucleotide reductase R1 protein. (a) Experimental X-ray scattering data (1), scattering from the single-envelope model (2) and from the two-envelope model (3). (b) Restored single-envelope model (top row) and two-envelope model (bottom row) merged with the atomic structure in the crystal (dots). Left orientation: twofold axis coincides with Z; right orientation is rotated 90° clockwise around X (see Plate Section).

(Uhlin and Eklund 1994), the single envelope representation (12 parameters, $\delta r = 3.4$ nm) yields a fair, low resolution description of the shape, whereas the parametrization of the monomer (20 parameters, $\delta r = 2.6$ nm) enhances the resolution and provides a better fit to the experimental data.

Of course, the method presented is aimed at determining the shape of proteins with unknown atomic structures, and it is being used successfully to this end (e.g. König et al. (1998); Macheroux et al. (1998); Schmidt et al. (1995)). The above tests were done to check its performance and limitations in a real experiment; the comparisons indicate that the *ab initio* shape determination yields an adequate description of the protein envelopes. One should, however, always bear in mind the limitations of this procedure. First, as $F(\omega)$ is assumed to be single-valued, complicated (e.g. U-like) shapes or those containing internal holes cannot be exactly represented. Second, the uniqueness of the restoration can be guaranteed only if the number of model parameters is comparable to that of the Shannon channels. This restricts the number of terms in series (11.22), and the omission of the higher harmonics with $l > L$ is compensated in the fitting procedure by an artificial enhancement of the lower ones. The deviations between the restored envelopes and the crystal structures in Figs 11.5 and 11.6 provide an idea of the magnitude of this effect.

11.3.2.3 Use of atomic models

Information about the atomic structure of the entire macromolecule or of its individual fragments in the crystal can significantly enhance the resolution of solution scattering studies. A necessary prerequisite is accurate evaluation of scattering patterns from the atomic models taking into account the influence of the solvent. Earlier methods (e.g. Lattman 1989; Ninio et al. 1972; Pavlov and Fedorov 1983) differently represented the particle volume inaccessible to the solvent, but did not account for hydration effects. It is now well established (e.g. Grossmann et al. 1993; Hubbard et al. 1988; Perkins et al. 1993) that the latter should be included to describe properly the experimental scattering patterns.

The program CRYSOL (Svergun et al. 1995) calculates the X-ray scattering curves from the atomic models taking into account the scattering from the solvation shell as follows. The macromolecule is surrounded by a hydration layer of thickness 0.3 nm with an adjustable density ρ_b which may differ from that of the bulk solvent ρ_s. The scattering from a particle in solution is

$$I(s) = \langle |A_a(s) - \rho_s A_s(s) + \delta\rho_b A_b(s)|^2 \rangle_\Omega, \quad (11.26)$$

where $A_a(s)$ is the scattering amplitude from the particle *in vacuo*, $A_s(s)$ and $A_b(s)$ are, respectively, the scattering amplitudes from the excluded volume and the hydration layer, both with unitary density, $\delta\rho_b = \rho_b - \rho_s$, and $\langle \rangle$ stands for the average over all particle orientations [Ω is the solid angle in reciprocal space, $s = (s, \Omega)$]. The multipole expansion of the scattering amplitudes is used to facilitate the spherical average in (11.26). Given the atomic coordinates, the program either fits the given experimental scattering curve using two free parameters, the excluded volume of the particle and the contrast of the hydration layer $\delta\rho_b$, or predicts the scattering pattern using the default values of these parameters.

The applications of CRYSOL to X-ray scattering data from proteins with known atomic structure (lysozyme, hexokinase, enolpyruvyltransferase, reverse transcriptase, aspartate transcarbamylase (see Section 11.7.1), ribonucleotide reductase, pyruvate decarboxylase, etc.) indicate that inclusion of the hydration shell significantly improves the agreement between the experimental and calculated X-ray scattering curves. The scattering density in the border layer is typically 1.05–1.25 times that of the bulk. Further, neutron scattering experiments in H_2O and D_2O solutions were performed for some of these proteins by Svergun et al. (1998a) and the data were analysed using the program CRYSON, a version of CRYSOL for neutron scattering. Taking advantage of the significantly different contrasts between the protein and the solvent for the X-rays and neutrons, it was demonstrated that the higher scattering density in the shell cannot be explained by disorder and mobility of the surface side chains in solution and is indeed due to a higher density of the bound solvent.

CRYSOL evaluates not only the scattering intensities but also the multipole components of the scattering amplitudes of the macromolecule or of its fragment(s). This opens the possibility of fast computation of the scattering patterns from assemblies of several macromolecules and thus for rigid body refinement of their structures against experimental solution scattering data. To illustrate this, let us consider a macromolecule consisting of two domains with known atomic structures. On fixing one domain (A) while moving and rotating the other (B), the scattering intensity of the particle is

$$I(s, \alpha, \beta, \gamma, \mathbf{u}) = I_a(s) + I_b(s) + 4\pi^2 \sum_{L=0}^{\infty} \sum_{m=-l}^{l} \mathrm{Re}[A_{lm}(s) C_{lm}^*(s)], \qquad (11.27)$$

where $I_a(s)$ and $I_b(s)$ are the scattering intensities from domains A and B, respectively, $A_{lm}(s)$ are partial amplitudes of the fixed domain A, $C_{lm}(s)$ are those of domain B rotated by the Euler angles α, β, γ and translated by a vector $\mathbf{u}$. The scattering amplitudes from both domains in reference positions are evaluated separately using CRYSOL. The algorithms of Svergun (1991), Svergun and Stuhrmann (1994) rapidly evaluate the amplitudes $C_{lm}(s)$ and thus the intensity $I(s, \alpha, \beta, \gamma, \mathbf{u})$ for arbitrary rotations and displacements of the second domain.

The program *ALM22INT* implementing algorithm (11.27) performs an automated six-dimensional search over the positional parameters to fit the experimental scattering curve from the complex. As visual criteria (interparticle contacts, position of rotation axes, etc.) often play an important role, this program was coupled with the graphics package *ASSA* for an interactive rigid body refinement. The fit to the experimental data is plotted in a separate window, and these plots can be kept on the screen to compare the results of different movements and rotations of the selected domains. Evaluation and display of a single fit for a current position of the domains takes about 1 s CPU time on a SUN SPARC-20 Workstation.

The above approach is illustrated by the quaternary structure determination of the extrinsic domain (ECF_1) of the *Escherichia coli* ATP Synthase in solution (Svergun *et al.* 1998b). ECF_1 is a 380 kDa complex consisting of five subunits in the stoichiometry $\alpha_3\beta_3\gamma\delta\varepsilon$ (Deckers-Hebestreit and Altendorf 1996). Its crystallographic structure at 0.28 nm resolution (Abrahams *et al.* 1994) reveals the $\alpha_3\beta_3$ subassembly as well as part of the γ subunit; the rest of the molecule remains unresolved, presumably due to disorder in the crystal. X-ray crystallography and NMR of individual subunits provided the atomic models of the ε subunit (Uhlin *et al.* 1997) and of the N-terminal domain of the δ subunit (Wilkens *et al.* 1996).

As the major part of the ECF_1 determined by Abrahams *et al.* (1994) possesses a quasi-threefold symmetry, the particle envelope was restored *ab initio* from the X-ray solution scattering data assuming a three-fold symmetry for the entire molecule at $L=5$ ($N_p=11$ free parameters, $\delta r=3$ nm). The crystal structure of the $\alpha_3\beta_3\gamma\delta$ subassembly is unequivocally placed in the low resolution model as illustrated in Fig. 11.7, and the protuberance along the threefold axis provides the space to accommodate the rest of the molecule. The atomic models of the δ and ε subunits were positioned with the help of the interactive modelling software in a way consistent with the results of the cross-linking (Capaldi *et al.* 1996) and fluorescence studies. The model in Fig. 11.7 yields the first direct evidence for the arrangement of the subunits in the native ECF_1 of ATP Synthase.

11.3.2.4 Singular value decomposition

Scattering curves can be considered as vectors with as many coordinates as data points. Any titration or kinetic study of a structural transition (see Section 11.6) thus leads to a two-dimensional data set, the two dimensions being the scattering angle and the variable parameter (titration) or time (kinetics). The minimum number of components (obtained as eigenvectors) required to account for the data within the experimental error in the least-squares sense is provided by singular value decomposition (SVD) analysis (Fowler *et al.* 1983; Provencher and Glöckner 1983), a method to determine the dimension of a set of vectors. The method has several advantages: (1) no hypothesis or approximation is involved in the analysis, which handles data independently of their physical nature (in our case scattering patterns); (2) it is particularly powerful in decomposing complex spectra where components share overlapping features; (3) since SVD is a linear operation, errors are easily taken into account and propagated to the resulting eigenvectors. It should be emphasized, however, that the eigenvectors do not represent the scattering curves of molecular species. Those can only be retrieved from the data when extra information is available. In the case of two components, where no intermediate species can be detected, the projection of each scattering curve upon each component is equivalent to the concentration of the corresponding species after proper scaling. The analysis thus directly yields the titration curve of the transition under study. Examples of applications of SVD are presented below (Section 11.7).

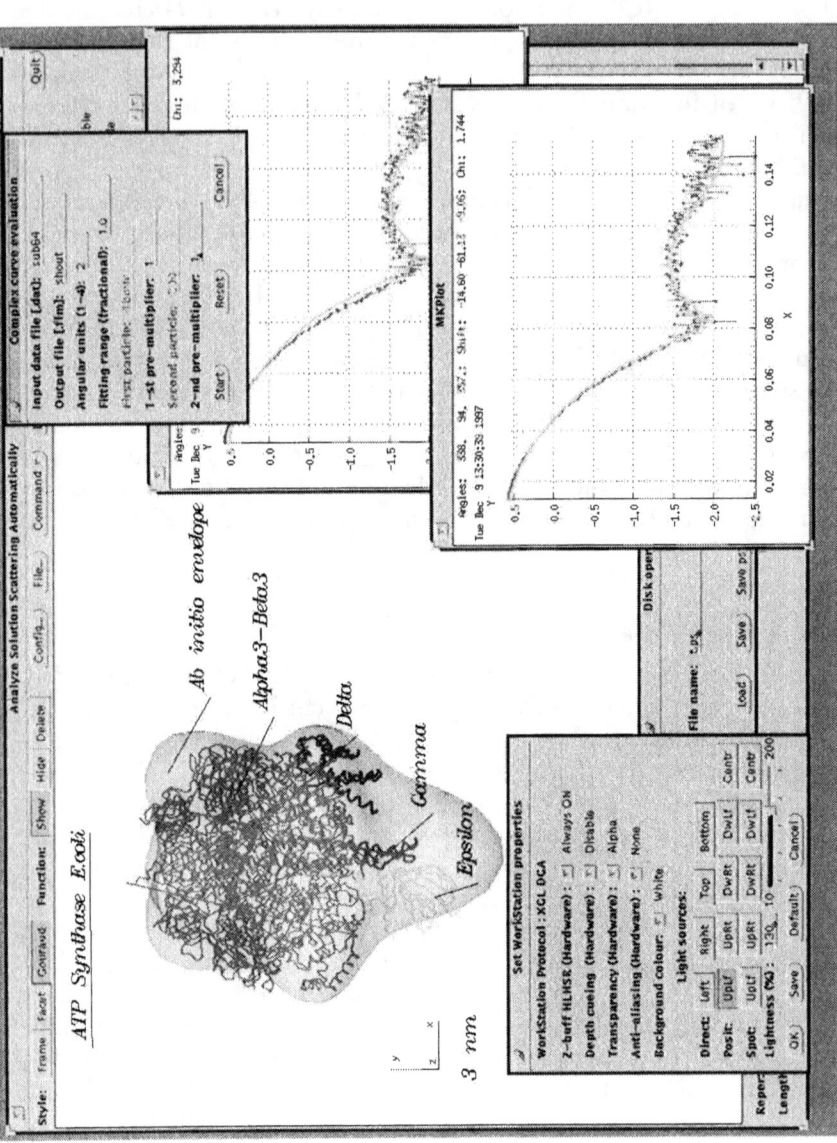

Fig. 11.7. Structure modelling of the ECF_1 ATP synthase. Semi-transparent envelope: low resolution model determined *ab initio* from the scattering data (threefold axis coincides with Y). The atomic structures of the domains are displayed as C_α traces (red: $\alpha_3\beta_3$ complex, magenta: γ subunit, green: ε subunit, blue: δ subunit). The domains are positioned interactively using ASSA and the two-dimensional plot in the bottom right corner shows the fit (blue curve) to the experimental data (red curve; the X-axis corresponds to $0.2\pi s$ in nm^{-1}) (see Plate Section).

11.3.3 Scattering by an extended chain

In the case of an unfolded protein, the description in terms of a compact, more or less globular particle is no longer valid, and models developed for polymers are better suited. We just briefly mention some of them with the corresponding expressions of SAXS intensity and their main features.

A Gaussian chain is described as the linear association of N monomers of length l with only short range interactions between adjacent units. Clearly, such an object will not adopt a single conformation in solution and its properties must be described in statistical terms. The distance between any pair (i,j) of units at a sufficient distance along the chain follows a Gaussian distribution. A particular case of a Gaussian chain is that of a free-linked chain where the angles between adjacent units exhibit no correlation. Debye (1947) established the expression for the scattering intensity of the Gaussian chain,

$$i(s)/i(0) = \frac{2(e^{-x} + x - 1)}{x^2}, \quad (11.28)$$

where $x = (2\pi s R_g)^2$. The scattering intensity depends on a single parameter, the radius of gyration R_g. The Guinier approximation still holds for a very expanded structure, but its range of validity is restricted to the immediate vicinity of the origin, and the Debye equation should be preferred to derive a reliable estimate of R_g (Petrescu et al. 1997). At large s, the scattering intensity has the following limit

$$\lim_{s \to \infty} [s^2 i(s)] = \frac{2[1 - 1/(s^2 R_g^2)]}{R_g^2}. \quad (11.29)$$

$I(s)$ varies as s^{-2} instead of the s^{-4} behaviour of the Porod law for globular objects. This marked difference in the asymptotic behaviour provides an easy and sensitive way of monitoring the degree of folding (the compactness) of a protein with a given parameter (temperature, pressure, pH, denaturing agent, etc.). It is most conveniently represented using the so-called Kratky plot of $s^2 i(s)$ vs s. The scattering curve of a globular particle will exhibit a distinctive bell-shaped curve with a maximum the position of which depends on the radius of gyration, while a Gaussian curve shows a plateau at larger s values.

The Gaussian chain model assumes that there are no intrachain interactions, which is the case in a so-called θ-solvent. However in a good solvent a polymer is fully solvated, which creates repulsive interactions between distant monomers along the chain, thereby preventing them from coming into contact. The polymer is said to have an excluded volume, since the solvation layer is excluded from the accessible configuration space. Finally, real polymers often display a local stiffness which Kratky and Porod proposed to describe using the worm-like chain model (Kratky and Porod 1949). The stiffness is characterized by the statistical length b, twice the persistence length, which is defined as the minimum length of a polymer segment the extremities of which are orientationally uncorrelated. Two regions of the scattering pattern can now be defined beyond the central domain (Calmettes et al. 1994):

(1) An intermediate region $3/(2\pi R_g) < s < 1.4/b$ where the scattering intensity is given by

$$I(s)/I(0) = Px^{-1/2\nu}. \qquad (11.30)$$

The pre-exponential factor P and the excluded volume exponent ν are constants whose values depend on the type of chain. Thus
- for a Gaussian chain $P = 2$ and $\nu = 0.5$;
- for a chain with excluded volume $P = 1.11$ (experimentally 1.3) and $\nu = 0.588$.

(2) A high s domain $s > 1.4/b$ where the local chain conformation is probed and where the cross-section of the chain must be taken into account via its radius of gyration R_c. The intensity scattered by a worm-like chain of contour length L without exclusion volume is then given by

$$\frac{I(s)}{I(0)} = \frac{1}{2\pi sL}\left(\pi + \frac{2}{3\pi sb}\right)e^{(-2\pi^2 R_c^2 s^2)}. \qquad (11.31)$$

This will give rise to an s^{-1} behaviour at large angles, hence to a linear increase in a Kratky plot until the exponential factor prevails. The case of a worm-like chain with excluded volume effect has recently been studied by Pedersen and Schurtenberger (1996).

In conclusion, the SAXS (or SANS) pattern contains a wealth of information about the denatured states of proteins, although the quality of experimental data may restrict the analysis.

11.4 Interactions between proteins in concentrated solutions

In the case of monodisperse solutions of spherical particles, the solution can be described as the convolution product of two functions: the distribution of the electron density contrast $\Delta\rho(r)$ and the distribution of the particles in the solution $d(r)$ (i.e. a set of delta functions positioned at the centre of each particle) (see Fig. 11.8). The Fourier transform of a convolution product is the product of the two Fourier transforms. Therefore, the intensity scattered by the concentrated solution is the product of the scattering curve of the particle in an ideal solution, called the form factor, times a function called the structure factor of the solution, which is due to the interactions between particles. Thus

$$I(c, s) = I(0, s) \cdot S(c, s). \qquad (11.32)$$

In the case of non-spherical but still globular particles or in the case of a slight polydispersity, eqn (11.32) still holds, though in a restricted s-range.

solution: described by a convolution

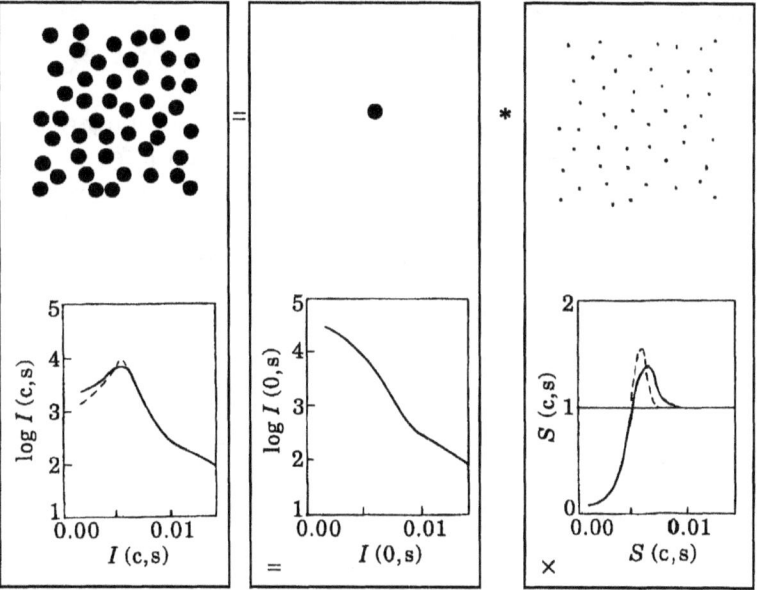

solution:		motive (protein)	*	lattice
$\Delta\rho\,(r)$	=	$\Delta\rho_p\,(r)$	*	$d\,(r)$
$F\,(c,s)$	=	$F\,(0,s)$	×	$\delta\,(c,s)$
$*I\,(c,s)$	=	$I\,(0,s)$	×	$S\,(c,s)$

(*) true for spherically symmetrical particles. In fact more general.

Fig. 11.8. Interactions between particles in non-ideal solutions. [After Tardieu and Delaye (1988).]

The form factor is obtained by recording the scattering pattern at low concentration under conditions chosen so as to render the interactions practically negligible (ideal solution). The structure factor of the solution under given conditions is the ratio of the scattered intensity to the form factor. Note that we hereby make the implicit assumption that the object under study does not change shape or size with concentration. This assumption is often verified with proteins (but beware of aggregation!), but can be a real problem with other kinds of objects such as micelles or vesicles. For examples of data analysis of interacting systems, the reader is referred to Tardieu (1994). Recently, SAXS experiments and numerical simulations using models developed in liquid state physics, were coupled to investigate the fluid–fluid phase separation observed with lysozyme and γ-crystallin solutions upon lowering the temperature

220 *SAXS by solutions of biological macromolecules*

(Malfois et al. 1996). Structure factors were measured as a function of temperature above the transition temperature and simulated by methods developed for liquid state physics, which make use of a pair particle potential model to describe the protein interactions. A model of attractive interactions depending on three parameters, protein diameter, potential depth and range was shown to account for both the X-ray structure factors measured at high temperature and for the low temperature phase separation. Although van der Waals forces could be the cause of the attractive interactions, other more specific effects also contribute to the protein phase diagrams. A clear and up to date presentation of this approach is to be found in Tardieu et al. (1999).

11.5 Instrumentation

The content of the two previous sections is independent of the X-ray source used, X-ray generator or synchrotron radiation emitted by a storage ring. The following description is that of the instrument ID2 (Bösecke 1992; Bösecke and Diat 1995, 1997) (also to be found on the ESRF web site) using the radiation from an undulator at the ESRF (Grenoble, France), but all instruments essentially use the same elements, even if significant differences exist between their actual implementation at various places (EMBL outstation in Hamburg, SRS in Daresbury, LURE-DCI in Orsay, the Photon Factory in Tsukuba, NSLS in Brookhaven, SSRL in Stanford, Elettra in Trieste, etc.).

A SAXS instrument essentially comprises the following elements (see Figs 11.9 and 11.10).

A monochromator selects in the emission from the undulator one well-defined wavelength. A cryogenically cooled monolithic Si(111) double-reflection monochromator yields a monochromatic reflected beam, the wavelength of which is given by Bragg's law, $2d \sin \theta = n\lambda$, where d is the interplanar spacing

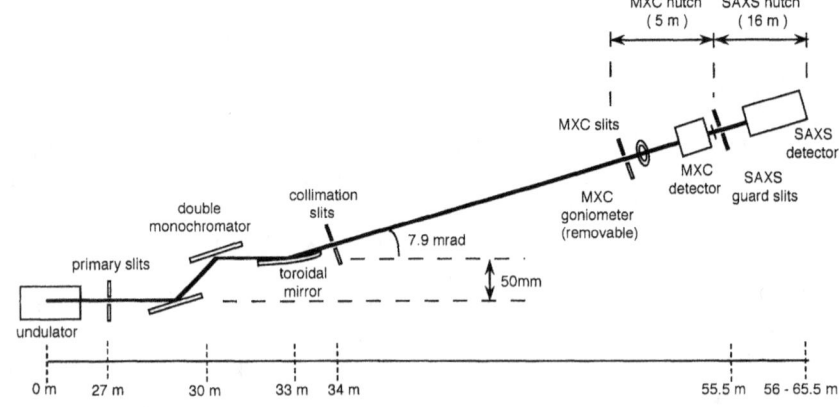

Fig. 11.9. General layout of the optical set-up of the high brilliance beam line ID2 at the ESRF (MXC: crystallography, SAXS: Small-Angle X-ray Scattering). [From Bösecke and Diat (1995).]

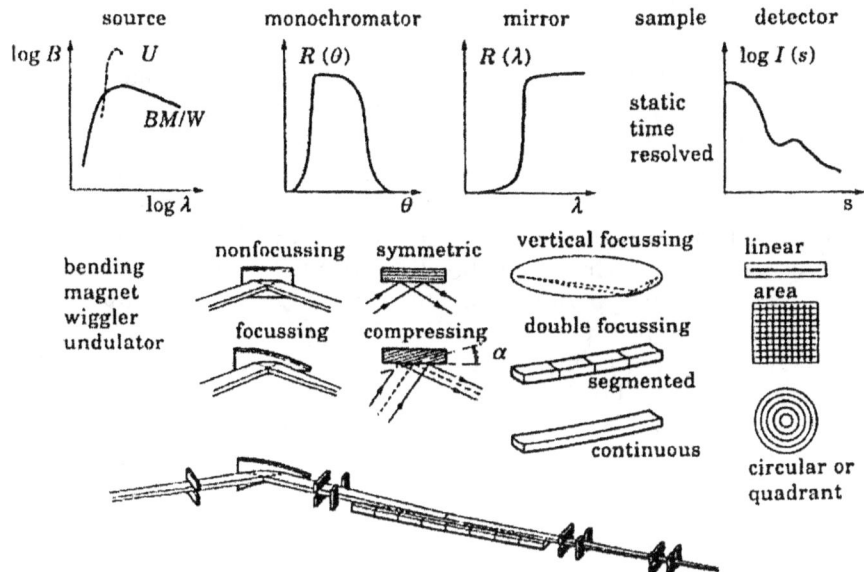

Fig. 11.10. Basic components of the systems for SAXS and example of a double-focusing monochromator–mirror camera. [From Koch (1988).]

and θ is the angle between the planes and the incident beam. Harmonics are eliminated by total reflection at glancing angle on a mirror, taking advantage of the decrease of the limiting angle for total reflection with decreasing wavelength. The vertical beam displacement between incident and reflected beams is 30 mm, which considerably reduces the background radiation from the storage ring. Some monochromators use a single Bragg reflection on a triangular bent focusing crystal (Lemonnier et al. 1978); see also Fig. 11.10.

Focusing in the two directions normal to the beam is provided by the Rh-coated toroidal mirror which, as mentioned above, also eliminates higher harmonics. Owing to the small beam divergence, the focusing parameters can be kept fixed for all experiments. The mirror characteristics were chosen so as to minimize the beam size at the detector position when it is farthest from the sample. The same result could be obtained using two elliptically bent mirrors orthogonal to each other (Kirkpatrick–Baez set-up). Alternatively, focusing in one direction can be achieved by sagittal focusing of one monochromator crystal.

Several pairs of slits made of a very absorbing material such as tungsten or tantalum are used to eliminate the parasitic scattering which is very intense around the monochromatic beam. This is crucial in order to reach very small angles. In the case of ID2, a first set of slits between the monochromator and the mirror can be used to reduce the beam cross-section at the sample position by a factor up to 10 000.

The SAXS camera proper consists of three sets of slits, two collimating slits, one positioned after the mirror at 34 m, and the second one at 50 m, while a guard slit is located directly before the sample at 54 m. Usually, the slit apertures are set to twice the full-width-at-half-maximum of the focused monochromatic beam. The slits confine the region around the primary beam in which diffuse parasitic scattering from the optics is visible. This defines the minimum size of the beam-stop and the minimum observable scattering angle. The sample position is fixed and the sample-to-detector distance can be adjusted at will between 0.75 and 10 m by automatically moving the detector inside a vacuum tube. Beam size on ID2 is typically 0.35 mm × 0.7 mm (V × H) at the sample position.

A sample holder, generally a thermostated quartz capillary, contains the solution under study. Alternatively, small cells with quartz or mica windows can be used with thickness typically between 10 and 20 µm. The optimum sample thickness is about 1 mm at $\lambda = 1.5$ Å.

The beam path is practically entirely evacuated to eliminate absorption and scattering by air. At LURE, a thermostated cell under vacuum has been designed, thereby eliminating two windows and the residual air path present when using conventional sample holders, and thus reducing considerably the parasitic scattering (Dubuisson et al. 1997).

A beam-stop made out of absorbing material absorbs the direct beam, preventing it from hitting the detector and damaging it because of its intensity (about 10 000–100 000 times more intense than the total scattered intensity, because of the low scattering power of solutions of biological macromolecules already mentioned).

An X-ray detector records the scattered intensity as a function of the scattering angle 2θ. This detector can be linear or two-dimensional, can count individual photons (proportional, position-sensitive detectors) or integrate the signal during exposure before reading is performed (image-plates, CCD) (Morse 1993).

For safety reasons, access to the experimental set-up is forbidden when the beam is on. The entrance into the experimental enclosure (generally known as the hutch) is possible only when beam shutters are closed, their opening being regulated by an interlock system. All movements of the various elements (slits, sample holder, detector, etc.) required by the instrument alignment and operation are thus entirely motorized and under remote control.

Data acquisition is performed via a set of VME (Bösecke and Diat 1997) [or CAMAC (Boulin et al. 1988)] electronic modules driven by a workstation. The time sequence (number and duration of patterns to be recorded) is defined by the experimentalist. Data acquisition can be synchronized with other operations (activation of a fast mixing device, T-jump triggering, etc.). Data are transferred onto disk before reduction and processing.

11.6 Questions and systems

The SAXS studies in solution provide *low resolution structural information*. We have seen in the paragraph on shape determination an illustration of the potential of the method. In the case of complex systems such as ribosomes, SAXS can be employed in conjunction with SANS to make use of contrast variation with mixtures of H_2O/D_2O and deuteration of a specific component (RNA or protein). This allows the distribution of each component within the global shape to be described (Svergun et al. 1997c,d). For X-rays, the contrast of electron density can also be varied, although in a more restricted range than with neutrons, but this requires the addition of electron-dense small molecules such as salts or sucrose; the complexities of interactions between the macromolecules and the small solute can further limit its use.

In these low resolution studies, the main advantage of SAXS is to provide direct information on the global structure in solution, free from the specific artefacts of either electron microscopy or crystallography and without the size limitations of NMR. In this context, the comparison between the solution and the crystal structure can be very informative. It can reveal perturbations caused by the intermolecular forces that keep the crystal together but which prevent certain domain movements that occur in solution [e.g. Grossmann et al. (1992)], or which even severely distort the structure, as illustrated below in the case of the R state of aspartate transcarbamylase from E. coli.

The strongest impact of SAXS is probably in the study of *structural transitions* ranging from gross conformational changes to assembly processes triggered by a variety of perturbations. The changes in the scattering patterns are related to changes in the structure or in the distribution of the various species in the case of polydisperse systems. The modification of the physico-chemical variable (pH, ionic strength, ligand concentration, temperature, pressure) can be made gradually through the transition, i.e. the scattering patterns are recorded *at equilibrium* at a number of intermediate values. This is a classical titration experiment. Alternatively, a fast perturbation is provoked and one follows the *relaxation of the system* to the new equilibrium state. This requires not only the very intense X-ray beam available at storage rings, but also instruments for rapid perturbation, such as a fast mixing apparatus ('stopped-flow') to mix a molecule with substrates or to create a pH or ionic strength jump (Tsuruta et al. 1989), a high pressure cell (Czelik et al. 1996) or field-jump devices (Koch et al. 1988). Finally, data acquisition capabilities must be fast compared to the rate of the process. At first and second generation facilities, the available flux is not sufficient to follow most conformational reactions in physiological conditions, where typical times are less than a second. A number of studies are thus performed at low temperature in the presence of an antifreeze agent to slow down the reaction (see below). The high flux delivered by instruments such as ID2 at third-generation machines may be sufficient to yield scattered intensities in excess of 10^8 photons per second, high enough for processes to be followed with a time resolution of the order of a millisecond and without the physicochemical and biochemical complications linked with the use of co-solvents. However, in spite of progress, there is to date no detector with the requisite capabilities.

A related problem is the *denaturation–renaturation of proteins*, which uses a similar experimental approach to detect and characterize putative intermediate states (molten globule, distinct stages in the un- or refolding of protein domains, etc.) (Damaschun et al. 1993; Kataoka et al. 1993) or the fully denatured state. This is illustrated below by some examples.

Finally, a last and very promising field is the study of the *interactions between proteins* in concentrated solutions. The original work of Tardieu's group on proteins of the eye lens has conclusively shown that some physiological properties of the eye lens such as its transparency or the existence of a correction for chromatic aberrations are simply due to non-specific protein interactions (Tardieu 1994; Tardieu and Delaye 1988). This approach has since been extended to other proteins in order to correlate the variations of interactions as a function of the physicochemical conditions with the solubility properties and, in the end, to use SAXS as a predictive tool for protein crystallization.

11.7 Some examples

We first present the case of the allosteric enzyme aspartate transcarbamylase from E. coli, since it has been the object of a variety of studies covering several of the applications listed above. This is followed by examples of (un)folding studies. A final paragraph will list references to a number of other systems studied by SAXS.

11.7.1 *The allosteric transition of aspartate transcarbamylase from* E. coli

Aspartate transcarbamylase (ATCase) catalyses the first committed step of the biosynthetic pathway of pyrimidines. Carbamyl phosphate (CP), which binds first, reacts with aspartate (A) to form carbamyl aspartate (CA) and inorganic phosphate [Fig. 11.11(a)]. The ATCase from E. coli is composed of two types of polypeptide chains, catalytic chains (C) with a molecular weight of 34 kDa and regulatory chains (R) with a molecular weight of 17 kDa. The enzyme comprises two trimers of C chains linked by three dimers of R chains with D3 symmetry [Fig. 11.11(b)]. The enzyme activity is regulated in two ways: first, it displays homotropic cooperativity with respect to A, since the active site, comprising residues from two adjacent C chains within a trimer, has a higher affinity for A in the presence of a high concentration of A than at low concentration. Furthermore, CTP, the final product of the biosynthetic pathway, inhibits the activity of the enzyme in a classical feedback inhibition mechanism, while ATP activates it. Both nucleotides bind to the same site on the R chain, about 60 Å away from the nearest active site.

The structures corresponding to the two states of activity of the enzyme in the absence of substrates (T state) and in the presence of saturating concentrations of substrates or analogues such as PALA (N-phosphonacetyl-L-aspartate) (R state) have been determined at high resolution by X-ray crystallography (Kantrowitz and Lipscomb 1988). The two crystal structures display large differences: in the T to R transition, the two catalytic trimers increase their separation along the threefold axis by about 11 Å and rotate each about 5° around the same axis, while the regulatory dimers rotate about 15° around the twofold axes (Fig. 11.11(c)). These changes in the quaternary structure are accompanied by tertiary structure modifications, most notably a closure of the active site cleft between the two domains of each C chain and a large displacement and reconfiguration of the so-called '240' loop, also from the C chain. As can be expected, this structural change is manifested by a change in the radius of gyration. Indeed,

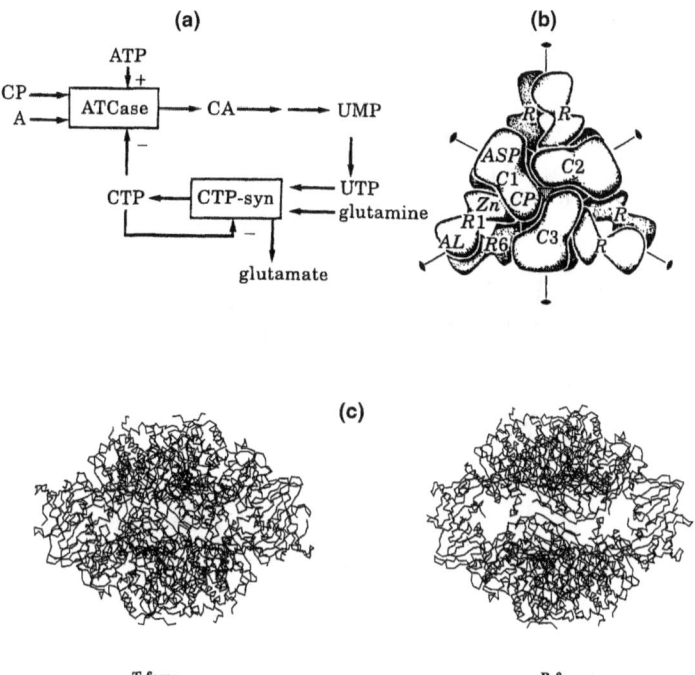

Fig. 11.11. (a) Schematic representation of the activity and regulation of ATCase from *E. coli*, at the start of the pyrimidine biosynthetic pathway (Koch and Vachette 1989). (b) Schematic representation of the quaternary structure of ATCase along the threefold axis. [Reprinted with permission from Krause *et al.* (1987); © Academic Press.] (c) α-carbon tracings of ATCase in both the T (less active) and the R (more active) forms. The threefold axis lies vertically in the plane of the paper, one twofold axis is perpendicular to the paper. [Reprinted with permission from Krause *et al.* (1987); © Academic Press.]

this parameter increases by about 5% upon the T to R transition. However, and much more interestingly, this structural change causes dramatic modifications in the X-ray scattering curve at larger angles (see the two extreme curves in Fig. 11.12). The changes, which can in parts reach 100%, are entirely due to the quaternary structure transition (Hervé *et al.* 1985). The X-ray scattering pattern is thus a sensitive and specific probe to study the modifications of the global conformation of the enzyme in response to a variety of stimuli: pH or temperature changes, addition of co-solvents, influence of substrate analogues, chemical or genetic modifications.

Structural properties of point mutants

Several point mutants of ATCase have thus been studied and the alterations in their structural properties have been correlated to modifications in their functional properties (Vachette *et al.* 1994). In one case, the total loss of homotropic cooperativity observed could be accounted for by the effect of CP alone: the addition of CP alone was enough to bring most of the mutant enzyme into

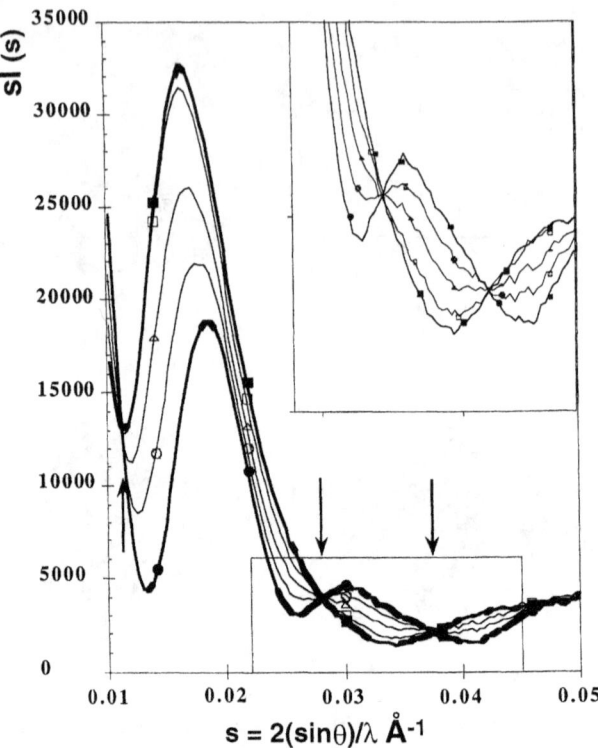

Fig. 11.12. X-ray scattering curves of ATCase in the two extreme forms T and R and in the presence of three subsaturating PALA concentrations expressed in mole of PALA/mole of active site (●: 0, ○: 0.20, △: 0.40, □: 0.60, ■: 2). These curves were recorded using a sector shape detector yielding integrated intensities $sI(s)$ along circular arcs. The arrows point towards the three crossing points. Inset: zoom on the framed region around the outermost two crossing points. [From Fetler and Vachette (1997).]

the R structure of high affinity, while its effect is much more limited on the wild-type enzyme (see below). Further addition of A was thus ineffectual, all sites having already a high affinity for A (Tauc *et al.* 1990). All cases have been encountered for the quaternary structure of the unligated mutant: T quaternary structure identical to that of the wild-type, structure close to T or close to R, population in equilibrium between two structures. Experiments involving various combinations of substrates and effectors have yielded a wealth of information on the structural consequences of those mutations, even if the interpretation of the structural and functional observations in terms of modified groups and bonds requires great caution.

Comparison between crystal and solution quaternary structure

Using the program CRYSOL discussed above, solution scattering curves were evaluated from the crystal structures of the T and R states of ATCase and compared with the experimental X-ray scattering patterns (Svergun *et al.* 1997a).

For the T state, the agreement obtained between the experimental and calculated curves is good, suggesting that the quaternary structures of the T state in solution and in the crystal should not differ. Note that if the hydration layer is not taken into account, the fit is much poorer with a systematic shift of the calculated curve to larger angles (see Fig. 11.13): in other words, without the hydration layer, the scattering object appears smaller than the actual one. In the case of the R state, large, systematic deviations between the experimental and the calculated curves are observed, most notably near the origin and around the first subsidiary maximum, which prove the existence of a significant difference between the quaternary structures of the R state in the crystal and in solution. The experimental curve of the R state was fitted by rigid body movements of the subunits in the crystal R structure which displace the latter further away from the T structure along the reaction coordinates of the T → R transition observed in the crystals. Taking the crystal R structure as a reference, it was found that in solution the distance between the catalytic trimers along the threefold axis is 3.4 Å larger and the trimers are rotated by 11° in opposite directions around the same axis; each of the three regulatory dimers is rotated by 9° around the corresponding twofold axis and displaced by 1.4 Å away from the molecular centre along this axis (see Fig. 11.14). The differences between the R state crystal structure and the model in solution are thus anything but marginal. The rationale behind this unexpectedly large difference is clear if one keeps in mind that ligand binding and conformational changes in allosteric enzymes involve only

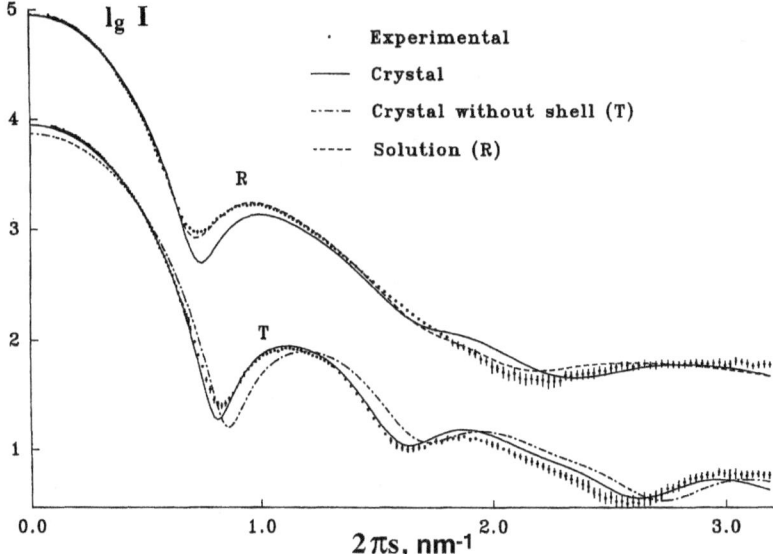

Fig. 11.13. Comparison of the solution scattering curves of ATCase with the scattering from the atomic structures for the T and R states. The curves from R state are shifted by one decade for better visualization. The curve evaluated for the T state corresponds to the model in Fig. 11.14(a), those for the R state to the models in Fig. 11.14(b) and (c). Error bars correspond to σ. [Reprinted with permission from Svergun *et al.* (1997a); ©Wiley-Liss Inc.]

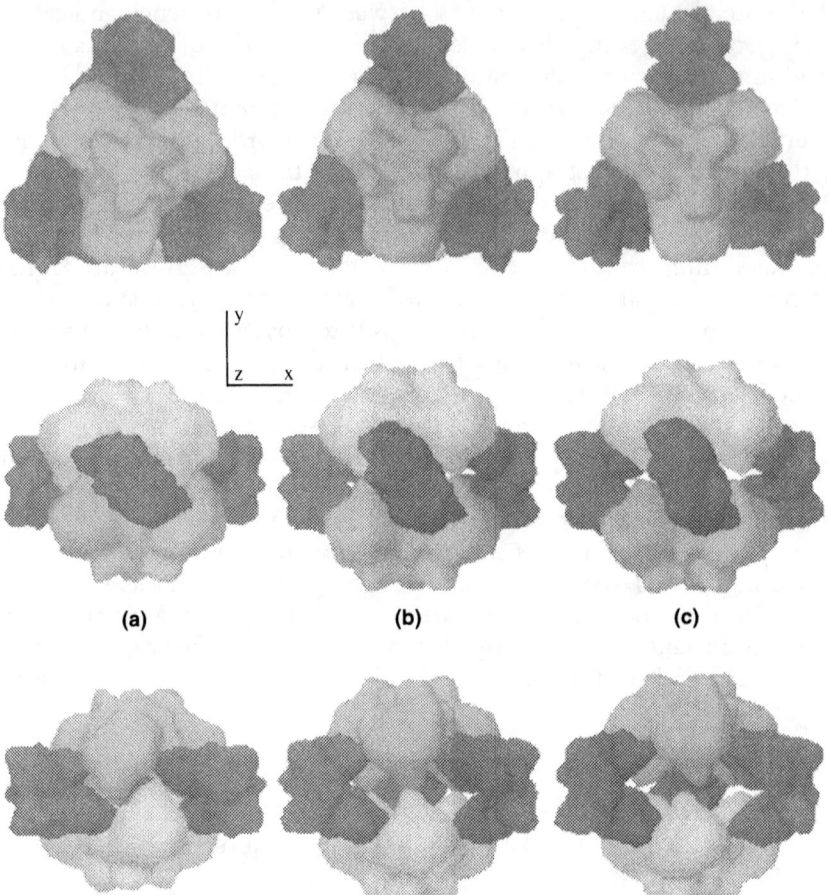

Fig. 11.14. Models of ATCase quaternary structures 'as seen' by X-rays at low resolution, since the envelope functions of the domains are represented using the spherical harmonics description for the scattering patterns. The two C trimers are coloured in yellow and cyan, the three R dimers in magenta. Top: View along the threefold axis (Z) from the side of the lower C trimer. Middle and bottom: View along one of the two fold axes (top view rotated along X-axis by 90° counterclockwise and clockwise, respectively). In each row, A and B are the crystal structures of the T and R states, respectively. C is the model of the R state in solution (see Plate Section). [Reprinted with permission from Svergun *et al.* (1997a); © Wiley-Liss Inc.]

low energy non-covalent interactions. The activation energy of the quaternary structure transition is thus also low and the crystal packing forces originating from non-covalent bonds between neighbouring molecules may distort the architecture of proteins selected for easy reorganization.

Equilibrium study of the structural transition with various ligands and effectors

Titration experiments can be performed, whereby a series of patterns (typically 20 curves) are recorded at different concentrations of substrate

(Fetler et al. 1995, 1997). The whole data set is then analysed using the SVD approach already described (Section 11.3.2.4). An experiment was performed using PALA (the bisubstrate analogue mentioned above) to provoke the transition. A subset of a few curves is presented in Fig. 11.12. The existence of three well-defined crossing points shown by arrows and enlarged in the inset is already very suggestive of a two-state transition. This is confirmed by the SVD analysis which shows that all curves can be approximated within experimental error by a linear combination of the first two eigenvectors only (Fig. 11.15). In other words, the transition triggered by PALA involves only the two extreme structures T and R, and no intermediate species can be detected. This supports the view of an equilibrium between the two forms which is shifted by PALA binding as proposed by Monod et al. in their model of a concerted transition (Monod et al. 1965). With only two components, the variation of the projection of each scattering curve onto the second component is directly related to the fraction of the corresponding species in solution. Thus, proper scaling, using the initial and the final state, yields the titration curve of the structural transition, i.e. the variation of the fraction of molecules in the R state vs the active site occupancy shown as a thin line in the inset in Fig. 11.15. Practically all molecules appear to be in the R state when only two-thirds of the active sites are occupied.

Similar titration experiments were performed using succinate, an analogue of aspartate, in the presence of saturating amounts of CP, the first physiological substrate (Fetler et al. 1997). Again, only two quaternary structures were detected. However, the scattering patterns of the unligated enzyme (T state) and of the ATCase−CP complex show small but significant differences. An SVD analysis shows that the latter is not a linear combination of the T and R patterns, and that a third quaternary structure is involved. This proves that CP modifies the quaternary structure of ATCase to a T′ state, close to but different from T. A last titration experiment by PALA in the presence of saturating amounts of CP again shows a two-state transition, but the titration curve is systematically shifted by 10–15% towards R (inset in Fig. 11.15, thick line). Therefore, CP changes the quaternary structure of ATCase to T′, a structure more readily converted to R by PALA (Fetler et al. 1997).

Time-resolved study of the structural transition

Finally, the signal associated with the transition is large enough that, after fast mixing of the enzyme and its substrates, the kinetics of the relaxation to the new equilibrium state, i.e. the transition to the R state of high activity, can be followed. This last aspect, the most spectacular, is possible only because of the high intensity of the X-ray beams available at storage rings [Fig. 11.16(a)]. However, even so, the reaction has to be considerably slowed down to the range of seconds, since indirect measurements suggest a characteristic time of 10–20 ms with the physiological substrates (Kihara et al. 1984). Experiments performed at the Photon Factory (Tsukuba, Japan), a second generation storage ring, were thus slowed down by working at low temperature (between −5 and −10 °C) in the presence of 20% or 30% ethylene glycol, conditions which have been

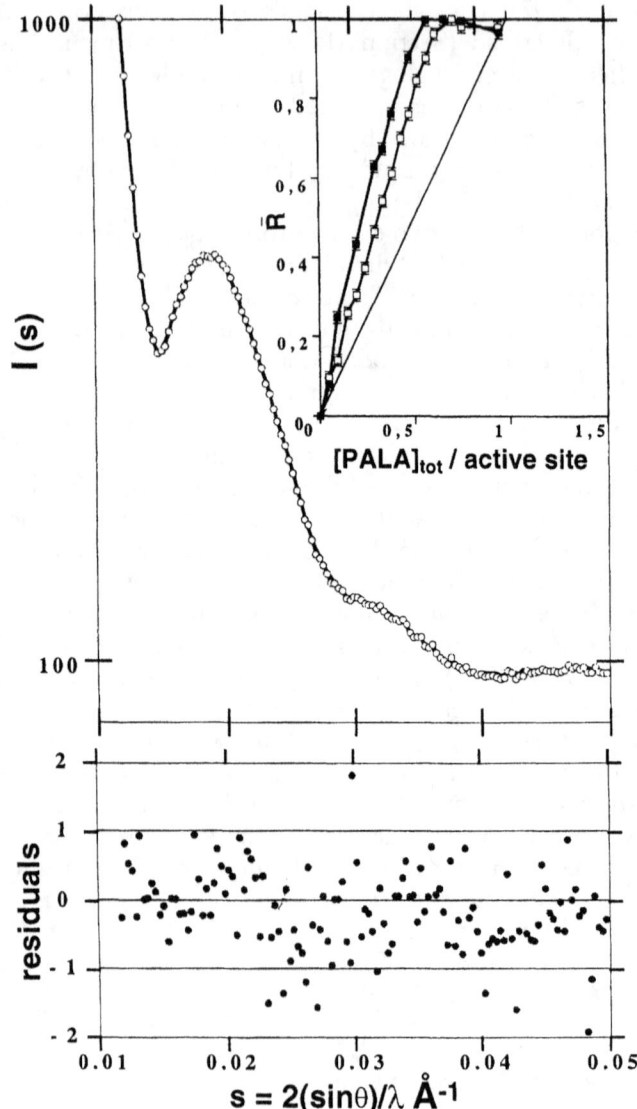

Fig. 11.15. Comparison of an experimental scattering pattern with its reconstruction using the first two eigenvectors with the corresponding normalized residuals below. This is from a study using a linear detector (Fetler *et al.* 1995). Inset: titration curves $\bar{R}$ vs $[PALA]_{tot}$ of ATCase with PALA alone (empty squares, thin line) and with PALA+ 5 mM CP (filled squares, thick line). [From Fetler *et al.* (1997).]

shown not to alter qualitatively the regulatory properties of ATCase. Measurements were performed using saturating CP together with various concentrations of succinate, a competitive inhibitor of aspartate. The scattering pattern can be seen to change progressively from that of the unligated enzyme (T state) to that

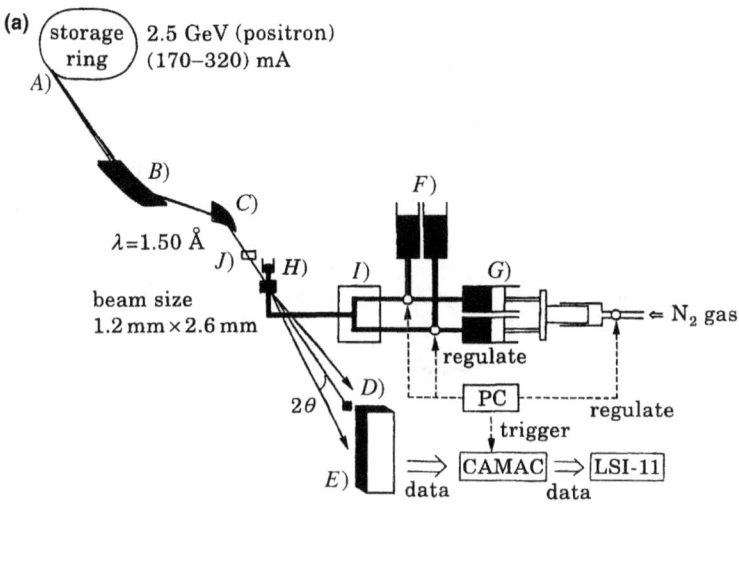

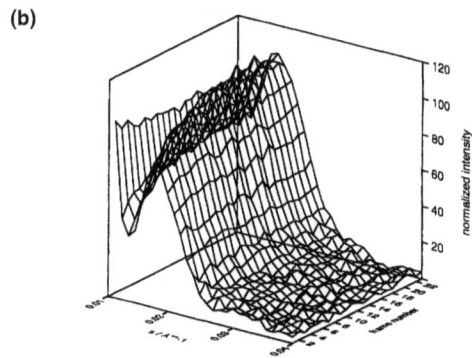

Fig. 11.16. (a) Stopped-flow apparatus on the small-angle installation BL-15A at the Photon Factory, Tsukuba, Japan (courtesy of H. Tsuruta); (b) Time-resolved scattering patterns during the course of the succinate reaction. [Reprinted with permission from Tsuruta et al. (1994); © American Chemical Society.] Two successive frames are separated by a time interval of 4 s.

of the PALA-saturated enzyme (R state) [Fig. 11.16(b)] (Tsuruta et al. 1994). Data were analysed in a way similar to the titration experiments mentioned above, and similarly no intermediate state could be detected during the transition [Fig. 11.17(a)], in agreement with the MWC model of a concerted transition. The variation of the overall rate constant of the transition k_{app} could be fitted with the expression derived from the MWC model and yielded parameter values in good agreement with the literature values, further supporting the model of a concerted transition [Fig. 11.17(b)] (Tsuruta et al. 1994).

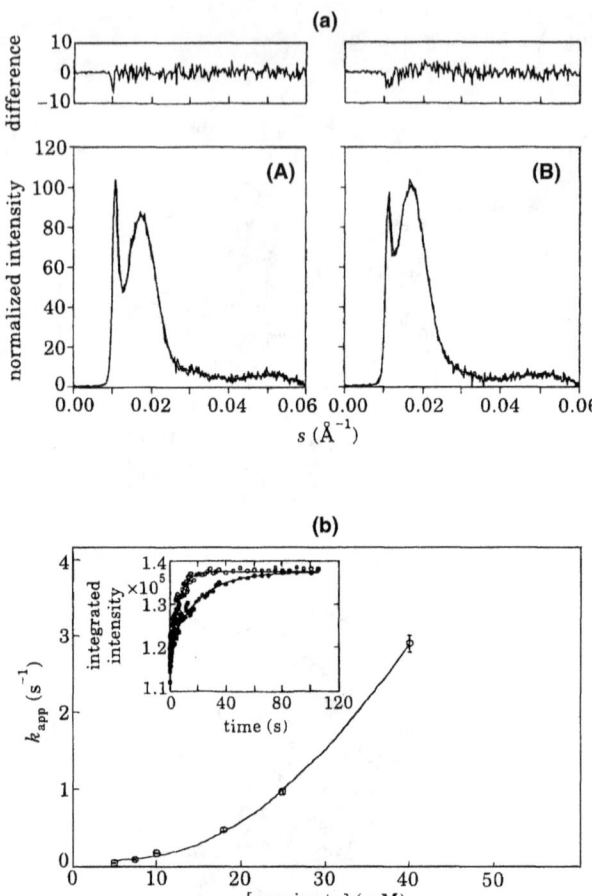

Fig. 11.17. (a) Comparison of two experimental curves and of their fit by linear combinations of T and R patterns (left-hand side: 39% R; right-hand side: 80% R) with the corresponding residuals on top of each frame. [Reprinted with permission from Tsuruta et al. (1994); © American Chemical Society.] (b) Variation of the apparent time constant of the transition as a function of succinate concentration. The curve is the least-squares fit using the MWC expression of k_{app}. [Reprinted with permission from Tsuruta et al. (1994); © American Chemical Society.]

11.7.2 Folding studies

A study of the equilibrium unfolding of hen egg lysozyme as a function of urea concentration at acidic pH has been performed using a combination of spectroscopic methods (UV absorption, far and near UV CD) and SAXS (Chen et al. 1996). Differences in the unfolding transition were observed as monitored by the radius of gyration and by far (222 nm) and near (298 nm) UV CD, suggesting the existence of a third unfolding species, in addition to the native and the unfolded states. This was confirmed by an SVD analysis of the set of SAXS patterns which showed clear evidence of a third basis component. The thermodynamic parameters of denaturation were estimated using a denaturant binding model and further used to reconstruct a scattering profile for the pure intermediate state. Simplified partially folded models, based on the crystal structure of hen lysozyme, support a working model for the intermediate, whose structure may be correlated with that of the kinetic intermediate found in the refolding pathway studied by NMR methods. Very recently, a collapsed state of lysozyme during refolding was characterized using the SVD analysis of time-resolved X-ray scattering patterns (Chen et al. 1998).

Another thermodynamic parameter, namely pressure, has been used in a recent study of staphylococcal nuclease folding and unfolding (Panick et al. 1998). A specially designed high pressure cell makes it possible to record the SAXS patterns of solutions under pressures of several kbar (Czelik et al. 1996). SAXS was used in conjunction with Fourier transform infrared (FTIR) spectroscopy under pressure, so as to follow the global structure (compactness) of the protein together with the secondary structure. Equilibrium experiments showed that a pressure of 3 kbar leads to an approximately twofold increase of the radius of gyration of the native protein together with a broadening of the distance distribution function, indicating the formation of an extended chain structure, which was shown by FTIR to retain some degree of β-like secondary structure. In addition, pressure-jump kinetics studies indicate that the changes in secondary structure and chain compactness are probably dependent upon the same rate-limiting step as changes in tertiary structure monitored by fluorescence.

11.7.3 Other examples

A number of other systems have been studied which cannot be discussed in any detail. Let us mention for instance the formation of microtubules from tubulin rings (Mandelkow et al. 1983) as well as the non-linear dynamic behaviour (oscillations) of microtubules (Mandelkow et al. 1988), actin polymerization (Sayers et al. 1985), the assembly of tobacco mosaic virus (Hiragi et al. 1990; Potschka et al. 1988) a comprehensive study of chromatin condensation (Koch et al. 1987), cold denaturation of phosphoglycerate kinase (Damaschun et al. 1993), the molten globule of cytochrome c (Kataoka et al. 1993) and apomyoglobin (Kataoka et al. 1995), the structure of highly denatured proteins (Petrescu et al. 1997), among many others.

Acknowledgements

P.V. is grateful to the Italian Physical Society for the permission to use parts of his contribution to the proceedings of the 128th course of the International School of physics 'Enrico Fermi' (Vachette 1994) and thanks D. Durand for fruitful discussions on protein folding studies. The X-ray solution scattering patterns from lysozyme, hexokinase, reverse transcriptase, ribonucleotide reductase and ATP synthase were collected as parts of ongoing projects at the EMBL Outstation in Hamburg. D.I.S. is grateful to M.H.J. Koch, C. Reißner,

L. Goobar-Larsson, S. Kuprin and G. Grüber for providing the experimental data. D.I.S. acknowledges support by the INTAS (Grants No. 95-1272 and No. 96-1115) and EU Biotechnology program (Grant BIO4-CT97-2143), and P.V. support by the CNRS.

Bibliography

Feigin, LA and Svergun, DI (1987). *Structure analysis by small angle X-ray and neutron scattering.* Plenum Press.
Glatter, O and Kratky, O (1982). *Small-angle X-ray scattering.* Academic Press, New York.
Guinier, A and Fournet, A (1955). 'The origins' (no recent edition): *Small angle scattering of X-rays.* Wiley, NY.
Lindner, P and Zemb, Th (eds), (1991). *Neutron, X-ray and light scattering: introduction to an investigative tool for colloidal and polymeric systems.* North-Holland.
The Proceedings of the SAS Conferences held every three years are usually published in the *Journal of Applied Crystallography*. The latest proceedings are in the *Journal of Applied Crystallography*, **30** (1997).

References

Abrahams, JP, Leslie, AGW, Lutter, R and Walker, JE (1994). *Nature*, **370**, 621–628.
Bennett, WS Jr and Steitz, TA (1980). *Journal of Molecular Biology*, **140**, 183–209.
Bernstein, FC, Koetzle, TF, Williams, GJB, Meyer, ERJ, Brice, MD, Rodgers, JR, Kennard, O, Shimanouchi, T and Tasumi, M (1977). *Journal of Molecular Biology*, **112**, 535–542.
Bösecke, P (1992). *Review of Scientific Instruments*, **63**, 438–441.
Bösecke, P and Diat, O (1995). *ESRF beamline handbook*, pp. 31–39. ESRF, Grenoble.
Bösecke, P and Diat, O (1997). *Journal of Applied Crystallography*, **30**, 867–871.
Boulin, C, Kempf, R, Gabriel, A and Koch, MHJ (1988). *Nuclear Instruments and Methods in Physics Research*, **A269**, 312–320.
Calmettes, P, Durand, D, Desmadril, M, Minard, P, Receveur, V and Smith, JC (1994). *Biophysical Chemistry*, **53**, 105–114.
Capaldi, RA, Aggeler, R, Wilkens, S and Grüber, G (1996). *Journal of Bioenergetics and Biomembranes*, **28**, 397–401.
Chen, L, Hodgson, KO and Doniach, S (1996). *Journal of Molecular Biology*, **261**, 658–671.
Chen, L, Wildegger, G, Kiefhaber, T, Hodgson, KO and Doniach, S (1998). *Journal of Molecular Biology*, **276**, 225–237.
Czelik, C, Malessa, R, Winter, R and Rapp, G (1996). *Nuclear Instruments and Methods*, **A368**, 847–851.
Damaschun, G, Damaschun, H, Gast, K, Misselwitz, R, Mueller, JJ, Pfeil, W and Zirwer, D (1993). *Biochemistry*, **32**, 7739–7746.
Debye, P (1947). *Journal of Physical and Colloid Chemistry*, **51**, 18–32.
Deckers-Hebestreit, G and Altendorf, K (1996). *Annual Review of Microbiology*, **50**, 791–824.
Dubuisson, JM, Decamps, T and Vachette, P (1997). *Journal of Applied Crystallography*, **30**, 49–54.
Fetler, L, Tauc, P, Hervé, G, Moody, MF and Vachette, P (1995). *Journal of Molecular Biology*, **251**, 243–255.
Fetler, L, Tauc, P and Vachette, P (1997). *Journal of Applied Crystallography*, **30**, 781–786.
Fowler, AG, Foote, AM, Moody, MF, Vachette, P, Provencher, SW, Gabriel, A, Bordas, J and Koch, MHJ (1983). *Journal of Biochemical and Biophysical Methods*, **7**, 317–329.

Garrigos, M, Mallam, S, Vachette, P and Bordas, J (1992). *Biophysical Journal*, **63**, 1462–1470.
Geissler, E (1994). In *Neutron and synchrotron radiation for condensed matter studies*, Baruchel, J, Hodeau, JL, Lehmann, MS, Regnard, JR and Schlenker, C (eds), Vol. III—Applications to soft condensed matter and biology, pp. 7–22, Les Editions de Physique, Springer Verlag, Les Ulis.
Glatter, O (1977). *Journal of Applied Crystallography*, **10**, 415–421.
Grossmann, JG, Neu, M, Pantos, E, Schwab, FJ, Evans, RW, Townes-Andrews, E, Lindley, PF, Appel, H, Thies, W-G and Hasnain, SS (1992). *Journal of Molecular Biology*, **225**, 811–819.
Grossmann, JG, Abraham, ZHL, Adman, ET, Neu, M, Eady, RR, Smith, BE and Hasnain, SS (1993). *Biochemistry*, **32**, 7360–7366.
Guinier, A and Fournet, G (1955). *Small angle scattering of X-rays*, Wiley, New York.
Hervé, G, Moody, MF, Tauc, P, Vachette, P and Jones, PT (1985). *Journal of Molecular Biology*, **185**, 189–199.
Hiragi, Y, Inoue, H, Sano, K, Kajiwara, K, Ueki, T and Nakatani, H (1990). *Journal of Molecular Biology*, **213**, 495–502.
Hubbard, SR, Hodgson, KO and Doniach, S (1988). *Journal of Biological Chemistry*, **263**, 4151–4158.
Kantrowitz, ER and Lipscomb, WN (1988). *Science*, **241**, 669–674.
Kataoka, M, Hagihara, Y, Mihara, K and Goto, Y (1993). *Journal of Molecular Biology*, **229**, 591–596.
Kataoka, M, Flanagan, JM, Tokunaga, F and Engelman, DM (1994). In *Synchrotron radiation in the biosciences*, Chance, B, Deisenhofer, J, Ebashi, S et al. (eds), Oxford University Press, Tokyo, pp. 187–194.
Kataoka, M, Nishii, I, Fujisawa, T, Ueki, T, Tokunaga, F and Goto, Y (1995). *Journal of Molecular Biology*, **249**, 215–228.
Kihara, H, Barman, TE, Jones, PT and Moody, MF (1984). *Journal of Molecular Biology*, **176**, 523–534.
Koch, MHJ (1988). *Die Makromolekulare Chemie, Macromolecular Symposia*, **15**, 79–90.
Koch, MHJ and Vachette, P (1989). *Synchrotron Radiation News*, **2**(2), 16–21.
Koch, MHJ, Sayers, Z, Vega, MC and Michon, AM (1987). *European Biophysics Journal*, **15**, 133–140.
Koch, MHJ, Dorrington, E, Kläring, R, Michon, AM, Marquet, R and Houssier, C (1988). *Science*, **240**, 194–196.
König, S, Svergun, DI, Volkov, VV, Feigin, LA and Koch, MHJ (1998). *Biochemistry*, **37**, 5329–5334.
Kozin, MB, Volkov, VV and Svergun, DI (1997). *Journal of Applied Crystallography*, **30**, 811–815.
Kratky, O and Porod, G (1949). *Recueil des Travaux Chimiques des Pays-Bas*, **68**, 1106–1122.
Krause, KL, Volz, KW and Lipscomb, WN (1987). *Journal of Molecular Biology*, **193**, 527–553.
Lattman, EE (1989). *Proteins Structure, Function and Genetics*, **5**, 149–155.
Lemonnier, M, Fourme, R, Rousseaux, F and Kahn, R (1978). *Nuclear Instruments and Methods*, **152**, 173–177.
Levelut, AM (1994). In *Neutron and synchrotron radiation for condensed matter studies*, Baruchel, J, Hodeau, JL, Lehmann, MS, Regnard, JR and Schlenker, C (eds), Vol. III—Applications to soft condensed matter and biology, pp. 69–80, Les Editions de Physique, Springer Verlag, Les Ulis.
Macheroux, P, Schönbrunn, E, Svergun, DI, Volkov, VV, Koch, MHJ, Bornemann, S and Thorneley, RNF (1998). *Biochemical Journal*, **335**, 319–327.
Malfois, M, Bonneté, F, Belloni, L and Tardieu, A (1996). *Journal of Chemical Physics*, **105**, 3290–3300.

Mandelkow, E, Mandelkow, EM and Bordas, J (1983). *Trends in Biochemical Sciences*, **8**, 374.
Mandelkow, EM, Lange, G, Spann, U, Jagla, A and Mandelkow, E (1988). *EMBO Journal*, **7**, 357–365.
Monod, J, Wyman, J and Changeux, JP (1965). *Journal of Molecular Biology*, **12**, 88–118.
Moore, PB (1980). *Journal of Applied Crystallography*, **13**, 168–175.
Morse, J (1993). In *Neutron and synchrotron radiation for condensed matter studies*, Baruchel, J, Hodeau, JL, Lehmann, MS, Regnard, JR and Schlenker, C (eds), Vol. I—Theory, instruments and methods, pp. 95–112, Les Editions de Physique, Springer Verlag, Les Ulis.
Ninio, J, Luzzati, V and Yaniv, M (1972). *Journal of Molecular Biology*, **71**, 217–229.
Panick, G, Malessa, R, Winter, R, Rapp, G. Frye, KJ and Royer, CA (1998). *Journal of Molecular Biology*, **275**, 389–402.
Pavlov, MY and Fedorov, BA (1983). *Biopolymers*, **22**, 1507–1522.
Pedersen, JS and Schurtenberger, P (1996). *Macromolecules*, **29**, 7602–7612.
Perkins, SJ, Smith, KF, Kilpatrick, JM, Volanakis, JE and Sim, RB (1993). *Biochemical Journal*, **295**, 87–99.
Petrescu, AJ, Receveur, V, Calmettes, P, Durand, D, Desmadril, M, Roux, B and Smith, JC (1997). *Biophysical Journal*, **72**, 335–342.
Porod, G (1951). *Kolloïd Zeitschrift*, **124**, 83–114.
Porod, G (1952). *Kolloïd Zeitschrift*, **125**, 51–57, 109–122.
Potschka, M, Koch, MHJ, Adams, ML and Schuster, TM (1988). *Biochemistry*, **27**, 8481–8491.
Provencher, SW and Glöckner, J (1983). *Journal of Biochemical and Biophysical Methods*, **7**, 331–334.
Sayers, Z, Koch, MHJ, Bordas, J and Lindberg, U (1985). *European Biophysics Journal*, **13**, 99–108.
Schmidt, B, König, S, Svergun, DI, Volkov, VV, Fischer, G and Koch, MHJ (1995). *FEBS Letters*, **372**, 169–172.
Shannon, CE and Weaver, W (1949). *The mathematical theory of communication*, University of Illinois Press, Urbana.
Stuhrmann, HB (1970a). *Zeitschrift für Physikalische Chemie Neue Folge*, **72**, 177–198.
Stuhrmann, HB (1970b). *Acta Crystallographica*, **A26**, 297–306.
Stuhrmann, HB (1994). In *Neutron and synchrotron radiation for condensed matter studies*, Baruchel, J, Hodeau, JL, Lehmann, MS, Regnard, JR and Schlenker, C (eds), Vol. III—Applications to soft condensed matter and biology, pp. 117–144, Les Editions de Physique, Springer Verlag, Les Ulis.
Svergun, DI (1992). *Journal of Applied Crystallography*, **25**, 495–503.
Svergun, DI (1993). *Journal of Applied Crystallography*, **26**, 258–267.
Svergun, DI (1994). *Acta Crystallographica*, **A50**, 391–402.
Svergun, DI (1997). *Journal of Applied Crystallography*, **30**, 792–797.
Svergun, DI and Stuhrmann, HB (1991). *Acta Crystallographica*, **A47**, 736–744.
Svergun, D, Semenyuk and Feigin, LA (1988). *Acta Crystallographica*, **A44**, 244–250.
Svergun, DI, Barberato, C and Koch, MHJ (1995). *Journal of Applied Crystallography*, **28**, 768–773.
Svergun, DI, Volkov, VV, Kozin, MB and Stuhrmann, HB (1996). *Acta Crystallographica*, **A52**, 419–426.
Svergun, DI, Barberato, C, Koch, MHJ, Fetler, L and Vachette, P (1997a). *Proteins Structure, Function and Genetics*, **27**, 110–117.
Svergun, DI, Volkov, VV, Kozin, MB, Stuhrmann, HB, Barberato, C and Koch, MHJ (1997b). *Journal of Applied Crystallography*, **30**, 798–802.

Svergun, DI, Burkhardt, N, Pedersen, JS, Koch, MHJ, Volkov, VV, Kozin, MB, Meerwink, W, Stuhrmann, HB, Diedrich, G and Nierhaus, KH (1997c). *Journal of Molecular Biology*, **271**, 588–601.
Svergun, DI, Burkhardt, N, Pedersen, JS, Koch, MHJ, Volkov, VV, Kozin, MB, Meerwink, W, Stuhrmann, HB, Diedrich, G and Nierhaus, KH (1997d). *Journal of Molecular Biology*, **271**, 602–618.
Svergun, DI, Richard, S, Koch, MHJ, Sayers, Z, Kuprin, S and Zaccai, G (1998a). *Proceedings of the National Academy of Sciences USA*, **95**, 2267–2272.
Svergun, DI, Aldag, I, Sieck, T, Altendorf, K, Koch, MHJ, Clark, RJ, Kane, DJ, Kozin, MB and Grüber, G (1998b). *Biophysical Journal*, **75**, 2212–2219.
Tardieu, A (1994). In *Neutron and synchrotron radiation for condensed matter studies*, Baruchel, J, Hodeau, JL, Lehmann, MS, Regnard, JR and Schlenker, C (eds), Vol. III—Applications to soft condensed matter and biology, pp. 145–160, Les Editions de Physique, Springer Verlag, Les Ulis.
Tardieu, A and Delaye, M (1988). *Annual Review of Biophysics and Bioengineering*, **17**, 47–70.
Tardieu, A, LeVerge, A, Malfois, M, Bonneté, F, Finet, S, Riès-Kautt, M and Belloni, L (1999). *Journal of Crystal Growth*, **196**, 193–203.
Tauc, P, Vachette, P, Middleton, SA and Kantrowitz, ER (1990). *Journal of Molecular Biology*, **214**, 327–335.
Taupin, D and Luzzati, V (1982). *Journal of Applied Crystallography*, **15**, 289–300.
Tsuruta, H, Nagamura, T, Kimura, K, Igarashi, Y, Kajita, A, Wang, ZX, Wakabayashi, K, Amemiya, Y and Kihara, H (1989). *Review of Scientific Instruments*, **60**, 2356–2358.
Tsuruta, H, Vachette, P, Sano, T, Moody, MF, Amemiya, Y, Wakabayashi, K and Kihara, H (1994). *Biochemistry*, **33**, 10 007–10 012.
Uhlin, U, Cox, GB and Guss, JM (1997). *Structure*, **5**, 1219–1230.
Uhlin, U and Eklund, H (1994). *Nature*, **370**, 533–539.
Vachette, P (1994). *International School of Physics 'Enrico Fermi', Course CXXVIII—Biomedical Applications of Synchrotron Radiation, Varenna on Lake Como*.
Vachette, P, Tauc, P and Kantrowitz, ER (1994). In *Synchrotron radiation in the biosciences*, Chance, B, Deisenhofer, J, Ebashi, S, Goodhead, DT, Helliwell, JR, Huxley, HE, Iizuka, T, Kirz, J, Mitsui, T, Rubinstein, E, Sakabe, N, Sasaki, T, Schmahl, G, Stuhrmann, HB, KW and Zaccai, Z (eds), pp. 145–157, Oxford University Press, Tokyo.
Wang, J, Smerdon, SJ, Jaeger, J, Kohlstaedt, LA, Friedman, J, Rice, PA and Steitz, TA (1994). *Proceedings of the National Academy of Sciences USA*, **91**, 7242–7246.
Wilkens, S, Dunn, SD, Chandler, J, Dahlquist, FW and Capaldi, RA (1996). *Nature Structural Biology*, **4**, 198–201.
Williams, CE (1993). In *Neutron and synchrotron radiation for condensed matter studies*, Baruchel, J, Hodeau, JL, Lehmann, MS, Regnard, JR and Schlenker, C (eds), Vol. I—Theory, instruments and methods, pp. 235–245, Les Editions de Physique, Springer Verlag, Les Ulis.

12

Small-angle neutron scattering

Giuseppe Zaccai

12.1 Small-angle scattering

Small-angle scattering (SAS) of X-rays or neutrons arises from scattering density fluctuations of the order 10–100 Å. In biological studies, samples are usually solutions of macromolecules in this size range. Depending on the experimental conditions such as macromolecular concentration, for example, SAS provides information on structural properties of individual particles as well as on interactions between them.

12.2 Neutrons and X-rays

Compared to X-rays, neutrons offer a choice of wavelength from below 1 Å to more than 10 Å, with low energies and negligible absorption in most matter. A significant advantage of neutrons is that they are scattered by atomic nuclei, yielding amplitudes independent of atomic mass unlike X-rays that are scattered by electrons; isotope effects (especially the large difference between the scattering amplitudes of hydrogen and deuterium) lead to the possibility of labelling different parts of a structure and provide powerful applications for experiments with low spatial resolution. The relatively low incident flux and high incoherent background from H in H_2O solutions, however, are the major disadvantages of neutron scattering. They are compensated for somewhat by the use of long wavelengths (that reduce the requirement for severe collimation to obtain high scattering vector resolution) and large beam cross-sections, and the favourable contrast and background conditions that can be obtained by using heavy water (D_2O) solvents.

12.3 Solutions of biological macromolecules

The structure–function relation in biology is now well established and the study and understanding of biological function pass through *experimental* determinations of macromolecular structure and interactions to varying degrees of spatial (and temporal) resolution.

Diffraction to a resolution of a few angstroms can only be observed from crystalline samples in which a very large number of atoms are essentially immobilized and scatter coherently. This is because atomic scattering cross-sections for X-rays or neutrons are very small ($\sim 10^{-24}$ cm^2). Studies of macromolecular structure and interactions performed *in solution* by X-rays or neutrons provide information at much lower resolution. Here too, and for the same reasons, a very large number of macromolecules is required; typical concentrations are $\sim$ mg/ml, or in the 10^{-4}–10^{-5} M range for 10–100 kD proteins, for example.

12.4 Biological macromolecules and their complexes have highly defined structures

In the 1930s, scientists saw, for the first time, an experimental example of radiation scattering by a sphere. It was known that the Fourier transform of a sphere had a series of maxima decreasing in height with increasing scattering vector. But these had never been observed. Their scattering angle values (and those of the minima in between) depend on the radius of the spheres and physics or chemistry could not provide samples of spheres that were sufficiently homogeneous (*monodisperse*) for such an observation to be made. Maxima from spheres of different sizes (a *polydisperse* solution) are at different angles and the resulting scattering curve is smeared out.

In the experiment referred to, the sample was a spherical virus. The sample was monodisperse, i.e. all spherical particles were strictly identical. This is because *biological systems are well defined*. In a pure virus solution, a pure protein solution or a pure solution of a given nucleic acid, all particles are identical: they have the same molecular weight, the same spatial structure.

If a protein were considered as a polymer, it would be one with a well-defined beginning, its N-terminal, a well-defined end, its C-terminal and a well-defined tertiary structure. This is an important difference with polymers usually studied in physical chemistry, for which statistical approaches must be applied because they have variable lengths and structures about mean values and configurations—a difference that is reflected in the data analysis approach to SAS.

12.5 SAS from a biological system

SAS from a solution of biological macromolecules can be interpreted only when

(a) the biochemistry of the system is well understood and
(b) it is well behaved.

In the case of a sample solution, this means that

(a) its composition is known accurately and
(b) all particles are identical or, if not, interact with each other in a way that can be described by a mathematical model.

Aggregation (proteins sticking together because they are not folded properly or because the solvent is not suitable), for example, is not a good thing; scattering from solutions showing aggregation cannot be interpreted (other than to conclude that there is aggregation). On the other hand, proteins interacting with each other to form functional dimers (or higher multimers), or with other proteins or nucleic acids to form functional complexes are usefully studied by scattering methods.

Table 12.1. SAS in the very low scattering vector limit

The assumption: No correlation in position or orientation of the particles in solution
The Guinier approximation (valid for $q \to 0$ up to $qR_g \sim 1$): $I(q) = I(0)\exp(-R_g^2 q^2/3)$
The experimental parameters: $I(0)$ and R_g^2 (from the intercept and slope of $\ln(I)$ vs q^2)
Their interpretation:

$$I(0) = \sum_i C_i A_i^2(0)$$

$$A_i(0) = \left(\sum b_\kappa - \rho^0 v_\kappa\right)_i$$

$$R_g^2 = \left\{\sum_i C_i A_i^2(0) R_{gi}^2\right\}/I(0)$$

$$R_{gi}^2 = \left\{\sum m_\kappa r_\kappa^2 \Big/ \sum m_\kappa\right\}_i$$

$$m_\kappa = b_\kappa - \rho^0 v_\kappa$$

I is scattered intensity; q is scattering vector modulus ($4\pi \sin\theta/\lambda$; 2θ is scattering angle, λ radiation wavelength); b_κ, v_κ are the scattering amplitude and volume, respectively, of the atom group κ in particle i; ρ^0 is solvent scattering density; R_{gi} is the radius of gyration of scattering amplitude difference with the solvent (m_κ) in particle i; r_κ is the distance between the atom group κ and the centre of mass of the m_κ distribution; C_i is the number concentration of particle i in the solution.

The parameters obtained from the scattering curve at small values of scattering vector modulus (q), in the Guinier approximation (Guinier and Fournet 1955) and their interpretation are given in Table 12.1.

12.6 SANS studies on biological systems in solution

SAS studies of macromolecular structure and interactions in solution can be performed by X-rays or neutrons. The theory behind the analysis of scattering curves, in the Guinier approximation, at small q values (Guinier and Fournet 1955) as well as in the larger q domain that can be interpreted in terms of distance distribution functions (Glatter 1980) is identical for the two types of radiation, provided the appropriate atomic scattering amplitudes are used in each case [see Chapter 11 on small-angle X-ray scattering (SAXS)]. The advantages of neutrons were outlined above. Low absorption leads to high penetration and negligible radiation damage to samples and to easy-to-design sample containers that can withstand extremes of temperature or pressure. The coherent neutron scattering amplitudes of most atoms have about the same absolute value ($\sim 10^{-12}$ cm) but the amplitude of ^{1}H is negative (-0.37×10^{-12} cm) while for D (0.667×10^{-12} cm) and the other atoms found in biological macromolecules these are positive. This makes the coherent amplitude close to *zero* for the light water molecule. For the first time in biophysics, solution experiments could be done as if in a 'vacuum'. As will be shown below, the solvent must always be taken into account in a hydrodynamic or scattering experiment in solution. The solvent term always appears as in Archimedes principle as a solvent density multiplied by a particle volume ($\rho^0 V$). In X-rays or analytical centrifugation, the electron density or mass density of solvent and macromolecules are quite close so that the solvent term plays an important role in the analysis. For a neutron experiment in H$_2$O, the term almost vanishes because of the low negative value of the scattering density of the solvent.

12.7 Contrast

The concept of contrast is valid only at low resolution, where groups of atoms in a volume can be considered together in a continuous distribution. A volume element v_k of scattering amplitude b_k, in a particle in solution, contributes to the scattering according to

$$m_k = (b_k - \rho^0 v_k) \tag{12.1}$$

(see also Table 12.1) where ρ^0 is the scattering amplitude density (or scattering density) of the solvent. Note the similarity to the buoyancy term in analytical centrifugation (Archimedes principle). This is related to the contrast of the volume element in the particular solvent. The contrast is defined as the difference in scattering amplitude density:

$$\rho = (b_k/v_k) - \rho^0. \tag{12.2}$$

Equations (12.1) and (12.2) are general and valid for both SANS and SAXS, for example. In the neutron case, scattering densities are denoted by the subscript N (ρ_N, ρ_N^0). Because of the natural neutron scattering density contrast values between important components of biological structures—proteins, lipids, nucleic acids, aqueous solutions—and the possibility of enhancing it by deuterium labelling (Fig. 12.1), neutron methods are very powerful for *low resolution* studies in crystals or in solution. It is difficult, for example, to distinguish protein from nucleic acid regions in a low resolution electron density distribution obtained from X-ray scattering of a protein–nucleic acid complex, because the difference in density between them is small. As shown below, however, by using neutrons and contrast variation, it is possible to enhance or suppress each component almost completely in chosen solvent conditions, making the interpretation very reliable.

Contrast variation consists in changing the contrast (or contribution) to the scattering of different components in a complex by adjusting solvent density (ρ^0 in Table 12.1). In general, however, comparing X-ray and neutron scattering curves is also equivalent to contrast variation because relative scattering amplitudes are different. This approach was used to study nucleoprotein complexes (Serdyuk *et al*. 1979).

12.8 Scattering density matching by D$_2$O : H$_2$O exchange. Fluctuations and resolution. Effects of D$_2$O

The value of ρ_N^0 for H$_2$O is close to -5×10^9 cm^{-2}. That of D$_2$O is close to 64×10^9 cm^{-2}. By mixing H$_2$O and D$_2$O in the solvent, ρ^0 varies linearly between

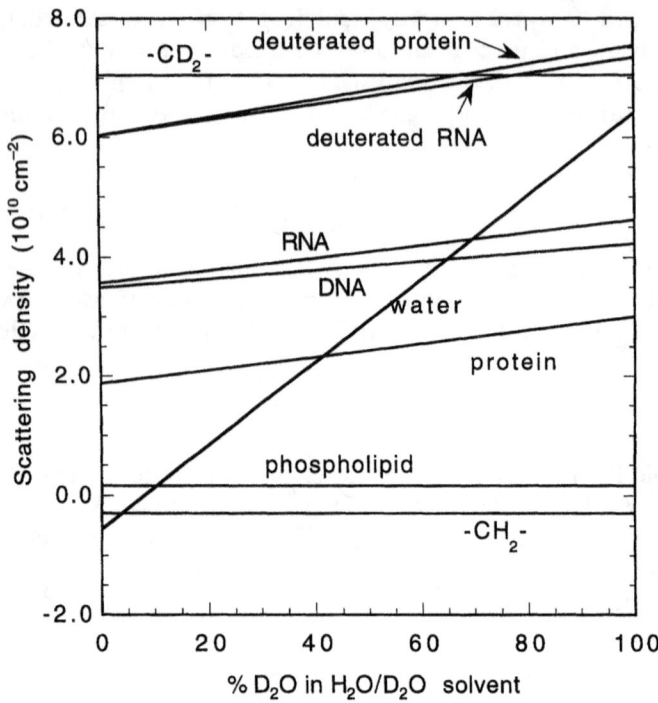

Fig. 12.1. Neutron scattering densities of various biological components as a function of D_2O/H_2O ratio in their solvent. It is assumed that labile hydrogen atoms exchange fully with the solvent.

a small negative value and a large positive value, passing through zero (at about 8% D_2O) (Fig. 12.1). The average scattering densities of the various components in Fig. 12.1 correspond, respectively, to different D_2O/H_2O ratios. The important word in the previous sentence is 'average'. Proteins, for example, are not homogeneous in scattering density. The scattering density of an amino acid residue depends on its non-labile hydrogen content (labile H will exchange with the solvent and increase the scattering amplitude in D_2O solvent). A table of amino acid scattering amplitudes and volumes is given by Jacrot (1976). At sufficiently low resolution, however, different parts of a protein cannot be distinguished separately, it will appear as homogeneous and it will be possible to match its scattering density by the appropriate D_2O/H_2O ratio—close to 40% D_2O. Nucleotides are more homogeneous in neutron scattering density than amino acids. Nucleic acids can be matched by the appropriate solvent (about 70% D_2O) at a resolution corresponding to a few nucleotides.

Solvent $H_2O:D_2O$ exchange is not always without effect on interactions and care must be exercised that the basic assumption of any contrast variation method is valid, i.e. that the system is behaving identically in all conditions examined. D_2O stabilizes hydrophobic interactions (Bonneté et al. 1994) and is known to favour aggregation especially in proteins.

12.9 The buoyancy term in hydrodynamics and SANS, molecular weights and solvent interactions

In densimetry or analytical centrifugation studies, there is a buoyancy term equivalent to eqn (12.1),

$$M_2(1 - \rho^0 \bar{v}_2) \tag{12.3}$$

where $M_2, \bar{v}_2$ are the molar mass and partial specific volume of the particle, respectively; ρ^0 is now the mass density of the solvent. The solvent term is relatively important in the expression. For example, $\bar{v}_2$ is close to 0.75 cm^3/g for a protein, $\rho^0 = 1$ for water so that the buoyancy term, $(1 - \rho^0 \bar{v}_2)$, is the difference between two numbers of similar magnitude. In this example, the relative error in $\bar{v}_2$ will be magnified by a factor of three in the final result.

Typical values for expression (12.1) applied to a gram of protein in H$_2$O scattering neutrons are $b_k = 14 \times 10^9$ cm/g, $\rho_N^0 = -5 \times 10^9$ cm^{-2}, $\bar{v}_k = 0.75$ cm^3/g. Not only is the solvent term much smaller than the protein term but also, because it is negative, the final expression is the sum of the two terms and the error is smaller than the relative error in the solvent term. In effect, because ρ^0 is small, and can even be made equal to zero, neutrons have provided for the first time the opportunity of studying a particle in solution with a solvent that is essentially transparent, as if the particle were in a vacuum. This makes the volume term negligible and is of particular interest for absolute value determination of macromolecular weight and macromolecule–solvent interactions in solution (Jacrot and Zaccai 1981; Block et al. 1982; Zaccai and Jacrot 1983; Zaccai 1986; Zaccai et al. 1989; Ebel et al. 1992, 1995; Bonneté et al. 1993). The equations underlying these studies are given in Table 12.2.

12.10 Structure and interactions in membrane proteins and complexes

Multicomponent complexes are studied by SANS by using solvent H$_2$O : D$_2$O exchange and/or specific deuteration by various ingenious approaches:

(1) Contrast variation by solvent exchange. There are continuing examples of studies that have made use of the natural contrast between complex components (Fig. 12.1), especially between protein (matched by about 40% D$_2$O) and DNA or RNA (matched by about 70% D$_2$O). Jacrot (1976) and Timmins and Zaccai (1988) provide good reviews of the method and early work on nucleoprotein complexes. The *in situ* association state of a membrane protein (the stoichiometry of polypeptides in the protein complex) is important and often difficult to establish. In a recent paper, SANS H$_2$O : D$_2$O contrast variation studies are described to assess the aggregation state of a membrane protein reconstituted into vesicles formed from deuterated lipids (Hunt et al. 1997). The specific deuteration was chosen so as to minimize the scattering density fluctuations between the lipid head groups and hydrocarbon chains.

Table 12.2. Solvent interactions in SANS

Three-component solution: component 1 = water; component 2 = macromolecule (non-diffusible across a dialysis membrane); component 3 = diffusible solute, e.g. salt

The thermodynamic interpretation of $I(0)$, the forward scattered intensity in terms of scattering density increment (Eisenberg 1981): $N_A I(0) = M_2 c_2 \, (\partial \rho_N / \partial c_2)_\mu^2$

For a particle which binds B_1 g/g of water, B_3 g/g of, for example, salt:

- Mass density increment (relevant in hydrodynamic experiments such as analytical centrifugation)

$$\left(\frac{\partial \rho}{\partial c_2}\right)_\mu = (1 + B_1 + B_3) - \rho^0 (\bar{v}_2 + B_1 \bar{v}_1 + B_3 \bar{v}_3)$$

- Neutron scattering density increment

$$\left(\frac{\partial \rho_N}{\partial c_2}\right)_\mu = (b_2 + b_1 B_1 + b_3 B_3) - \rho_N^0 (\bar{v}_2 + B_1 \bar{v}_1 + B_3 \bar{v}_3)$$

N_A is Avogadro's number; M_2 is the molar mass of the macromolecule; c_2 is the mass concentration of macromolecule; the density increments are defined at constant chemical potential μ of components 1 and 3; ρ, ρ_N are mass and neutron scattering density, respectively; ρ^0, ρ_N^0 are the mass and neutron scattering densities of the solvent, respectively; $\bar{v}_i$ is the partial specific volume of component i; b_i is the neutron scattering amplitude per gram of component i. Because the relative contributions of the various components are very different for hydrodynamic experiments, on the one hand, and SANS, on the other, the two sets of experiments are strongly complementary for the determination of B_1 and B_3, the solvent binding parameters.

(2) Label triangulation of deuterium-labelled components in a complex particle. The distances between the centres of mass of the components and sometimes even their shapes *in situ* are obtained from the analysis of scattering curves obtained for particles with different label combinations. For example, the positions of all the proteins in the 30S subunit of the *E. coli* ribosome have been derived in this way (Moore 1988). In a beautiful study, by using a special genetic construction, the quaternary structure of repressors bound to a control region of DNA was determined (Lederer *et al.* 1989). Recently, conformational changes upon formation of the GroEL–GroES chaperonin complex from *E. coli* were studied in solution by this method (Stegmann *et al.* 1998).
(3) Triple isotopic substitution. Pavlov *et al.* (1991) and Serdyuk *et al.* (1994) on elongation factor EF-Tu from *E. coli* and its complexes.

Dilute samples (or data extrapolation from concentration series) must be used in the solvent exchange method to avoid interparticle interference or aggregation which would make the analysis unreliable. In the label triangulation and triple isotopic substitution methods, concentrated samples can be used; the effects of interparticle interference are compensated for by the analysis itself.

12.11 Complexes can also be studied without labelling

If a complex can be 'built' up sequentially, the mutual arrangement of components could be calculated from a neutron experiment without using isotope labelling. This is a very powerful approach when complementary biochemical information is available on the structure of the components, e.g. the study on complement proteins (Zaccai *et al.* 1990; Esser *et al.* 1993).

12.12 Studies in special solvents

Highly concentrated solvents that are strongly absorbing in X-ray scattering can be essentially transparent in SANS. This is the case of guanidinium chloride, which is used extensively in protein denaturation and folding studies. SANS was used by Calmettes *et al.* (1994) to characterize the structure of denatured protein states in such solvents. Interactions of proteins from extreme halophiles (organisms that live in high salt environments) with water and ions were also characterized by SANS in multimolar salt solvents (Ebel *et al.* 1992, 1995; Bonneté *et al.* 1993).

12.13 Hydration around nucleic acids and proteins

The $H_2O:D_2O$ exchange and SANS provide a unique opportunity to study hydration around macromolecules. Nucleic acids are polyelectrolytes and a hydration layer around them, of density higher than that of the bulk solvent, was discovered by SANS, through its contribution to the radius of gyration measured in D_2O (Li *et al.* 1983). Recently, a 'dense' hydration layer was shown to exist around proteins in solution as well by comparing SANS scattering curves in H_2O and D_2O solvents, X-ray scattering curves and using existing high resolution crystal structure coordinates in the interpretation (Svergun *et al.* 1998). In this study the entire scattering curves were analysed and not just the two parameters (the radius of gyration and forward scattered intensity) obtained from the Guinier approximation. In fact, a hydration shell about 3 Å thick influences a scattering curve to the higher values of q and can even be observed qualitatively through the shift of the first subsidiary maximum. Thus, the first subsidiary maximum in the scattering curves of the three soluble proteins studied by Svergun *et al.* (1998) appeared at progressively higher q values, respectively, for SAXS, SANS in H_2O and SANS in D_2O. The dense hydration shell is essentially invisible to neutrons in H_2O; it makes the particle appear larger to X-rays, and smaller to neutrons in D_2O because its contrast is positive compared to the negative contrast of the protein in this solvent. The analysis of the full scattering curves was done by using two powerful programmes, CRYSOL developed for SAXS (Svergun *et al.* 1995) and its modification for SANS, CRYSON. Hydration layers around proteins and nucleic acids can be analysed in terms of electrostriction around charges and packing at the macromolecular surface. These are examples of contrast variation to study solvent components that can be distinguished through a difference in density with bulk solvent. In other examples, the hydration layer around proteins in glycerol or alcohol was characterized by SANS because of its difference in composition with bulk solvent (Lehmann and Zaccai 1984; Zaccai 1986).

12.14 Interacting functional systems in solution

Beautiful examples of SANS studies of interacting nucleic acid−protein systems in solution are the studies on elongation factors and amino-acyl tRNA (Antonsson et al. 1986; Bilgin et al. 1998) and on aminoacyl-tRNA synthetases and tRNA (Dessen et al. 1978; Giegé et al. 1982). The titrations of methionyl tRNA synthetase (MetRS) by tRNA^{met} in Dessen et al. (1978) shown in Figs 12.2 and 12.3 are an excellent illustration of the method. MetRS is a dimer that binds two tRNA molecules in an anticooperative fashion in buffer with 10 mM MgCl$_2$. The anticooperativity is removed in 50 mM MgCl$_2$. The $I(0)$ and R_g values from SANS experiments (Table 12.1) were plotted as a function of protein : tRNA ratio. The $A(0)$ values (Table 12.1) of protein and tRNA, respectively, were calculated from their compositions and volumes. In the H$_2$O solvent, both contribute with tRNA dominating, in 77% D$_2$O the tRNA is essentially matched out and only protein contributes, in D$_2$O, both contribute but protein dominates (Fig. 12.1). The $I(0)$ in any condition can be calculated exactly from these values and an interaction model that yields number concentrations for each component in the solution: protein dimer, protein monomer, tRNA, various complexes, etc. The lines through the data points in Figs 12.2 and 12.3 are for different interaction models. This permits a quantitative test of each model against the data. In Fig. 12.2, the panel corresponding to $I(0)$ in H$_2$O, the anticooperative binding appears clearly by the break of slope at the 1 : 1 protein : tRNA point. When, in 50 mM MgCl$_2$, two tRNA molecules are bound with high affinity, the corresponding $I(0)$ plot breaks at the 1 : 2 protein : tRNA ratio (Fig. 12.3). Note that the full curve is fitted quantitatively by the calculation and not just the break point. The nature of the change between 10 and 50 mM MgCl$_2$ appeared in the R_g dependence (see the panels in Figs 12.2 and 12.3): the dimer R_g increased significantly, presumably separating the two tRNA binding sites. A slight dissociation of the dimer upon binding tRNA is concluded from the drop in intensity in the 77% D$_2$O panel in Fig. 12.3; only the protein is 'visible' in this solvent. This dissociation is absent in Fig. 12.3, presumably because no strain is imposed by tRNA binding upon the dimer in 50% MgCl$_2$, the two sites being far enough from each other so as not to interfere. SANS experiments such as those of Dessen et al. (1978) have allowed a full characterization of structural aspects of the interactions in solution between the amino acylating enzymes and tRNA that would not be possible by any other method. They allowed an understanding of other biochemical data and paved the way for high resolution structural studies (Giegé et al. 1982).

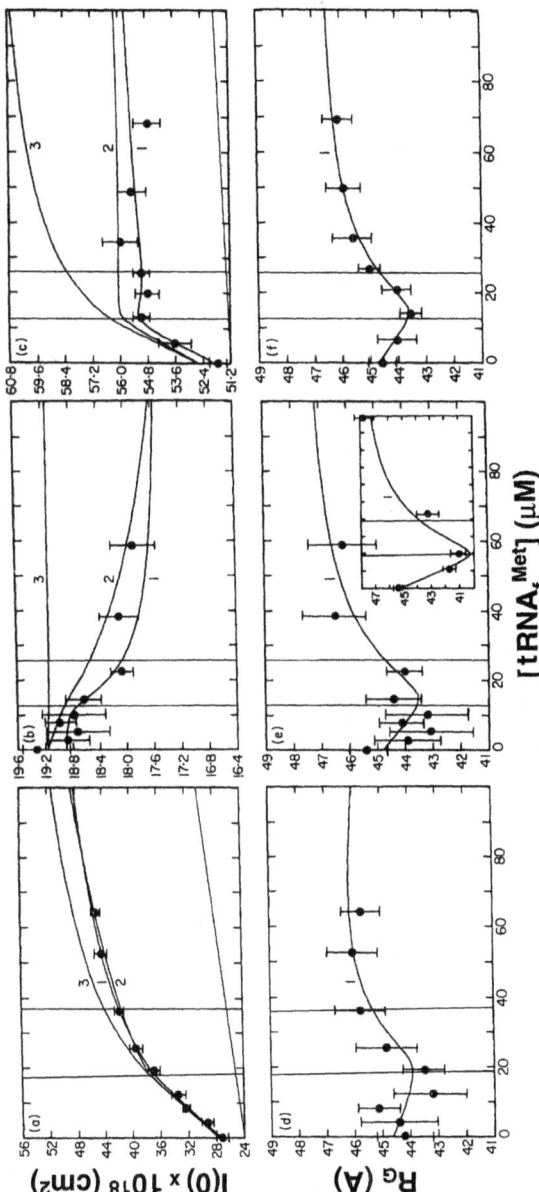

Fig. 12.2. $I(0)$ and R_g values from SANS during titration of *E. coli* methionyl-tRNA synthetase by tRNA^{met} in buffer with 10 mM MgCl$_2$, where the protein anticooperatively binds two tRNA molecules. Protein concentration was maintained constant. The faint vertical lines correspond to protein : tRNA ratios of 1 and 2. The lines 1, 2, 3 are from calculations for different interaction models. [From Dessen *et al.* (1978).]

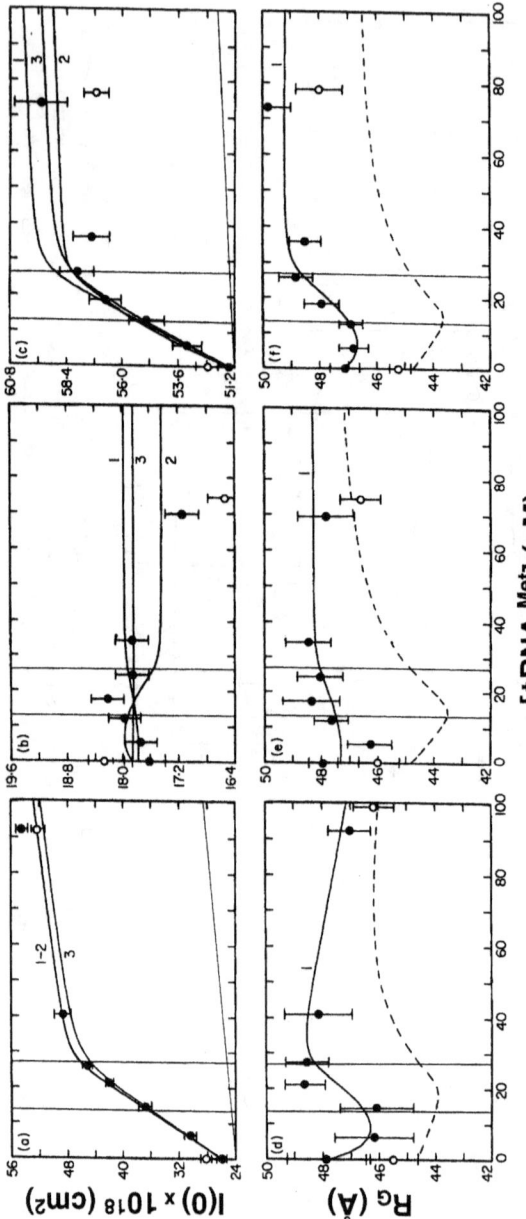

Fig. 12.3. $I(0)$ and R_g values from SANS during titration of *E. coli* methionyl-tRNA synthetase by tRNA^{met} in buffer with 50 mM MgCl$_2$, where the protein binds two tRNA molecules with equal and high affinity. The faint vertical lines correspond to protein : tRNA ratios of 1 and 2. The dotted curve recalls the data from 10 mM MgCl$_2$ (Fig. 12.2). The lines 1, 2, 3 are from calculations for different interaction models. [From Dessen *et al.* (1978).]

12.15 Conclusion

SANS approaches have great potential for future scientific studies that, for success, depend mainly on sufficient investment of effort in sample preparation, in order to have a well-defined system and structural problem.

The least effort required is for a solution study of a macromolecule, in which a milligramme or even less of pure material could allow a very useful characterization at the molecular level with a short experiment. At the other end of the scale, an experiment on an interacting system involving specific deuterium labelling is likely to require a very sophisticated effort in molecular biology and biochemistry. In early work, SANS was mainly a technique to characterize, albeit at very low resolution, the structure of systems that did not crystallize and where there was little hope of a high resolution structure. There has now been tremendous progress in the determination of high resolution structures, even of large systems, from synchrotron radiation crystallography and SANS is bridging the structure–function gap by evolving towards studies on complex interacting systems where the known high resolution structures of components can be used very usefully in the interpretation of scattering curves under different conditions.

References

Antonsson, B, Leberman, R, Jacrot, B and Zaccai, G (1986). *Biochemistry*, **25**, 3655–3659.
Bilgin, N, Ehrenberg, M, Ebel, C, Zaccai, G, Sayers, Z, Koch, MHJ, Svergun, D, Barberato, C, Volkov, V, Nissen, P and Nyborg, J (1998). *Biochemistry*, **37**, 8163–8172.
Block, M, Zaccai, G, Lauquin, G and Vignais, P (1982). *Biochemical and Biophysical Research Communications*, **109**, 471–477.
Bonneté, F, Ebel, C, Zaccai, G and Eisenberg, H (1993). *Journal of the Chemical Society Faraday Transactions*, **89**, 2659–2666.
Bonneté, F, Madern, D and Zaccai, G (1994). *Journal of Molecular Biology*, **244**, 436–447.
Calmettes, P, Durand, D, Desmadril, M, Minard, P, Receveur, V and Smith, JC (1994). *Biophysical Chemistry*, **53**, 105–114.
Dessen, P, Blanquet, S, Zaccai, G and Jacrot, B (1978). *Journal of Molecular Biology*, **126**, 293–313.
Ebel, C, Guinet, F, Langowski, J, Urbanke, C, Gagnon, J and Zaccai, G (1992). *Journal of Molecular Biology*, **223**, 361–371.
Ebel, C, Altekar, W, Langowski, J, Urbanke, C, Forest, E and Zaccai, G (1995). *Biophysical Chemistry*, **54**, 219–227.
Eisenberg, H (1981). *Quarterly Reviews of Biophysics*, **14**, 141–172.
Esser, AF, Thielens, NM and Zaccai, G (1993). *Biophysical Journal*, **64**, 743–748.
Giegé, R, Lorber, B, Ebel, JP et al. (1982). *Biochimie*, **64**, 357–362.
Glatter, O (1980). *Acta Physica Austriaca*, **52**, 243-256.
Guinier, A and Fournet, G (1955). *Small angle scattering of X-rays*, Wiley, New York.
Hunt, JF, McCrea, PD, Zaccai, G and Engelman, DM (1997). *Journal of Molecular Biology*, **273**, 1004–1019.
Jacrot, B (1976). *Reports on Progress in Physics*, **39**, 911–953.
Jacrot, B and Zaccai, G (1981). *Biopolymers*, **20**, 2414–2426.
Lederer, H, Tovar, K, Baer, G, May, RP, Hillen, W and Heumann, H (1989). *EMBO Journal*, **8**, 1257–1263.

Lehmann, MS and Zaccai, G (1984). *Biochemistry*, **23**, 1939–1942.
Li, ZQ, Giegé, R, Jacrot, B, Oberthur, R, Thierry, JC and Zaccai, G (1983). *Biochemistry*, **22**, 4380–4388.
Moore, PB (1988). *Nature*, **331**, 223–227, and references therein.
Pavlov, MY, Rublevskaya, IN, Serdyuk, IN, Zaccai, G, Leberman, R and Ostanevich, YM (1991). *Journal of Applied Crystallography*, **24**, 243–254.
Serdyuk, IN, Grenader, AK and Zaccai, G (1979). *Journal of Molecular Biology*, **135**, 691–708.
Serdyuk, IN, Pavlov, MYu, Rublvskaya, IN, Zaccai, G and Leberman, R (1994). *Biophysical Chemistry*, **53**, 123–130.
Stegmann, R, Nieba-Axmann, S, Marakova, E, Rössle, M, Hermann, T, May, RP, Wiedemann, A, Plückthun, A and Heumann, H (1998). *Journal of Structural Biology*, **121**, 30–40.
Svergun, DI, Barberato, C and Koch, MHJ (1995). *Journal of Applied Crystallography*, **28**, 768–773.
Svergun, DI, Koch, MHJ, Kuprin, S, Richard, S and Zaccai, G (1998). *Proceedings of the National Academy of Sciences (USA)*, **95**, 2267–2272.
Timmins, PA and Zaccai, G (1988). *European Biophysical Journal*, **15**, 257–268.
Zaccai, G and Jacrot, B (1983). *Annual Review of Biophysics and Bioengineering*, **12**, 139–157.
Zaccai, G (1986). *Methods in Enzymology*, **127**, 619–629.
Zaccai, G, Cendrin, F, Haik, Y, Borochov, N and Eisenberg, H (1989). *Journal of Molecular Biology*, **208**, 491–500.
Zaccai, G, Aude, CA, Thielens, NM and Arlaud, GJ (1990). *FEBS Letters*, **269**, 19–22.

13

Structure and dynamics of biological membranes

Georg Büldt, Ramona Schlesinger, Eva Pebay-Peyroula and Hans Jürgen Sass

Biological membranes are the barriers of cells and organelles against the outside medium. The two major components of membranes are lipids and proteins. Lipids are amphiphilic molecules, which in an aqueous medium spontaneously form double-layered structures. The amphiphilic character of the lipids with polar head-groups and apolar hydrocarbon chains is responsible for the two-dimensional geometry of biological membranes. Integral membrane proteins, which span the lipid bilayer, mediate the transport of matter and signals. Soluble proteins can be associated with one side of a membrane by electrostatic interactions or through a lipid anchor covalently bound to the protein. A membrane can also contain protein anchors for filamentous structures, which mechanically stabilize the cell.

The first part of this chapter describes X-ray and neutron diffraction experiments on pure lipid bilayer systems, whereas the second part presents experiments concerning the structure and dynamics of membrane proteins. Scattering experiments on biological membranes or on pure bilayer systems are normally carried out on aggregates such as single shell vesicles or multibilayer systems. Diffraction experiments on membranes have two levels in the degree of difficulty. Recording diffraction patterns and simply measuring positions of diffraction peaks provide geometrical information of bilayer systems such as the thickness of the membranes, packing of lipid chains, secondary structure of membrane proteins and membrane protein packing. Any non-specialist can collect this information without spending too much effort. Measuring integrated intensities of diffraction peaks, solving the phase problem and calculating density maps in order to solve structures is much more complicated.

13.1 Pure lipid bilayers

Although in a biological membrane proteins modify the interactions between membranes as well as between membrane components, it is the amphiphilic architecture of a lipid molecule that is responsible for its geometric arrangement in the form of bilayers. In addition, the inherent ability of most lipids to take up a large number of conformations and their segmental mobility leads them to form an excellent two-dimensional solvent for membrane proteins. Therefore, to have a clear understanding of the forces that are responsible for the various lipid and bilayer properties, an investigation of pure lipid bilayer systems is necessary. X-ray and neutron diffraction are the most powerful tools for characterizing lipid and bilayer morphology and for following up their changes due to alterations in the number of components and their concentrations. It should be noted that ions and other molecules in the water phase adjacent to the bilayer surface influence the physical state of membranes and should therefore be considered as a component of the bilayer.

As a well-written introductory article for the diffraction method applied to bilayer systems, the paper of Frank and Lieb (1981) is recommended.

13.1.1 Diffraction experiments on oriented films of multilamellar stacks of bilayers

With respect to the basic principles of a diffraction experiment, there is no large difference between neutrons and X-rays. As the flux of neutron sources is several orders of magnitude smaller than for X-rays, the sample volume in a neutron diffraction experiment is typically of the order of $10 \times 20 \times 1\,\text{mm}^3$ and therefore much larger than in the X-ray case. In the following, we describe the experiments carried out with X-rays. The same type of experiments can be done with neutrons, after modifying the instrumentation with respect to beam size, scattering volume and detectors.

In a first experiment, a sonicated dispersion of about 4 mg of the lipid 1,3-DPPC (1,3-dipalmitoyal-glycero-2-phosphocholine) in 0.2 ml H_2O is spread on a glass slide and allowed to evaporate slowly. After evaporation of the bulk water, a multibilayer film is formed with predominant orientation parallel to the bilayer surface. The slide with the lipid film is aligned in an X-ray beam that hits the film tangentially (Fig. 13.1A). In quantitative experiments the sample should be controlled with respect to temperature and relative humidity, which is achieved by a sample chamber with windows transparent for the X-ray beam.

In Fig. 13.1B three diffraction patterns of this lipid film are shown. The equidistant intensity spots on the meridian of the X-ray film prove that we have a one-dimensional periodicity along the direction perpendicular to the lipid film. From Bragg's law

$$|\mathbf{G}| = |\mathbf{s}| \quad \text{with} \quad |\mathbf{G}| = \frac{h}{d} \quad \text{and} \quad |\mathbf{s}| = \frac{2}{\lambda}\sin\theta;$$

we obtain

$$\frac{h}{d} = \frac{2}{\lambda}\sin\theta \quad \text{with} \quad h = 1, 2, 3, \ldots,$$

where d is the repeat distance of this one-dimensional lattice, which is the thickness of the lipid bilayer plus the water layer. The Bragg angle θ is calculated from $\tan 2\theta = 1/L$ (see Fig. 13.1A).

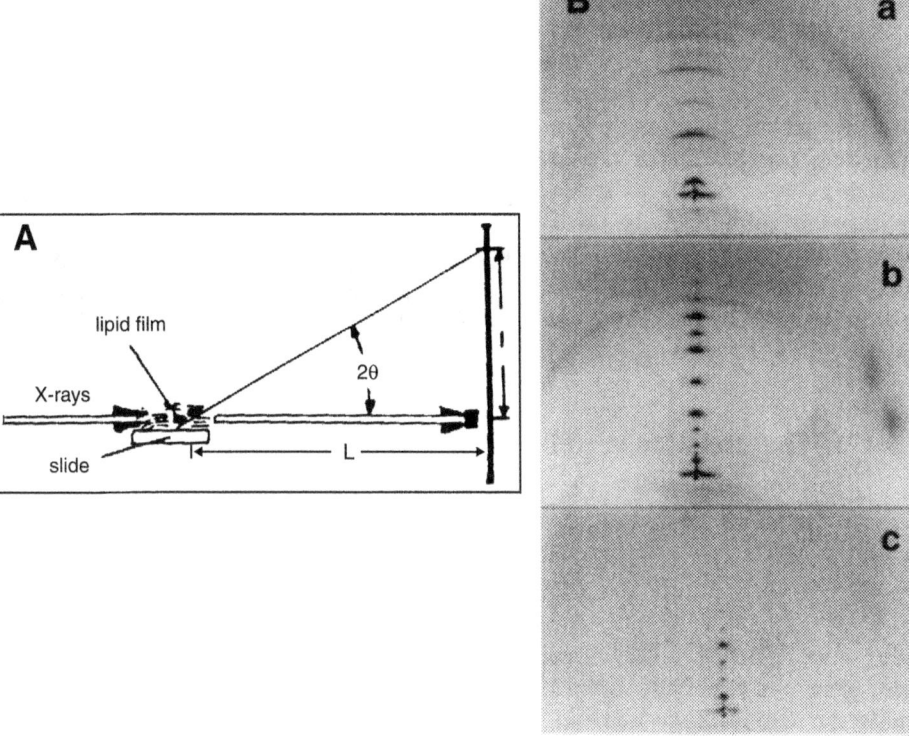

Fig. 13.1. (A) Experimental set-up for an X-ray diffraction experiment on bilayer stacks oriented on a glass slide. The incident beam is absorbed by a small lead beam stop in front of the X-ray film (L: sample-to-film distance, l: distance of a diffraction spot on the film from the origin of the diffraction pattern). (B) Diffraction pattern of oriented bilayer stacks of 1,3-DPPC (a) in the gel phase, not annealed (broad mosaic spread), at 22 °C, about 6% H_2O content; (b) same conditions as in (a), but after the annealing procedure (narrow mosaic spread); (c) in the fluid phase, 70 °C, about 8% H_2O content. [Photographs from Büldt and de Haas (1982).]

The diffraction spots in Fig. 13.1B(a) are extended arcs. The intensity distribution over an arc is directly proportional to the number of domains in the lipid film with different orientations relative to the glass side. The angular distribution of all domains in the sample is called the mosaic spread, which in (a) is of the order of 25° FWHM (full-width-at-half-maximum). The mosaic spread for this lipid film can be reduced by an annealing procedure, by heating up the sample to 110 °C for several hours and then cooling it down slowly (Büldt et al. 1979). The procedure results in an enhanced intensity of diffraction orders, in this case by a factor of 20 as seen in (b). The diffraction patterns (a) and (b) were collected at 22 °C where DPPC is in the so-called gel phase, which is characterized by hydrocarbon chains in all-*trans* conformation. When the temperature is raised to 70 °C, i.e. above the first order phase transition temperature of DPPC, the sample is in the fluid phase with hydrocarbon chains forming *gauche+* and *gauche−* conformations that result from rotations around C–C bonds by +120° or −120°. The diffraction pattern of the fluid phase is shown in (c). The Bragg peaks are further apart, which means that the bilayer plus water layer become smaller. This is the result of a reduction in the thickness of the hydrocarbon chains due to segmental disorder. We also observe a decrease in intensity that is quite dramatic for diffraction spots greater than $h=6$. This is caused by the Debye–Waller factor, which is a Gaussian function centred on $s=0\,\text{Å}^{-1}$ with a width depending on the disorder and motion of lipid segments in the bilayer.

Off-axis to the equator at $1/s = 4.2\,\text{Å}$, one single reflection spot is visible in (a) and (b). This reflection originates from the packing of hydrocarbon chains. If we assume a hexagonal arrangement, the observed reflections can be indexed by $h, k = 1, 0$. Using Bragg's law

$$\frac{2}{\sqrt{3}a}\sqrt{h^2 + hk + k^2} = \frac{2}{\lambda}\sin\theta,$$

the lattice constant a, which is the distance between two neighbouring chains, is determined to be $a = 4.85\,\text{Å}$. The off-axis position of the chain-packing reflection is caused by a tilt of hydrocarbon chains with respect to the bilayer normal (Tardieu et al. 1973). This reflection becomes more pronounced at smaller mosaic spread (b) and is very weak in the fluid phase. The increase of disorder in this phase causes a broadening and a shift of the chain reflection from 4.2 to 4.5 Å.

13.1.2 *Diffraction experiments on powders of multilamellar stacks of bilayers*

For sample preparation, 7.5 mg of DPPC is mixed with 2.5 mg H_2O and placed in a glass capillary of 1 mm diameter with 0.01 mm wall thickness. The capillary is closed by wax and aligned perpendicular to an X-ray beam. The X-ray film shows a pattern of concentric equidistant rings centred around the origin of the diffraction pattern. The ring system originates from the diffraction of multilamellar bilayer stacks in the powder. The chain packing gives a ring at 4.2 Å. The lamellar reflections in the diffraction pattern of the fluid phase, which can be obtained by heating the capillary with hot air, are shifted away from the centre, which means that the thickness of the bilayer plus water become smaller. The chain reflection is broadened and shifts to 4.5 Å, as observed for the oriented sample.

13.1.3 *Lipid polymorphism*

It was demonstrated in the previous sections how X-ray diffraction is used to obtain the geometrical parameters of multilamellar stacks of bilayers. However, lipids and even a single kind of lipid can aggregate in many different structures depending on the water content and temperature (Fig. 13.2). These forms exhibit different degrees of long range and short range orders and different kinds of

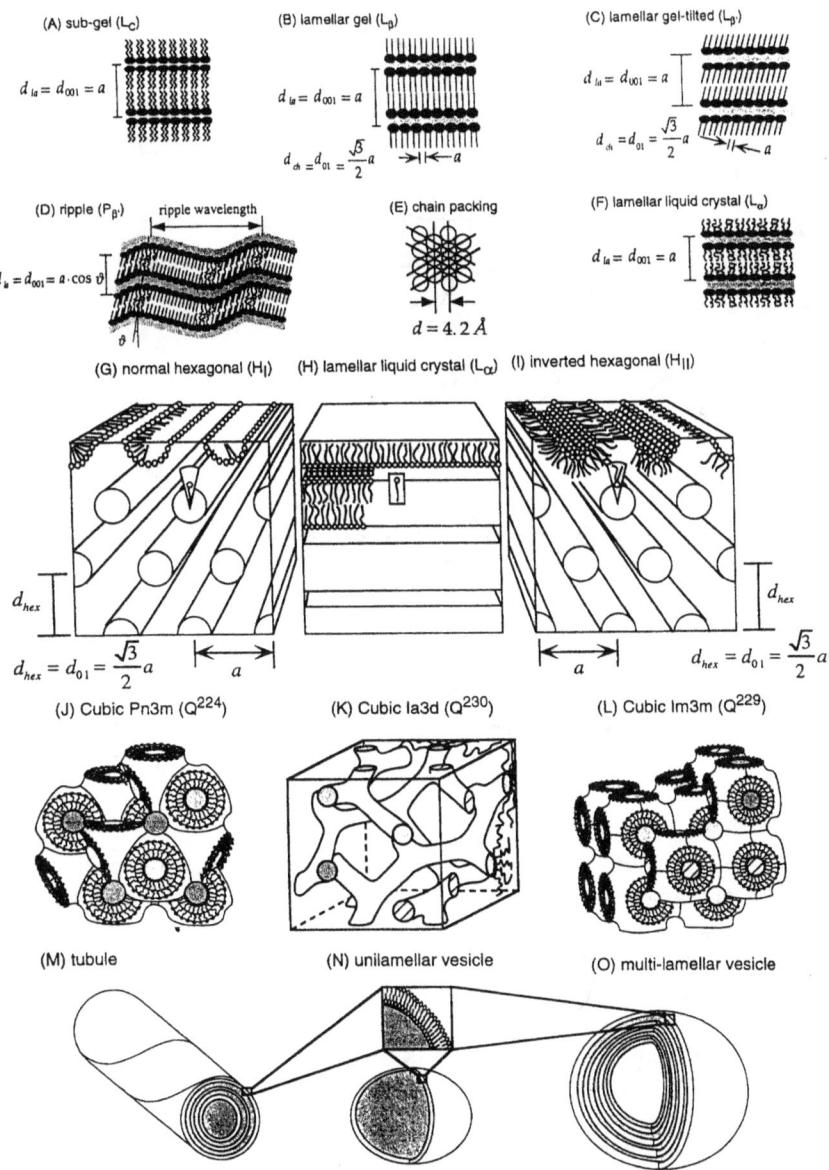

Fig. 13.2. Schematic view of the various mesophases, crystalline forms, and states of aggregation adopted by membrane lipids. (A) Lamellar subgel phase (L_c); (B) lamellar gel phase (untilted chains, L_β); (C) lamellar gel phase (tilted chains, $L_{\beta'}$); (D) ripple phase ($P_{\beta'}$); (E) cross-sectional view of the hydrocarbon chains in a hexagonal close-packed arrangement (view is down the long axis of the chains); (F) lamellar liquid crystalline phase (L_α); (G) normal hexagonal phase (H_I); (H) lamellar liquid crystalline phase (L_α); (I) inverted hexagonal phase (H_{II}); (J) inverted cubic phase (space group Pn3m, number 224); (K) inverted cubic phase (space group Ia3d, number 230); (L) inverted cubic phase (space group Im3m, number 229); (M) tubule; (N) unilamellar vesicle; (O) multilamellar vesicle. The labeled elements refer to parameters or dimensions commonly used in identifying structural features of phases. [Figure from Caffrey and Cheng (1995).]

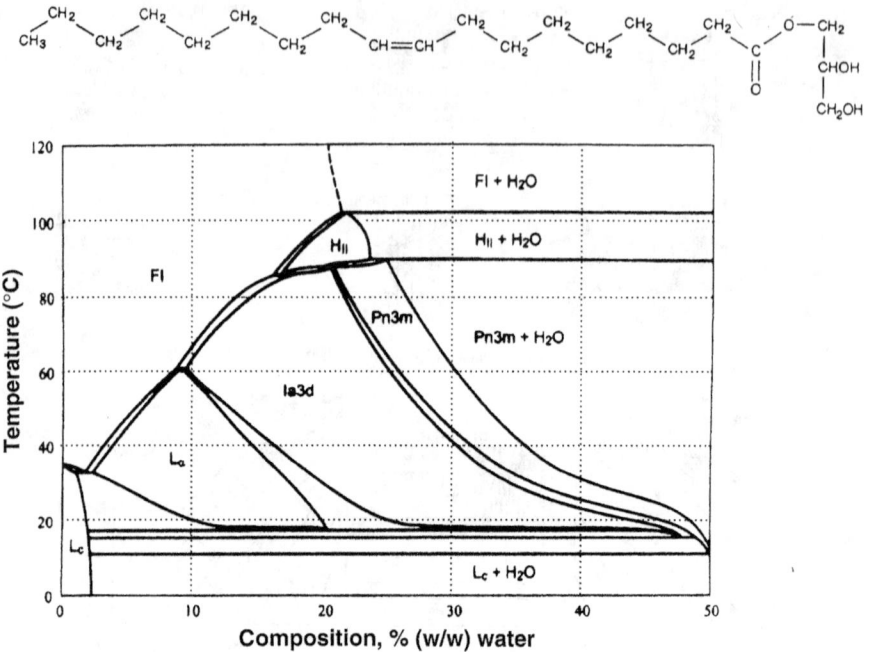

Fig. 13.3. Chemical formula of 1-monooleoyl-glycerol (MO or monoolein) and its phase diagram. [Phase diagram from Rummel et al. (1998).]

periodicities. Figure 13.3 depicts the chemical structure and the phase diagram of 1-monooleoyl-glycerol (MO or monoolein). The hydrocarbon chain, which contains one double bond at the 9th carbon atom, is attached to the first carbon atom of glycerol. The different geometries of these structures as well as their boundaries in the phase diagram were mainly determined by X-ray diffraction. An interesting feature of the cubic phases is that they form two continuous networks of water channels which are not connected to each other. The cubic lipid phases became of interest for crystallizing membrane proteins since in 1996 Landau and Rosenbusch succeeded in crystallizing bacteriorhodopsin in the cubic phase Pn3m of MO (see Section 13.2).

13.1.4 The conformation of a phospholipid molecule in a bilayer as studied by neutron diffraction

In the previous sections, we have derived the geometrical information on bilayer systems. In this section it will be shown how coordinates of certain segments in a lipid molecule in the bilayer are obtained, applying methods of one-dimensional crystallography to the one-dimensional periodicity of multilamellar stacks of bilayers. Structure factors, derived from lamellar diffraction spots, give only profile information of the bilayer, that is the projection of density on to the bilayer normal. It is not possible to see the positions of certain segments in the bilayer profile in the gel or fluid phase using X-ray diffraction. The fluctuations of individual segments along the bilayer normal are quite large so that neighbouring segments overlap. Therefore, it is impossible to determine the average position of a segment. This problem was solved by neutron diffraction in combination with selective deuteration of a molecular group, thus enhancing the local contrast. In this way the average position and the extent of fluctuations of a segment were determined from a difference of two density maps with and without the deuterated group. In general, one great advantage for the application of neutrons over X-rays in biological structures is the possibility of contrast enhancement by isotope exchange, which normally is a very small disturbance for the structure. For biological material, in most cases, the difference in coherent scattering length between hydrogen ($b = -0.374 \times 10^{-12}$ cm) and deuterium ($b = +6.67 \times 10^{-12}$ cm) is used. A neutron diffraction experiment on an oriented lipid film is performed in a similar way to an X-ray experiment. However, the area of the lipid film in a neutron diffraction experiment has to be quite large (about 25×50 mm^2). The diffraction pattern of such a film is shown in Fig. 13.4.

The experimental details are given in the figure legend. The differences in relative height of the diffraction orders demonstrate the sensitivity of the

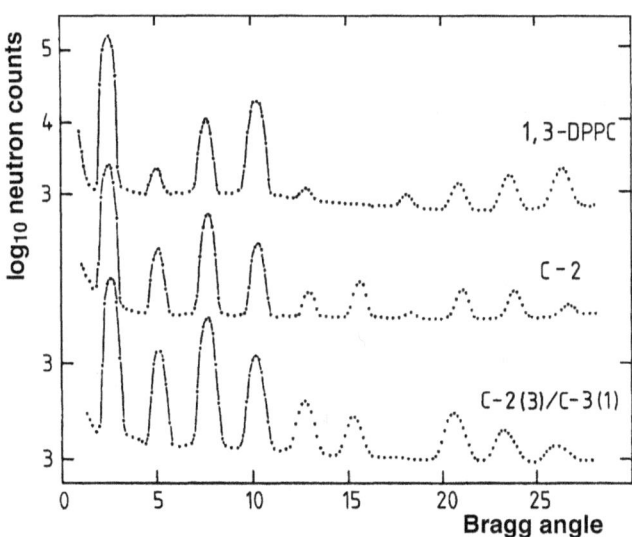

Fig. 13.4. Log plots of neutron counts vs the Bragg angle (θ) for 1,3-DPPC in H$_2$O measured on the diffractometer D16 at the ILL Grenoble ($\lambda = 4.5$ Å): upper plot: non-deuterated 1,3-DPPC; pattern in the middle: 1,3-DPPC deuterated at the first methylene segment or at the second carbon atom (C-2) in both chains; third plot: the same lipid, deuterated at the second carbon atom in chain 3 (C-2(3)) and at the third carbon atom in chain 1 (C-3(1)). All three patterns are taken in the gel phase at 20 °C and 6% H$_2$O content, lamellar spacing 52.2 Å. [From Büldt and de Haas (1982).]

method with respect to label positions, e.g. the change in label position to a neighbouring segment leads already to a remarkable change in the relative intensities (Fig. 13.4).

The absolute values of the structure factors are derived from the integrated intensities of diffraction spots. Since pure lipid bilayers are centrosymmetric, the phase problem reduces to a distinction between phase angles 0 or π. This problem can be solved by H_2O/D_2O exchange experiments using the water layer between membranes as an isomorphous replacement site.

Figure 13.5A shows the profile of DPPE (1,2-dipalmitoyl-*sn*-glycero-3-phosphoethanolamine), selectively deuterated in the first methylene segments of both chains, together with the profile of the non-deuterated DPPE. The resulting difference profile (lower trace) shows two peaks at positions 17.4 and 21.1 Å. Although the profile is quite noisy, the numbers can be determined very

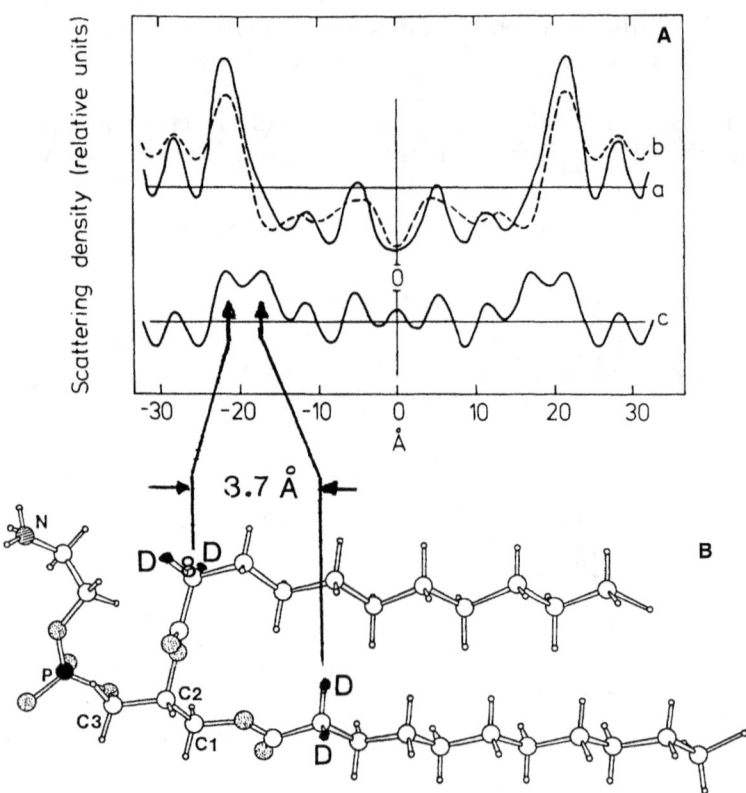

Fig. 13.5. (A) Neutron scattering length density profiles calculated from 10 structure factors of (a) DPPE deuterated at the C-2 position, (b) of the non-deuterated sample, and (c) the difference of these profiles. [Plot from Büldt and Seelig, (1980), with permission.] (B) Molecular model of a phospholipid in a bilayer according to the distance of the first methylene segments in both hydrocarbon chains. [Figure from Büldt (1994).]

accurately from a refinement procedure on the structure factors. The main result of this profile is that the first methylene segments in both hydrocarbon chains are shifted with respect to each other by 3.7 Å along the bilayer normal (Büldt and Seelig 1980).

The consequence of this experimental finding on the molecular model of a phospholipid is illustrated in Fig. 13.5B. The glycerol backbone is oriented perpendicular to the membrane plane with the sn-1 chain following this direction, whereas the sn-2 chain becomes parallel to sn-1 after a sharp bend. Using the deuterium magnetic resonance this conformation was shown to be general for all phospholipids in all phases and membranes (J. Seelig and A. Seelig 1980) and even in its crystal structure where it was first discovered (Hitchcock *et al.* 1975).

13.2 The light-driven proton pump bacteriorhodopsin

A wide range of methodical approaches is available for the investigation of membrane proteins with neutrons or X-ray synchrotron radiation. A number of membrane proteins have now been crystallized. The first two structures solved had fundamentally different compositions in secondary structure. The reaction centre showed α-helical rods traversing the membrane (Deisenhofer *et al.* 1985) whereas a β-barrel forms a wide pore in porin (Weiss *et al.* 1991).

In this section the integral membrane protein bacteriorhodopsin (bR) is chosen as an excellent example for demonstrating how information is obtained from different biophysical methods to understand the function of this membrane protein on the basis of its structure and dynamics. There are only a few such proteins which have attracted the interest of a large number of research groups.

13.2.1 *The photocycle and proton translocation*

Bacteriorhodopsin is densely packed in hexagonal two-dimensional lattices, the so-called purple membranes (PM) in the plasma membrane of the *Halobacterium salinarum*. It was found by Oesterhelt and Stockenius (1973) to contain the chromophore, retinal, covalently bound via a Schiff base to Lys 216, which enables this protein to pump protons across the cell membrane to the outside medium (Fig. 13.6). The photocycle of bR (Fig. 13.7) is linked to this active transport of protons by delivering part of the energy of the absorbed photon to this process. In the ground state (bR_{568}) of the photocycle the retinal is in an all-*trans* conformation, the Schiff base being protonated ($pK \sim 13$) and positively charged. Upon absorption of a photon, the retinal goes via an excited state to a 13-*cis* retinal conformation, which results in a dramatic pK down-shift to 3. Hence the Schiff's base proton is taken up by Asp 85 (Fig. 13.6) and another proton appears at the extracellular surface of bR. These steps pass through intermediate states of the protein which can be monitored by colour changes of the chromophore (Fig. 13.7) that originate from the altered electric field at the chromophore position and the different isomerization states of the retinal. At this point of the photocycle, bR has reached its M_{412} intermediate which is easily detected by the blue shift of the absorption maximum to 412 nm. In order

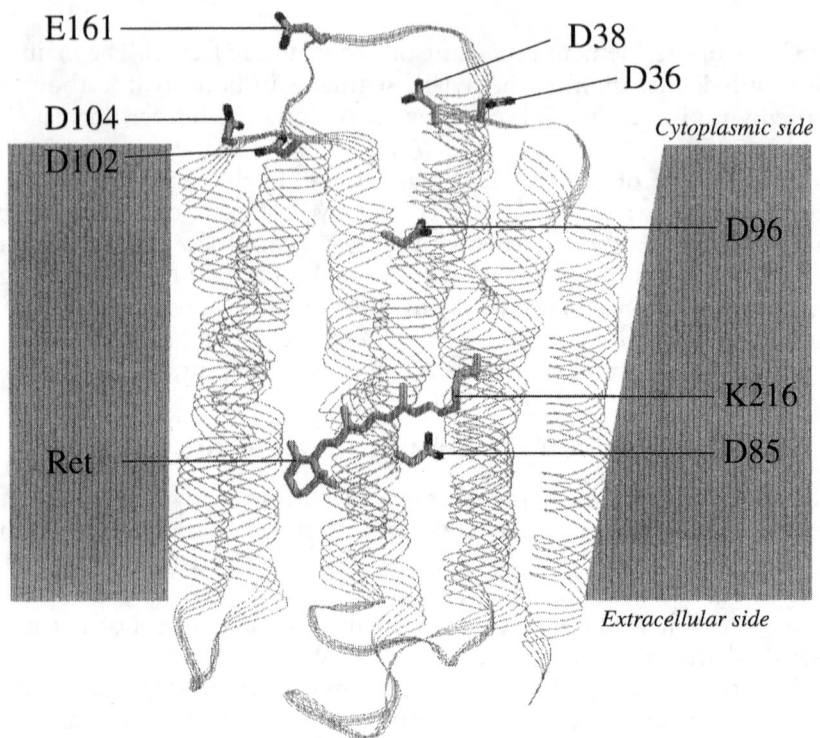

Fig. 13.6. Schematic picture of the seven transmembrane α-helices of bR, displaying the retinal in its all-*trans* conformation connected to K216 and several negatively charged amino acids.

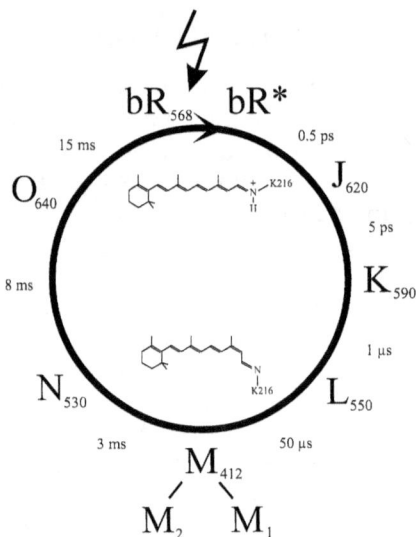

Fig. 13.7. The photocycle of bR showing the light-adapted ground state bR_{568} and several intermediates with their absorption maxima and lifetimes. The conformation of all-*trans* and 13-*cis* retinals are also depicted.

to complete the cycle and to return to its initial state, the Schiff base has to be reprotonated via the protonated Asp 96 (Fig. 13.6). This amino acid will then be reprotonated and the retinal is driven back to its all-*trans* conformation.

It should be noted that this active proton translocation can take place against a proton gradient. It seems clear that the vectoriality of the pump is achieved by controlled changes of the pK from different groups, the accessibility of certain positions in the structure for protons and the position of water molecules to bridge the gaps in the proton translocation pathway. What drives these changes in pK, accessibility and water positions? So far, experiments reveal three observations: the retinal isomerization is followed by a charge redistribution, which then drives conformational changes within the protein.

In the following we would like to present some structural details that give an understanding of these processes. Diffraction methods using X-rays, neutrons and electrons have contributed a lot to the current understanding of the proton pumping mechanism. Further, important details were obtained, especially from FTIR spectroscopy and site-directed mutagenesis.

13.2.2 *The arrangement of bR molecules and α-helices in the PM*

The characterization of the aggregation state of bR in the PM was given by X-ray diffraction on these natural two-dimensional lattices (Blaurock 1975; Henderson 1975). These experiments are best understood by comparing the density in real space with the intensity distribution in reciprocal space. Figure 13.8 illustrates in (A) a two-dimensional protein lattice in real space and its counterpart in reciprocal space. The $h-k$ plane, which corresponds to the plane of two-dimensional periodicity in real space, has a distinct intensity distribution at reciprocal lattice points. Perpendicular to these points in the l-direction, continuous intensity distributions are obtained, the so-called Bragg rods, which correspond to the direction with no periodicity in real space. This is the situation for electron diffraction in electron microscopy where a diffraction pattern from a single protein lattice can be recorded (Henderson and Unwin 1975), since the interaction of an electron with matter is much larger than for X-rays or neutrons. The latter case is illustrated in (B), where several millions of lattices form a film of PMs. In this film, lattices have small but different angles to the supporting surface, which determines their mosaic spread. Within each stack, lattices are randomly oriented with respect to each other relative to an axis perpendicular to the membrane plane. The corresponding reciprocal lattices are also randomly oriented with respect to the l-axis and form a system of concentric cylinders, which have a continuous intensity distribution in the l-direction. Only the Bragg rod at $h, k = 0, 0$, which is on the l-axis, is related to the Fourier transform of the projected density on the membrane normal (the profile) and is therefore periodic in real space, showing distinct intensity spots in reciprocal space. In scattering geometry (i), where the incident beam runs parallel to the PMs, the Ewald sphere or 'sphere of reflection' intersects the reciprocal space tangentially to the l-axis at the origin. The intersection picture is shown on the X-ray film with the lamellar spots on the meridian and the intersection lines of the

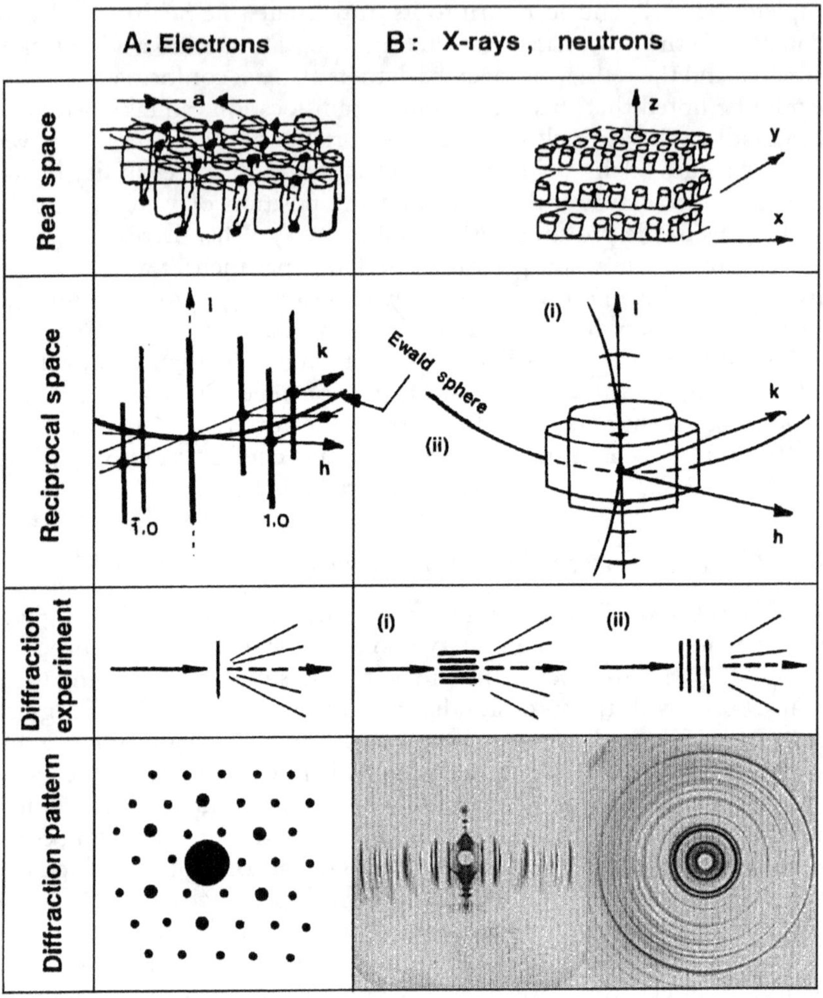

Fig. 13.8. (A) Density in real space and intensity distribution in reciprocal space of a single two-dimensional protein lattice (PM) and an electron diffraction experiment. (B) Intensity distribution of a stack of two-dimensional lattices randomly oriented with respect to the z-axis. X-ray films show diffraction pattern of stacks of PMs in two different scattering geometries (i) and (ii). [Figure from Büldt (1994).]

Ewald sphere with the concentric cylinders on the equator. In configuration (ii), with the incident beam perpendicular to purple membranes, the Ewald sphere intersects the reciprocal space tangentially to the $h-k$ plane at the origin, resulting in concentric rings on the X-ray film. From similar experiments on oriented PM films, Blaurock (1975) and Henderson (1975) determined the geometry of the lattice and the predominant type of secondary structure (α-helices). Even the orientation of the membrane spanning the α-helices was determined from the 1.5 Å reflection on the meridian. The intensity distribution of this reflection perpendicular to the meridian gave an indication of the deviation of the individual α-helices from the membrane normal.

13.2.3 The electron microscopic structure of bR and the refinement of this structure by neutron and X-ray diffraction

The basic picture of the tertiary structure, describing the three-dimensional arrangement of α-helices spanning the membrane, was given in several papers on low dose electron microscopy [e.g. Unwin and Henderson (1975); Henderson and Unwin (1975)]. The knowledge of the amino acid sequence (Ovchinnikov et al. 1979) allowed the prediction of a folding model by determining seven α-helical stretches in the polypeptide chain, which were assigned to the seven α-helices seen in the electron microscope structure (Engelman et al. 1980). In the following years, neutron diffraction was the method of choice to prove this model using bR molecules containing certain perdeuterated amino acids (Engelman and Zaccai 1980; Trewhella et al. 1983) and partially deuterated α-helices (Trewhella et al. 1986; Popot et al. 1989). This technique was also successful in the localization of perdeuterated or partially deuterated retinal within the protein (Jubb et al. 1984; Heyn et al. 1988). Heavy-atom labelled retinal analogues incorporated into the protein confirmed these results by X-ray diffraction (Büldt et al. 1991). In early neutron diffraction experiments the hydration of PMs was studied (Zaccai and Gilmore 1979) and more recently the location of the proton channel in bR was determined by neutron diffraction in combination with H_2O/D_2O exchange experiments (Papadopoulos et al. 1990).

In 1990, cryo-electron microscopy provided an additional breakthrough providing a structure of bR at 3.5 Å resolution in the $X-Y$ plane and 7 Å in the Z-direction. This picture allowed for the first time, including all the information from other methods mentioned above, the construction of a molecular model of bR (Henderson et al. 1990). This model was further improved by cryo-electron microscopy data of Grigorieff et al. (1996) and Kimura et al. (1997).

13.2.4 Structural investigations on photocycle intermediates using PMs

For a complete understanding of the function of a protein such as bR, it would be desirable to follow the structural changes in space and time with high resolution parallel to the working cycle. Since the intensity scattered from a single molecule is too low, taking into account the limits set by radiation damage, an ensemble of molecules has always to be considered. In this situation information about the structural changes during the working cycle can be obtained in two ways, either by trapping intermediate states or by time-resolved detection of the scattered intensity after excitation. Both alternatives have been successfully carried out in the case of bR.

13.2.4.1 Trapping of the M state in wild-type bR

After the observation that a proton leaves bR at the extracellular side during the transition from L to M, the M intermediate was considered to be a strategic state for the pumping process. It seemed that the knowledge of the M state structure would give some insight into the mechanism of proton translocation. One of the first trapping experiments of the M intermediate at low temperatures was performed by Glaser et al. (1986) using electron diffraction for detecting the M state structure. These experiments showed no intensity changes in the resolution region from 60 to 5 Å and small changes between 5 and 3.5 Å. Therefore neutron diffraction experiments were undertaken by Dencher et al. (1989) with the aim to observe changes in the distribution of water molecules compared to the ground state. As it was known that guanidine hydrochloride (GuaHCl) at high pH slows down the decay of the M state, a stack of several PM films was soaked in a buffer containing GuaHCl at pH 9.4 and illuminated at +8 °C. The films became yellow, indicating complete transformation to the M state, which was then preserved at liquid nitrogen temperatures in a cryostat. Neutron diffraction patterns of the ground state and the M intermediate showed clear differences in the reflections of up to 9% in $\sum|\Delta I|/\sum I$ in the resolution region of 60–7 Å. By comparing diffraction patterns between films in H_2O and D_2O it was deduced that at least 80% of the differences resulted from changes in the protein structure and only a minor contribution could originate from a redistribution of water molecules. These observations indicate that small changes in the tertiary structure of bR take place during the photocycle. These results are in contradiction to the electron diffraction experiments of Glaser et al. (1986).

13.2.4.2 Trapping the M state in the mutant Asp 96 Asn

It was observed by optical spectroscopy that the bR mutant Asp 96 Asn is characterized by a large decrease in the decay rate of the M state with increasing pH. Koch et al. (1991) performed X-ray diffraction experiments on films of this mutant under continuous illumination at room temperature and pH 9.6. They found intensity changes between the bR_{568}- and M_{412}-state structures (Fig. 13.9B) similar to those observed in the neutron diffraction measurements.

13.2.4.3 Time-resolved X-ray diffraction experiment on the bR mutant Asp 96 Asn

In order to examine how these structural changes correlate with relaxation processes in the photocycle and pumping cycle of bR, the structural transition from the M state to the ground state was followed by time-resolved X-ray diffraction using intense synchrotron radiation (Koch et al. 1991). The time course of flash-induced changes for three reflections at neutral pH is illustrated in Fig. 13.10. The changes in individual reflections before and immediately after the light flash are consistent both in amplitude and in direction with the steady-state experiments. A comparison of these structural relaxation times with optical decay rates of intermediate states in the photocycle indicated that the observed structural changes decay with the transition from the N state to the ground state. In functional terms this means that the structural changes relax after the reprotonation of the Schiff base.

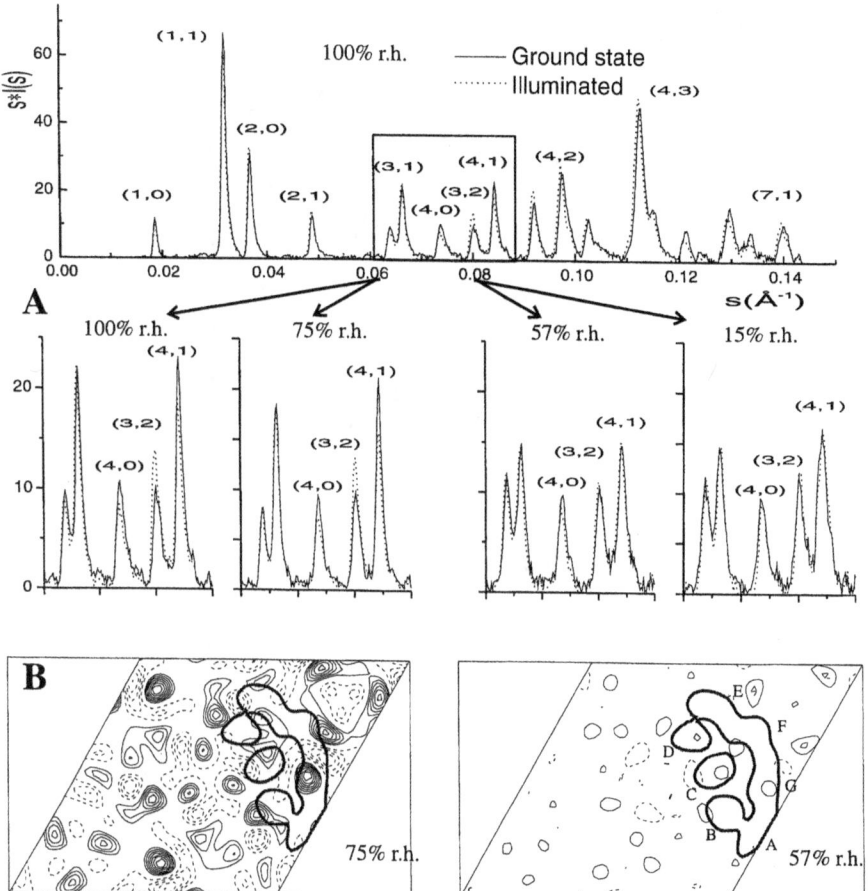

Fig. 13.9. (A) X-ray diffraction patterns of bR D96N light-adapted PMs (pH 9.6, room temperature) at different hydration levels in the absence of light (solid line) and under steady-state illumination (dotted line). $s = 2 \sin \theta / \lambda$, with Bragg angle θ and wavelength $\lambda = 1.5$ Å. The lower panels represent the reflection range (2,2) to (4,1) at different RH on an expanded scale. (B) Difference electron density maps (M state minus light-adapted ground state) of the bR D96N PMs at different hydration levels. Left: 57% RH, M_1 state. Right: 75% RH, M_2 state. The bold contour outlines the bR monomer, individual helices are marked by upper case letters from A to G. Continuous lines correspond to positive, dashed lines to negative electron density levels. Contour levels are scaled to each other in both maps.

13.2.4.4 M splits into two states M_1 and M_2

Up to this point, the relaxation of tertiary structural changes had been followed by time-resolved X-ray experiments. The onset of these changes could not precisely be attributed to the photocycle intermediates. A first hint of an answer to this problem was obtained from X-ray diffraction experiments on bR mutant Asp 96 Asn (pH 9.6) at different hydration levels (Sass *et al.* 1997). PM films equilibrated at different relative humidities (RHs) (15%, 57%, 75% and 100%) were

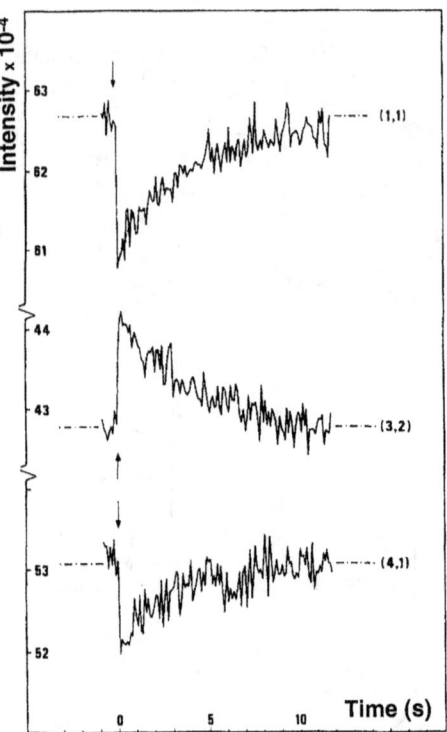

Fig. 13.10. Time course after a short light flash (∼ 0.75 ms FWHM) at $t = 0$ s of the integrated intensity including background under the (1,1), (3,2) and (4,1) reflections for PMs with mutated bR D96N at pH 9.6 (100 mM carbonate buffer, 500 mM KCl). The data were collected with a time resolution at 100 ms.

transformed to the M state by continuous illumination. Films, equilibrated at RHs above 60% showed the known changes in the tertiary structure, whereas films below 60% RH displayed only very small changes in their diffraction patterns (Fig. 13.9A). The corresponding difference density maps at high humidity show positive difference density at helices B, F and G (Fig. 13.9B). No significant difference peaks can be seen below 60% RH.

These experiments demonstrated that two types of M states with and without structural changes compared to the ground state exist. Reconsidering the earlier experiments of Glaeser et al. (1986), it is very probable that the M_1 intermediate was accumulated and therefore no changes in the tertiary structure were observed. If one assumes that both M states would occur in the photocycle also at high hydration but in a sequential order, the changes in the tertiary structure would develop within the M intermediate between M_1 and M_2 and would relax after the N state.

The concept of two M states arose from the evaluation of time-resolved spectroscopy data in the visible wavelength region (Varo and Lanyi 1990). An irreversible step was assumed at this position in the photocycle between M_1 and M_2, acting as a switch by changing the proton accessibility of the Schiff base from the extracellular to the cytoplasmic side and thus creating the vectoriality of the proton pump.

With respect to the function of bR, Thiedemann et al. (1992) were able to show that proton pumping was found only in samples with hydration levels above 60% RH. These results indicate that the observed structural changes are necessary for proton translocation and that at least part of these changes may form the switch that changes the accessibility to the Schiff base.

13.2.4.5 *Charge-controlled conformational changes*

Further support for this interpretation of the proton translocation mechanism was obtained from investigations on the bR mutant Asp 38 Arg (Sass *et al.* 1998). X-ray diffraction experiments on samples at pH 6.7 do not show changes in the tertiary structure of bR whereas measurements at pH 9.6 display a diffraction pattern showing the characteristic large structural changes. The interpretation of these experiments is that at pH 6.7 the M_1 state is trapped under illumination whereas at pH 9.6 the M_2 state is accumulated. Assuming a sequential order of M_1 and M_2 independent of pH, with individual pH-dependent relaxation times, it seems that for this mutant also the observed large structural changes are necessary for vectorial proton pumping. In addition, these results give a clear indication that the changes in the tertiary structure are driven by alterations in the charge distribution of the protein, which follow photoisomerization.

At pH 6.7 the substitution of an aspartic acid by an arginine makes the charge pattern at the cytoplasmic side more positive either directly by the positive charge of the arginine or indirectly but more effectively since another positive charge is no longer compensated by the interaction with the aspartic acid. This new charge pattern, more positive at the cytoplasmic side than in wt bR, could interfere with the charge variation resulting from the deprotonation of the Schiff base and therefore slow down the onset of the large structural rearrangements. This would result in the accumulation of the M_1 state. Since no large structural changes are detectable under this condition and, if one assumes a sequential order of M_1 and M_2, the general conclusion for wt bR can be drawn that a charge redistribution around the Schiff base and at the extracellular side of the molecule results in an altered force field within bR which drives the large structural changes.

At a pH above 9, another group on the cytoplasmic side might be deprotonated and compensate, at least partly, for the added positive charge of the arginine. This would allow the structural changes associated with the transition to the M_2 state to take place.

13.2.5 *The crystal structure*

13.2.5.1 *Crystallization of bR*

Bacteriorhodopsin was one of the first integral membrane proteins to be crystallized in three dimensions (Michel and Oesterhelt 1980). These crystals showed only poor diffraction up to 8 Å resolution. Considerable progress was made by Schertler et al. (1993) using the surfaces of freshly formed benzamidine crystals as nucleation sites. These plate-shaped crystals gave diffraction to 3.6 Å in the a- and b-directions and 4.2 Å in the c-direction.

A novel concept for crystallization of membrane proteins in a lipidic cubic phase was reported by Landau et al. (1996). They crystallized bR in a monoolein–water cubic phase and obtained microcrystals of space group $P6_3$ with a unit cell of $a = b = 61.76$ Å; $c = 104.16$ Å, $\alpha = \beta = 90°$ and $\gamma = 120°$. Although the exact mechanism of crystallization is still unknown, it seems probable that bR, once inserted into the continuous three-dimensional curved lipid bilayer of the cubic phase, diffuses within this bilayer to form a nucleation particle from which well-ordered crystals can grow. They started with microcrystals of up to 30 µm in diameter and 5 µm in thickness. These small crystals already gave diffraction to 2 Å resolution on a specially designed beamline, the microfocus beamline ID13 at the ESRF Grenoble (Pebay-Peyroula et al. 1997). Using a highly focused beam with a diameter of about the same size as the crystal, the signal-to-noise ratio increased significantly.

The cell constant in the a- and b-directions of 61.76 Å together with the hexagonal space group gave an indication for an arrangement of bR molecules similar to the PM. The crystal can be described as a stack of PM-like two-dimensional lattices with the stack axis parallel to the screw axis, which is along the c-direction of the crystal (Fig. 13.11.) The crystallization conditions in the lipidic cubic phase were continuously improved, so that crystals larger than

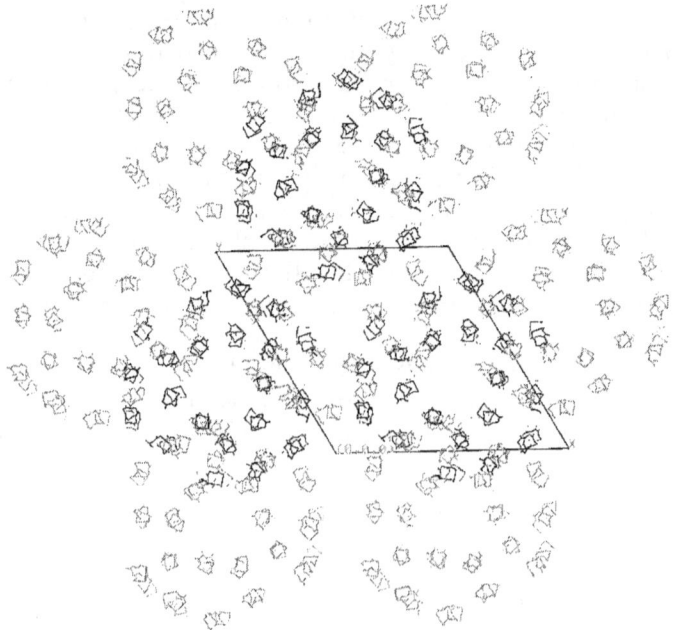

Fig. 13.11. Crystal packing of the two bR layers projected on the a–b plane (see Plate Section).

100 μm in diameter became available. The crystals showed a merohedral twinning which had to be taken into account (Luecke et al. 1998). Meanwhile, two additional crystal structures of bR were published using different crystallization strategies (Essen et al. 1998; Takeda et al. 1998). The structure obtained by Essen et al. nicely illustrates how original lipids from the PM in the protein boundary interact with amino acid side-chains of bR.

13.2.5.2 *The ground state crystal structure of bR*

The crystal structure of bR was solved by molecular replacement on the basis of the electron microscopic structure (Grigorieff et al. 1996) to 2.4 Å resolution (Pebay-Peyroula et al. 1997). The interactions between the PM-like protein layers are limited. Protein–protein contacts exist only between loops BC (loop between helix B and helix C) and loop EF (Pebay-Peyroula et al. 1997). Intertrimer protein contacts were not found. The most important new features of these maps are that the conformations of functionally important amino acid side-chains become more reliable and that several water molecules are localized in the proton translocation channel. From the H-bond network involving the water molecules and side-chains, the proton release from the Schiff base to the extracellular medium could be understood. However, the reprotonation of the Schiff base from the cytoplasm still remains unclear. The four X-ray crystal structures of the ground state of bR will very soon be refined using higher resolution data from better crystals; therefore, we hesitate to present a molecular model here.

Without going through all the still existing uncertainties of the elements of the proton wire through bR, it is evident that only high resolution structures of several important intermediates like L, M_1, M_2 and N will give the desired information for modelling a more probable mechanism for proton translocation through bR.

References

Blaurock, AE (1975). *Journal of Molecular Biology*, **93**, 139–158.
Büldt, G (1994). In *Neutron and synchrotron radiation for condensed matter studies*, Baruchel, J, Hodeau, JL, Lehmann, MS, Regnard, JR, Schlenker, C, eds, Les Editions de Physique, Springer, Les Ulis and Heidelberg, Ch. X, Fig. X.4., 161–173.
Büldt, G and de Haas, GH (1982). *Journal of Molecular Biology*, **158**, 55–71.
Büldt, G and Seelig, J (1980). *Biochemistry*, **19**, 6170–6175.
Büldt, G, Gally, HU, Seelig, J and Zaccai, G (1979). *Nature*, **271**, 182–184.
Büldt, G, Konno, K, Nakanishi, K, Plöhn, H-J, Rao, BN and Dencher, NA (1991). *Photochemistry and Photobiology*, **54**, 873–879
Caffrey, M and Cheng, A (1995). *Current Opinion in Structural Biology*, **5**, 548–555.
Deisenhofer, J, Epp, O, Miki, K, Huber, R and Michel, H (1985). *Nature*, **318**, 618–624.
Dencher, NA, Dresselhaus, D, Zaccai, G and Büldt, G (1989). *Proceedings of the National Academy of Sciences USA*, **86**, 7876–7879.
Engelman, DM and Zaccai, G (1980). *Proceedings of the National Academy of Sciences USA*, **77**, 5894–5898.
Engelman, DM, Henderson, R, McLachlan, AD and Wallace, BA (1980). *Proceedings of the National Academy of Science USA*, **77**, 2023–2027.

Essen, L, Siegert, R, Lehmann, WD and Oesterhelt, D (1998). *Proceedings of the National Academy of Sciences USA*, **95**, 11673–11678.
Frank, NP, and Lieb, WR (1981). In *Liposomes*, Knight, CG, ed. Elsevier, Cambridge, pp. 243–270.
Glaeser, RM, Baldwin, J, Ceska, TA and Henderson, R (1986). *Biophysical Journal*, **50**, 913–920.
Grigorieff, N, Ceska, TA, Downing, KH, Baldwin, JM and Henderson, R (1996). *Journal of Molecular Biology*, **259**, 393–421.
Henderson, R (1975). *Journal of Molecular Biology*, **93**, 123–128.
Henderson, R, Baldwin, JM, Ceska, TA, Zemlin, F, Beckmann, E and Downing, KH (1990). *Journal of Molecular Biology*, **213**, 899–929.
Henderson, R and Unwin, PNT (1975). *Nature*, **257**, 28–32.
Heyn, MP, Westerhausen, J, Wallat, I and Seiff, F (1988). *Proceedings of the National Academy of Sciences USA*, **85**, 2146–2150.
Hitchcock, PB, Mason, R, Shipley, GG (1975). *Journal of Molecular Biology*, **94**, 297–299.
Jubb, JS, Worcester, DL, Crespi, HL and Zaccai, G (1984). *EMBO Journal*, **3**, 1455–1461.
Koch, MHJ, Dencher, NA, Osterhelt, D, Plöhn, H-J, Rapp, G and Büldt, G (1991) *EMBO Journal*, **10**, 521–526.
Kimura, Y, Vassylyev, DG, Miyazawa, A, Kidera, A, Matsushima, M, Mitsuoka, K, Murata, K, Hirai, T and Fujiyoshi, Y (1997). *Nature*, **389**, 206–211.
Landau, EM and Rosenbusch, JP (1996). *Proceedings of the National Academy of Sciences USA*, **93**, 14 532–14 535.
Luecke, H, Richter, HTh and Lanyi, JK (1998). *Science*, **280**, 1934–1937.
Michel, H and Oesterhelt, D (1980). *Proceedings of the National Academy of Sciences USA*, **77**, 1283–1285.
Oesterhelt, D and Stoeckenius, W (1973). *Proceedings of the National Academy of Sciences USA*, **70**, 2853–2857.
Ovchinnikov, Yu, Abdulaev, N, Fergira, M, Kiselev, A and Lobanov, N (1979). *FEBS Letters*, **100**, 219–224.
Papadopoulos, G, Dencher, NA, Zaccai, G and Büldt, G (1990). *Journal of Molecular Biology*, **214**, 15–19.
Pebay-Peyroula, E, Rummel, G, Rosenbusch, JP and Landau EM (1997). *Science*, **277**, 1676–1681.
Popot, J-L, Engelman, DM, Gurel, O and Zaccai, G (1989). *Journal of Molecular Biology*, **210**, 829–847.
Rummel, G, Hardmeyer, A, Widmer, Ch, Chiu, ML, Nollert, P, Locher, KP, Pedruzzi, I, Landau, EM and Rosenbusch, JP (1998). *Journal of Structural Biology*, **121**, 82–91.
Sass, HJ, Schachowa, IW, Rapp, G, Koch, MHJ, Oesterhelt, D, Dencher, NA and Büldt, G (1997). *EMBO J.*, **16**, 1484–1491.
Sass, HJ, Gessenich, R, Koch, MHJ, Oesterhelt, D, Dencher, NA, Büldt, G and Rapp, G (1998). *Biophys. J.*, **75**, 399–405.
Schertler, GFX, Bartunik, HD, Michel, H and Oesterhelt, D (1993). *Journal of Molecular Biology*, **234**, 156–164.
Seelig, J and Seelig, A (1980). *Quarterly Reviews of Biophysics*, **13**, 19–61.
Takeda, K, Sato, H, Hino, T, Kono, M, Fukuda, K, Sakurai, I, Okada, T and Kouyama, T (1998). *Journal of Molecular Biology*, **283**, 463–474.
Tardieu, A, Luzzati, V and Reman, FC (1973). *Journal of Molecular Biology*, **75**, 711–733.
Thiedemann, G, Heberle, J and Dencher, NA (1992). *Colloque NSERM/John Libbey Eurotext Ltd*, **221**, 217–220.

Trewhella, J, Anderson, S, Fox, R, Gogol, E, Khan, S, Engelman, D and Zaccai, G (1983) *Biophysics Journal*, **42**, 233–241.
Trewhella, J, Popot, J-L, Zaccai, G and Engelman, DM (1986) *EMBO Journal*, **5**, 3045–3049.
Unwin, PNT and Henderson, RJ (1975). *Journal of Molecular Biology*, **94**, 425–440.
Varo, G and Lanyi, J (1990). *Biochemistry*, **29**, 2241–2250.
Weiss, MS, Kreusch, A, Schiltz, E, Nestel, U, Welte, W, Weckesser, J and Schultz, GE (1991) *FEBS Letters*, **280**(2), 379–382.
Zaccai, G and Gilmore, DJ (1979). *Journal of Molecular Biology*, **132**, 181–191.

14

Fibre and muscle diffraction

John M. Squire

PART 1: DIFFRACTION FROM HELICES

14.1 Introduction

Many natural and synthetic materials (e.g. plastics, wood, textiles, bone, nucleic acids, proteins, polysaccharides) are made of polymer molecules. Polymers comprise repeating subunits joined together end-to-end like beads on a string to form very long molecules. In some cases (e.g. some proteins) these polymers fold up to form globular structures which are conveniently studied by X-ray crystallography. However, in other cases (e.g. DNA, collagen, cellulose, some synthetic polymers), the molecules occur in extended forms in which they twist regularly into helical structures. Even in the case of some globular proteins (e.g. actin, tubulin) the globular molecules themselves aggregate in a regular fashion to produce helical macromolecular assemblies (in these cases, actin filaments and microtubules).

A three-dimensional crystal is one of nature's ways of packing together identical motifs so that they are all in exactly equivalent (i.e. identical) environments. What is obtained is a minimum energy configuration and therefore one that is highly stable. An exactly analogous situation occurs with polymer molecules or molecular aggregates. In fact, there is only one type of configuration that a polymer molecule *can* adopt so that all of the monomers along the chain have identical situations. As shown in Fig. 14.1, this configuration is a helix. Two apparently simple structures, a straight chain molecule [Fig. 14.1(a)] and a circular molecule [Fig. 14.1(c)], can both be considered as special cases of the helix, one with infinite pitch and one with zero pitch.

Because the helical conformation is so common in synthetic polymers and in biological systems, analysis of the structures of these systems by X-ray diffraction requires an understanding of the special features of helical diffraction theory. From a study of this theory, it is found that a number of key helical parameters can be determined easily by inspection of the observed 'fibre' diffraction patterns. The aim of this brief chapter is to provide both a qualitative feel for the main features of helical diffraction patterns, and also the mathematics involved, for those who wish to go into the subject more deeply. Examples of helical diffraction patterns will be discussed to illustrate the theory, including patterns from some synthetic polymers, from DNA and from particular biological systems, with a special focus in Part 2 on muscle filaments. Muscle is a particularly interesting fibre structure. Not only do intact muscles provide extremely rich fibre X-ray diffraction patterns, but muscles can be activated to produce tension while still in the X-ray beam and changes in the observed diffraction patterns

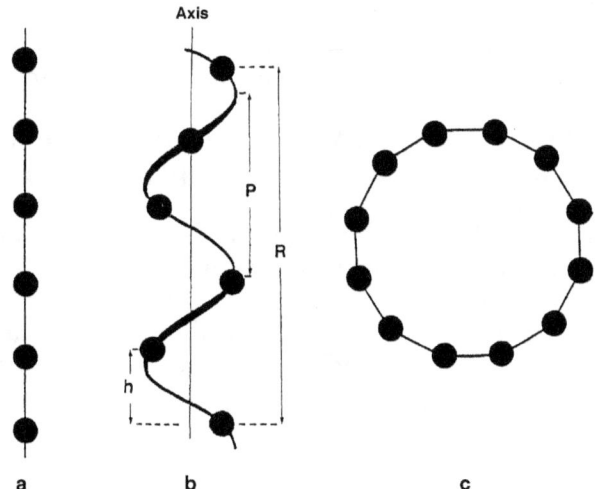

Fig. 14.1. A way to produce equivalence for the subunits on a polymer is for it to fold into a helix (b). Straight chain (a) and circular structures (c) are special cases of this.

provide direct evidence on the molecular movements involved in the production and regulation of muscular activity. Using synchrotron X-ray sources and modern high speed fast readout area detectors (usually multiwire proportional counters) muscle molecular dynamics can now be studied with a time-resolution of about 1 ms to 100 μs. This is a remarkable application of the X-ray diffraction technique.

Additional recommended texts on diffraction from fibres and helical structures are Cochran *et al.* (1952), Fraser and MacRae (1973), Harford and Squire (1997), Holmes and Blow (1965), Hukins (1981), Rosenbaum *et al.* (1971), Squire (1997b) and Vainshtein (1966).

14.2 Theory of diffraction from helical structures

14.2.1 General features

The monomers on a helix can be thought of as analogous to the steps on a spiral staircase. Each is exactly equivalent to every other monomer (or step) on the spiral. The distance climbed vertically by each monomer (step) is called the subunit axial translation or unit height [h, Fig. 14.1(b)], the distance climbed in exactly one turn of the spiral is called the pitch (P) and that climbed to reach a monomer (step) which is exactly over the first (lowest) is called the axial repeat (R). If there happens to be exactly a whole number of monomers (steps) in one turn of the helix (i.e. $P/h = N$; N is an integer) then R and P are clearly the same.

What we need to find is the form of the diffraction pattern from a three-dimensional array of helical molecules in which the monomer contains several or many atoms. This may appear to be a formidable task. However, the approach

274   Fibre and muscle diffraction

here is to give both a non-mathematical description of the procedure, making liberal use of the ideas in what is known as the 'convolution theorem', and also to provide the appropriate mathematics. Here and later two-dimensional optical diffraction patterns will be used as analogues to illustrate the three-dimensional diffraction ideas that are presented. Most of the optical diffraction illustrations are taken from Squire (1981) and the mathematics presented here is largely based on that in Harford and Squire (1997). We will see that the three-dimensional crystal of helical molecules can be thought of as convolutions and products of rather simple structures, for which the form of the diffraction pattern can readily be deduced.

Figure 14.2 illustrates the ideas contained within the convolution theorem. The convolution theorem stated simply is that if a function C can be expressed as the convolution of two simpler functions A and B, then the diffraction pattern from C is simply the *product* of the diffraction patterns of A and B. Using $G(A)$ to represent 'the diffraction pattern of' or, strictly, 'the Fourier transform of' A, the convolution theorem can be written as follows:

$$\text{if } A^*B = C \text{ then } G(A) \times G(B) = G(C), \tag{14.1}$$

where the symbol * is used to represent 'convoluted with'. So what does 'convoluted with' mean? It literally means 'folded together with', but in simple terms it means that function A is picked up and placed successively on every point in B and the results are all added together. Thus a crystal C may be thought of as a

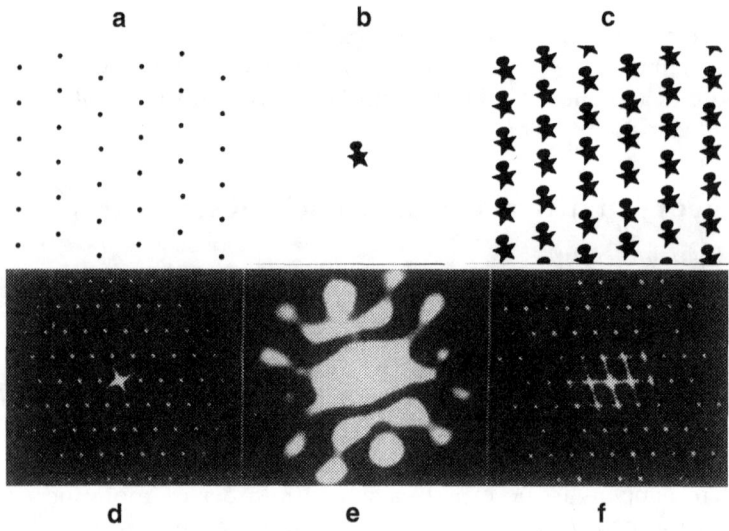

Fig. 14.2. Representation of the concept of convolution and the consequences of the convolution theorem. (a) A lattice, (b) a motif (a baby), (c) the convolution of (a) with (b); a baby has been put on every point in the lattice in (a). This means that the transforms (d) and (e) of (a) and (b) need to be multiplied together to give the transform (f) of (c). In effect the motif (baby) transform has been 'sampled' by the reciprocal lattice in (d).

convolution of *A*, the motif (repeating unit, asymmetric unit) in the crystal unit cell, with *B*, the space lattice that defines the unit cell shape. This is illustrated in Fig. 14.2. Figure 14.2(a) is a two-dimensional lattice of points, analogous to the lattice points in a Bravais space lattice. Figure 14.2(b) is the motif, in this case the outline of a baby. Figure 14.2(c) is what is obtained by putting a baby motif on every point in (a); a two-dimentional lattice of babies is obtained. Figure 14.2(d)–(f) are the optical diffraction patterns obtained from opaque masks with apertures as illustrated in (a)–(c). The diffraction pattern from (a) is another lattice; it is actually the reciprocal lattice of the real space lattice in (a). Figure 14.2(e) is the diffraction pattern from the single baby in (b). Figure 14.2(f) is the *product* of the diffraction patterns in (d) and (e). This is obtained by placing the centres of these two diffraction patterns on top of each other, and then at each point in the overlapping diffraction patterns multiplying the two patterns together. Thus if there is zero intensity in one pattern, it will be zero in the product as well. Where both patterns have non-zero intensity, the resultant intensity will also be non-zero, with the actual magnitude of the resultant depending on the product of the two diffraction patterns at that point. The effect in (f) is of a 'baby' diffraction pattern that is only seen at the points corresponding to the reciprocal lattice points in (d). We say that the baby transform has been 'sampled' by the lattice transform.

Note finally that the convolution theorem works backwards as well. If a function *C* is the *product* of two other functions *A* and *B*, then the diffraction pattern (*G*) from *C* is the *convolution* of the two diffraction patterns from *A* and *B*. Thus

$$\text{if } A \times B = C \text{ then } G(A)^*G(B) = G(C). \tag{14.2}$$

14.2.2 A single helical turn

The starting point in the mathematics is the general expression for the Fourier transform $G(S)$ which corresponds to diffraction in the direction defined by the vector S in reciprocal space. $S = S_1 - S_0$ is a chord on the Ewald sphere [Fig. 14.3(a)] and S_1 and S_0 are each vectors of the same length $(1/\lambda)$ defining the radius of the Ewald sphere. (Note that some texts use k, k' and Δk for S_0, S_1 and S, and the vector length is sometimes put as $2\pi/\lambda$ rather than $1/\lambda$.)

If the density (scattering power) of a volume element dV at the end of vector r is $\rho(r)$, then the Fourier transform is

$$G(S) = \int \rho(r) \exp[2\pi i r \cdot S] \, dV \tag{14.3}$$

Here $\rho(r)$ dV represents the amount of scattering and $\exp[2\pi i\, r \cdot S]$ represents the relative phase of that scattering relative to scattering from a volume at the origin of r.

As described above, when considering a helical molecule or assembly of molecules it is convenient to build up the final transform in stages by considering relatively simple structures. As a starting point we can consider

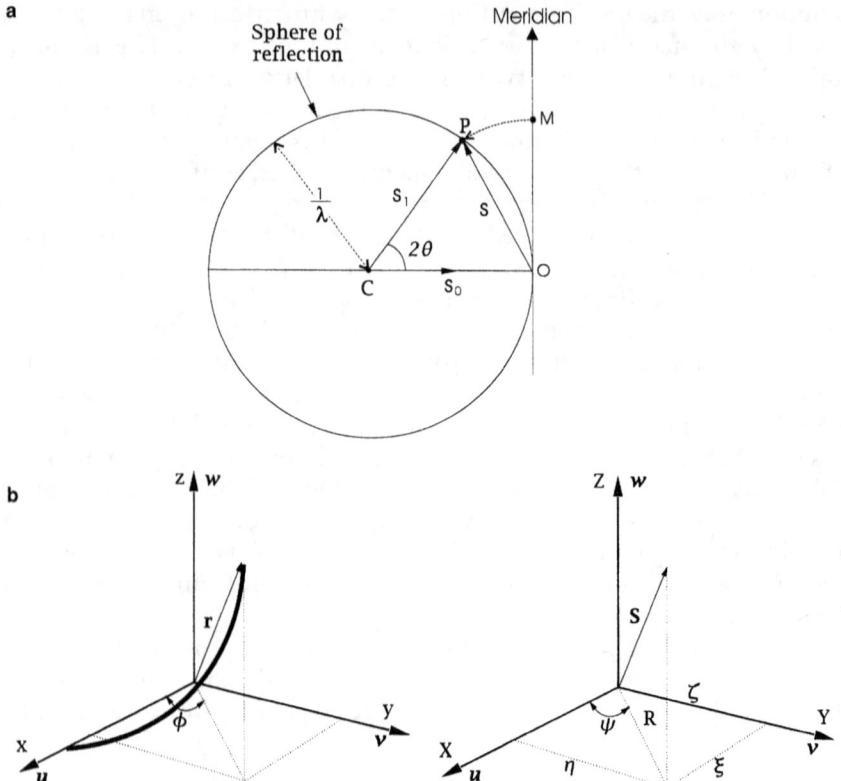

Fig. 14.3. (a) Geometry of the Ewald sphere. S_0 is the incident beam direction. O is at the origin of reciprocal space. S_1 is parallel to the direction of a diffracted beam and $S = S_1 - S_0$. The line OM is the meridional direction in a fibre diffraction pattern where the fibre is perpendicular to the incident beam. To record the meridional reflection M properly, the fibre must be tilted to bring M onto the Ewald sphere. (b) Geometry in real space (left) of a helical wire (bold line) in terms of orthogonal axes x, y and z. Right is the corresponding geometry in reciprocal space showing how S can also be defined in terms of directions x, y and z.

diffraction from one turn of a simple uniform helical wire, as illustrated in Fig. 14.3(b), (left). This can be described in cylindrical coordinates by the expression

$$x = r \cos \phi, \qquad y = r \sin \phi, \qquad z = z,$$

where x, y and z are positions in Cartesian coordinates defined by unit vectors $\boldsymbol{u}$, $\boldsymbol{v}$ and $\boldsymbol{w}$ and r is the radius of the helix (i.e. the projected length of $\boldsymbol{r}$ onto the x–y plane). A point on the wire can therefore be defined by the vector $\boldsymbol{r}$ such that

$$\boldsymbol{r} = r \cos \phi \, \boldsymbol{u} + r \sin \phi \, \boldsymbol{v} + z \boldsymbol{w}.$$

The pitch P of the helix is then $\phi = 2\pi(z/P)$ and

$$\mathbf{r} = r\cos[2\pi(z/P)]\mathbf{u} + r\sin[2\pi(z/P)]\mathbf{v} + z\mathbf{w}. \tag{14.4}$$

If the scattering direction is $\mathbf{S}$, where [Fig. 14.2(c)],

$$\mathbf{S} = \xi\mathbf{u} + \eta\mathbf{v} + \zeta\mathbf{v}, \tag{14.5}$$

then from Fig. 14.3b(right) we have $\xi = R\cos\Psi$ and $\eta = R\sin\Psi$. And then from eqns (14.4) and (14.5) we have

$$\mathbf{r}\cdot\mathbf{S} = [r\cos(2\pi z/P)]\xi + [r\sin(2\pi z/P)]\eta + z\zeta.$$

Putting these all together gives

$$\mathbf{r}\cdot\mathbf{S} = [rR\cos(2\pi z/P)\cos\Psi] + [rR\sin(2\pi z/P)\sin\Psi] + z\zeta,$$

and finally for $G(\mathbf{S})$ [eqn (14.3)],

$$G(\mathbf{S}) = \rho\int \exp\{2\pi i[rR\cos(2\pi z/P)\cos\Psi + rR\sin(2\pi z/P)\sin\Psi + z\zeta]\}\,dz,$$

giving [recall $\cos A\cos B + \sin A\sin B = \cos(A-B)$]

$$G(\mathbf{S}) = \rho\int \exp\{2\pi i[rR\cos(2\pi z/P - \Psi)]\}\exp(2\pi i z\zeta)\,dz, \tag{14.6}$$

where ρ is the scattering power per unit length along z in a direction defined by $\mathbf{S}$.

14.2.3 Repeated helical turns

A continuous helical wire can be considered as a convolution of a single helical turn with a periodic function of spacing P along the z-axis (Fig. 14.4). If the full helical wire is a convolution of a single turn with a function that repeats axially after a distance P, then the Fourier transform will be obtained from the product of the transform of a single turn and the transform of the periodic array. The latter is a series of intensity planes oriented perpendicular to the helix axis and separated in reciprocal space by $1/P$; the transform will only be non-zero on planes in reciprocal space defined by $\zeta = n/P$. The intensity distribution along these planes will be defined by the single turn transform at those positions. Equation (14.6) then becomes

$$G(\mathbf{S}) = \rho\int \exp\{2\pi i[rR\cos(2\pi z/P - \Psi)]\}\exp(2\pi i z[n/P])\,dz.$$

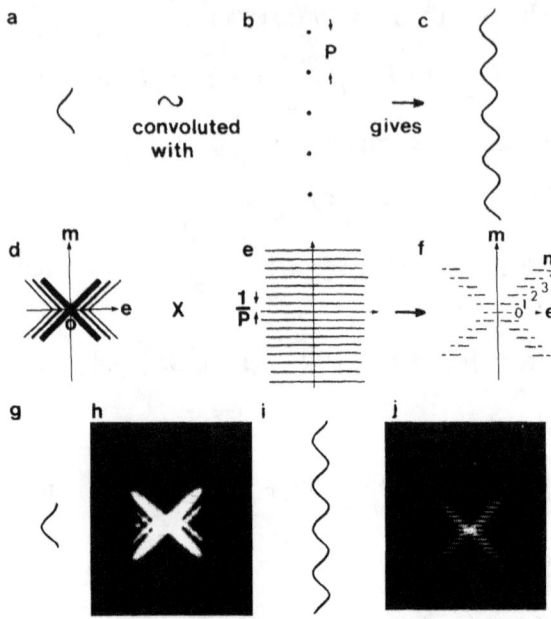

Fig. 14.4. A single helical wire (c) can be thought of as a convolution of one turn of the helix (a) with a row of points P apart as in (b). The diffraction pattern (f) from (c) is then the product of the diffraction patterns [d; see Fig. 14.5(b)] and (e), respectively, from (a) and (b). Optical diffraction simulations of (a), (d), (c) and (f) are shown in (g)–(j). [From Squire (1981).]

At this point, a convenient trick is to multiply both sides by the factor $\exp(-n\Psi)$. Thus

$$G(\mathbf{S})\exp(-n\Psi) = \rho \int \exp\{2\pi i[rR\cos(2\pi z/P - \Psi)]\}$$
$$\times \exp(in[2\pi z/P - \Psi])\,\mathrm{d}z.$$

Because, by definition,

$$\int \exp(Xi\cos\phi)\exp(in\phi)\mathrm{d}\phi = 2\pi i^n J_n(X),$$

where $J_n(X)$ is a Bessel function of order n and argument X [Fig. 14.5(a)]. If $X = 2\pi Rr$ and $\phi = (2\pi z/P - \Psi)$ so that $\mathrm{d}\phi = (2\pi/P)\,\mathrm{d}z$, then we have

$$G(\mathbf{S}) = G(R, \Psi, n/P) = [P\rho \exp\{in(\pi/2 + \Psi)\}]J_n(2\pi Rr). \qquad (14.7)$$

[Note that $\exp(in\pi/2) = i^n$.]

14.2.4 Appearance of the continuous helix transform

In the diffraction pattern from the continuous helix [see Fig. 14.4(f)], which has non-zero intensity only when $\zeta = n/P$, these axial positions are termed layer-lines and are numbered consecutively from $n = 0$. The layer-line number is usually written l and corresponds to the Miller index l for crystals, when the helix axis is along the crystallographic c axis. Note also that the transform is cylindrically symmetrical in amplitude (it is independent of Ψ). Bessel functions are characterized by being zero when $X = 0$, except for the zero order Bessel functions, which have a maximum at this argument [Fig. 14.5(a)]. Other orders have a first maximum for small values of X and, as X increases, there are smaller (subsidiary) maxima. Also as the order n increases, the first maximum occurs at a larger value of X. Since Bessel functions of increasing order n describe the helical transform on layer-lines of increasing number l, the general form of the intensity in Fig. 14.4(f) is that of a 'cross' of intensity centred on the origin of the pattern—it is known as the 'helix cross-pattern' and is shown in terms of Bessel functions in Fig. 14.5(b).

As a qualitative indication of the analysis so far, Fig. 14.4(g)–(j) illustrate the derivation of the transform of a continuous helix using the two-dimensional

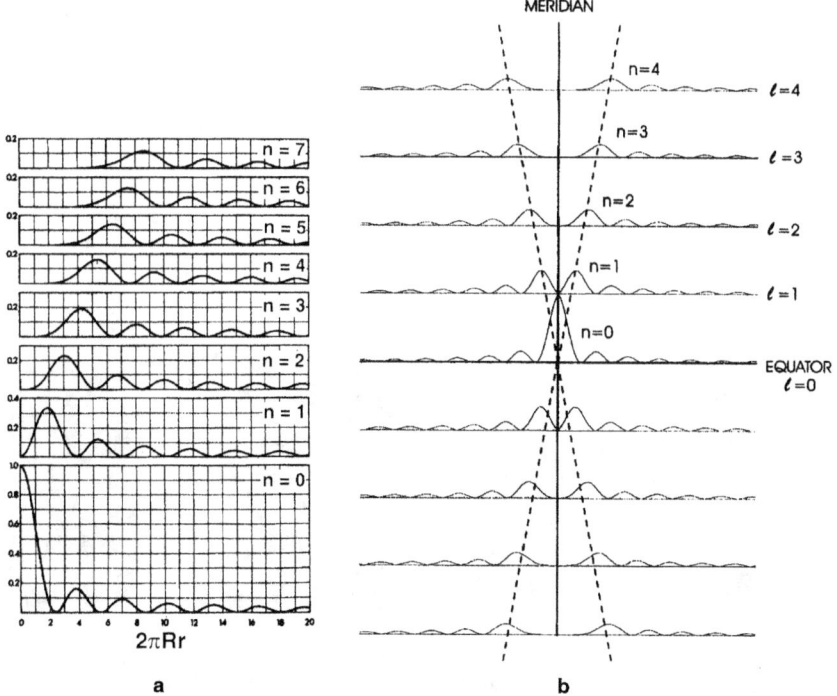

Fig. 14.5. (a) The amplitudes squared of low order Bessel functions with varying argument $X = 2\pi Rr$ [from Fraser and MacRae (1973)]. (b) Central part of a helix cross pattern from a continuous helical wire in terms of Bessel orders n which are the same as the layer-line number l.

analogue of a sine wave and its transform. Application of the convolution theorem is illustrated there too. A full sine wave is the convolution of a single repeat (one wavelength) with a periodic array along the axis of the sine wave and spaced by the wavelength. The wavelength of the sine wave is the two-dimensional analogue of the pitch of the helix.

Note that the Bessel function argument $X = 2\pi Rr$ is smaller if the helix radius r is smaller. Since the first peak in an nth order Bessel function occurs at a defined value of X, helices of small radius give Bessel peaks at large values of R in reciprocal space and vice versa. The positions R of peaks along the layer-lines can therefore be used to find the radius r of the helix if the appropriate Bessel order is known.

14.2.5 The transform of a discontinuous helix of points

To extend the analysis, it is useful to make use of the second form of the convolution theorem [eqn (14.2)]. A discontinuous helix of points (Fig. 14.6) can be

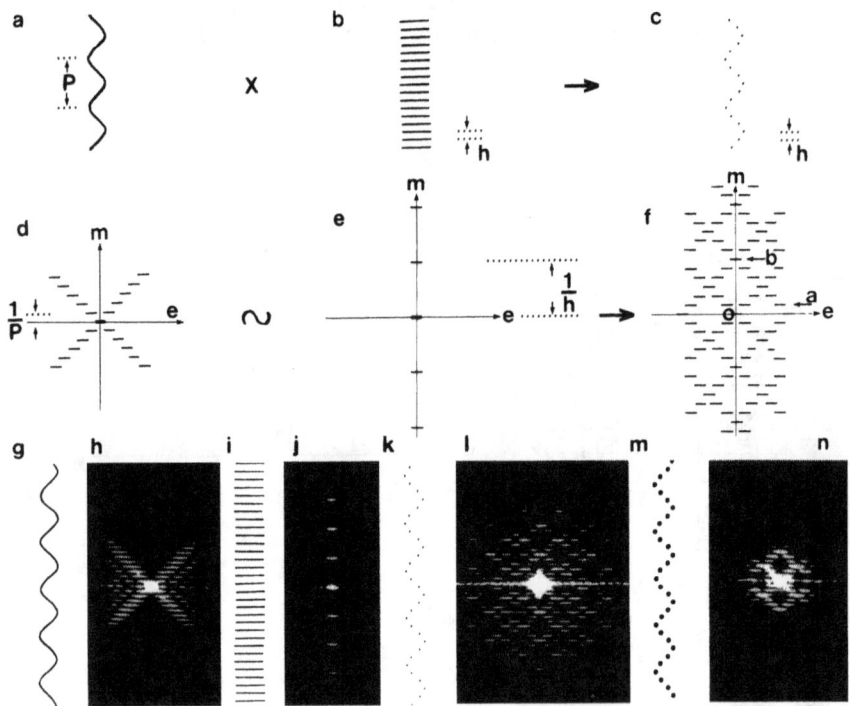

Fig. 14.6. A discontinuous helix (c) can be thought of as the product of a continuous helix (a) and a set of density planes (b) perpendicular to the helix axis in (a) and spaced h apart. The diffraction pattern of (c) is the convolution of the diffraction patterns (d) and (e) from (a) and (b), respectively. The optical analogue of this process is given in (g)–(l). If the points on the helix in (c) or (k) are replaced by atoms of finite size (m), then the transform in (f) or (l) needs to be multiplied by the transform of the atoms as in (n).

considered to be the product of the continuous helical wire that we have already considered and a set of thin density planes of interplanar spacing h oriented perpendicular to the helix axis, where h is the subunit axial translation or unit height of the monomers on the helix [Fig. 14.1(b)]. If the discontinuous helix is a product of two functions, then we need to form the convolution of the transforms of the two functions. This will give us the transform of the discontinuous helix. One of the transforms that we need is the helix cross-pattern [Figs 14.4(f), 14.5(b)]. The other is simply the transform of a single set of density planes. This is the three-dimensional analogue of the transform of a two-dimensional grating; it is a series of spots spaced along the z^* axis in reciprocal space at distances $\zeta = m/h$ from the origin. The final transform is then the convolution of the helix cross pattern with these points along z^*; something obtained simply by placing helix crosses with their centres on every point m/h from the origin and adding the result [Fig. 14.6(f)]. The transform is now only non-zero on a series of planes of spacing

$$\zeta = n/P + m/h, \tag{14.8}$$

where m is an integer (positive, negative or zero) and n is the Bessel function order. If the helix is such that there are exactly N points in K turns of the helix (often termed an N/K helix), then the structure has an exact axial repeat c, where $c = KP$. In the full transform the layer-line numbers now refer to this long repeat c. The ratio N/K is clearly the same as P/h and eqn (14.8) can be written in the form

$$\zeta = n/P + m/[P(K/N)],$$

and from this

$$KP\zeta = nK + mN = l,$$

where l is now the layer-line number for the discontinuous helix transform. Successive layer-lines of number l are now orders of the repeat c. In this way, if the symmetry N/K of the helix is known, the orders n of Bessel functions that contribute to a given layer-line l can be determined. Note that it is quite common for a layer of the helix cross pattern centred on the origin to coincide axially with other layers from $m = \pm 1, \pm 2$, etc. A single layer-line can have an intensity distribution that is the sum of contributions from several different Bessel function orders.

Once again, the ideas in this section are summarized by the two-dimensional optical diffraction analogues in Figs 14.6(g)–(l). The analysis for a particular non-integral helix (5/2 symmetry, $c = 2P$) is given in Fig. 14.7.

14.2.6 Helices of atoms and molecules

The very simplest helical molecule must have at least one atom on each of the points along the discontinuous helix discussed in the previous section. This can be thought of as being obtained by convoluting an atom (assumed to have spherical symmetry) with the helix of points. The result is the multi-cross pattern of Fig. 14.6(f) being multiplied by the transform of the atom, the atomic

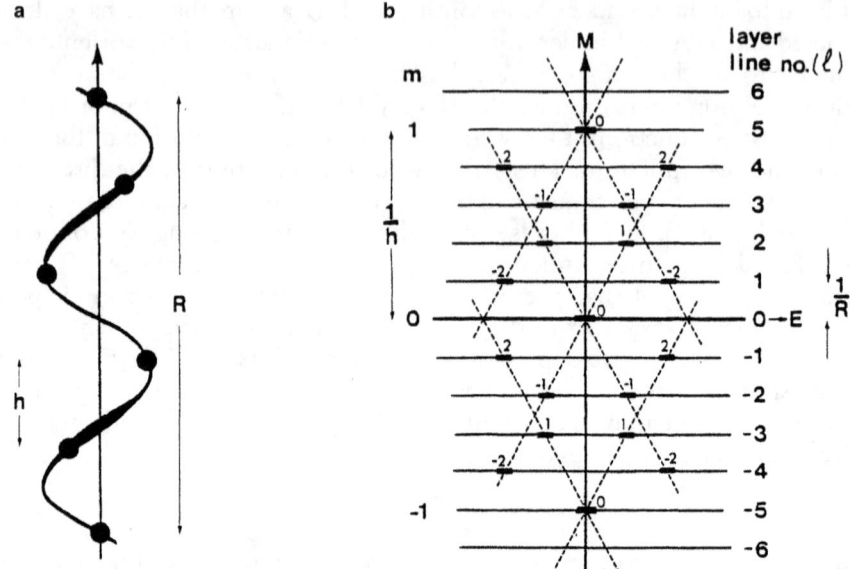

Fig. 14.7. (a) Simple representation of a discontinuous, non-integral, helix (5/2 symmetry; $N=5$, $K=2$) of axial repeat c ($=2P$), together with (b) its diffraction pattern consisting of helical crosses as in Fig. 14.4(f) at different points m along the meridian. [From Squire (1981).]

scattering factor $f(S)$. This is simulated in Fig. 14.6(k)–(n). Of course, real polymers will have several atoms in each repeating unit. Supposing that there are j atoms per repeating unit (monomer, residue) then the full structure can be considered as j discontinuous helices of the same pitch P, and the same number of repeating units N in K turns of the helix, but each helix can have a different radius r_j, azimuthal position ϕ_j and axial position z_j. A translation of the discontinuous helix in Fig. 14.6(c) results in the transform being multiplied by $\exp(2\pi i(\mathbf{r} \cdot \mathbf{S}))$, where $\mathbf{r} = z_j \mathbf{w}$ and $\mathbf{S}$ is as in eqn (14.5). The factor $\exp(2\pi i(\mathbf{r} \cdot \mathbf{S}))$ then becomes $\exp\{2\pi i(lz_j/c)\}$ since $\zeta = l/c$. Application of a rotation means that the transform is now sampled at $(R, \Psi - \phi_j, \zeta)$ instead of the previous sampling at (R, Ψ, ζ). The transform for all the atoms therefore becomes

$$G(R, \Psi, l/c) = \sum_j \sum_n f_j J_n(2\pi R r_j) \exp\{i[n(\pi/2 + \Psi - \phi_j) + 2\pi l z_j/c]\}.$$

(14.9)

The generation of this transform is illustrated in Figs 14.8(a)–(c); in Fig. 14.8(c) there is a complicated distribution of intensity along each layer-line.

14.2.7 Crystalline arrays of helical structures

Helical molecules or aggregates can pack together in three dimensions to form a crystalline array in which the c-axis of the unit cell is the same as the axial

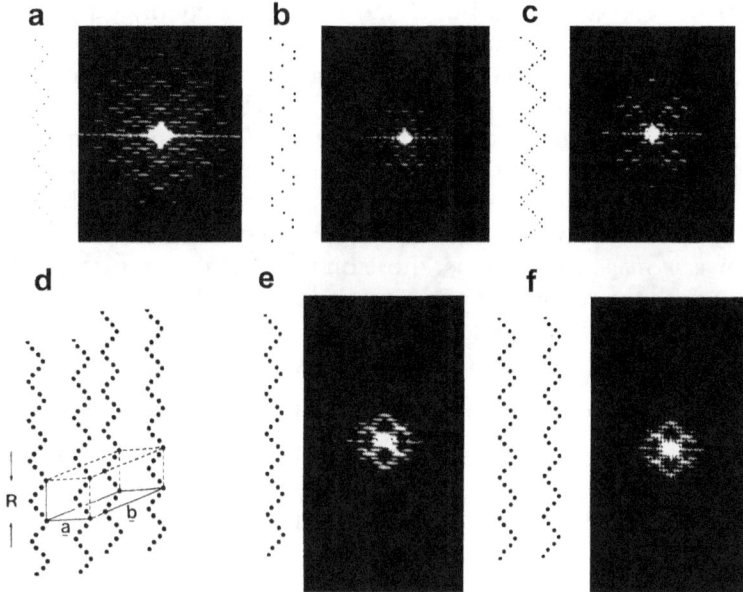

Fig. 14.8. Different helices of atoms in a helix of multiatom monomers (c) with their transforms are represented by (a) and (b). The transform in (c) is the vector sum of the transforms in (a) and (b). A crystalline array of helical molecules is represented in (d) as a convolution of the molecule with a two-dimensional array with unit cell vectors **a** and **b**. The transform in (f) from a simple array of two molecules shows sampling along vertical row-lines of the pattern from a single helix in (e).

repeat c of the helix. Using the convolution theorem again, the final structure can be considered as the convolution of the whole helical molecule with a two-dimensional array defined by lattice vectors **a** and **b** [Fig. 14.8(d)]. Note, however, that the unit cell may contain more than one helical molecule with different relative axial positions $z_q w$ and rotations ϕ_q for the qth helix. With these extra molecules at translations along the **a** and **b** axes of the unit cell for the qth helix of $x_q u$ and $y_q v$, the final transform (structure factor) is

$$F(h, k, l) = \sum_q \sum_j \sum_n f_j J_n(2\pi R r_j) \exp\{i[n(\pi/2 + \Psi - \phi_j) + 2\pi l z_j/c]\}$$
$$\times [\exp 2\pi i(h x_q + k y_q + l z_q)] \exp(-in\phi_q). \qquad (14.10)$$

A simple example of the sampling of the helical transform by the reciprocal lattice is shown in Figs 14.8(d)–(f). Note that the so-called 'fibre diffraction patterns' are usually described by eqn (14.9) when there is no regular three-dimensional packing of helices but there is preferred orientation of all the molecules parallel to the helix axis direction. In the case of crystalline fibres, the basic transform is described by eqn (14.10), but apart from preferred orientation along the c-axis, it is usually assumed that the crystallites have random orientations around the helix (c) axis. Equation (14.9) describes a continuous helical transform in which the observed intensities are governed by the diffraction from a single molecule. Equation (14.10) describes a sampled helical transform where the relationships of adjacent molecules need also to be considered.

Note finally some nomenclature. The layer-line $l=0$ (e.g. in Fig. 14.7), which passes through the middle of the diffraction pattern in a direction perpendicular to the c-axis, is known as the *equator*. The c-axis direction in the diffraction pattern ($h=0$, $k=0$) is known as the *meridian*. Reflections that lie on the meridian are known as *meridional* reflections; those on the equator as *equatorial* reflections. Reflections with their intensity peaks centred close to, but not on, the meridian are often called *off-meridional* reflections.

14.3 Diffraction from synthetic polymers and DNA

The simulated diffraction pattern in Fig. 14.9(a) is from crystalline polyethylene. It consists of a series of well-sampled, equally spaced, layer-lines, with strong intensity on the meridian only on the even order layer-lines. This shows immediately that the structure is helical with two 'residues' per turn of the helix. Polyethylene is conventionally written $[CH_2-CH_2]_n$, but since the two parts of the repeating unit are the same, the monomer in the helix is actually CH_2. The carbon backbone forms a kind of planar zig-zag structure, but that is exactly what a 2/1 (sometimes also written 2_1) helix is. The diffraction pattern simulated in Fig. 14.9(b) from polypropylene shows well-sampled, equally spaced, layer-lines with strong meridional intensity on every third layer-line. In this case the

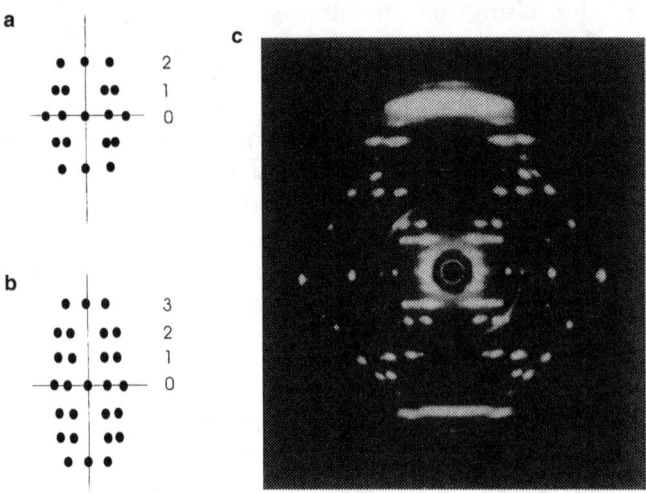

Fig. 14.9. (a) Representation of the diffraction pattern from oriented, crystalline polyethylene. (b) As for (a) but from polypropylene. (c) Diffraction pattern from the B form of DNA, with the DNA fibre tilted as in Fig. 14.3(a) to bring the strong meridional reflection on the 10th layer-line onto the Ewald sphere.

helix has three monomers per turn, each monomer being CH_2–$CHCH_3$. The diffraction pattern in Fig. 14.9(c) is from the B form of DNA. It comprises a series of equally spaced layer-lines, but the diffraction pattern does not appear symmetrical across the equator. This is because the DNA fibre has been tilted around an axis perpendicular to the X-ray beam, as in Fig. 14.3(a), in order to bring the first meridional reflection onto the Ewald sphere, thus properly satisfying the Bragg condition. In this case the first meridional reflection happens to be on the 10th layer-line. The pattern is in some ways deceptive in that there are equally spaced layer-lines, but some of them happen to have very low intensity. If the 'missing' layer-lines are also counted then the fact that the strong meridional is on the 10th layer-line becomes more obvious.

By determining the angle of diffraction out to the first meridional reflection, or out to the middle of the first layer-line, and using those angles in Bragg's law, the values for the subunit axial translation h and the pitch P can be calculated. Figure 14.10 shows that if the physical spacing between the centre of the diffraction pattern and the reflection of interest is S and if the specimen-to-film (detector) distance is D, then the corresponding Bragg spacing d is given by a combination of Bragg's law ($n\lambda = 2d\sin\Theta$) and the camera geometry ($\tan 2\Theta = S/D$) as

$$d = \lambda/2 \sin(\tfrac{1}{2}\tan^{-1}(S/D)). \quad (14.11)$$

[Note that it is usual to put $n=1$ in this, since *a priori* the value of n is unknown. If the true value of n is needed, then having found the different values of d assuming that $n=1$, the true value of n can usually be deduced from the relationship between different d spacings.] In the case of form B of DNA, the values of P and h determined using eqn (14.11) turn out to be 34 and 3.4 Å.

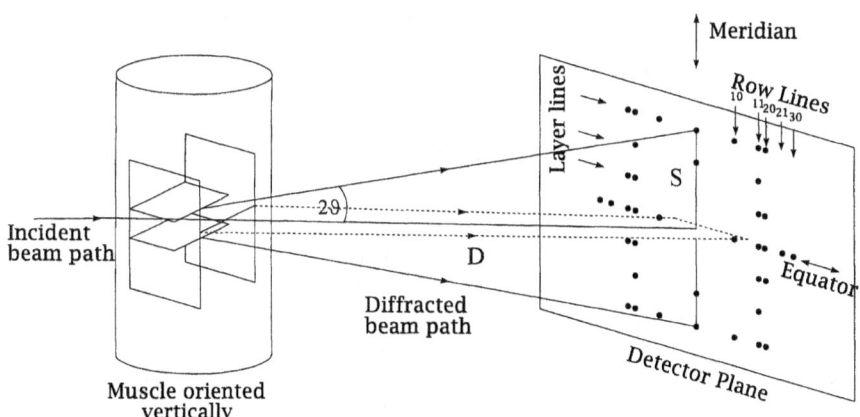

Fig. 14.10. Illustration of the experimental geometry of a rotating crystal or fibre diffraction pattern being recorded on a detector, showing row-lines and layer-lines. Locations of the meridian and equator are indicated. Here S is the distance from the centre of the recorded diffraction pattern of the reflection diffracted at angle 2Θ, when the detector (e.g. film) is D from the diffracting object. Courtesy of Dr J. Barry.

Another feature of the DNA molecule that can be obtained readily from the diffraction pattern is the approximate radius of the strong scatterers in the helix. The monomers in DNA comprise a sugar–phosphate backbone and 'base' side-chains. The phosphate groups are by far the strongest X-ray scatterers. From a knowledge of Bessel functions [Fig. 14.5(a)] it is found that the first peak in, say, Bessel function J_3 occurs when $X = 2\pi Rr$ is about 3.4. By measuring the radial position R of the main intensity on the third layer-line, which in this case is related to a third order Bessel function which peaks at $X = 3.4$, then the formula $3.4 = 2\pi Rr$ can be used to give the helix radius r. In this case the main peak is at about the same radius as the layer-line separation $(3/34 \text{ Å}^{-1})$ from the equator. From this one finds that r is about 7.5–8 Å. So, from an inspection of the diffraction pattern and a few rather simple calculations, we have found that the DNA molecule in the B form has a pitch of 34 Å, a subunit axial translation of 3.4 Å, has 10 nucleotides (base, plus sugar and phosphate) per turn and that the phosphate group is at a radius of about 7.5–8 Å. A study of the equatorial reflections, which provide information about the lateral molecular packing, would also have shown that the intermolecular distance in a quasi-hexagonal lattice is about 18–20 Å, giving a molecular diameter of the same value. From this, one can deduce that the phosphate groups at a diameter of 15–16 Å are near the outside of the helix. Clearly, in order to determine the structure properly, one needs to apply the full molecular transform [eqn (14.10)], but a great deal has been learnt even without that. The fact that the first layer-line is weak whereas the second is relatively strong suggests a pseudo-halving of the axial repeat at low resolution, consistent with the presence of two helices on a common axis but not exactly equivalent. In fact the well-known DNA double helix has two antiparallel chains twisting around a common axis.

PART 2: DIFFRACTION FROM MUSCLE

14.4 Muscle structure and diffraction from actin filaments

Striated muscles comprise overlapping arrays of myosin and actin filaments which together form a muscle sarcomere which is usually about 2.0–2.5 μm long. Similar sarcomeres then repeat throughout the muscle to produce the large masses that are familiar to us [see an overview of muscle anatomy in Luther et al. (1995)]. Muscle structure is illustrated schematically in Fig. 14.11. It can be seen in Fig. 14.11(b) that changes in muscle (sarcomere) length are associated with relative sliding of the actin and myosin filaments without the filaments themselves changing much in length. Projections, myosin cross-bridges, on the myosin filaments are thought to interact with actin filaments in a cyclical manner, powered by the hydrolysis of ATP, to produce muscular force and filament sliding (Huxley 1969; Squire 1997a). In this part of the chapter, the general ideas about diffraction from helical structures, given in Part 1, are illustrated by their application to the structures of the myosin and actin filaments in muscle, both of which are helical.

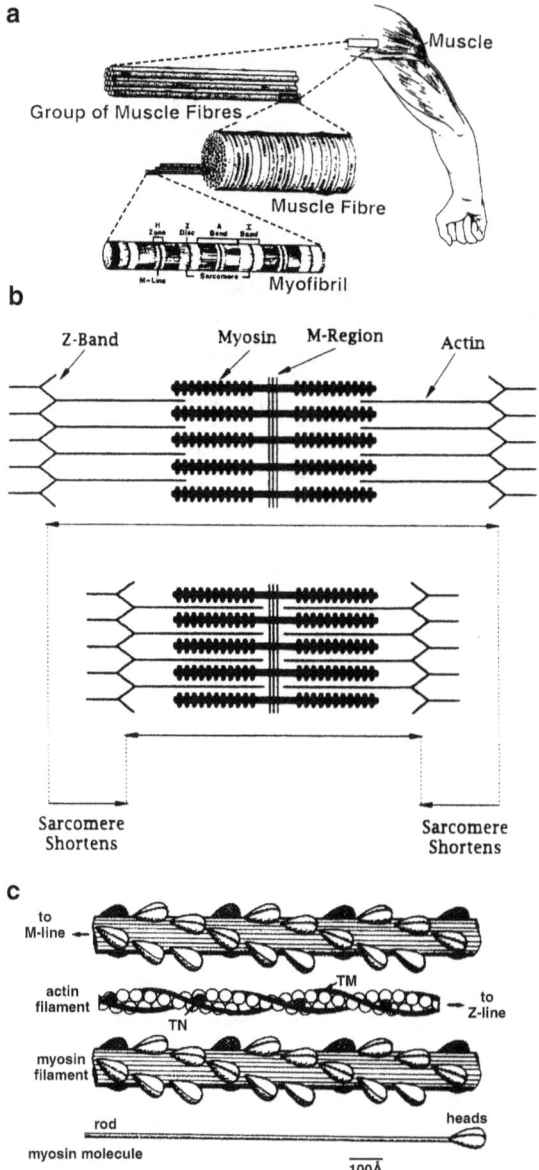

Fig. 14.11. (a) Levels of structural heirarchy within vertebrate striated muscles [after Bloom and Fawcett (1975)]. Typical fibre diameters are 20–100 μm, and myofibrils are 1–2 μm across. (b) The structure of the vertebrate striated muscle sarcomere in terms of overlapping myosin and actin filaments and illustration of the change in filament overlap that occurs when the sarcomere shortens [after Offer (1974)]. (c) Schematic illustrations of the myosin rod and myosin filament, showing the rods forming the filament backbone and the heads in a nearly helical array on the myosin filament surface. Also shown is an actin filament containing globular actin monomers (empty circles), coiled-coil tropomyosin strands (TM, dark lines) and the troponin complex (TN, filled circles). [From Squire (1983)]. Note the alternative names used for the same structure: Z-line, Z-band and Z-disc are all same, the M band is the central part of the M region.

As shown in Fig. 14.11(c), actin filaments [details in Al-Khayat et al. (1995)] contain helical arrays of globular actin molecules. The symmetry is that of a helix with ($N=$) 13 actin monomers in ($K=$) 6 turns of the helix. The pitch (P) is just over 59 Å, the subunit axial translation (h) is about 27.5 Å and the true repeat (c) is about 360 Å (about 6×59 Å). Because there are very nearly two actin monomers per helix turn ($N/K=13/6$), the appearance of the filament [Fig. 14.11(c)] is that of two long, slowly twisting strands of actin monomers. These are the so-called long-period or long-pitched helices. The helix of pitch 59 Å is often referred to as the 'genetic' helix.

The transform of a 13/6 helix can be deduced from the analysis in Part 1. The 59 Å genetic helix, taken to be continuous, is left-handed [Fig. 14.12(a)]. To generate the appropriate discontinuous helix, the continuous helix in Fig. 14.12(a) needs to be multiplied by a set of density planes [Fig. 14.12(b)] of spacing 27.5 Å (the subunit axial translation h). The long-pitched strands then become very clear as in Fig. 14.12(c). The transform of the genetic helix consists of a series of layers

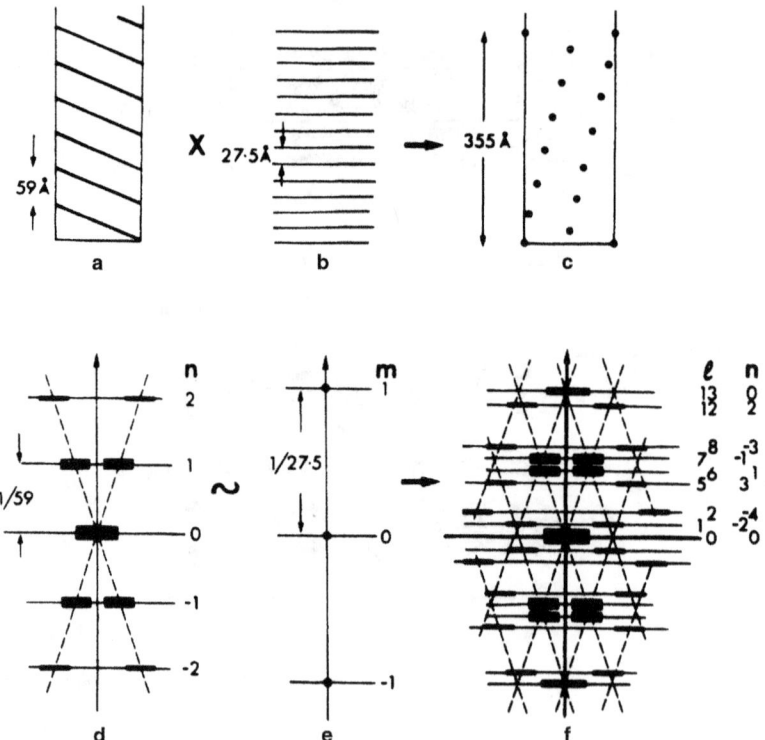

Fig. 14.12: Generation of the Fourier transform of the F-actin helix. (a) Radial projection of a continuous helix of pitch 59 Å. When this is multiplied by a set of density planes of spacing 27.5 Å (b), a discontinuous helix is generated (c), which has 13 residues in six turns of the left-handed helix in (a). (d)–(f) The transforms of (a)–(c) and indicate the strong layer-lines in (f) from the actin filament, together with the Bessel orders that describe the intensity distribution. [From Squire (1981).]

of intensity at axial positions along the z^*-axis (meridian) in reciprocal space equal to $\zeta = n/59 \text{ Å}^{-1}$ (cf. Fig. 14.6). The transform of the density planes spaced 27.5 Å (h) apart is a series of points along the z^*-axis [Fig. 14.12(e)] at positions $\zeta = m/27.5 \text{ Å}^{-1}$. These transforms need to be convoluted to give the transform of the 13/6 helix, as in Fig. 14.12(f). As can be seen here, layer-lines close to the equator are $l = 1$ with intensity described by Bessel order -2, $l = 2$ described by Bessel order -4 and so on. There is then a series of layer-lines described by low order Bessel functions around $l = 5$–8: 5 has $n = 3$, 6 has $n = 1$; 7 has $n = -1$; 8 has $n = -3$. The first and second layer-lines with spacings 1/360 and 2/360 Å^{-1} figure largely in discussions of actin filament structure, as do the sixth and seventh layer-lines at 1/59 Å^{-1} (i.e. 6/360 Å^{-1}) and 1/51 Å^{-1} (i.e. 7/360 Å^{-1}). The latter are commonly referred to as the 59 and 51 Å reflections. The first meridional reflection ($m = 1$) is at 1/27.5 Å^{-1} (1/h = 13/360 Å^{-1}).

The validity of this analysis is shown in Fig. 14.13, which reproduces both an optical diffraction pattern from a two-dimensional mask of circular apertures arranged to simulate a projection of a 13/6 helix and also part of a muscle X-ray diffraction pattern showing clear actin layer-lines up to the 13th at 27.5 Å.

All actin filaments in muscle contain tropomyosin molecules as well as F-actin, and some contain troponin as well [reviewed in Squire (1997a); Squire and Morris (1998)]. Tropomyosin molecules are two-chain coiled-coil α-helical structures of diameter about 20 Å and length 400 Å. They are thought to lie end-to-end along the two long-pitched strands of the F-actin helix [Fig. 14.11(c)]. Since they produce essentially continuous helical strands of pitch 2×360 Å, they should contribute a single cross pattern at $m = 0$, with layer-lines spaced axially along z^* by $\zeta = n/720 \text{ Å}^{-1}$. However, since there are two coaxial strands, the actual repeat is halved from 720 to 360Å; only even order layer-lines are seen. Therefore there is no layer-line at 1/720 Å^{-1}. The intensity from tropomyosin on the first actin layer-line at $\zeta = 1/360 \text{ Å}^{-1}$ is described by a second order Bessel function ($n = 2$), that on the second (third) actin layer-line(s) by $n = 4$ (6), etc. There is little contribution beyond this from tropomyosin.

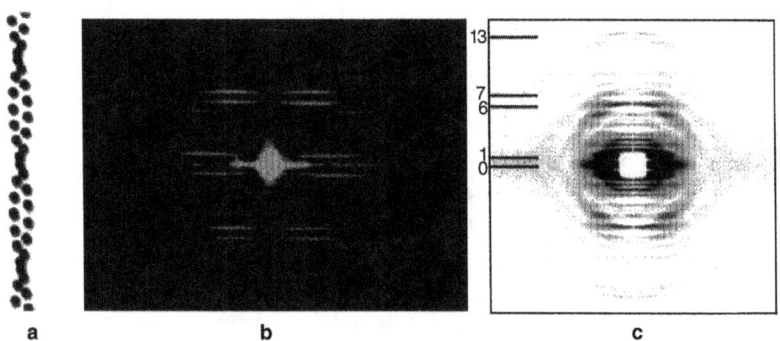

Fig. 14.13. (a) A set of points with the same symmetry as a 13/6 actin filament together with the optical diffraction pattern from it (b). For comparison, (c) is a recorded X-ray diffraction pattern from fish muscle labelled with myosin heads and showing the actin layer-line pattern. (JJ Harford, unpublished data.)

290   *Fibre and muscle diffraction*

The axial repeat of the slightly overlapping, roughly 400 Å long, tropomyosin molecules is about 385 Å. When the troponin complex is present, it attaches specifically to each tropomyosin molecule and therefore shows the same 385 Å axial repeat [Fig. 14.11(c)]. Because in many vertebrate muscles this is slightly different from the repeat of the actin helix, pairs of troponin molecules on opposite strands of tropomyosin gradually appear to rotate around the actin filament axis. The diffraction patterns from troponin tend to have contributions mainly along the meridian at axial spacings that are of the orders of 385 Å, although troponin may also make a contribution to the low order actin layer-lines. Note that the discussion so far has referred to actin filaments in vertebrate striated muscles. In some muscles, such as those in the asynchronous flight muscles of insects (Reedy 1968) the actin filament itself is slightly untwisted to have exactly 28 subunits in 13 turns of the helix, giving a pitch P of 385 Å that coincides with the troponin repeat. This may be related to some of the special contractile properties of these muscles (Pringle 1967). The diffraction patterns from actin filaments with 28/13 symmetry are very similar to those shown in Figs 14.12 and 14.13.

14.5   Diffraction from myosin filaments

Myosin filaments in different muscles, although apparently being variations on a single theme, have different helical symmetries (Squire 1981). It is therefore necessary to treat them individually. Figure 14.14 shows radial projections of the surface lattices of myosin heads on some representative filaments. Of particular

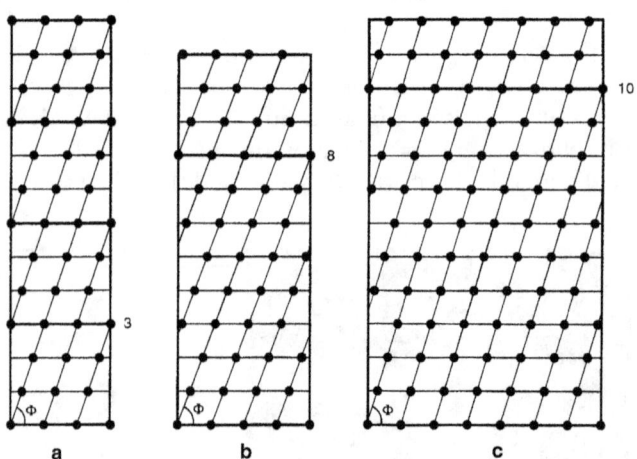

Fig. 14.14. Myosin filament nets for (a) vertebrate striated muscle, (b) insect flight muscle (*Lethocerus*) and (c) scallop striated muscle. To generate the appropriate helix in each case, imagine cutting around the outline of each helical net and then bending the paper into a cylinder to bring the left- and right-hand edges of each image together. Note that the numbers to the right of each figure show the multiples of 143–145 Å at which the structure repeats. Each 143 Å repeat is a crown of heads; there are three crowns in the 429 Å repeat of vertebrate myosin filaments (a). Note that despite the different numbers of strands and the different axial repeats, the lattice cell of angle ϕ is similar in each case. [From Chew and Squire (1995).]

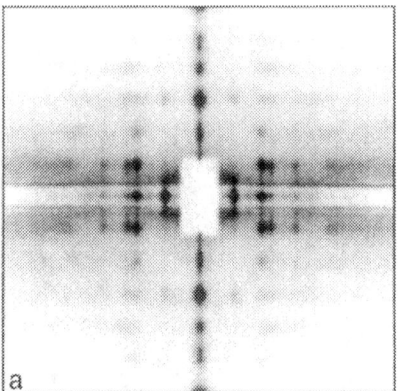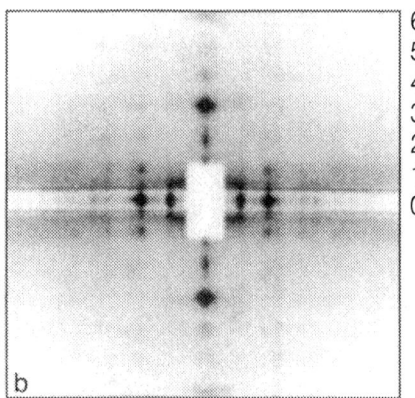

Fig. 14.15. Low angle X-ray diffraction patterns recorded on a multiwire proportional counter on line 2.1 at the Daresbury Laboratory from bony fish muscle in the resting state (a) and the plateau of a tetanus (b). Note the high degree of sampling of the myosin layer-lines (0–6) and the meridional reflections on layer-lines other than multiples of 3 that would be 'forbidden' for a perfectly helical structure as in Fig. 14.12(a). [From Harford and Squire (1992).]

interest are the myosin filaments in vertebrate striated (i.e. skeletal and cardiac) muscles [Fig. 14.14(a)]. The myosin diffraction pattern from resting fish muscle is shown in Fig. 14.15(a). Vertebrate myosin filaments have three-fold rotational symmetry (Squire 1972, 1981; Luther and Squire 1980) and the helical arrays can be described as three-start helices with ($N=$) 9 subunits in ($K=$) 1 turn of the helix. The helix pitch (P) is 1287 Å and the subunit axial translation (h) is 143 Å. Because there are three coaxial helices of myosin heads, the true repeat of the structure is 1287/3 Å = 429 Å. Layers 1 and 2 of the helix cross are missing, as are *any* orders of 429 Å that are not multiples of 3, as illustrated in Fig. 14.16. This means that the first layer-line ($l=1$) at $\zeta = 1/429$ Å^{-1} has intensity described by a third order Bessel function from $m=0$, together with a further contribution of order -6 from $m=+1$. There are also several further potential contributions of higher Bessel function orders from crosses centred at other m values. Likewise, the second layer-line at $\zeta = 2/429$ Å^{-1} has Bessel orders $+6$ from $m=0$, -3 from $m=+1$, and so on. The equator and layer-lines (l) which are multiples of 3, all have meridional reflections (J_0 Bessel orders from cross centres), together with orders $+9$ and -9, $+18$ and -18, etc. from other adjacent helix crosses.

The myosin heads [Figs 14.11 and 14.14(a)] occur in rings spaced approximately 143 Å apart axially along the myosin filament backbone. These have been termed 'crowns' of heads. In thick filaments of vertebrate striated muscle, there are three pairs of myosin heads in each crown and successive crowns are related by a 143 Å axial shift and a 40° azimuthal rotation around the filament axis (Squire 1972). In fact, these crowns in vertebrate muscle are not exactly equivalent. There are small perturbations to the basic helix of myosin heads that make three successive crowns slightly different (Huxley and Brown 1967; Harford and Squire 1986). This set of three crowns then repeats at 429 Å intervals to give the full filament. The effect of this slight non-equivalence of the crowns

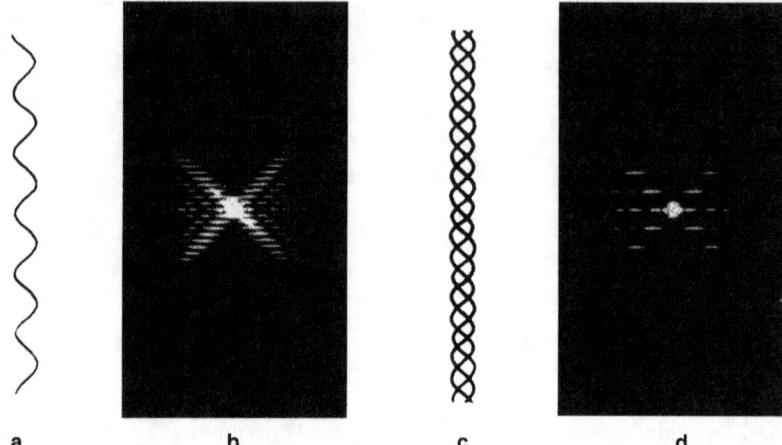

Fig. 14.16. If a helical structure is multistranded as in (c), which shows a three-start helix, then the transform (d) has some of the layer-lines missing relative to the transform (b) from a single-start helix (a) of the same pitch (P). With a three-stranded continuous helix, every third layer-line is missing since the axial repeat has become P/3. [From Squire (1981).]

is to introduce the so-called 'forbidden' meridional reflections on layer-lines 1, 2, 4, 5, 7, etc. of the 429 Å repeat [see Fig. 14.15(a) and Squire et al. (1982)]. These reflections are 'forbidden' in the sense that they are not part of the transform of the perfect undistorted helix.

As far as is known, such a perturbation does not occur on the myosin filaments of insect asynchronous flight muscle. Here the myosin filaments have four-fold rotational symmetry with four pairs of myosin heads per crown (Reedy 1968; Morris et al. 1991). In this case the helix, shown in Fig. 14.14(b), repeats after eight 145 Å-spaced crowns so that c is 1160 Å. The strong layer-lines are at $l = 3$ ($n = 4$; $\zeta = 3/1160 = 1/387\,\text{Å}^{-1}$), $l = 5$ ($n = -4$; $\zeta = 5/1160 = 1/232\,\text{Å}^{-1}$), $l = 8$ ($n = 0$; $\zeta = 8/1160 = 1/145\,\text{Å}^{-1}$), and so on. Myosin filaments in other (invertebrate) muscles have higher rotational symmetries, such as seven-fold in scallop striated muscle [Fig. 14.14(c)].

14.6 Whole muscle diffraction patterns

Diffraction patterns from whole muscle are a combination of the diffraction contributions from actin and myosin filaments, together with any other minor components of the muscle that are massive enough and ordered enough to contribute to specific peaks [e.g. the Z-band; Harford et al. (1994)]. In some muscles (e.g. bony fish muscles, insect flight muscles) the myosin and actin filaments are ordered well enough that the typical filament layer-lines are sampled by the reciprocal lattice corresponding to the ordered unit cell in the muscle. Fish muscles have their myosin filaments arranged with good rotational register so that the layer-lines that are of the order of 429 Å repeat are well sampled along

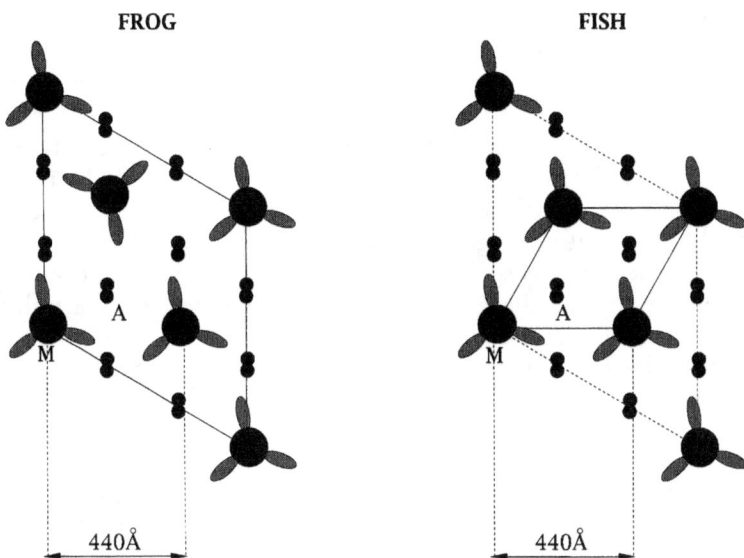

Fig. 14.17. Filament lattices in the overlap region of the A-bands in various vertebrate striated muscles: (a) superlattice muscles (higher vertebrates, e.g. frog, rabbit, human); (b) simple lattice in bony fish muscles. In each case the large black circles (M) represent myosin filaments and the pairs of small filled circles (A) represent actin filaments and indicate their relative orientations. The solid outlines show the unit cells in each case. In (a) this is a statistical superlattice unit cell containing three myosin filaments, whereas in (b) there is a simple unit cell containing only one myosin filament per cell. The shaded ellipses indicate the three-fold symmetry of the cross-bridge arrangements on the myosin filaments. [After Squire (1981).]

axially aligned *row-lines* [Fig. 14.15(a)]. The radial positions (R) of these row-lines are determined by the lattice spacing of the hexagonal array of filaments in the A-band (Fig. 14.17). The first few orders from the hexagonal lattice, corresponding to the largest d spacings and therefore the smallest angles of diffraction, can be determined from the expression for the interplanar spacing d_{hk0} in a conventional hexagonal lattice ($a = b \neq c$, $\alpha = \beta = 90°$, $\gamma = 120°$):

$$d_{hk0} = a\sqrt{(3/4)}/\sqrt{(h^2 + hk + k^2)}.$$

From the largest spacings downwards, the reflection indices hk are 10, 11, 20, 21, 30, etc. On a particular layer-line l, the reflections on these row-lines are the $10l, 11l, 20l, 21l, 30l$, and so on. On the equator ($l = 0$), the reflections are the 100, 110, 200, etc. The equatorial reflections provide information on the appearance of the muscle (or fibre) looking down the fibre axis (i.e. they show what a thick cross-section might look like).

In bony fish muscle the myosin filaments are quite well ordered in three dimensions in the sarcomere [Fig. 14.17(b)] and good sampling of the myosin layer-lines is apparent [Fig. 14.15(a)]. However, in the muscles of vertebrates, apart from the bony fish (e.g. human, chicken, rabbit, frog, etc.), the myosin filaments are not simply ordered as in fish muscles and, as a consequence, the myosin layer-lines are either totally unsampled or they are only partially sampled (Huxley and Brown 1967). These higher vertebrate muscles have their myosin filaments arranged in a statistical superlattice [Fig. 14.17(a); Luther and Squire 1980]. This means that, although a kind of superlattice unit cell can be defined with a side equal to $a\sqrt{3}$ (a, the intermyosin filament spacing, is the length of lattice vector a of the simple hexagonal lattice), the contents of one unit cell are not necessarily identical to those of its neighbours—there is both order and disorder in this structure. The superlattice is such that, in a view (i.e. projection) down the filament axis, all the myosin filaments in the unit cell appear to be similar at low resolution. The equator of the diffraction patterns from frog and other superlattice muscles is therefore similar to that from fish muscles, in that it relates to the basic hexagonal lattice of side a. However, on layer-lines other than those where l is a multiple of 3, the row-line sampling relates to the superlattice unit cell of side $a\sqrt{3}$. Even then, this sampling is only partial; some underlying unsampled transform remains. The general effect of any kind of disorder on a diffraction pattern is to reduce the intensity of sampled peaks, sometimes to broaden them, but always to increase the relative intensity of the underlying unsampled 'background' diffraction pattern (Fraser and MacRae 1973; Hukins 1981; Vainshstein 1966). The presence of strong unsampled scattering underlying the observed sampled diffraction peaks is good evidence for disorder.

Note finally that, even if the myosin and actin filaments show no significant rotational register, provided they are still packed in the basic hexagonal A-band lattice with their axes parallel, they will appear to be well ordered in a view down the filament axis. The equator of the diffraction pattern will then be sampled as described above, even though all the other layer-lines may be completely unsampled. Partial ordering of the filaments will produce intermediate effects. For example, perfect axial register of the myosin filaments will give meridional reflections such as that at 143 Å that are extremely narrow across the meridian. If there is no axial order, these layer-lines will be totally unsampled and the 143 Å reflection will be broad across the meridian. With partial axial ordering, the meridional reflections will be of intermediate widths across the meridian. This width can therefore be used as a measure of the degree of ordering of the specimen, once broadening due to the inherent width of the X-ray beam has been accounted for.

14.7 Example of detailed analysis: modelling the myosin head array in fish muscle

In order to solve directly the observed low angle X-ray diffraction data from fish muscle myosin filaments [Fig. 14.15(a)], as stripped by using the CCP13 fibre diffraction software [Fig. 14.18; Denny (1996)], it is necessary to set up a model with the known myosin head shape (Rayment et al. 1993) with appropriate positional parameters as variables and then to search over and refine these parameters to optimize the fit between observed and calculated data. In the case of fish muscle, a simple hexagonal unit cell can be described in terms of parameters defining

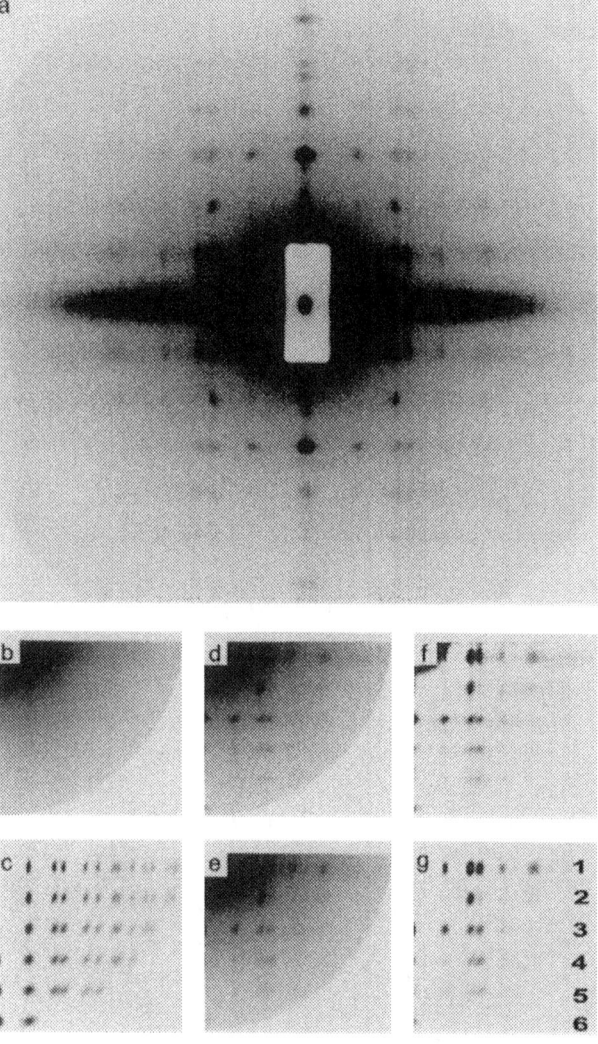

Fig. 14.18. (a) A resting diffraction pattern from plaice fin muscle showing the myosin layer lines 1–6 and various steps (b)–(g) in fitting this diffraction pattern using CCP13 software (Denny 1996). (c) A NOFIT representation of the shapes and positions of the Bragg peaks in (a) but not their intensities. (b) The background obtained by fitting the intensities between the NOFIT peaks. (f) The original data with the background in (b) removed and (g) the Bragg peaks modelled as in (c) but with the intensities fitted. Comparison can be made between the original data (d) and the modelled data (e), which is the sum of (b) and (g). Note that data in (a) were quadrant-folded to give (d), and (b)–(g) all represent the bottom-right quadrant of (a); the undiffracted beam is at the top-left corner of each figure in (b)–(g). [From Hudson et al. (1997).]

tilt, slew and rotation of the heads (Hudson et al. 1997). The filament can be assumed to retain a three-fold rotational symmetry, but the heads on different crowns may have different configurations. The whole filament structure must also be allowed to rotate within the unit cell about its own long axis, say by an angle β. This makes 22 parameters as being the smallest number that can be used for a sensible search. Amplitudes and phases of reflections h, k, l in the general Fourier transform expression (Hudson et al. 1997) were computed from

$$F(h, k, l) = \sum f_j(h, k, l) \exp[2\pi i(hx_j + ky_j + lz_j)]$$

in terms of these parameters. Note that here the myosin head shape was built up from spheres, so $f_j(h, k, l)$ in this case is the spherical scattering factor

$$f_j(h, k, l) = (4/3)W\pi a_j^3 \{3[\sin(X) - X\cos(X)]\}/X^3,$$

where $X = 2\pi a_j \rho$, a_j is the sphere radius, W is its density and ρ is the distance from the origin in reciprocal space to the point h, k, l in the transform that is being calculated.

The computed amplitudes (F_c) were then compared with the observed amplitudes (F_o) using a suitable 'goodness of fit' R-factor. A useful R-factor is

$$R = \frac{\sum_{i=1}^{N}(I_i^o - I_i^c)^2/\sigma_i^2}{\sum_{i=1}^{N}(I_i^o)^2/\sigma_i^2},$$

where I^o is the observed intensity, I^c is the calculated intensity, N is the number of intensities and σ_i denotes the standard deviation associated with I_i^o. Simulated annealing procedures (Press et al. 1992; Hudson et al. 1997) can be used to bring the parameters close to giving a good R-factor fit, after which a downhill simplex routine will rapidly bring the parameters to give the minimum R-factor. The simulated annealing approach has the advantage that the search can escape from local minima and should converge to the principal minimum.

Carrying out this procedure for the diffraction pattern from the plaice fin muscle [Fig. 14.15(a)], as stripped by using the CCP13 procedures discussed in Fig. 14.18 to give the data in Fig. 14.19(a), produced the simulated diffraction pattern in Fig. 14.19(b), with an R-factor of about 3% for 56 fitted reflections to a resolution of about 65 Å. Although head pairs on the three non-equivalent crowns were not constrained to be the same, in fact this best model [Fig. 14.19(c)] makes two crowns rather similar and the third not totally dissimilar. The 'forbidden' meridional reflections are nicely reproduced by this model, as are features such as the marked intensity difference between the closely adjacent 112 and 202 reflections.

Further optimization of the modelling required the introduction of the myosin filament backbone which makes a small contribution to the inner end of the third (143 Å) myosin layer-line (Chew and Squire 1995). In addition, the contribution of the so-called 'extra' proteins, such as C protein (Bennett et al. 1986; Rome et al. 1973) and titin (Labeit and Kolmerer 1995), needs to be considered. These are presented in detail elsewhere (Squire et al. 1998).

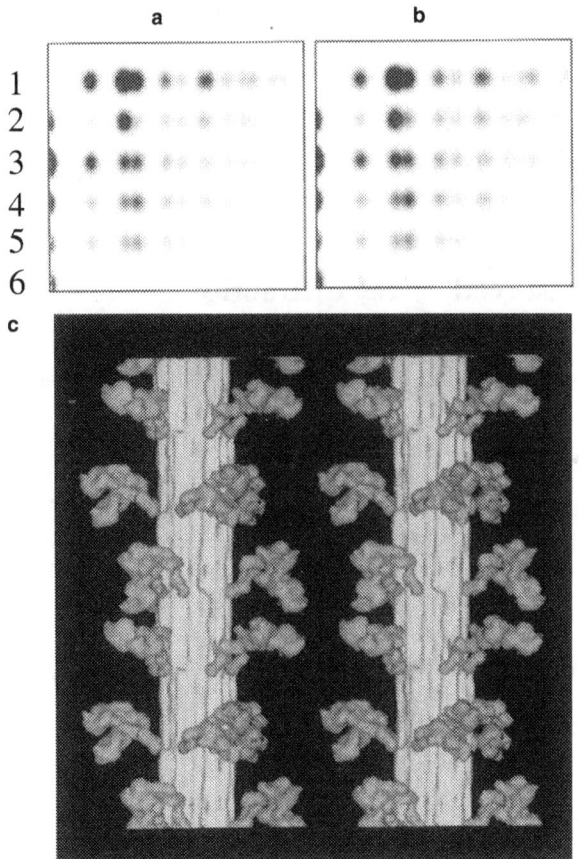

Fig. 14.19. (a) The observed and (b) the modelled intensities from the optimized myosin filament structure in (c). Each pattern corresponds to the same quadrant of the diffraction pattern as in Fig. 14.18(g). (c) The 'best' model of the myosin head array in fish muscle obtained by simulated annealing and further local refinement of a parametrized filament structure (Hudson *et al.* 1997). [The filament backbone is from Chew and Squire (1995).]

PART 3: FUTURE DEVELOPMENTS
14.8 High resolution diffraction studies

New methods of fibre preparation, including the inducement of orientation by strong magnetic fields and also simply leaving liquid crystals to gradually anneal over a period of years, are producing fibres of unprecedented order. Recent examples of this are flagella which diffract out to 0.9 nm resolution (Yamashita *et al.* 1998) and TMV crystalline fibres which diffract to about 0.15 nm (D Caspar, personal communication). With this kind of resolution, some of the conventional approaches of protein crystallography, modified for application to fibres, can be successfully applied to solve the structures to atomic detail. Molecules like DNA have already been studied to high resolution and, with the application of neutron diffraction, linked to contrast variation and use of deuterated water, it is possible to define the positions of ordered water molecules and some ions to very high precision (Pope et al. 1996). Actin filaments, which are ordered aggregates of actin monomers for which the atomic arrangement is known (Kabsch et al. 1990), can be solved to quite high resolution by searching and refinement (Holmes et al. 1990).

14.9 Time-resolved X-ray diffraction from dynamic systems

One of the most important applications of synchrotron radiation in muscle research is the use of fast, time-resolved X-ray diffraction data from active muscle [Fig. 14.15(b)] or muscle undergoing rapid length or tension changes (Bagni et al. 1994; Bershitsky et al. 1996; Huxley et al. 1983; Harford and Squire 1992; Irving et al. 1992; Lombardi et al. 1995; Martin-Fernandez et al. 1994; Wakabayashi et al. 1993; Yagi and Matsubara 1989). Studies of this kind are too numerous to summarize here. The reader is referred to a number of reviews where such studies are discussed (Harford and Squire 1990, 1997; Lowy and Poulsen 1987; Squire 1981; Wakabayashi and Amemiya 1991). Suffice it to say that time-resolved X-ray diffraction studies of muscle, which is probably the key example of a dynamic biological system that can be studied on a macroscopic scale, are starting to probe different stages of the myosin head cycle on actin filaments involved in force production and movement; they are playing a central part in getting to this 'holy grail' of muscle research (Squire et al. 1993).

14.10 New synchrotron sources and detectors

Some time-resolved diffraction studies, for example of muscle, are limited by the flux available from conventional, second generation, synchrotron sources [see Harford and Squire (1997)]. Muscles are also composed of many parallel fibres between which there may be small variations. It is also desirable to monitor the muscle sarcomere length during the imposition of fast mechanical interventions and this is much more easily done with single muscle fibres. There is therefore a clear need for synchrotron sources that are of high brilliance; a high flux in a small beam cross-section. The new third generation sources such as the ESRF, the APS and Spring-8 are such high brilliance machines. It is therefore possible with beamlines on these machines to record good diffraction patterns from muscle single fibres or other small fibrous specimens (e.g. keratin in a single hair; collagen in a single strand of tendon). The one over-riding problem with such studies is that there are very few detectors available that can cope both with high flux and fast readout for time-resolved studies. For static diffraction patterns, image plates can be very effective, but these are of little use if time resolutions of $10\,\mu s$ to $1\,ms$ are needed. The best area detectors available for this are the multiwire proportional counters (MWPC) (Lewis 1994). Current standard

MWPCs have a count rate that at best is about 1 MHz, but is usually limited to about 200–300 kHz. This is very much lower than the rate at which photons are being delivered to the specimen, even with second generation synchrotrons like the Daresbury SRS, so in the past it has been necessary to attenuate the beam so as not to saturate the detectors. A new multiwire detector known as RAPID has now been developed at CCLRC Daresbury. This is a very high resolution, low noise detector that has a count rate up to 10 MHz; it has not yet saturated on Daresbury beamlines. Currently, line 16.1 at Daresbury with the RAPID detector is the best combination available worldwide for fast low angle time-resolved diffraction studies of any system, whether biological, such as muscle, or synthetic, such as polymer systems. The seemingly ideal combination of third generation source beamlines with the RAPID detector still seems to be some way off. When eventually they come together, experiments with unprecedented time-resolution and quality will become a reality rather than a dream.

Acknowledgements

The author is pleased to acknowledge the grant support of the MRC, EPSRC (formerly SERC), BBSRC, CCLRC and Wellcome Trust for various parts of the work reported here. Various figures were generously provided by Drs J.J. Harford, J.S. Barry, M. Chew, L. Hudson and G. Offer. The help of Sarah Tomlin in preparing the text and figures in a suitable format for reproduction is much appreciated.

References

Al-Khayat, HA, Yagi, N and Squire, JM (1995). *Journal of Molecular Biology*, **252**, 611–632.
Bagni, MA, Cecchi, G, Griffiths, PJ, Maeda, Y, Rapp, G and Ashley, CC (1994). *Biophysical Journal*, **67**, 1965–1975.
Bennett, PM, Craig, R, Starr, R and Offer, G (1986). *Journal of Muscle Research and Cell Motility*, **7**, 550–567.
Bershitsky, S, Tsaturyan, A, Bershitskaya, O, Mashanov, G, Brown, P, Webb, M and Ferenczi, MA (1996). *Biophysical Journal*, **71**, 1462–1474.
Bloom, W and Fawcett, DW (1975). *A textbook of histology*, 10th edn, (Philadelphia, PA: Saunders).
Chew, MWK and Squire, JM (1995). *Journal of Structural Biology*, **115**, 233–249.
Cochran, W, Crick, FHC and Vand (1952). *Acta Crystallographica*, **5**, 581–586.
Denny, R (1996). *Fibre Diffraction Review*, (CCP13 Newsletter, http://www.dl.ac.uk/CCP/CCP13).
Fraser, RDB and MacRae, TP (1973). Conformation in fibrous proteins and related synthetic polypeptides, Academic Press, New York and London.
Harford, JJ and Squire, JM (1986). *Biophysical Journal*, **50**, 145–155.
Harford, JJ and Squire, JM (1990). In *Molecular mechanisms in muscular contraction*, JM Squire, ed. Macmillan Press, Vol. 13, pp. 287–320.
Harford, JJ and Squire, JM (1992). *Biophysical Journal*, **63**, 387–396.
Harford, JJ and Squire, JM (1997). *Reports on Progress in Physics*, **60**, 1723–1787.

Harford, JJ, Luther, PK and Squire, JM (1994). *Journal of Molecular Biology*, **239**, 500–512.
Holmes, KC and Blow, DM (1965). *The use of X-ray diffraction in the study of protein and nucleic acid structure*, Wiley, New York.
Holmes, KC, Popp, D, Gebhard, W and Kabsch, W (1990). *Nature*, **347**, 44–49.
Hudson, L, Harford, JJ, Denny, R and Squire, JM (1997). *Journal of Molecular Biology*, **273**, 440–455.
Hukins, DWL (1981). *X-ray diffraction by disordered and ordered systems*, Pergamon Press, Oxford and New York.
Huxley, HE (1969). *Science*, **164**, 1356–1366.
Huxley, HE and Brown, W (1967). *Journal of Molecular Biology*, **30**, 383–434.
Huxley, HE, Simmons, RM, Faruqi, AR, Kress, M, Bordas, J and Koch, MHJ (1983). *Journal of Molecular Biology*, **169**, 469–506.
Irving, M, Lombardi, V, Piazzesi, G and Ferenczi, MA (1992). *Nature*, **357**, 156–158.
Kabsch, W, Mannherz, HG, Suck, D, Pai, EF and Holmes, KC (1990). *Nature*, **357**, 37–44.
Labeit, S and Kolmerer, B (1995). *Science*, **270**, 293–296.
Lewis, R (1994). *Journal of Synchrotron Radiation*, **1**, 43–53.
Lombardi, V, Piazzesi, G, Ferenczi, MA, Thirlwell, H, Dobbie, I and Irving, M (1995). *Nature*, **374**, 553–555.
Lowy, J and Poulsen, FR (1987). In *Fibrous protein structure*, Squire, JM and Vibert, PJ, eds, Academic Press. pp. 451–494.
Luther, PK and Squire, JM (1980). *Journal of Molecular Biology*, **141**, 409–439.
Luther, PK, Munro, PMG and Squire, JM (1995). *Micron*, **26**, 431–459.
Martin-Fernandez, ML, Bordas, J, Diakun, G, Harries, J, Lowy, J, Mant, GR, Svensson, A and Towns-Andrews, E (1994). *Journal of Muscle Research and Cell Motility*, **15**, 319–348.
Morris, EP, Squire, JM and Fuller, GW (1991). *Journal of Structural Biology*, **107**, 237–249.
Offer, G (1974). In *Companion to biochemistry*, Bull, AT et al., eds. pp. 623–671. Longmans, London.
Pope, LH, Shotton, MW, Forsyth, VT, Langan, P, Grimm, H, Rupprecht, A, Denny, R and Fuller, W (1996). *The Fibre Diffraction Review* (CCP13 Newsletter **5**), 34-38 (http://www.dl.ac.uk/CCP/CCP13).
Press, WH, Teukolsky, SA, Vetterling, WT and Flannery, BP (1992). *Numerical recipes in C*, Cambridge University Press, Cambridge, UK and New York.
Pringle, JWS (1967). *Progress in Biophysics and Molecular Biology*, **17**, 1–60.
Rayment, I, Rypniewski, WR, Schmidt-Base, K, Smith, R, Tomchick, DR, Benning, MM, Winkelmann, DA, Wesenberg, G and Holden, HM (1993). *Science*, **261**, 50–58.
Reedy, MK (1968). *Journal of Molecular Biology*, **31**, 155–176.
Rome, E, Offer, G and Pepe, FA (1973). *Nature*, **244**, 152–154.
Rosenbaum, G, Holmes, KC and Witz, J (1971). *Nature*, **230**, 434–437.
Squire, JM (1972). *Journal of Molecular Biology*, **72**, 125–138.
Squire, JM (1981). *The structural basis of muscular contraction*, Plenum Press, New York and London.
Squire, JM (1983). *Trends in Neuroscience*, **6**, 409–413.
Squire, JM (1997a). *Current Opinion in Structural Biology*, **7**, 247–257.
Squire, JM (1997b). In *Current methods in muscle physiology*, Sugi, H, ed, pp. 241–285. Oxford University Press, Oxford.
Squire, JM and Morris, E (1998). *FASEB Journal*, **12**, 761–771.
Squire, JM, Harford, JJ Edman, A-C and Sjostrom, M (1982). *Journal of Molecular Biology*, **155**, 467–494.
Squire, JM, Harford, JJ and Morris, EP (1993). *Muscle—The Movie, Image Processing*, Spring 1993, 22–23.

Squire, JM, Cantino, M, Chew, M, Denny, R, Harford, JJ, Hudson, L and Luther, PK (1998). *Journal of Structural Biology*, **122**, 128–138.
Vainshstein, BK (1966). *Diffraction of X-rays by chain molecules*, Elsevier, Amsterdam.
Wakabayashi, K and Amemiya, Y (1991). In *Handbook on synchrotron radiation*, Vol. 4, Ebashi, S, Koch, M and Rubenstein, E, eds, Elsevier Science Publishers BV, Tokyo, pp. 597–678.
Wakabayashi, K, Saito, H, Moriwaki, N, Kobayashi, T and Tanaka, H (1993). In *Mechanism of myofilament sliding in muscle contraction*, Sugi, H and Pollack, GH, eds, Plenum Press, New York, pp. 451–460.
Yagi, N and Matsubara, I (1989). *Journal of Molecular Biology*, **208**, 359–363.
Yamashita, I, Hasegawa, K, Suzuki, H, Vonderviszt, F, Mimiri-Kiyosue, Y and Namba, K (1998). *Nature Structural Biology*, **5**, 125–132.

Selected topics

15

Biological spectroscopy using low energy (VUV/UV) synchrotron radiation

G.R. Jones and I.H. Munro

15.1 Introduction

Synchrotron radiation (SR) storage rings provide the most highly collimated and intense continuous sources of X-rays, soft X-rays and extreme vacuum ultraviolet (UV) radiation. At longer wavelengths, in the UV, visible and infrared the photon flux of SR is often surpassed by alternative (usually less expensive) sources, such as lasers, arc lamps, blackbody sources, etc. However, there are inherent properties of synchrotron light that make it the preferred source for many experiments. Synchrotron light from bending magnets provides a wavelength continuum of highly collimated, linearly polarized radiation, emanating from a tiny source (of from 10^3 to $\sim 10^6 \, \mu m^2$ in cross-section) of electrons or positrons with a well-understood time structure and long term stability. This unique ensemble of properties is further augmented at third generation synchrotron radiation sources by the much higher flux and brightness available from insertion devices such as undulators or multipole wigglers.

The main low energy SR techniques that are directly relevant to the area of structural biology are time-resolved fluorescence spectroscopy and vacuum ultraviolet (VUV) circular dichroism (CD and MCD), both of which will be described in detail. However, low energy SR is now also being exploited in other areas of research, such as the study of the effects of ionizing radiation on biological samples, the study of specific vibrations in hydrated systems (such as proteins) by the chemical mapping of biological specimens using Fourier transfer infrared microscopy and, of course, biological imaging at shorter wavelengths using soft X-ray microscopy.

15.2 Time-resolved fluorescence spectroscopy

Fluorescence is the light emitted when an electronically excited state in a molecule rapidly relaxes to a state of lower energy (within a time range of from picoseconds to microseconds). As an analytical tool fluorescence has been widely exploited in many areas of biology, because of its simplicity, specificity, selectivity, sensitivity and often low cost. This is illustrated by its use in microscopy, enzyme assays, electrophoresis gels and blots, etc. In principle, fluorescence should be an extremely powerful technique in structural biology because every fluorescent group will be sensitive to the pH, ionic strength, specific ions, or the polarity within its immediate environment. Also, since the fluorescence relaxation process occurs within the same time-scale of the segmental and Brownian motions of many biological macromolecules and their complexes ($\sim$ from picoseconds to nanoseconds), it can be used to measure their size and mobility.

When steady-state fluorescence measurements are used in structural biology, however, the usefulness of fluorescence is significantly diminished because these measurements cannot resolve the fluorescence lifetime τ. The fluorescence lifetime is the average time following the absorption of a photon, for which molecules exist in the excited state before emitting a photon. Measuring the time course of fluorescence decay requires more sophisticated equipment than for steady-state fluorescence, but has the advantage of providing direct information on a macromolecule's size, flexibility, environment and degree of molecular aggregation.

When used as a hydrodynamic technique, time-resolved fluorescence offers tremendous advantages, in that it can be measured over a vast range of concentrations (molar to nanomolar). Other hydrodynamic techniques such as X-ray solution scattering, laser light scattering, and analytical ultra-centrifugation are more limited. This advantage is particularly important in the determination of thermodynamic and kinetic parameters such as binding constants and free energies, or distinguishing between Arrhenius and non-Arrhenius behaviour, or even the type of binding in macromolecular complexes, i.e. ionic, hydrogen bond, etc.

15.3 The fluorescence process

The fluorescence behaviour of aromatic or heterocyclic molecules is best explained with the help of the modified Jablonski diagram [Fig. 15.1(a)]. It can be broken down into three stages:

Stage 1. The absorption of the normally UV photon which converts the fluorophore from its ground state S_0 to an excited state S_1 in a time scale so fast (10^{-15} s) that the atomic nuclei have not had time to change their positions. This state is sometimes called the Franck–Condon state.

Stage 2. From S_1 the fluorophore relaxes to a lower energy state (still assigned S_1). The energy is lost via a number of processes such as vibrational energy loss, losses due to excited state reactions and, most importantly to the structural biologist, the loss of energy incurred by the positional rearrangement of the excited fluorophore with respect to its environment. This occurs because the excited state is electronically very different from the ground state of the molecule.

Stage 3. The emission of the fluorescence photon.

The whole absorption and fluorescence process imposes two important properties. Firstly, the emission of the fluorescence occurs at longer wavelengths than the absorption (the Stokes shift) due to stage 2, thereby imparting a high sensitivity to fluorescence measurements since the emitted photons can easily be distinguished from excitation wavelength background scatter by wavelength

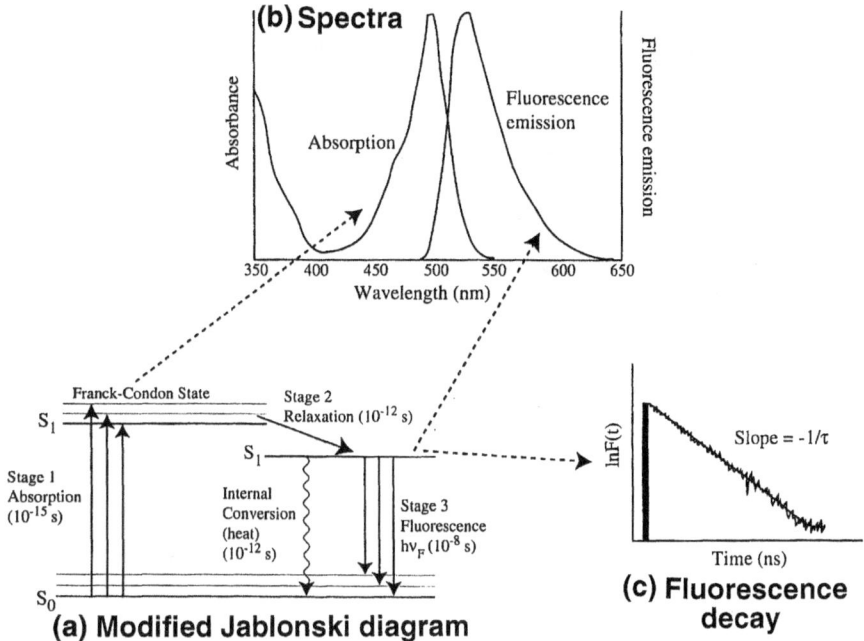

Fig. 15.1. Schematic diagram illustrating fluorescence processes.

selection with a monochromator [Fig. 15.1(b)]. The second property is that the fluorescence lifetime [Fig. 15.1(c)], which usually occurs on a timescale from 10^{-8} to 10^{-9} s, reflects the changes caused by the rearrangement of electron orbit and momentum in the excited state and is therefore very sensitive to the fluorophore's environment in this time domain. This turns out to be ideally suited to report on certain molecular rearrangements, e.g. the segmental motions along the backbone of protein and the rotation of fluorescent amino acid side-chains. Not every excited state species emits a fluorescence photon as there are many strongly competing non-radiative processes for their deactivation, such as vibrational relaxation by internal conversion, collisional quenching and intersystem crossing where the spin of an electron is changed (sometimes leading to the emission of phosphorescence). However, most effective fluorescence probes have quantum efficiencies of fluorescence between 30% and 90%. A full description of fluorescence spectroscopy may be found in Lakowicz (1983).

15.4 Fluorescent molecules

There are a huge number of naturally occurring fluorescent species in organic materials, the amino acids tryptophan and tyrosine being particularly useful. When these are used to report on proteins they are termed intrinsic probes because they are both fluorescent and, of course, occur naturally in proteins. Often it is more useful to chemically bind a small fluorescent molecule to a macromolecule, selected to measure a particular property. These are termed

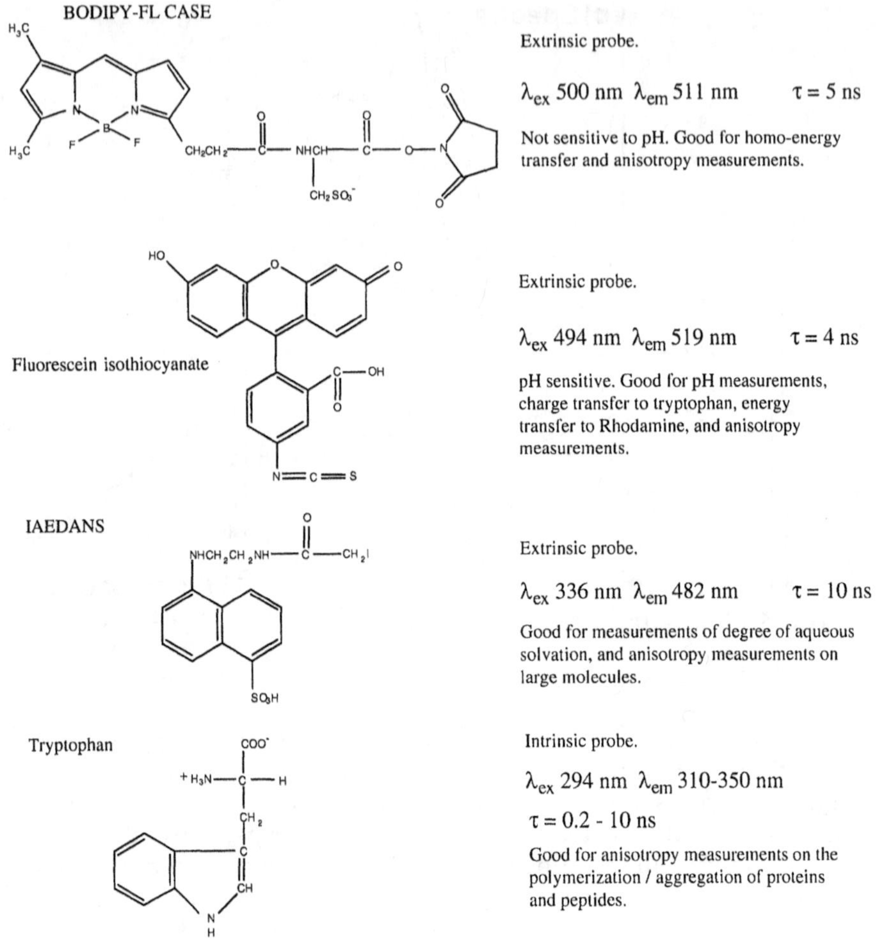

BODIPY-FL CASE

Extrinsic probe.

λ_{ex} 500 nm λ_{em} 511 nm $\tau = 5$ ns

Not sensitive to pH. Good for homo-energy transfer and anisotropy measurements.

Fluorescein isothiocyanate

Extrinsic probe.

λ_{ex} 494 nm λ_{em} 519 nm $\tau = 4$ ns

pH sensitive. Good for pH measurements, charge transfer to tryptophan, energy transfer to Rhodamine, and anisotropy measurements.

IAEDANS

Extrinsic probe.

λ_{ex} 336 nm λ_{em} 482 nm $\tau = 10$ ns

Good for measurements of degree of aqueous solvation, and anisotropy measurements on large molecules.

Tryptophan

Intrinsic probe.

λ_{ex} 294 nm λ_{em} 310-350 nm

$\tau = 0.2 - 10$ ns

Good for anisotropy measurements on the polymerization / aggregation of proteins and peptides.

Fig. 15.2. Examples of fluorescent probes.

extrinsic probes. Extrinisic probes that have excitation wavelengths away from the natural fluorescence of the sample are especially useful when many intrinsic fluorophores are present, in order to reduce the complexity of the fluorescence signal. Some examples of popular extrinsic and intrinsic probes are given in Fig. 15.2.

15.5 Measurement of the fluorescence lifetime

There are two general methods routinely used to measure the time evolution of fluorescence, the pulse coincidence method and phase modulation technique (Lakowicz 1983). Each has advantages and disadvantages over the other. The phase modulation technique enables the fluorescence lifetime to be determined from the phase shift between the emitted fluorescence and the (usually sinusoidally modulated) exciting light or, alternatively, by measuring the ratio of the intensity of modulation (AC/DC signal) between the excitation light and the fluorescence. These methods have been used at LURE (Orsay), DAPHNE (Rome), and ALADDIN (USA), but they are technically difficult to undertake using SR sources (De Stasio et al. 1991). This is because the usual acceleration frequencies of storage ring sources are high in order to maintain good control of the electron or positron beam. The SRS at Daresbury Laboratory, for example, runs at a radio frequency of around 500 MHz. This means that when the machine is running in multibunch mode with 160 bunches of circulating electrons, the equation

$$\tan \phi = 2\pi \omega \tau, \qquad (15.1)$$

where ϕ is the shift in phase angle and ω the modulation frequency, shows that, for optimal phase shifts between say 2° and 70°, the corresponding fluorescence lifetimes would vary between 0.01 and 0.8 ns. This is not a very useful window for lifetime measurements in structural biology although in principle at least, it offers scope for the measurement of very fast processes (e.g. in photosynthesis). Unfortunately, in practice, it is extremely difficult to make high precision phase shift measurements in the few hundred MHz region and the techniques that are available (e.g. using variable delay lines) are not adaptable to the longer lifetimes of primary interest. The phase method could be very useful if the bunch frequency in the storage ring were to be varied, on request, in the range 3–500 MHz. This is also feasible, but is an unrealistic prospect even for the most up-to-date sources because of the uncertainties associated with stable machine operations and the much lower mean circulating currents achievable in the 'few bunch' modes. Storage rings can be operated in a variety of these modes, e.g. single bunch, two bunch mode or other hybrid modes (see Table 15.1), but for these modes, it is actually pulse lifetime measurements that have proved most useful.

Time-structured modes also operate at MaxI (Lund) and Bessy (Berlin). The pulse lifetime measurements are best described by considering a population, N_0 of fluorophores to be excited by an infinitely short pulse of light. As time, t passes, after excitation the number of excited species $N(t)$ decays by

$$dN(t)/dt = -(E_F + E_{NF})N(t), \qquad (15.2)$$

where E_F is the decay rate of fluorescence and E_{NF} is the sum of the decay rate of all other competing process. Integrating eqn (15.2) gives

$$N(t) = N_0 e^{-t/\tau}, \qquad (15.3)$$

where the lifetime of the excited state is given as $\tau = 1/(E_F + E_{NF})$.

The fluorescence intensity F must be proportional to the excited population N and, therefore, decays exponentially as

$$F(t) = F_0 e^{-t/\tau} \qquad (15.4)$$

Table 15.1. Some relevant characteristics of SR sources

Source	Beam mode	Bunch length (FWHM) (ps)	Intensity (mA)	Energy	Frequency (MHz)	Beam lifetime (h)
SRS (Daresbury)	Single bunch	160	30	2 GeV	3.1	>20
UVSOR (Japan)	Single bunch	170	100	750 MeV	5.65	5
ESRF (Grenoble)	Single bunch	30	15	6 GeV	0.337	5
	Hybrid mode	30	7	6 GeV	2.97	7
	Hybrid 2 mode	30	14	6 GeV	0.143	7
	Hybrid 4 mode	30	20	6 GeV	2.86	7
APS (Argonne)	Single bunch	50–100	8.5	7 GeV	0.77	10
NSLS (Brookhaven)	Single bunch	350	100	2.584 GeV	1.67	20
X-ray	Five bunches	350	200	2.584 GeV	8.33	20
NSLS (VUV)	Seven bunches	320	950	800 MeV	5.88	5
ALS (Berkeley)	Two bunch mode	55	40	1–1.9 GeV	3.05	2.5
ELETTRA (Trieste)	Single bunch	20–50		2 GeV	1.1	11
DORIS II (Hamburg)	Two bunch mode	150	50	4.45 GeV	2.08	8
	Five bunch mode		120	4.45 GeV	5.21	8
SUPER-ACO (LURE)	Two bunch mode	600	200	800 MeV	8.33	>5

[as shown in Fig. 15.1(c)] or

$$F(t) = \sum_i a_i e^{-t/\tau_i} \qquad (15.5)$$

when the decay is multi-exponential. This would occur if the fluorophore is distributed in i different environments within the fluorescence lifetime or if i different fluorophores are measured.

Pulse lifetime methods measure the whole fluorescence time decay directly. Early methods of recording the fluorescence decay involved sampling the fluorescence intensity after the excitation flash, by energizing the detector at a series of set times after excitation, or by gating the output of the detector to bin the detected fluorescence into individual time-frames. These, when placed in sequence, built a profile of the fluorescence decay intensity. Such box-car methods have, however, been largely superseded by the single photon counting technique.

The principle behind single photon counting [for a more detailed description see Munro (1980); Munro and Schwentner (1983); O'Connor and Phillips (1984)] is to record the time delay between the excitation light pulse and the first (single) fluorescent photon that is detected repeatedly, so that a histogram of the fluorescence decay is built up. A diagram of the equipment required to achieve this is shown in Fig. 15.3. The heart of the instrument is the time-to-amplitude converter (TAC), which on command of a start signal, specially shaped by a device called a discriminator, commences ramping the voltage in a special capacitor until another signal arrives at the stop channel. This signal curtails the voltage ramp and, after a set time, the TAC outputs the voltage accumulated by the capacitor. This voltage pulse is fed to an amplitude-to-digital converter, ADC and the digitized signal. The amplitude of the pulse from the

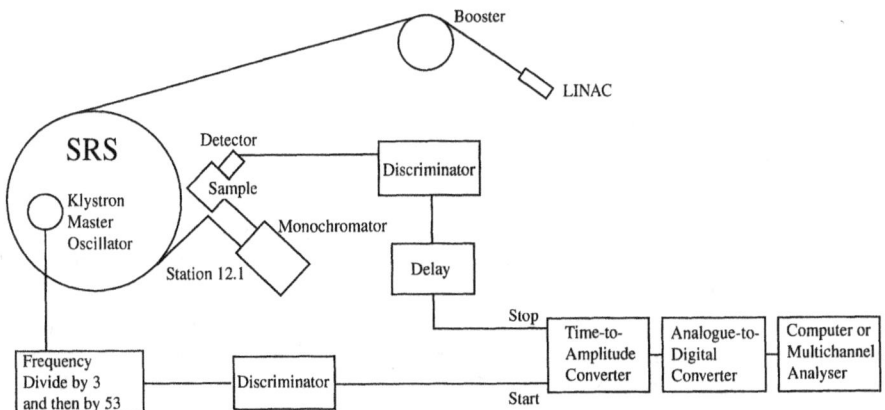

Fig. 15.3. Schematic of single photon counting equipment.

TAC is therefore proportional to the time lapse between the start and stop pulse. If the TAC is started by the excitation flash and stopped by an emitted fluorescence photon and this is repeated many times (the precision of the measurement is defined by the total number of counts; 10^3 is a lower limit and 10^5–10^6 would be typical) a profile of the fluorescence decay may be built up by accumulating the digitized voltage signals in a storage device that histograms counts vs time, such as a computer or multi-channel analyser. If a stop signal is not detected within a few microseconds then the TAC is reset without an output to the ADC.

When high frequency light sources are used, e.g. SR and some laser sources, it is usual to delay the fluorescence-generated electronic pulse from the photomultiplier or microchannel plate detector, so that it can be used as the start signal of the TAC. This delay is achieved with a length of RF-screened cable, and this is called the delay line (the speed of electronic pulses through a wire cable is about 23 cm/ns and this is used to delay the arrival of the excitation pulse until after the fluorescence pulse has been detected). The signal from the excitation flash is therefore used to stop the TAC and a reverse fluorescence decay profile is accumulated. This arrangement is commonly used because the conditions of the experiment must in any case be adjusted so that statistically during the TAC ramp there is only time for one fluorescence photon to be counted. To achieve this, the fluorescence detection rate is adjusted by an attenuator (iris diaphragm or similar device) so as to be less than one fluorescence photon per 20 excitation flashes. If two photons were to arrive at the TAC during the ramp, it would only detect the first, thus skewing the counting statistics of the fluorescence decay profile to shorter lifetimes. SR sources in running modes for time-resolved fluorescence operate at 1–10 MHz and therefore the fluorescence count rates must be limited to 10–50 kHz to ensure that the measured decay curve reproduces the actual decay to better than 1%. Usually, the time-resolved fluorescence instruments are equipped with fast photodiodes to detect the excitation flash. This is unnecessary at SR sources where suitable signals to provide a TAC stop can be derived from the Klystron master oscillator of the storage ring RF system as shown in Fig. 15.3.

15.6 Why SR for time-resolved fluorescence?

Table 15.2 illustrates why synchrotron radiation is a suitable pulsed source for single photon counting fluorescence lifetime measurements. SR is the only modulated light source that provides high intensities, high frequencies, short pulse-widths, excellent pulse-to-pulse and extreme short- and long-term stability, and which is continuously tunable. The excitation wavelength is selected simply via a monochromator. This contrasts with lengthy reconfiguration time required by tunable lasers. The most striking and unique advantage is that the light pulse profile emitted from an SR source is absolutely identical in time position and shape at absolutely all wavelengths. This makes SR ideally suited to coping with the throughput of work demanded at national

Table 15.2. A review of pulsed sources for time-resolved fluorescence

Source	Frequency	Pulse width	Wavelength	Long term stability	Pulse-to-pulse reproducibility
Synchrotron radiation (SRS)	3.1 MHz	200 ps	Continuously tunable	Very good	Excellent
Nitrogen flash lamp	1–100 kHz	1 ns	Lines between 300 and 400 nm	Poor due to electrode wear	Moderate
Deuterium flash lamp	1–100 kHz	1 ns	Continuous but low intensity	Poor due to electrode wear	Moderate
Titanium sapphire solid state laser	4 MHz	100 fs	200–1090 nm By changing interference filters and mirrors plus returning	Good	Poor
Argon ion pumped dye laser	4 MHz	10 ps	310–700 nm By changing dyes, approx. a week to retune	Moderately good	Poor
Optical parametric amplifier laser	1 kHz	1 ps	220–2200 nm By changing interference filters and mirrors plus a few days to retune	Moderate	Poor

and international facilities. Not all SR sources run time-resolved fluorescence experiments, but Table 15.1 shows that they are well suited to the task. Time-resolved fluorescence spectroscopy work is currently carried out at the SRS at Daresbury, the Super ACO at LURE, the NSLS at Brookhaven and UVSOR at Okazaki.

15.7 Analysis of time-resolved fluorescence decay

There are two reasons why time-resolved fluorescence data require careful analysis. Firstly, the assumption made in Section 15.5 that the excitation source pulse is infinitely short is in reality incorrect! Even if the excitation source is very short, in practice, processes in the detector have the effect of broadening this pulse. The response of the instrument (i.e. the overall time profile of the excitation pulse combined with the response of the detection system) is easily measured from the light pulses scattered from a non-fluorescent specimen, such as a colloid (silica). The instrumental response can be deconvoluted from the fluorescence data to give the true fluorescent decay. An advantage of SR is that the transit time of the photomultiplier is wavelength dependent, and with the continuous spectrum of SR, the scatter used to determine the instrumental response function can be at the same wavelength as the fluorescence maximum. The second reason results from the heterogeneity of the fluorescence decays of macromolecules of interest to the structural biologist. In these cases, a multi-exponential decay kinetics often indicates numerous environments for the probe or a complex quenching environment.

There are many methods of analysis available [reviewed in O'Connor and Phillips (1984)], but three methods, broadly based on least-squares reduction have gained prominence. These methods avoid the difficult problems of deconvolution (i.e. analytical methods) by instead reconvoluting the instrumental response with trial sums of exponential functions to fit the experimental data. The least-squares convolution method does just that. The approach is to analyze the experimental decay data with the minimum number (usually chosen) of acceptable exponents (lifetimes), by minimizing the weighted sum of the squares of the deviations of the experimental points from the calculated fitting function. The quality of fit is described in terms of a reduced chi-squared parameter which is usually acceptable only when it closely approaches unity. Another method (the maximum entropy method) (Livesey and Brochon 1987) makes no attempt to limit the number of lifetimes, but searches for a distributed set of decays logarithmically spaced across the time domain. Decay function amplitudes are recovered from an entropy-like function with the imposed constraint that the reduced chi-squared value must be near to unity. This method requires the highest quality (i.e. high number of total counts) data to avoid spurious oscillations. A third method, which is growing in popularity is the global analysis method (Beechem and Gratton 1989). This method is particularly powerful when attempting to analyse several data sets that have a common 'target' parameter.

15.8 Applications of time-resolved methods for structural biology

There are many examples where time-resolved fluorescence methods have been exploited in the field of structural biology. Here we will focus on some where the intrinsic probe tryptophan has been used to investigate the protein structures. These studies often acquire a greater significance when the protein structure is known, but this is not a prerequisite.

The development of site-directed mutagenesis techniques has enabled experimenters to investigate the fluorescence lifetime behaviour of tryptophan residues located at different selected sites within a large biomolecule. For example, by systematically replacing first one and then two of the three tryptophan residues of lactate dehydrogenase from *B. stearothermophilus* with tyrosine, Waldman *et al.* (1986) were able to assign each lifetime of the tri-exponential decay fit to the fluorescence of the wild-type protein to each one of the three individual tryptophyl residues. This was interpreted to mean that each tryptophyl had a very different but homogenous environment to generate its fluorescence lifetime. The lifetimes ranged from 0.3 to 7.4 ns whereas the segmental motion of a freely rotating tryptophyl side-chain in water would be of the order of 10 ps. This indicated that the longest lifetime, emitted by trp-203, reflected its particularly highly rigid environment. Similarly, a mono-exponential lifetime of 5.2 ns (Fig. 15.4) for the single trp-72 of the soluble peptide of the proton-translocating transhydrogenase from *R. rubrum* (Diggle *et al.* 1995) is indicative of this probe being deeply buried in a rigid and solvent inaccessible environment. By observing the fluorescence quenching of trp-72 by NADH, the authors were able to specify that its binding site must be on a specific domain of the peptide.

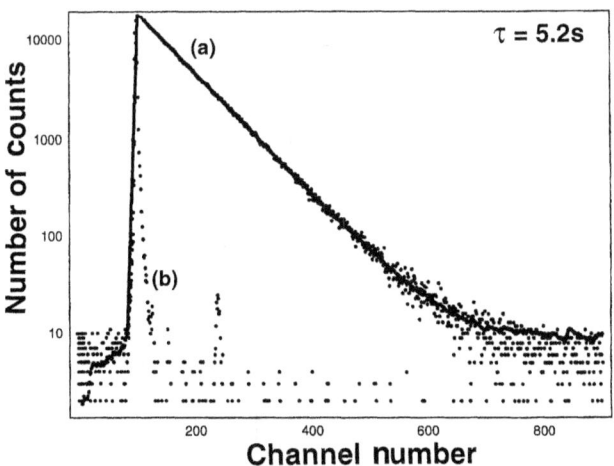

Fig. 15.4. Fluorescence decay of tryptophan in transhydrogenase from *R. rubrum*, by the single photon counting method: (a) instrumental response, (b) line shows the least-squares fit of a mono-exponential decay.

The combination of site-directed mutagenesis and SR fluorescence lifetime analysis has been used by Kuipers et al. (1991). In this work, the conformational dynamics of pancreatic phospholipase A_2 was examined by substituting the single tryptophyl residue of the wild-type protein with a phenylalanyl residue and then inserting a tryptophyl at each of two other locations, one of which was situated at the calcium-binding site. The phe residue was not fluorescent at the particular excitation wavelength selected for these measurements and this amino acid was chosen in order to retain the hydrophobicity at the site of the substituted trp, therefore providing minimal perturbation to the mutant protein. Using the maximum entropy method and the published X-ray crystallographic structure, these authors were able to show that the specific distortions of the structure, following the binding of calcium, resulted in an increase in the flexibility of residues at its binding site. The binding of the micellar substrate to the phospholipase had the opposite effect, that of making the residues at the N-terminal region more rigid.

Figure 15.5 gives another example where the maximum entropy method has been used to investigate the dynamics of a known structure (Blandin et al. 1994). The lifetime distribution obtained by maximum entropy method shows good agreement with the least-squares analysis and was used to probe the active loop of cardiotoxin from *Naja nigricollis* via a single trp-11. The study shows that the very complex pattern of the returned lifetimes for the emission of this single chromophore originates exclusively from conformational fluctuations, which reflects the relatively external and unconstrained location of the residue. This is a good example of the complementary use of techniques in structural biology. Here the spectroscopic technique is used to provide further structural information that may not be readily available from the diffraction because of the unknown effects of crystal packing.

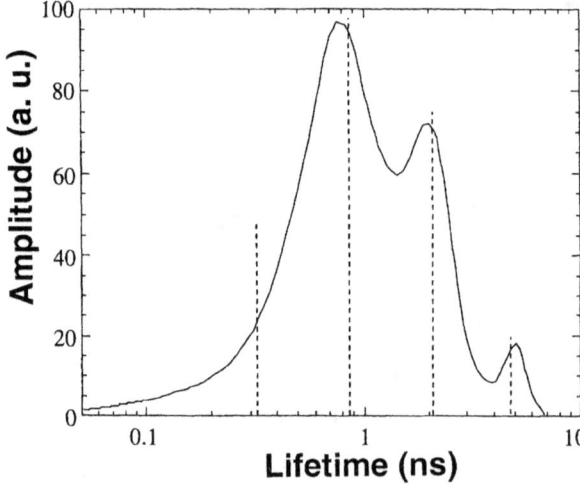

Fig. 15.5. The maximum entropy recovered lifetime distribution from *N. nigricollis* cardiotoxin (line); least-squares-fitted values (dashed lines). [Redrawn from Blandin et al. (1994).]

15.9 Instrumental developments

15.9.1 *Simultaneous spectral and lifetime collection*

Sutherland's group at the Brookhaven National Laboratory has developed an instrument (Kelly *et al.* 1997) that simultaneously records fluorescence lifetime decays over a large number of wavelengths using spectrographic techniques combined with microchannel plate array detectors. The spectrograph disperses the fluorescence spectrum across the array detector so that whole spectrum can be collected at once. This technique combines the two most environmentally sensitive parameters of fluorescence probes into one measurement and greatly improves the efficiency and power of the method.

15.9.2 *Microvolume fluorescence lifetime measurements*

Confocal microscopy (Pawley 1995) is a novel imaging technique in which a 'point source' is used to illuminate the sample and where a 'point detector' limits the volume within the sample that can be observed. Such a confocal system gives a real advantage in resolution over a conventional bright field microscope, which is specifically related to the size and shape of the apertures chosen. In the best cases, the lateral resolution is approximately equal to one-third of the wavelength and the axial resolution approximately equal to the wavelength of illumination. The pin-hole (1) in Fig. 15.6(a) defines the size of the focal spot at the specimen, and pin-hole (2) limits the volume of the specimen observed by the detector. This enhancement in resolution is obtained, of course, at the expense of loss of field of view and therefore the microscope has to use a scanning mirror system to scan the excitation beam across the specimen and create an image. The confocal method has the advantages over bright field microscopes of improved spatial resolution; improved image clarity, because image planes above and below the focal plane are eliminated; and the ability to achieve three-dimensional images (by moving the specimen in the Z-direction and collecting a series of separate 'optical slices').

Using SR as the light source (van der Oord 1996), microvolume time-resolved fluorescence measurements, as shown in Fig. 15.6(b) become feasible. In this case, the microscope was not used in a scanning mode. Instead, in the Daresbury SYCLOPS instrument (Gerritsen *et al.* 1994), it is straightforward to resolve the fluorescence lifetimes from the selected (in this case a $3\,\mu m^3$) volume within a living cell. This has been used, for example, to study the distribution of steroid hormones and their environments in Leydig cells. Results have shown that the concentration of hormone is much reduced in the nucleus of the cell compared with the cytoplasm. In principle, this selected volume could be used to undertake a full range of time, polarization and spectroscopically resolved experiments on wet samples at wavelengths in the far UV, without any significant radiation effects at the specimen.

15.9.3 *Real-time time-resolved microfluorimetry*

This is a newly developed technique at Daresbury's SRS which uses the pulsed SR in single bunch mode at 3.1 MHz to excite a specimen placed on the stage of

318   *Biological spectroscopy using low energy (VUV/UV) SR*

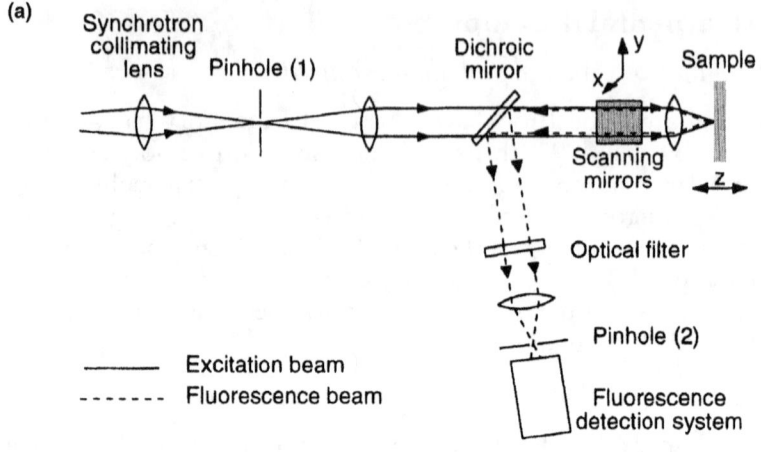

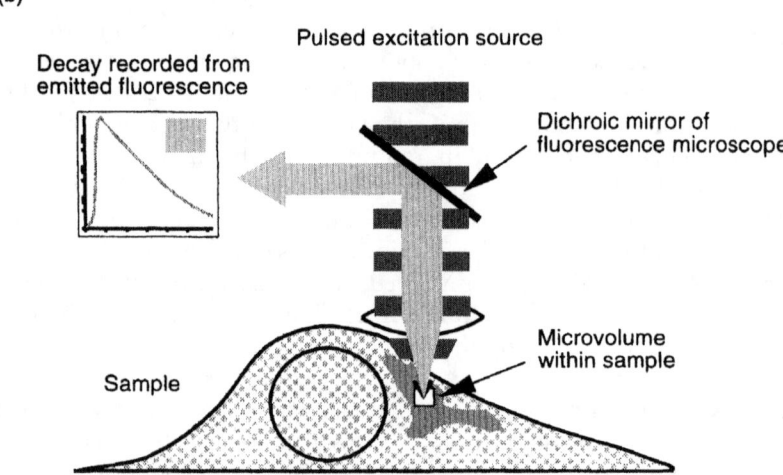

Fig. 15.6. Microvolume lifetime measurements using SR confocal microscopy.

an inverted optical microscope. This instrument is designed to continually record fluorescence lifetime data with a real-time resolution of about 1 s. The time-resolved microfluorimeter has been used to follow the endocytosis of epidermal growth factor in cultures of human epithelial cells, by following the quenching of fluorescence lifetime. The fluorescence lifetime is a very good measure of charge transfer between the BODIPY-labelled hormone and the hormone's receptor throughout the endocytotic process as it is not subject to the artefacts caused by sample photobleaching, trivial fluorescence transfer, and inner filter effects as is the equivalent steady-state measurement. This real-time

time-resolved measurement provides the occupancies and state of dimerization of the EGF receptors.

15.10 Time-resolved fluorescence anisotropy

Synchrotron light is linearly polarized in the plane of the storage ring. When this light is used to excite a fluorescent molecule, the fluorescence emitted will also be polarized since plane polarized light preferentially excites those molecules that have their absorption dipoles orientated in that plane, as shown in Fig. 15.7(a). For very small molecules, rotating freely in non-viscous solvents, however, this polarization is not preserved because Brownian rotation within the period of the (radiating) excited state lifetime quickly diminishes any 'memory' of the polarization direction of the excitation light. For biological macromolecules the time-scale for rotation of the molecule is often similar to the excited-state lifetime. This introduces a component(s) of the motion of the molecule, the fluorescence anisotropy, into the fluorescence decay curve. The time-resolved fluorescence anisotropy is measured by recording the polarized emitted fluorescence using transmission polarizers to select only light either parallel ($I_{\parallel}$), or perpendicular ($I_{\perp}$) to the direction of polarization of the excitation light.

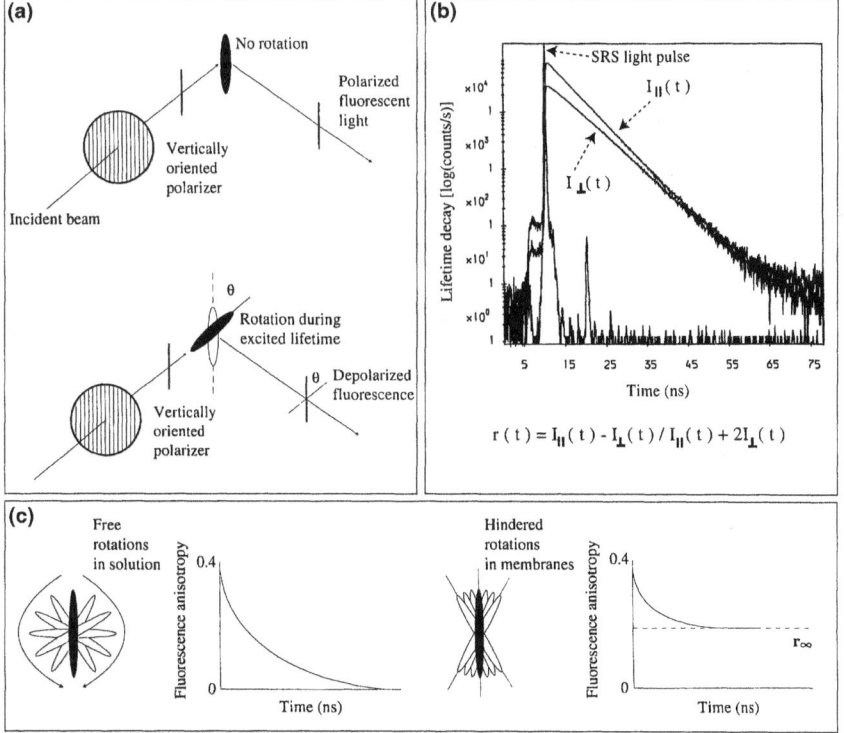

Fig. 15.7. Principles of time-resolved fluorescence anisotropy.

This is done using a simple rotating polarizer [Fig. 15.7(b)]. When molecular rotation occurs during the excited-state lifetime of the molecule then the decay curve of $I_\parallel$ is different to that of $I_\perp$ and fluorescence anisotropy is present [Fig. 15.7(c)]. Molecular motions, whether free in solution or hindered by being buried in, e.g. a membrane, can be extracted from these data. When undertaking fluorescence lifetime measurements on anisotropic samples, the anisotropy contribution to the fluorescence can be removed by placing a polarizer in the emission orientated at 54.6° (the so-called magic angle) to the plane of the polarized excitation to ensure that polarization selectivity is eliminated. This is because a complete loss of anisotropy occurs at this angle (Lakowicz 1983, p. 118).

The measurement of steady-state fluorescence anisotropy also uses the depolarization of fluorescence to provide information on the molecular size. These measurements can be useful but depend upon, and require, a knowledge of the excited-state lifetime for a correct interpretation. For example, if the fluorescence lifetime is shorter than expected then depolarization has less time to occur, and the anisotropy will correspondingly be greater resulting in an overestimation of the macromolecule's molecular weight.

15.11 Theory of time-resolved decays of fluorescence anisotropy

The time-resolved fluorescence anisotropy [$r(t)$] is calculated from the time-resolved decays of the parallel and the perpendicular components of fluorescence emission [Fig. 15.7(b)] using

$$r(t) = \frac{I_\parallel(t) - I_\perp(t)}{I_\parallel(t) + 2I_\perp(t)} = \frac{D(t)}{I_0(t)}. \tag{15.6}$$

For a single fluorophore the total intensity $I_0(t)$ is expected to decay as a single exponential and $D(t)$ is the difference between the parallel and perpendicular components, i.e.

$$I_0(t) = I_0 e^{-t/\tau}. \tag{15.7}$$

For continuous (steady-state) excitation the parallel and perpendicular components of the emission are given by

$$I_\parallel = \tfrac{1}{3} I_0 (I + 2r), \tag{15.8}$$

$$I_\perp = \tfrac{1}{3} I_0 (1 - r), \tag{15.9}$$

where r is the steady-state anisotropy. For molecules whose rotations are symmetric (isotropic) and unhindered, the anisotropy, following an infinitely sharp excitation pulse (δ-pulse), decays as a single exponential:

$$r(t) = r_0 e^{-t/\phi} = r_0 e^{-6Rt}, \tag{15.10}$$

where ϕ is the rotational correlation time of the fluorophore, R is its rotational rate ($6R = \phi^{-1}$), and r_0 is the anisotropy observed in the absence of rotational diffusion.

When the rotational motions of the fluorophore are hindered (e.g. if it is embedded in a membrane), the anisotropy does not decay to zero. In such cases, a limiting anisotropy ($r_\times$) is observed at times that are long compared to the fluorescence lifetime [Fig. 15.7(c)]:

$$r(t) = (r_0 - r_\times)e^{-t/\phi} + r_\times. \tag{15.11}$$

The maximum value for r_0 is 0.4 when the absorption and emission dipoles are parallel and -0.33 when perpendicular to each other. The time-resolved fluorescence anisotropy would be multi-exponential for a strongly asymmetric macromolecule and this would be of the form

$$r(t) = r_0 \sum_i a_i e^{-t/\phi_i}, \tag{15.12}$$

where a_i is related to number populations of species with i different correlation times. Of course, as for the measurement of fluorescence lifetime (see Section 15.7), the excitation flash is not a δ-pulse and therefore deconvolution of the instrumental response is essential. This is usually achieved using the least-squares method.

For a spherical molecule yielding a single exponential decay (15.10), the magnitude of the rotational correlation time for the fluorophore is governed by the viscosity (η) and temperature (T) of the solution and by the volume of the rotating unit (V), and is given by

$$\phi = \frac{\eta V}{RT}, \tag{15.13}$$

where V is related to the molecular weight by the partial specific volume of globular proteins (~ 0.73 ml/g).

15.12 Time-resolved fluorescence anisotropy at SR sources

A determination of the rotational correlation time from the time-resolved fluorescence anisotropy has two main advantages over steady-state measurements (continuous illumination). Firstly, the measurement is independent of the fluorescent lifetime and therefore provides a direct measure of the rotational rate of the fluorescent molecule. Secondly, if the rotational dynamics of the probe are complex then additional information can be obtained from the multi-exponential nature of the anisotropy decay kinetics. The time-resolved fluorescence anisotropy can be performed over a large concentration range making association constants of $10^9 \, M^{-1}$ and above measurable, because it possesses the inherent high sensitivity of the fluorescence methods. There are many advantages of synchrotron radiation for time-resolved measurements over other pulsed sources. The high repetition rate, combined with the high fluxes available (up to 10^9 photons per pulse are feasible in the best case) provides for high sensitivity and speedy acquisition times. The degree of linear polarization of SR is 100% at X-ray wavelengths and remains high (around 90%) in the UV. The advantage of being able to select any wavelength required for excitation is combined with the invariance of the time pulse profile with wavelength (although, of course, the response of the detector itself may have a wavelength–time dependence). For many samples, the first excited state S_1 can have other excited states ($S_2, S_3, \ldots$) in close proximity in terms of energy/wavelength. When measuring rotational correlation times it is important that excitation of fluorophore occurs uniquely to one excited state, usually S_1. Tryptophan is a good example of a molecule that has two excited states lying very close to each other. The experimental data are modified according to the states selected, thus giving rise to uncertainty in the data analysis.

The instrumentation for time-resolved fluorescence anisotropy is identical to that for fluorescence lifetimes, except for the addition of a rotatable emission polarizer. Data are collected cyclically for fixed periods with the emission polarizer orientated parallel and then perpendicular to the direction of polarization of the excitation beam. The typical total counts for a good anisotropy analysis would be around 500 000 for the parallel decay curve.

15.13 Applications of time-resolved fluorescence anisotropy in structural biology

Generally, time-resolved fluorescence anisotropy has been used to measure the size and shape of macromolecules, the rigidity of the interior of supramolecular structures, e.g. membranes, and the fluidity of the intramolecular environment. A good example of the power of the technique in the measurement of molecular dimensions is the work of Jackson *et al.* (1992). These workers were able to distinguish the monomer, dimer and tetramer states of L-lactate dehydrogenase from *B. stereothermophylus* by measurement of their time-resolved fluorescence anisotropies. Application of the Stokes–Einstein equation (15.13) and the analysis of the pre-exponential terms [a_i in eqn (15.12)] allowed them to determine the ratio of concentrations of the individual species in solution. From this the dissociation constant of the tetramer to the dimer and thus measurement of the stability of the tetramer is straightforward. Mutants of L-lactate dehydrogenases were used to identify important sequences at the tetramerization binding site.

An example of how time-resolved fluorescence anisotropy can be used to infer the shape of a macromolecule is shown in Fig. 15.8. The fluorescence of

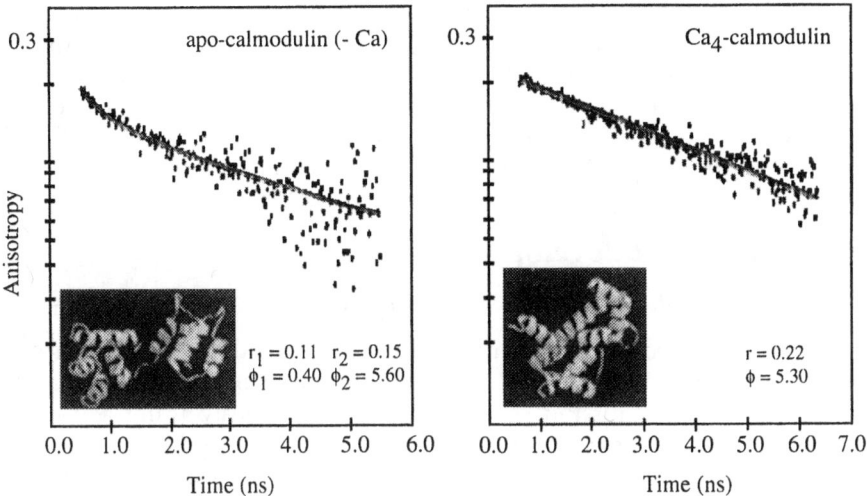

Fig. 15.8. Changes in the time-resolved fluorescence anisotropy of calmodulin in the presence and absence of calcium.

calmodulin arises from its four tyrosyl residues. In the absence of calcium, the calmodulin fluorescence anisotropy decay is bi-exponential, which is indicative of its 'dumbell-like' structure (i.e. the molecule can be characterized by two rotational correlation times, around its long and also its short axis). The addition of calcium to a solution of calmodulin transforms the decay to one that is monoexponential, thus indicating a more spherical (single correlation time) structure (Bayley *et al.* 1988). The two correlation times (ϕ) of the calcium-free calmodulin can be used to calculate a mean 'aspect ratio' for the molecule.

Time-resolved fluorescence anisotropy can also be used to characterize the viscosity of the environment surrounding a probe. In the interior of a membrane bilayer, viscosity can be related to a variety of parameters, such as the degree of order in the lipid hydrocarbon chains or to a notional 'membrane fluidity'. By introducing a lipid-soluble fluorescence probe, such as 1,6-diphenyl-1,3,5-hexartriene (DPH) into a suspension of natural membranes, membrane fluidity measurements can be undertaken under a variety of conditions. The time-resolved fluorescence anisotropy from probes buried in membranes is characterized by the presence of a limiting anisotropy $r_\times$ [eqn (15.11)]. The second-rank order parameter, $\langle P_2 \rangle$, of the fatty acid chains of the membrane may be calculated from the $r_\times$ and r_0 by the equation

$$\frac{r_\times}{r_0} = \langle P_2 \rangle^2. \tag{15.14}$$

This has been used to study, for example, the homeoviscous adaptation of membranes in deep sea fish (Behan *et al.* 1992). The order parameter measured for membranes from the brain myelin of deep sea fish is considerably less (more fluid) at atmospheric pressure than for the myelin of shallow water species. Time-resolved fluorescence studies have shown that by applying a hydrostatic pressure equivalent to the natural habitat of the deep sea fish, the measured order parameter increases to a value similar to that of the shallow water species, thus indicating that myelin order is highly conserved across a wide range of habitat conditions despite the profound ordering effects of high hydrostatic pressure.

15.14 Instrument developments

15.14.1 *Time-resolved fluorescence anisotropic spectra*

The fluorescence Omnilyser is an instrument constructed by Sutherland's group at the NSLS Brookhaven facility. This instrument measures the parallel and perpendicular polarized emission components simultaneously using a Wollaston polarizer and the data are stored in a two-dimensional histogramming memory. Using the spectrographic equipment described in Section 15.5 the instrument collects anisotropy information across the entire wavelength spectrum simultaneously. This instrument is ideal for the measurement of fragile or photosensitive biological specimens. It rapidly (simultaneously) collects all the data required to study systems with multiple fluorophores, or those systems undergoing chemical reactions, physical rearrangements, or energy transfer during the lifetime of the emitting species.

15.14.2 *Time-resolved fluorescence anisotropy in the time domain*

This instrument is used for the continuous measurement of time-resolved anisotropies using microfluorimetry (Martin-Fernandez et al. 1998). By positioning a rotatable polarizer in the emission path of the real-time time-resolved microfluorimeter (described in Section 15.9) it is possible to follow the changes in molecular motion of the fluorophore with a time resolution of approximately 60 s (determined by the minimum length of time required to assemble a useful data set). Figure 15.9 shows an example of an experiment where the second-rank order parameter from the fluorescence of fluorescein-labelled EGF is continuously monitored during endocytosis of the hormone by epithelial cells. An increase in the order parameter indicates higher environmental constraints to free rotation.

15.15 Circular dichroism

Circular dichroism (CD) (Rodger and Bengt, 1997) is a form of optical activity resulting from the differential absorption between left- and right-handed circularly polarized light by a solution of chiral molecules. Molecules are termed chiral when the arrangement of their covalent bonds is such that they cannot be superimposed on their mirror images. In other words, the molecule has 'handedness' or intrinsic asymmetry. CD measurements provide important information on the electronic structures of molecules and when applied to larger

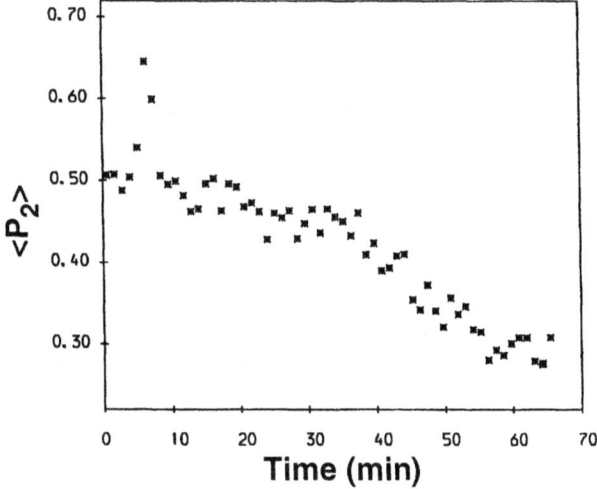

Fig. 15.9. Real time change in the $\langle P_2 \rangle$ order parameter from the fluorescence of EGF labelled with fluorescein during the endocytosis of EGF by epithelial cells.

(biological) macromolecules, such as proteins, nucleic acids and carbohydrates, can yield unique information on their secondary structure (Johnson 1990; Fasman 1996). This is because tetravalent carbon, which normally forms the asymmetric centres of molecules, giving L-amino acids and D-sugars, etc., confers its handedness to the backbone structures of biological macromolecules. CD can also be used to probe the interactions of small achiral molecules with macromolecules, because their binding often induces CD in the ligand. As CD is a phenomenon associated with the absorption process it occurs in an extremely short time-scale (10^{-15} s) and is well suited to study intermediate structures in rapid reactions and short residence binding processes. CD spectra are relatively quick and easy to measure and are usually performed in aqueous solutions. As a result, CD has become recognized as an important structural technique for biological molecules, which complements the more detailed information provided by X-ray crystallographic analysis, X-ray small and wide angle scattering and diffraction and NMR spectroscopy. In some situations it is the most informative structural technique available.

Circularly polarized light is the sum of two parallel electromagnetic waves, offset by 90°, and can most easily be envisaged in terms of the magnitude and direction of the electric vector of an electromagnetic wave at fixed times along the direction of propagation of the light wave. The upper wave in Fig. 15.10 is circularly polarized. The polarization is right-handed and the electric vector does not change in magnitude but does change in direction around the direction of propagation. The lower curve, which shows linear polarization, is identical to the superposition of two opposite light rays of circularly polarized light of identical phase and amplitude. In this case a projection of the combined amplitudes perpendicular to the direction of propagation produces a line. When

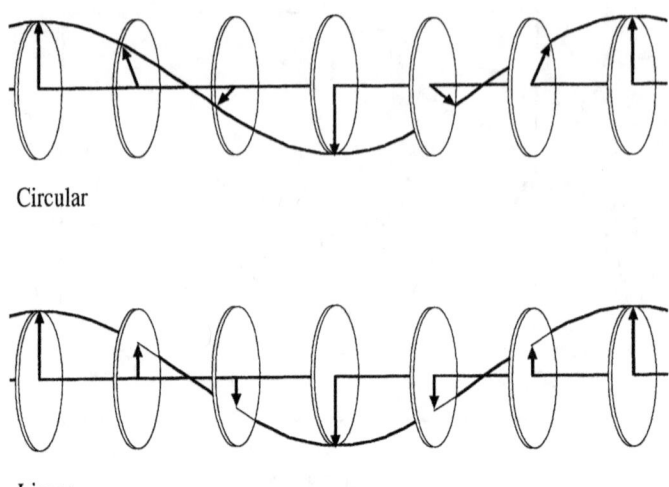

Fig. 15.10. Snapshots of the electric vector for circularly and linearly polarized light.

circularly polarized light at a given wavelength passes through a solution of chiral molecules absorbing at that wavelength, either the left- or right-hand circular polarization can be preferentially absorbed, as shown in Fig. 15.11(a). The CD signal [Fig. 15.11(b)] is derived from the difference between the absorbance of right-handed polarized light and that of the left-handed circularly polarized light. CD is, however, a very small effect when compared to the magnitude of the absorbance (typically less than 10^{-3}) at the same wavelength at which it has to be measured.

The CD signal arises because neither the ground state nor the excited state of the chiral molecule has a reflection plane. Therefore, the rearrangement of electrons during an electronic excitation, i.e. the transition moment of the electrons, will not be linear but roughly helical in character, thus accounting for the differential absorption for left- and right-handed circularly polarized light. Different electronic transitions of the same molecule may involve electron distributions of different handedness; therefore the molecule can have both positive and negative CD signals. The objective in large biomolecular CD is to identify how the electrons that give rise to the CD signal are involved in forming the three-dimensional structure of the molecule. As an example, the transitions of the amide bonds forming the polypeptide linkages of proteins, occurring in the wavelength region between 160 and 240 nm are shown (Fig. 15.12). The electron orbitals involved are diagrammatically represented and the position and polarity of the corresponding CD bands are given for helical and β sheet conformations. It is apparent that the CD of these electronic transitions is extremely sensitive to the angular arrangement of the atoms surrounding the asymmetric carbon atom of the amino acid residues (Fig. 15.13). The differing arrangements of bond angles along the peptide backbone is, of course, what determines its secondary

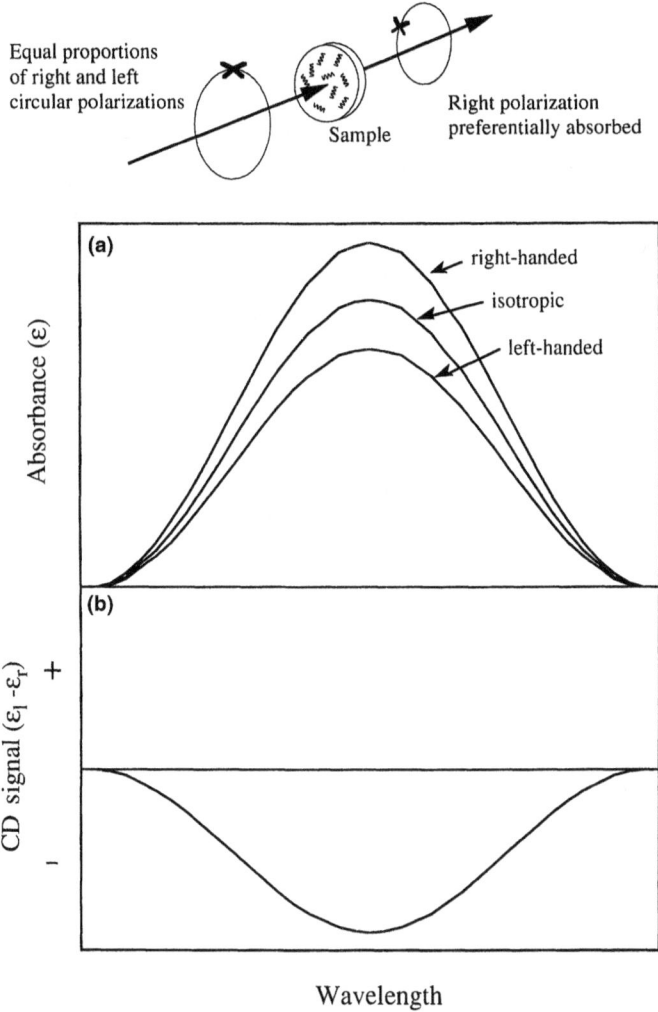

Fig. 15.11. The absorption (a) and CD (b) of an idealized chiral molecule.

structure. The resulting CD spectra for pure helix, β sheet, β turn and random polypeptides show remarkable differences (random is sometimes termed 'other', or 'aperiodic' structures because 'random coil' is sometime used to mean 'unstructured' polypeptide, which is usually not the case for this type of secondary structure in proteins). Most proteins are, of course, a mixture of these elements of secondary structure types. However, the relative proportions of the secondary structural types can be extracted from the CD spectrum with remarkable accuracy from such a quick and simple measurement, as long as the signal quality is good and covers a sufficiently extensive range of the available spectrum to enable spectral (and therefore structural) distinctions to be made.

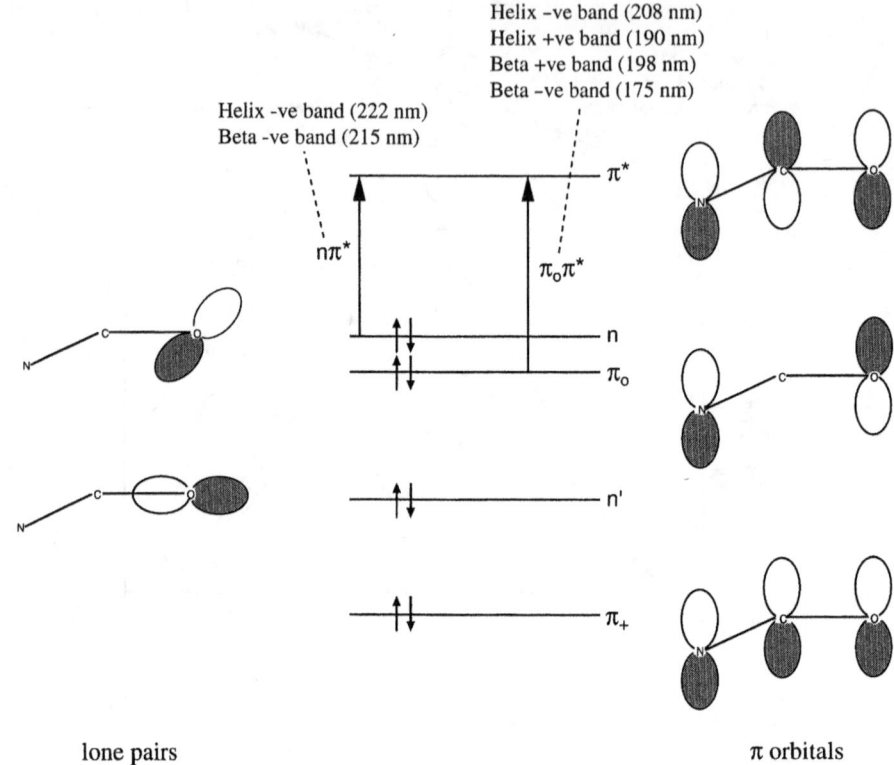

Fig. 15.12. Molecular orbitals and electronic transitions of the amide group.

15.16 CD measurements using SR

15.16.1 *Advantages of SR*

Linearly polarized light is a prerequisite for the production of circularly polarized light by the modulation method. Conventional CD instruments use a quartz or magnesium fluoride polarizer to linearly polarize the unpolarized output from a xenon arc lamp at the expense of the loss of over 60% of the available flux. SR provides inherently linear polarized light at high photon fluxes and has a major advantage over conventional sources at short wavelengths. This means that using SR, CD data can be collected to lower wavelength limits with improved signal-to-noise ratio, giving more precise secondary structural information than is possible by using instruments with conventional light sources. This is particularly important for proteins and carbohydrates since the majority of the absorption—and hence of the CD spectral information—lies at shorter wavelengths than 200 nm. Collection times for CD spectra using SR, are also considerably shorter than on conventional instruments and offer much higher signal-to-noise ratios. In a comparison of data quality (Fig. 15.14), the

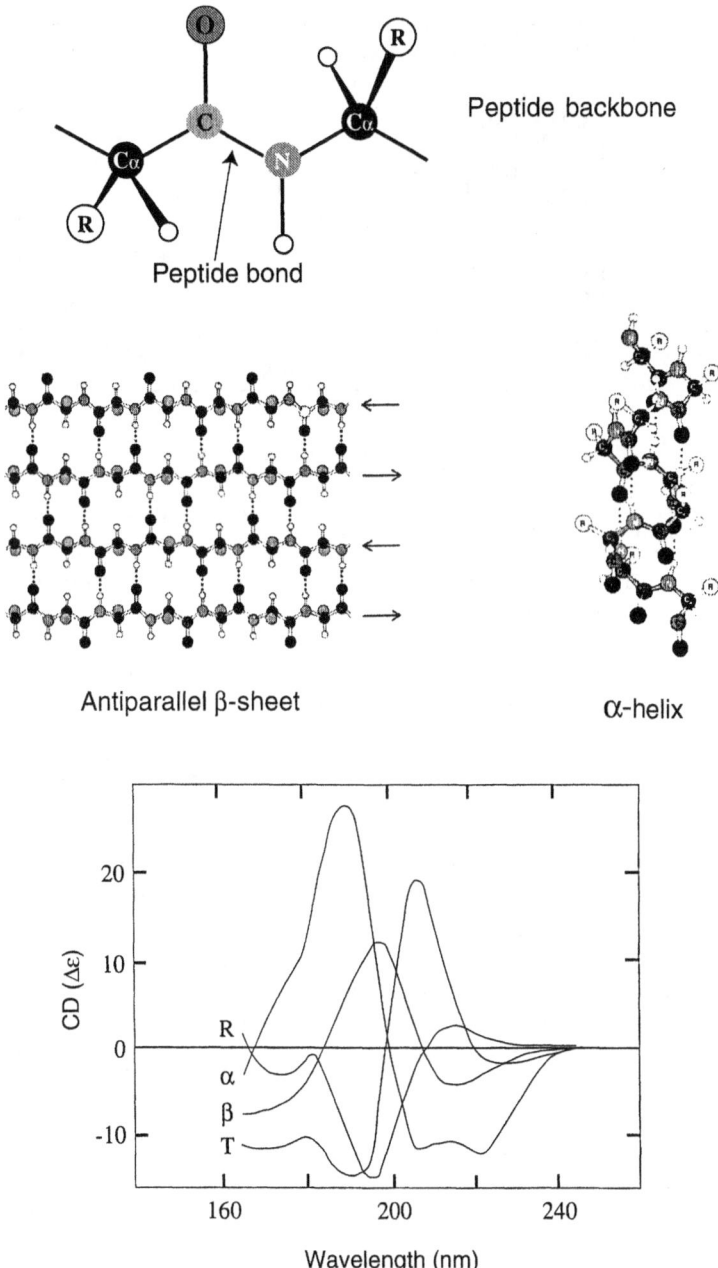

Fig. 15.13. The application of VUV CD for protein secondary structure, where α is the CD spectrum of α-helices, β of β-sheet, T of β-turns, and R of other structures.

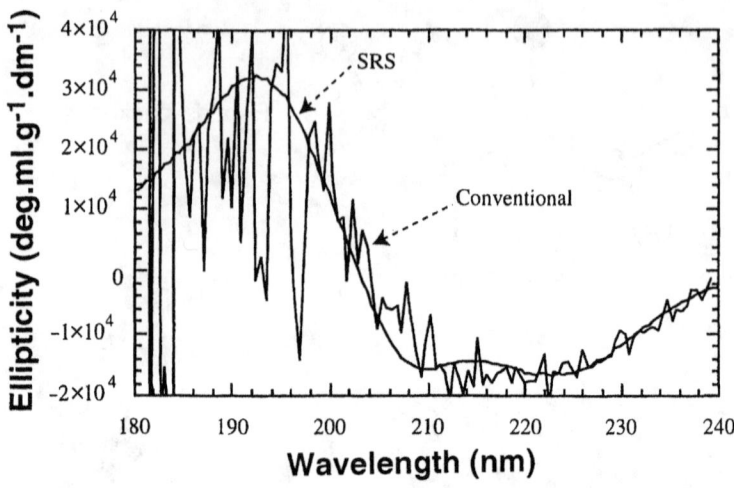

Fig. 15.14. Typical conventional and SR CD data sets taken at equivalent sampling times.

CD spectrum of myoglobin taken at the SRS compared to that taken on a conventional instrument for equivalent acquisition time illustrates this point.

The improved signal-to-noise ratio over conventional instruments is very important when undertaking time-resolved CD measurements in the millisecond time regime, especially at low wavelengths. The output flux of light sources employed in conventional instruments decreases dramatically and continuously for wavelengths below 220 nm. Below this wavelength range the transmission of practically all optical elements of the apparatus also decreases. Figure 15.15 shows typical absorbance curves (pathlengths of 10 µm have been used for the liquids) for what may be commonly required in a time-resolved CD experiment and it must be remembered that the CD is a small signal that has to be separated out from large absorbances. If the absorbance is too high then little light reaches the detector and the signal quality is inadequate. Oxygen, in one form or another, is mainly responsible for high absorbances in the measurement system at wavelengths less than 200 nm. Molecular oxygen within the optical paths in the instrument has to be eliminated, usually by purging the equipment with nitrogen gas. The absorption of water vapour, of quartz and of many varieties of commercial silica (used for cell windows) are the major constraint for measurements at wavelengths below 185 nm. The use of lithium fluoride windows in place of quartz and the smallest possible path lengths within the cell are vital considerations. At longer wavelengths, some buffer salts and compounds such as the denaturants, urea and guanidinium hydrochloride are very highly absorbing and also contribute to reduced signal quality. It is important to remember that the sample itself absorbs quite strongly at lower wavelengths, although without absorption there is of course no CD. It is sometimes advisable to perform two or more wavelength scans at different concentrations to achieve the best results. All these effects conspire to reduce the quality of data collected in the far UV and near VUV with conventional instruments.

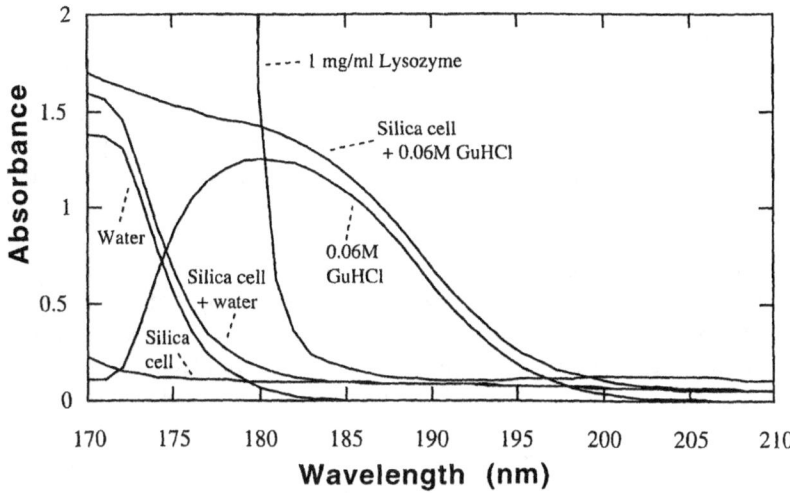

Fig. 15.15. Comparison of absorbance spectra for typical components of useful CD measurements. A 10 μm cell path length was used for the liquids.

While SR measurements suffer from the same effects, they have the tremendous advantage that the source flux is high and increasing, in principle at least, with decreasing wavelength in the VUV.

The signal-to-noise ratios of CD measurements can be maximized by keeping sample absorbance at an optimal level. The highest signal-to-noise ratios are obtained when $A = 0.869$ Fasman 1996 (p. 647). The maximum absorbance at which the CD measurement can be made usefully is dependent on the initial photon flux and the sensitivity of the detector. However, once 'single photon counting' sensitivity has been reached by the detection system, then only an increased photon flux can extend the absorbance range. To extend the measurement of CD for a sample by an absorbance unit of 1 requires an increase of photon flux of 10-fold. The development of 'third generation' SR sources will produce fluxes 10^2–10^5 times greater than is available, for example, at the SRS where CD measurements are already being successfully made. Therefore, VUV SRCD provides a tremendous opportunity for the future development of this technique.

Synchrotron light from bending magnets is linearly polarized in the plane of the ring. There is a contribution from the orthogonally polarized components above and below this plane yielding intrinsically left- and right-handed circular components of SR light (but in general it is elliptically polarized). However, in the VUV wavelength region, it is more efficient and convenient to exploit the linearly polarized component for CD measurements on biological molecules. This is easily converted and switched between left- and right-circularly polarized light at a frequency of 50 kHz using a standard photoelastic modulating crystal with wavelength compensation control. As an alternative to this, left- and right-switchable circularly polarized light will soon be available from helical undulators.

15.16.2 Instrumentation

The basic elements of the SR measurement are similar to that for the conventional instrument (Fig. 15.16) except, of course, that there is no requirement for the polarizer. In Fig. 15.16 arrows indicate the light path and wavy lines indicate electronic signals. The light first passes through a magnesium fluoride prism or grating scanning monochromator (gratings are used for SR work) and then, if necessary, through a polarizer to produce linearly polarized light. The long axis of the retarder crystal is orientated at 45° to the direction of polarization and the light then becomes circularly polarized prior to passing through the sample. The detector (usually a fast photomultiplier tube) detects the transmitted light and the signal is delivered to a lock-in amplifier. A signal from the modulator to the amplifier isolates the oscillating CD signal with minimum noise input from other frequencies and the CD signal is then fed to a computer for integration and analysis. There are two feedback systems required for the CD spectrum to be recorded successfully. The first is that the crystal retarder has to be programmed to one quarter-wavelength retardation to change the phase of one of the perpendicular components of linear polarized light by 90°, resulting in circular polarization. This retardation must, of course, be changed along with the wavelength. The wavelength retardation is controlled by a signal originating from the monochromator through the data acquisition system. A second feedback system takes account of the very large changes in the DC levels of the detector signal resulting from changes in the absorbance of the sample during the scan. In order to achieve this the DC controller is used to alter the high voltage supply to the photomultiplier tube (detector) so that the output is at a constant current (e.g. $\sim 10\,\mu A$) even though the light intensity falling on it will

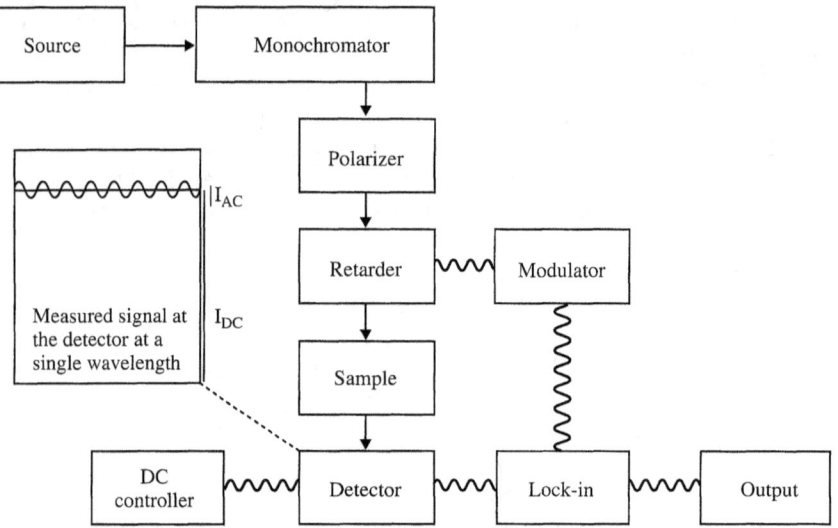

Fig. 15.16. Block diagram of a modulation CD instrument, with the signal measured at the detector (inset).

change considerably. The absorbance of the sample during the scan is therefore directly proportional to the voltage supplied to the detector and the CD signal proportional to the AC component as indicated by I_{AC} in Fig. 15.16. By monitoring the voltage level applied to the photomultiplier, and some measure of the intensity of the incident light (such as the current circulating in the storage ring), the absorption spectrum of the sample can be determined during the same scan used to measure the CD (Sutherland et al. 1982).

15.16.3 Data analysis

There are several computer programs freely available (to non-commercial organizations) for the analysis of the secondary structure of proteins (Greenfield 1996). CONTIN (Provencher 1982) is based on a general purpose constrained regularization designed to invert noisy linear algebraic and integral equations. SELCON (Sreerama and Woody 1993) is a self-consistent procedure based on the singular value decomposition method. Another method is convex constraint analysis, which uses volume minimization to deconvolute the CD spectra (Perczel et al. 1991). As an alternative to the linear statistical methods used by these three programs, neural networks have been used for the analysis of CD data (Bohm et al. 1992). All these methods use 'basis sets' (CD data sets for a number of proteins of known secondary structure from X-ray crystallography or NMR) to analyse the unknown CD spectrum. When analysing extended SRCD data, caution must be used because the basis data sets supplied with these programs do not usually extend much below 190 nm (178 nm at best). Basis sets used at the SRS extend to 168 nm.

15.17 Application of SRCD in structural biology

The potential of VUV SRCD has only recently been recognized. This can be best illustrated for carbohydrates, where almost the entire CD spectrum is inaccessible to the conventional instrument (Fig. 15.17). The CD analysis of protein secondary structure has in the past also been hampered by the dearth of protein structures readily available for the analysis programs. This situation has improved considerably with the almost exponential increase in the numbers of protein structures established from X-ray crystallography and NMR analyses (see the PDB, at http://www.pdb.bnl.gov/). The group at the NSLS Brookhaven has been active in demonstrating the advantages of VUV SRCD (Sutherland et al. 1992). This group has shown the analytical power of extended CD spectra in, for example, accurately characterizing the percentage of secondary structural elements (including parallel and antiparallel sheet) in the water soluble cloned constant antigen from the tick-borne spirochete of Lyme disease (France et al. 1992). Using these data they were able to undertake detailed stability studies and, in combination with protein prediction, they were able to identify the amino acid residues most likely to be the antigenic determinants.

Other groups have exploited the analytical power of extended VUV SRCD in the investigation of important biological processes. For example, Qi et al.

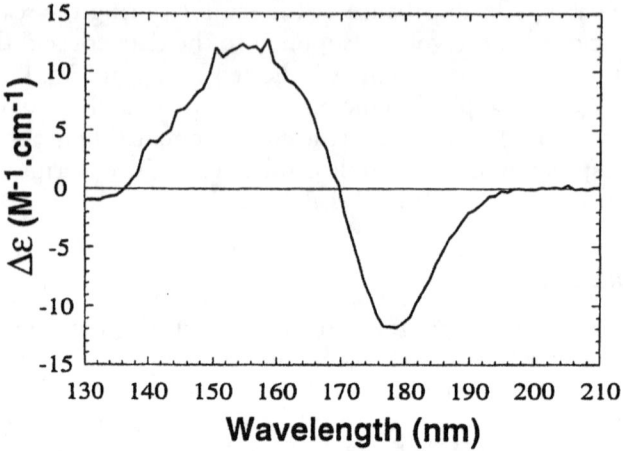

Fig. 15.17. SRCD spectrum of a film of α-carrageenan.

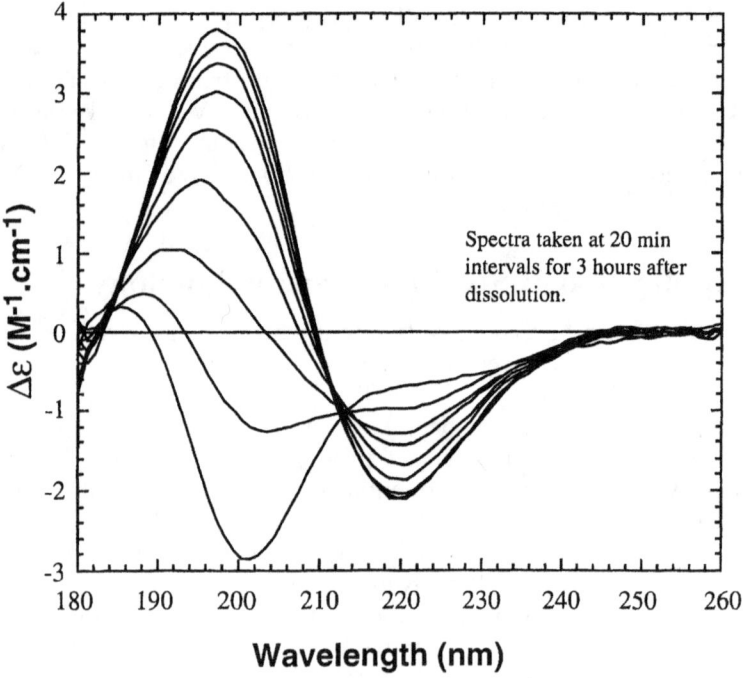

Fig. 15.18. Polymerization of LRRN peptide monitored by SRCD.

(1997) were able to show that the important and elusive protein folding intermediate, the proposed 'molten globule state', which state β-lactoglobulin had previously been claimed to adopt at temperatures in excess of 70 °C, had lost too much of its secondary structure by 65 °C to be so designated. Symmons *et al.*

(1997) have exploited the rapid acquisition times of VUV SRCD to investigate β-sheet formation in real time in a leucine-rich repeat peptide (Fig. 15.18) that mimics the formation of amyloid fibrils of prions and Alzheimer's disease β-protein. These are both examples of experiments that are not possible with other structural techniques such as X-ray crystallography or NMR. A further area of research, that has not yet been exploited by VUV SRCD is membrane–protein interactions. Both synthetic and natural bilayer membranes are usually CD-neutral, making any change in protein structure easily detectable. Unlike the conventional CD, the VUV SRCD instrument can usually measure highly turbid samples, a particular problem when using optical techniques in membrane investigations.

15.18 Instrumental developments

The most recent and exciting development in VUV SRCD is in the millisecond time-resolved instrument and attendant advances in rapid mixing technologies. Figure 15.19 shows how a three-dimensional plot of CD spectra vs time can be constructed after collecting numerous single wavelength millisecond time courses, in this case, the re-folding of Staph. nuclease by pH jump from pH 3 to 7. The spectra can be analysed to give time-courses for the re-folding of the protein's secondary structural types. This, for the first time, allows experimenters to study the kinetics of the folding of specific structures within a protein in the absolute terms of secondary structure.

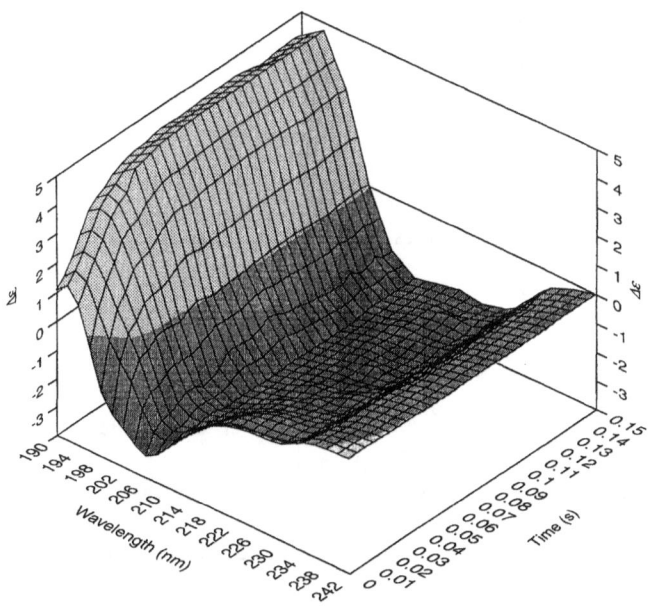

Fig. 15.19. Multiwavelength SRCD measurement on the folding of Staph. nuclease.

15.19 Future prospects for SRCD

Within the near future it should become possible to obtain an efficient and optimized medium resolution very high flux beam line for SRCD spectroscopy in the near VUV—still a rarity. It will also become possible to use a helical undulator within a storage ring to provide left- or right-switchable circularly polarized light. Although the relatively slow switching time for present day undulators (around 1 min or more) may change the way in which fast-time studies are undertaken, the gain in flux will be around 10 000! The prime measurements for the future will probably be to fully characterize all elements of the secondary structure for proteins, including modular proteins and develop special techniques for analysing peptides. These should stimulate *ab initio* theoretical analysis of peptides and proteins, including their side-chain interactions. It will also be important to characterize the structural conformations of carbohydrates and nucleic acids and their complexes below 180 nm and to study membrane proteins to the lowest possible wavelengths. Given the availability of reliable fast mixing (stopped-flow) and especially laser temperature-jump methods, it should be straightforward to extend wet CD structural studies to well below ~ 1 ms. VUV SRCD should therefore make a significant contribution to the identification of the elusive intermediate states postulated for the crucial processes of protein folding and macromolecular assembly.

Acknowledgements

The authors would like to acknowledge the enormous amount of help provided by Mrs Patricia Broadhurst, Dr David Clarke and other members of the VUV–IR Facility Group at the SRS in the preparation of this manuscript.

References

Bayley, P, Martin, S and Jones, GR (1988). *FEBS Letters*, **238**, 61.
Beechem, JM and Gratton, E (1989). *Proceedings of SPIE*, **909**, 70.
Behan, MK, MacDonald, AG, Jones, GR and Cossins, AR (1992). *Biochimica et Biophysica Acta*, **1103**, 317.
Blandin, P, Mérola, F, Brochon, J-C, Trémean, O and Ménez, A (1994). *Biochemistry*, **33**, 2610.
Bohm, G, Muhr, R and Jaenicke, R (1992). *Protein Engineering*, **5**, 191.
Diggle, C, Hutton, M, Jones, GR, Thomas, CM and Jackson, JB (1995). *European Journal of Biochemistry*, **228**, 719.
De Stasio, G, Zema, N, Antonangeli, F, Savoia, A, Parasassi, T and Rosato, N (1991). *Review of Scientific Instruments*, **62**, 1670.
Fasman, GD (1996). *Circular dichroism and the conformational analysis of biomolecules*, Plenum Press, New York.
France, LL, Kieleczawa, J, Dunn, JJ, Hind, G and Sutherland, JC (1992). *Biochimica et Biophysica Acta*, **1120**, 59.
Gerritsen, HC, van der Oord, CJR, Levine, YK, Munro, IH, Jones, GR, Shaw, DA and Rommerts, FFG (1994). *SPIE*, **2137**, 238.

Greenfield, N (1996). *Analytical Biochemistry*, **235**, 1.
Jackson, RM, Gelpi, JL, Emery, DC, Wilks, HM, Moreton, KM, Halsall, DJ, Sleigh, RN, Behan-Martin, M, Jones, GR, Clarke, AR and Holbrook, JJ (1992). *Biochemistry*, **31**, 8307.
Johnson WC (1990). *Proteins*, **7**, 205.
Kelly, LA, Trunk, JG, Polewski, K and Sutherland, JC (1995). *Review of Scientific Instruments*, **66**, 1496.
Kelly, LA, Trunk, JG and Sutherland, JC (1997). *Review of Scientific Instruments*, **68**, 2279.
Kuipers, OP, Vincent, M, Brochon, J-C, Verheij, HM, de Haas, GH and Gallay, J (1991). *Biochemistry*, **30**, 8771.
Lakowicz, JR (1983). *Principles of fluorescence spectroscopy*, Plenum Press, New York.
Livesey, AK and Brochon, JC (1987). *Biophysical Journal*, **52**, 693.
Martin-Fernandez, ML, Tobin, MJ, Clarke, DT, Gregory, CM and Jones, GR (1996). *Review of Scientific Instruments*, **67**, 3716.
Martin-Fernandez, ML, Tobin, MJ, Clarke, DT, Gregory, CM and Jones, GR (1998). *Review of Scientific Instruments*, **69**, 540.
Munro IH (1980). *Synchrotron radiation research*, Winick, H and Doniach, S, (eds), Plenum Press, New York, Chapter 8.
Munro, IH and Schwentner, N (1983). *Nuclear Instruments and Methods*, **208**, 819.
O'Connor, DV and Phillips, D (1984). *Time-correlated single-photon-counting*, Academic Press, London.
Pawley, JB (1995). *Handbook of biological confocal microscopy*, Plenum Press, New York.
Perczel, A, Hollosi, M, Tusnady, G and Fasman, GD (1991). *Protein Engineering*, **4**, 669.
Provencher, SW (1982). *Computer Physics Communications*, **27**, 229.
Qi, XL, Holt, C, McNulty, D, Clarke, DT, Brownlow, S and Jones, GR (1997). *Biochemical Journal*, **324**, 341.
Rodger, A and Bengt, N (1997). *Circular dichroism and linear dichroism*, Oxford University Press, Oxford.
Sreerama, N and Woody, RW (1993). *Analytical Biochemistry*, **209**, 32.
Sutherland, JC, Keck, PC, Griffin, KP and Takacs, PZ (1982). *Nuclear Instruments and Methods*, **195**, 375.
Sutherland, JC, Emrick, A, France, LL, Monteleone, DC and Trunk, J (1992). *Biotechniques*, **13**, 588.
Symmons, MF, Buchanan, SGStC, Clarke, DT, Jones, G and Gay, NJ (1997). *FEBS Letters*, **412**, 397.
van der Oord, CJR, Jones, GR, Shaw, DA, Munro, IH, Levine, YK and Gerritsen, HC (1996). *Journal of Microscopy*, **182**, 217.
Waldman, ADB, Clarke, AR, Wigley, DB, Hart, KW, Chia, WN, Barstow, D, Atkinson, T, Munro, IH and Holbrook, JJ (1986). *Biochimica et Biophysica Acta*, **913**, 66.

16
X-ray microscopy

C.J. Buckley

16.1 Introduction

Monochromatic X-rays interact with specimens in a very different way than light, infrared radiation or electrons. The highly specific interaction of these X-rays provides for elemental and chemical state analysis of samples in a range of environments including the wet state. The short wavelength of X-rays places the fundamental limit of spatial resolution more than two orders of magnitude smaller than that of light. However, microscopy using X-rays has only recently been possible. The current advances in this area are due to the emergence of high resolution X-ray optics and X-ray sources with sufficient brilliance. Due to the availability of these resources, it is likely that the next few years will see a considerable expansion in the use of X-ray microscopy to tackle problems in the biological and material sciences.

This article sets out some of the motivations behind the use of X-ray microscopy and comparisons are made with competing and complementary techniques. In particular, the analytic capability of X-ray microscopy is discussed. The configuration and main contrast modes are described and particular attention is given to X-ray microscopy in the transmission mode. Here, the X-ray interactions that give rise to elemental and chemical state maps are described and the method by which these maps may be calculated is provided. Finally, two examples of the use of X-ray microscopy for chemical state mapping are given together with appropriate references and comment on the future directions for this emerging microscopy technique.

16.2 X-ray microscopy in context

Several factors affect the choice of microscope for an examination of a specimen. These include: required sample condition (live, hydrated, frozen, dry, fixed, embedded and sectioned), the spatial resolution required, the contrast source (native, or via dyes), the pixel time, analysis capabilities and specimen damage. Table 16.1 summarizes some of the basic attributes of microscopes for a number of radiation types. It must be noted, however, that the table is only a guide and that a true comparison can only be made between techniques based on the composition/condition of individual specimens.

The principal attributes of X-ray microscopy are that elemental and chemical state maps may be made without stains on relatively thick specimens at a dose level that does not destroy the morphology or composition of the specimen

Table 16.1. Comparison of X-ray microscopy with conventional microscopies

Microscopy	Spatial resolution	Contrast origin	Specimen thickness	Pixel time	Specimen condition
Light: confocal	> 0.15 μm	Native or fluromarkers	Tens of μm	Real time	Live
Light	> 0.25 μm	Native or histochemial stains	Tens of μm	Real time	Embedded sections
TEM	> 1 nm	Via high Z stains	< 0.2 μm	Real time	Ultrathin sections
Cryo-STEM	> 1 nm	Native	< 0.2 μm	Near real time	Ultrathin cryo-sections
Infrared	> 1 μm	Native	—	10 s	Bulk
X-ray transmission	> 30 nm	Native	A few μm	0.1–20 ms	Hydrated or sectioned
X-ray fluorescence	> 50 nm	Native	< 20 μm	> 100 ms	Hydrated, sectioned or bulk

(for review, see Kirz et al. 1995). Also, specimens may be wet and imaged at atmospheric pressure at spatial resolutions down to 0.03 μm. In the case of X-ray induced fluorescence microscopy, the detection limits can be as low as 1 ppb (part per billion).

16.3 Analytical microscopies that use native contrast

The ability to gain substantial information about small specimens is greatly enhanced when the spatial information is combined with analytic information. This can include the elemental composition or information about the bonding state of specific elements providing the distribution of particular molecular species. Electron microscopy provides the most widely used of the analytic techniques through electron probe microanalysis (EPMA) (Kitsugi et al. 1995). In this technique, electrons excite atoms in the specimen which fluoresce characteristic X-rays. Here the spatial resolution is set by the excitation range of the scattered and secondary electrons and is typically 50 nm. EPMA provides a convenient method for obtaining the elemental composition of specimens and has a sensitivity for elements with medium to high atomic number of about 1000 ppm (parts per million).

Chemical state images may be obtained using transmission electron microscopes (TEMs) or scanning transmission electron microscopes (STEMs) which have an electron energy filter or analyser after the specimen to select and measure the energy losses of electrons as they interact with the specimen. The techniques that use electron energy loss are known as energy-filtered TEM and electron energy loss spectroscopy TEM (Jeanguillaume et al. 1996). These techniques are very successful when applied to specimens with medium to high atomic number where detection limits are more than an order of magnitude higher than EPMA, but have less application on low atomic number specimens containing organic molecular species. In these cases, specimen damage is a problem as the electrons impart a considerable dose to the specimens, which can easily disrupt the covalent bonds of the organic molecules. Infrared (IR) microscopy (Boskey et al. 1992) can provide chemical state information at a spatial resolution of a few microns. However, IR sources are intrinsically low in brightness, which has the consequence that pixel times are prohibitively long and therefore the technique is largely limited to selected area spot analysis.

16.4 X-ray microscopes

There are several types of X-ray microscopes. These may be categorized by contrast origin and configuration as described below. The refractive index for most materials is very close to 1 and, for this reason, X-rays are focused either by reflection at small grazing angles of incidence or by diffraction elements. The highest spatial resolution has so far been obtained by diffraction-based focusing devices known as Fresnel zone plates, which have produced images with a resolution limit of about 20 nm. A zone plate is a circular diffraction grating and the effective numerical aperture is set by the width of the finest ring and the wavelength of the X-rays (see Fig. 16.1). So far, the spatial resolution obtainable

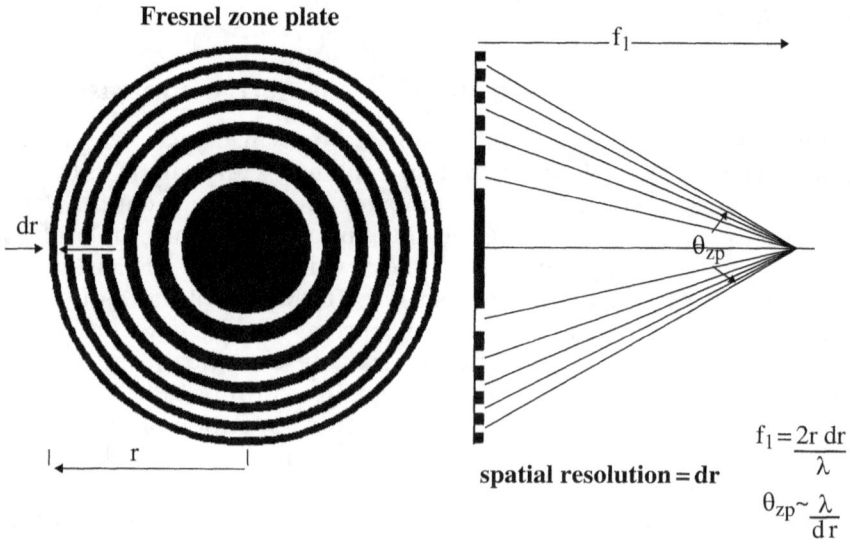

Fig. 16.1. Diagram of a Fresnel zone plate X-ray 'lens'. The spatial resolution is limited by the width of the smallest zone that can be manufactured.

has been limited by the width of the finest zone that can be manufactured with sufficient thickness to give a practical diffraction efficiency.

The spatial resolution obtained using grazing angle reflection lenses tends to be limited by aberrations. The resolution is typically a factor of 10 worse than that achievable with zone plates, but the efficiency is considerably higher.

16.4.1 Microscope configurations

16.4.1.1 Photoemission X-ray microscopes

In these microscopes the sample is held in a vacuum and electrons are liberated from the surface by X-rays. Here, the X-ray energy may be varied in order to obtain contrast from a given element or chemical state present in the specimen. Alternatively, the X-ray energy may be fixed, and the energy of the emitted electrons analysed to provide equivalent information. Configurations may be in a scanning system, where the specimen is scanned relative to an X-ray probe (H. Zhang et al. 1996), or in a non-scanning, full field configuration, where electron optics are employed to form the image (Tonner et al. 1998). The principal advantage of this form of imaging is that it is sensitive to only a few atomic layers at the surface of the specimen, while its main disadvantage is that the specimen must be kept in an ultrahigh vacuum.

16.4.1.2 X-ray-induced fluorescence microscopes

Many scanning electron microscopes are equipped with 'X-ray analysers' to perform EPMA. These devices measure the energy of X-rays that are emitted from the specimen as the electron beam in the electron microscope excites the atoms in the specimen. An energy tuneable X-ray probe has the advantage that its energy may be chosen to be that which excites the element of interest in a highly specific way. This means that the signal from the targeted element is largely free from background signal (bremsstrahlung radiation) which is an unavoidable consequence of stimulation with electron beams. The greatly reduced background radiation means that very low level signals can be detected and the detection limit for X-ray-induced fluorescence is about 1 ppb, about a factor of 1000 times more sensitive than is obtained with electron probes (Sparks 1980). Microscopes of this type are now being actively developed (Kaulich et al. 1998; Yun et al. 1998), and are providing hitherto unavailable information about specimens with very low concentrations of some elements. The principal advantages of X-ray probe fluorescence microscopes are the high sensitivity, and the fact that wet specimens several microns in thickness may be imaged at atmospheric pressures. A disadvantage of this form of microscopy is that imaging times may be long for some elements at low concentrations, with pixel times of several seconds.

16.4.1.3 Transmission X-ray microscopes

Transmission X-ray microscopes have a number of configurations to take advantage of a range of contrast mechanisms including absorption (covered in

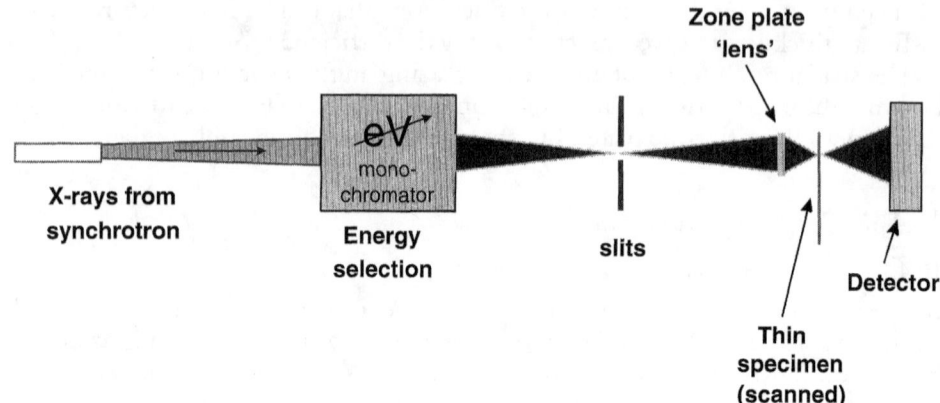

Fig. 16.2. Schematic diagram of a scanning transmission X-ray microscope.

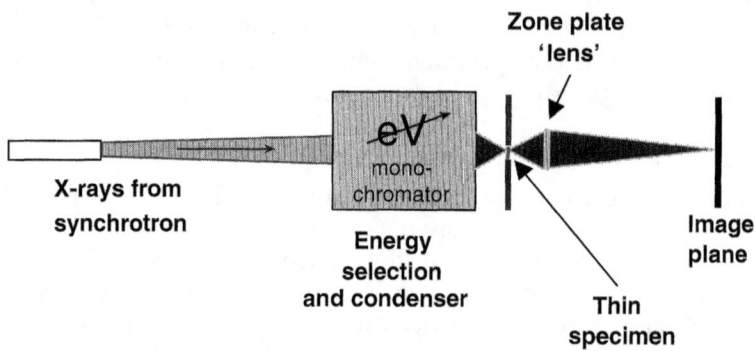

Fig. 16.3. Schematic diagram of a non-scanning (full field) transmission X-ray microscope.

detail below), phase (Schneider *et al.* 1998; Morrison *et al.* 1998), and luminescence (Jacobsen *et al.* 1993). These microscopes exist in scanning (Jacobsen *et al.* 1991) and non-scanning (Schmahl *et al.* 1995) configurations (see Figs 16.2 and 16.3).

The optical element responsible for image or probe formation is a modified Fresnel zone plate (see Figs. 16.1 and 16.4). The image resolution is fundamentally set by the width of the finest ring that can be fabricated accurately. Current zone plates are produced with a finest zone width of about 30 nm; however, 19 nm zone plates have been fabricated (Schliebe *et al.* 1998). The non-scanning microscope has primarily been used to examine wet and frozen specimens (Schnieder *et al.* 1998; Meyer-Ilse *et al.* 1998; Abraham *et al.* 1998) in both absorption and phase contrast at X-ray energies between 250 and 550 eV. The specimens may be in excess of 1 μm thick and the spatial resolution obtained is about 30 nm. This provides higher spatial resolution than light microscopy on thicker specimens that can be used in electron microscopy.

Fig. 16.4. A zone plate made by electron beam lithography (see Plate Section).

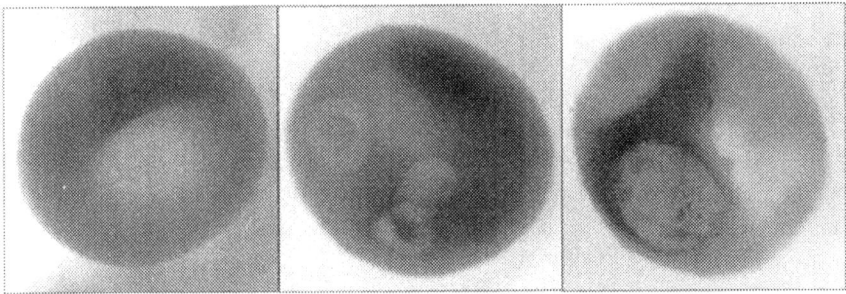

Fig. 16.5. Life cycle of malaria in human red blood cells. The images show from left to right an uninfected blood cell, a newly infected cell, and a 12-hour-old parasite. Image sizes are 7 × 7 μm (see Plate Section).

An example of transmission imaging of malaria-infected human blood cells is shown in Fig. 16.5. Here, one of the most interesting outcomes of the X-ray microscopy studies of intact, unstained red cells was the frequent observation of the existence of a tubular structure. This structure was found surrounding the parasite and protruding into the red blood cell cytosol. This tubulo-vesicular membrane network is thought to have a role in the import of nutrients to the parasite.

The scanning transmission X-ray microscope has been used to combine elemental and chemical state contrast on a variety of specimens. By using absorption differences provided by spectral absorption features (discussed below), it is possible to distinguish molecular species such as DNA and protein (X. Zhang *et al.* 1996) in a quantitative fashion and without the use of stains. This is a highly useful attribute and its application via X-ray microscopy to problems in biology and materials science has accelerated in the last two years. The origin of absorption fine structure and the method of obtaining elemental and chemical state contrast is described below.

16.5 Obtaining elemental and chemical state contrast using X-ray probes via X-ray absorption fine structure

The X-ray absorption spectrum of an element shows a substantial increase in absorption when the X-ray energy is sufficient to promote an electron from one of its orbitals to either an unoccupied orbital or to the vacuum continuum. These absorption features can be classified by three main processes:

- Promotion of an electron into a higher orbital
- Ionization of a bound electron
- Scattering of the ionized electron by atoms in the vicinity of the ionized atom

When the X-ray energy is the difference between an inner shell orbital and an unoccupied orbital a sharp excitation peak may occur in X-ray absorption. These excitation peaks can be further classified as originating from excitations to unoccupied atomic orbitals (Rydberg states), and molecular orbitals formed by chemical bonding (π^* and σ^* states). See Fig. 16.6.

An example of the absorption features due to electron promotion is shown in Fig. 16.7(a). This is an absorption spectrum of CO_2 gas formed using the NSLS synchrotron X-ray source on the X1a beam line. The spectrum shows the overall ionization step (often referred to as an absorption edge) and a series of peaks near the edge (referred to as near-edge X-ray absorption fine structure or NEXAFS). These peaks are due to the promotion of electrons into Rydberg and molecular orbital states. Not present in this spectrum are any peaks due to the scattering of ionized electrons by adjacent atoms, which occurs at higher

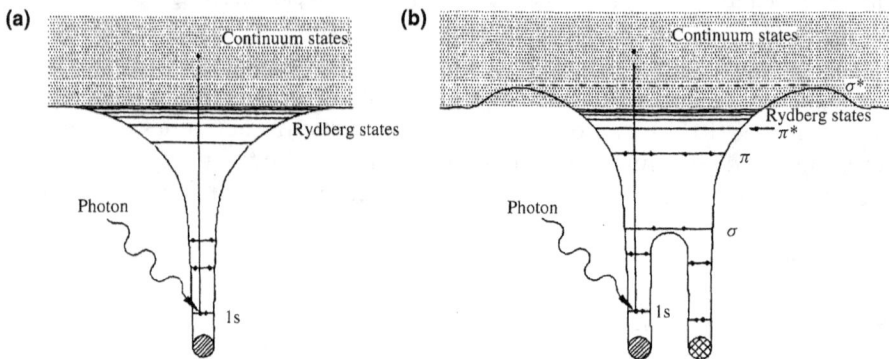

Fig. 16.6. Schematic potentials of left; an isolated carbon atom and right; a carbon based diatomic molecule. The diatomic molecule has the π and σ molecular bonds and the π and σ unoccupied molecular orbitals. When the incoming X-ray photon has an energy equivalent to the energy difference between the 1s and π^* states, an electron will be promoted and give rise to a sharp absorption peak. The energy difference between these two levels is affected by the identity of the diatomic molecule and the environment in which the molecule is located. Thus the energy position of the absorption peak can be used to identify the molecule which gives rise to the peak.

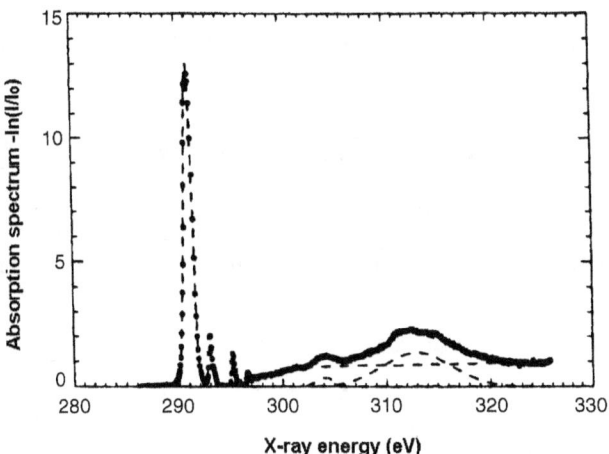

Fig. 16.7(a). X-ray absorption spectrum of gaseous CO_2.

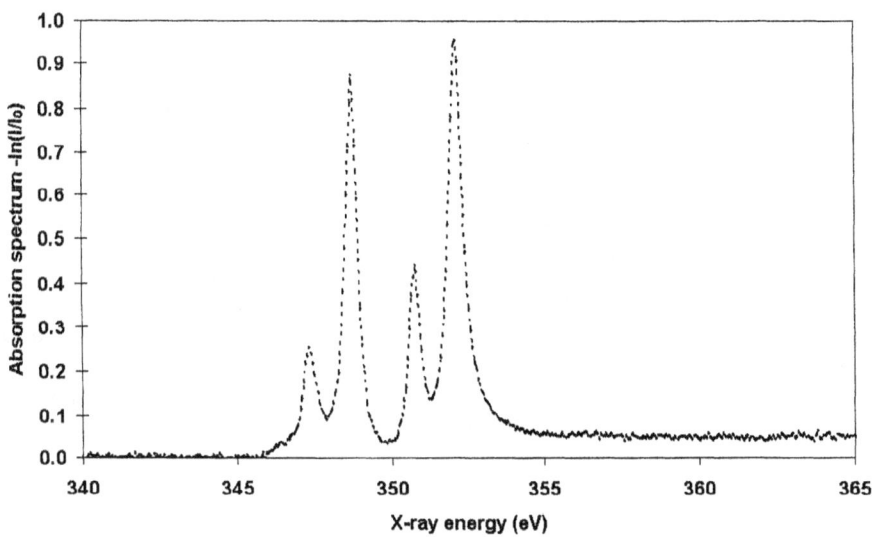

Fig. 16.7(b). X-ray absorption spectrum of CaO.

energies. Figure 16.7(b) shows an absorption spectrum of calcium oxide (also from X1a at the NSLS). This spectrum shows some NEXAFS peaks and an overall ionization step. However, the origin of the peaks in the two spectra is different, as explained in the following section.

The scattering of the ionized electron by atoms in the vicinity of the ionized atom gives rise to fine structure on the absorption spectrum at energies higher than the absorption edge. These features are known as extended X-ray absorption fine structure (EXAFS). They are prominent at the K absorption edge of medium to high atomic number elements and require X-ray energies above 1 keV. As high brightness X-ray sources become available at these energies, chemical state imaging via the EXAFS signal will become a possibility.

16.6 The origin and use of NEXAFS features

NEXAFS peaks are classified as described above as arising from promotion into orbitals not involved in molecular bonds and those that are. Both processes have peaks whose energy is affected by the environment of the atom or molecule.

16.6.1 Molecules with ionic bonds

In the case of ionically bonded molecules, Rydberg peaks dominate the near-edge spectrum. Figure 16.7(b) shows a spectrum of calcium oxide. The considerable difference in absorption with energy in these sharp peaks provides for an elemental mapping of calcium (Buckley et al. 1995). Here the peaks that straddle the edge originate from transitions between the 2p and 3d atomic orbitals of calcium. The energy of the outer orbitals is a function of the local electrostatic field. Thus the energy position of absorption peaks themselves is a function of the net potential experienced by the orbital. While this is dominated by the net nuclear and bound electron fields, nearby ions can impart a significant effect. The shift in the peak positions caused by the ions that surround the atom of interest (such as calcium) may be used to provide chemical information (Buckley et al. 1995) and hence obtain images of mixed mineral phases.

16.6.2 Molecules with covalent bonds

New orbitals are created by the sharing of electrons in a covalent bond due to the combined field of the constituent nuclei. The σ orbitals are symmetric about the axis joining the constituent atoms of the molecule, while the π orbitals are perpendicular to this axis. However, inspection of the CO_2 spectrum in Fig. 16.7(a) shows that a strong peak is present at an energy below the Rydberg peaks. This peak results from the transition of a carbon 1s electron into the lowest unoccupied π orbital (a π^* orbital). The energy of the peak is lower than the difference between the energies of the 1s and π orbitals involved. This is because as the electron is liberated from the 1s state, the π state experiences a greater nuclear field and its energy is pulled lower, creating a net smaller transition energy for the 1s electron. This π^* orbital is quite stable as it is pulled below the ionization threshold and has a long lifetime. By the Heisenberg principle, the absorption peak is narrow in energy. The sigma bonds are less strongly bound to the molecule and are therefore less stable and have broader peaks by the same principle.

The molecular orbitals are loosely bound to the atoms and their energy is not only affected by the combination of atoms that form the bond, but also by fields due to other neighbouring atoms. In this way the energy of the π^* and σ^* absorption peaks at a particular atomic absorption edge are sensitive to both the other bonding atoms and the local chemical environment surrounding the bond. The localization of the energy position of the π^* peak due to the nature of the bonding atom and the bond environment provides a very useful handle for accessing absorption by specific chemical species (Stohr 1992).

16.7 Forming elemental and chemical state maps using absorption contrast differences

The optical density (OD) at of the specimen at a point at a given energy is found by taking the natural logarithm of the transmission,

$$OD = -\ln\left(\frac{I}{I_0}\right), \qquad (16.1)$$

where I_0 is the incident X-ray intensity and I is the transmitted intensity. In terms of the property of the specimen at a given energy, this optical density is a sum of the densities of the individual components and their mass absorption coefficients times the specimen thickness, i.e. for a specimen containing n components the optical density is given by

$$OD = \sum_{i=1}^{n} \mu_i \rho_i t, \qquad (16.2)$$

where μ_i, ρ_i and t are the mass absorption coefficient, density and thickness of the ith component respectively. If, for example, a specimen is a blend of two substances (a and b) and the transmission is measured at two different energies, then the following equations can be written:

$$OD_1 = \mu_{a1}\rho_a t + \mu_{b1}\rho_b t, \qquad (16.3)$$
$$OD_2 = \mu_{a2}\rho_a t + \mu_{b2}\rho_b t. \qquad (16.4)$$

There are two equations and two unknowns and thus the mass thickness of a and b can be determined if the mass absorption coefficients are known. If the thickness of the specimen is known or measured, then the densities (in g cm^{-3}) of a and b can be determined. The accuracy of the mass thickness depends primarily on how accurately the mass absorption coefficients are known, while the signal-to-noise ratio is determined by both the signal-to-noise ratio of the transmission measurement and the extent to which the mass absorption coefficients of the components are numerically different from each other at a given energy.

For a specimen containing n components, it is necessary to measure the transmission at a minimum of n energies. It is often useful to image at more energies than there are principal components. The non-square matrices of the simultaneous equations will then be solved by singular value decomposition (Press et al. 1988), where every combination of the over-determined set of equations is applied to obtain a least-squares fit to every value of mass thickness, thus improving the accuracy of the result. This analysis has been applied to two problems given below as examples of quantitative chemical state X-ray microscopy.

16.8 Examples of chemical state microscopy via the transmission signal in X-ray microscopy

16.8.1 *Measurement and distribution of collagen, mineral and hydrated volume in osteoporotic bone*

In the study of osteoporotic bone, the distribution and relative amounts of calcium-based mineral and collagen (protein) are of interest (Buckley *et al.* 1998), as collagen forms the template for mineralization. Both collagen and calcium mineral were quantitatively mapped in normal and osteoporotic mouse bone. The mouse femoral neck bones from these mice were embedded and sectioned to a thickness of $\sim 0.2\,\mu\text{m}$. These sections were mapped in the X-ray microscope. The specimens were treated as being composed of three principal components, calcium mineral, protein and embedding resin. The mass absorption coefficients were obtained from spectra (e.g. Fig. 16.8) and the mass thicknesses of these components were mapped without stains using the method of singular value decomposition as described above. In total, seven different X-ray energies were used to produce the maps. An example of the mineral formation around cells in bone is shown in Fig. 16.9 which shows maps of embedding resin, collagen and calcium mineral. The work and its conclusions have been published (Buckley *et al.* 1997).

16.8.2 *Measurement and distribution of DNA and protein in sperm*

The packing of DNA in sperm is achieved by the proteins protamine1 and protamine2. The distributions and total amount of these proteins to that of DNA have implications in infertility (Balhorn *et al.* 1998). The mass absorption coefficients of protein representative of protamine and DNA were obtained from

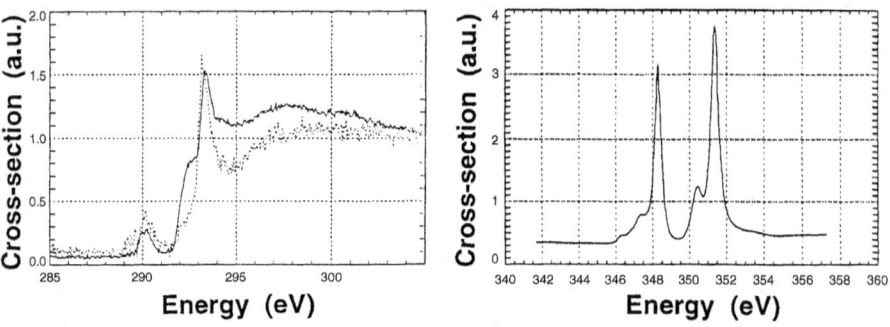

Fig. 16.8. Left: X-ray absorption spectrum of collagen and embedding medium at the carbon K absorption edge. Right: X-ray absorption spectrum of calcium hydroxy apatite at the calcium L absorption edge.

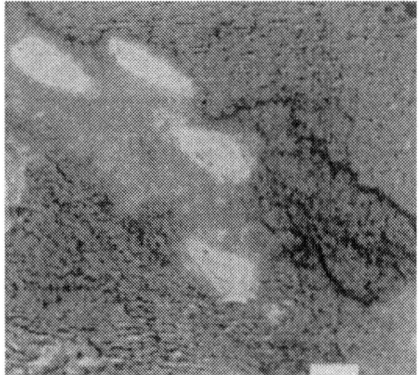

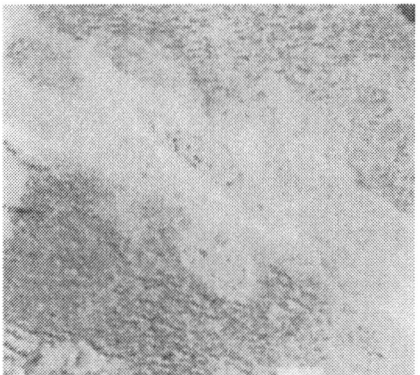 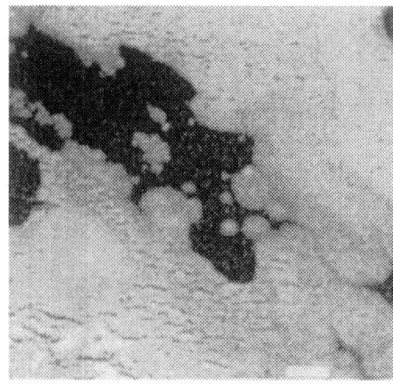

Fig. 16.9 Top: Embedding medium map of mineralizing tissue. This is the area between the cortical bone and the mineralizing cartilage. The embedding medium substitutes hydrated volume, and provides a useful means of highlighting cells in relation to the organic and mineralizing matrix. The scale bar represents 5 μm and the brightness scale is proportional to mass thickness. Lower left: Protein map. The high concentration of protein surrounding the cells is the collagen matrix exuded by the cells. The collagen forms the template for mineralization. Lower right: Calcium map. The more mature collagen matrix is on the upper right and lower left, and shows considerable mineralization. Initial mineralization islands can be seen in the vicinity of the cells. The structure is typical of mineralizing cartilage.

spectra (see Fig. 16.10) and these were used with the SVD method to obtain the protein and DNA maps of frozen dehydrated bull sperm (shown in Fig. 16.11). A total of six different X-ray energies were used to create the maps. This work and its conclusions have been published (X. Zhang *et al.* 1996).

16.9 Conclusions and future directions

The above applications were made possible by the availability of a bright source of soft X-rays. However, higher energy X-rays are also very useful for the study

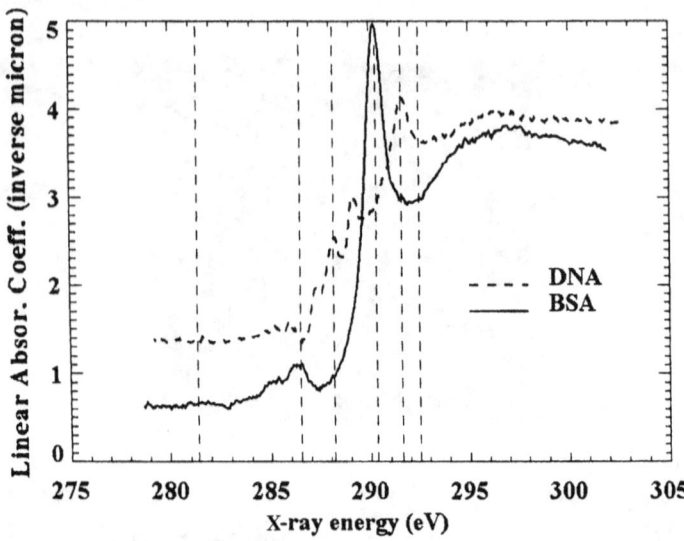

Fig. 16.10. X-ray absorption spectrum of DNA and BSA (standard protein) at the carbon absorption edge. The sharp peaks are π^* transitions from C=C, C=N and C=O double bonds.

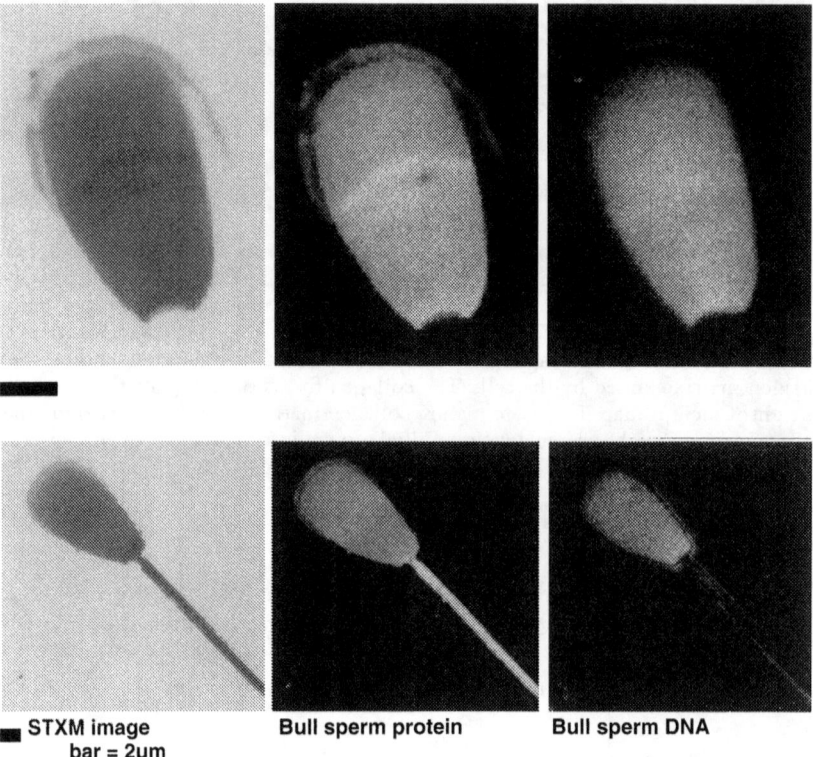

Fig. 16.11. Distributions of DNA and protein in bull sperm. The distribution of DNA in the sperm head is quite uniform in the mid-region, while the protein distribution varies considerably.

of the chemical composition of samples. For example, higher energy X-rays have been used in the study of biological mineralization. There is considerable interest and conjecture on the initial mineral phases in both healthy and diseased mineralized tissue. EXAFS on bulk samples has been used to investigate the structure of calcium compounds in biological systems (Hasnain et al. 1983) using the calcium K-edge at an X-ray energy of 4 keV. However, in order to relate these phases to the biological mechanisms responsible for their creation, it is necessary to identify the phases in relation to the micromorphology. When completed, the microscopy beam line at the ESRF (Susini et al. 1998) will be very well suited to this application as the energy range 0.28–6 keV will be covered with high spectral brilliance. The use of the EXAFS signal in transmission combined with a resolution of about 0.1 µm would permit a range of biological and materials investigations with spatially resolved chemical state information for elements and specimen thicknesses that are currently not accessible. Further, detection of EXAFS variations using the fluorescence signal is potentially several orders of magnitude more sensitive than detecting the transmitted intensity. Excitation of the fluorescence signal with a monochromatic X-ray source also gives lower detectable limits and specimen damage than excitation with charged particles. The limits should be low enough that intercellular ionic concentrations can be mapped. The wide energy range that will be available at the ESRF will provide several X-ray signals that can be obtained from the one specimen in different imaging modes. This combination will facilitate a unique and comprehensive mapping of both the composition and chemistry of the organic and inorganic components of bone and many other tissues.

References

Abraham, J et al. (1998). *X-ray microscopy and spectromicroscopy*, Springer, Vol. I, pp. 13–24.
Balhorn, R et al. (1998). *X-ray microscopy and spectromicroscopy*, Springer, Vol. II, pp. 29–46.
Boskey, AI, Pleshko, M, Doty, SP and Mendelsohn, R (1992). *Cells and Materials*, **2**, 20–220.
Buckley, CJ (1995). *Review of Scientific Instruments*, **66**, 1322–1324.
Buckley, CJ et al. (1995). *Review of Scientific Instruments*, **66**, 1318–1321.
Buckley, CJ et al. (1997). *Journal de Physique IV (France)*, **7**, C2, 82–90.
Buckley, CJ et al. (1998). *X-ray microscopy and spectromicroscopy*, Springer, Vol. II, pp. 47–56.
De Groot, FMF et al. (1990). *Physical Review B*, **41**(2), 928–937.
Hasnain, S et al. (1983). *EXAFS and near edge structure*, A, Bianconi, L, Incoccia and S, Stipcich (eds), Springer-Verlag, Berlin, p. 330.
Jacobsen, C et al. (1991). *Optics Communications*, **86**(3–4), 351–364.
Jacobsen, C et al. (1993). *Journal of Microscopy—Oxford*, **172**, 121–129.
Jeanguillaume, C, Tence, M, Zhang, L and Ballongue, P (1996). *Cellular and Molecular Biology*, **42**(3), 439–450.
Kaulich, B et al. (1998). *Microscopy and Microanalysis*, **4**(Suppl. 2), 374–375.
Kirz, J et al. (1995). *Quarterly Reviews of Biophysics*, **28**(1), 33–130.
Kitsugi, T, Yamamuro, T, Nakamura, T, Oka, M, Kokubo, T, Okunaga, K and Shibuya, T (1995). *Calcified Tissue International*, **56**(4), 331–335.

Meyer-Ilse, W et al. (1998). *Microscopy and Microanalysis*, **4**(Suppl. 2), 352–353.
Morrison, G et al. (1998). *X-ray microscopy and spectromicroscopy*, Springer, Vol. I, pp. 85–94.
Press, WH et al. (1988). *Numerical recipes*, Cambridge University Press.
Schliebe, T et al. (1998). *X-ray microscopy and spectromicroscopy*, Springer, Vol. IV, pp. 3–12.
Schmahl, G et al. (1995). *Review of scientific instruments*, **66**, 1285–1286.
Schneider, G et al. (1998). *X-ray microscopy and spectromicroscopy*, Springer, Vol. I, pp. 111–116.
Sparks, C (1980). In *Synchrotron radiation research*, H, Winick and S, Doniach (eds) Plenum Press, New York, pp. 459–512.
Stohr, J (1992). *NEXAFS spectroscopy*, Springer-Verlag, Vol. 25.
Susini, J et al. (1998). *X-ray microscopy and spectromicroscopy*, Springer, Vol. I, pp. 45–54.
Tonner, BP, Dunham, D, Droubay, T and Pauli, M (1997). *Journal of Electron Spectroscopy and Related Phenomena*, **84**(1–3), 211–229.
Yun, W et al. (1998). *Microscopy and microanalysis*, **4**(Suppl. 2), 362–363.
Zhang, H et al. (1996). In *X-ray microscopy and spectromicroscopy*, J, Thieme, G, Schmahl, D, Rudolph and E, Umbach (eds), Vol. II, pp. 143–148.
Zhang, X et al. (1996). *Journal of Structural Biology*, **116**, 335–344.

17
Crystallography of viruses and very large macromolecules

David I. Stuart, Jonathan M. Grimes and Elizabeth E. Fry

17.1 Background

Crystallography, using synchrotron radiation, can reveal the internal structures of very large macromolecules and macromolecular complexes. The limitation is the requirement for well-ordered three-dimensional crystals. Thus it is very difficult to determine the structure of a macromolecular complex which does not have a single, well-defined structure or which is difficult to obtain in a pure and stable state. Since the late 1970s, when technical and methodological advances opened the way to the determination of the structures of complete virus particles, virus crystallography has led the way in the analysis of complex structures. Crystallography is an extraordinarily powerful tool in the pursuit of a fuller understanding of the biological functions of viruses [see Chiu *et al.* (1997) for a recent compendium]. However, an intact virus structure determination is not a trivial undertaking and entails a significant scaling up in most respects compared to a traditional crystallographic analysis of a protein. This is because the number of data required to perform an analysis of a crystal by X-ray diffraction (at a fixed resolution) is proportional to the volume of the asymmetric unit of the crystal. Thus for a virus of particle mass 70 MDa (the approximate size of the largest particle thus far solved at reasonable resolution) there are 5000 times as many data to be measured as for, say, lysozyme (the first enzyme structure determined). Not only are there 5000 times as many X-ray reflections to be measured, but these reflections will each be, on average, about 5000 times weaker than those for lysozyme. Virus crystallography makes considerable demands on the qualities of X-ray sources, the performance and sensitivity of X-ray detectors and the computer hardware and software necessary to deal with this enormous increase in data. Nevertheless, the field is now sufficiently mature that it has become quite routine to solve the structure of a relatively small icosahedral virus. Here we give a brief introduction to the methods and take an example from our current work. Before embarking on the crystallographic analysis of a virus, the reader is referred to, e.g. Arnold *et al.* (1987), Luo *et al.* (1989), Fry *et al.* (1993) and Rossmann (1989) for a detailed treatment of various methods.

Viruses are some of the simplest biological systems that possess an evolutionary history. Their genetic simplicity is reflected in the simplicity of structure. A key aspect of the protein coat (capsid) which surrounds and protects the nucleic acid of a virus is its symmetry. There is a biological imperative behind the symmetry, namely the greater genetic 'efficiency' when many copies of just a very few proteins may be used to construct a massive shell. The viral genome, since it is a linear stream of information, cannot properly follow such symmetry. In fact, the RNA or DNA genome is usually almost completely invisible in a crystallographic analysis of an intact virus. However, in some cases, e.g. bean pod mottle virus (Chen et al. 1989), the symmetry of the coat proteins is reflected, in part, in the structure of the nucleic acid of the genome. Perhaps the most striking visualization of the genome is for bluetongue virus (BTV) (Gouet et al. 1998), which will be discussed below. Although symmetry is a common feature of viruses, there is a tremendous degree of structural variation. For example, the capsid may be spherical, bullet-shaped or rod-shaped. Viruses with an outer lipid membrane surrounding the protein coat (the enveloped viruses) have yielded, thus far, essentially useless crystals (Harrison et al. 1992). Similarly, although rod-like viruses have been studied by fibre diffraction [e.g. Namba et al. (1989)], only the isometric viruses have so far yielded to single crystal analysis, and are thus the focus of this chapter.

Neutron scattering, via contrast variation, can yield low resolution information about the relative distribution of the protein and nucleic acid components. Whilst electron microscopy has become an increasingly powerful tool (Böttcher et al. 1997), the results have not yet led to atomic models of viruses. However, at lower resolution the results can complement crystallographic analyses (see below).

17.1.1 Icosahedral symmetry

Isometric viruses are made from protein shells with closed symmetry. It has long been established that the most complex of such structures is the icosahedron. After DNA, Crick and Watson (1956) turned their attention to virus structure and realized that 60 identical protein subunits (which correspond to the triangular building blocks of an icosahedron) could be used to self-assemble a virus particle with icosahedral symmetry. They proposed this as the underlying symmetry of spherical viruses. They were right. However, many viruses are more complex than this and a major conceptual advance came when Caspar and Klug (1962) made a leap from exact to approximate symmetry. They introduced an idea termed quasi-equivalence. In simple terms, this proposes that the basic triangular icosahedral building block is broken down into a number (denoted T) of equal sub-triangles, so that the icosahedral shell is composed of $60 \times T$ chemically identical subunits. This idea explains the architecture of many virus particles but has some serious shortcomings when one considers the details of the molecular interactions (this is discussed further below). Irrespective of the further complexity arising from quasi-equivalence, the presence of icosahedral (532) symmetry leads to a minimum of five-fold non-crystallographic redundancy (for those that crystallize with the virus lying on a point of 23 crystallographic symmetry). This redundancy is of enormous importance in a crystallographic analysis, providing extremely powerful constraints which facilitate many aspects of the analysis, to the extent that routine *ab initio* phasing is potentially achievable.

17.1.2 Crystallization and crystal preparation

Crystallization conditions for viruses and other complex assemblies are not notably different from those for simpler proteins. The main rule is to show respect and not to subject these rather fragile structures to harsh conditions, so a thorough knowledge of the system gives the best chance of success. Particle aggregation is one of the main phenomena that prevent crystallization. Electron microscopy is a useful tool for checking this, in addition to methods such as dynamic light scattering. With viruses, highly ordered crystals are often obtained once the optimum crystallization conditions are found. This may be attributable to their isometric nature and is a dramatic reflection of the perfection with which these systems assemble. There is a tendency towards high symmetry space groups, often with some of the capsid symmetry axes coinciding with those of the crystal. A measure of the perfection of the crystals is the tiny mosaic spread that the crystals often exhibit (Fig. 17.1).

There are few reports of successful collection of virus data at cryogenic temperatures. For disease security reasons, virus crystals may have to be premounted in quartz capillary tubes, ruling out the most popular methods of cooling crystals. We have investigated alternative methods and found that the use of hand-pulled tapered quartz tubes with extremely thin walls that fit snugly around the crystal allows rather controlled experiments to determine freezing conditions. A freezing/annealing protocol has been established that allows crystals of foot-and-mouth disease virus to be frozen without a large increase in mosaic spread (allowing a complete data set to better than 2 Å resolution to be collected from a single crystal) (Stuart et al., unpublished). Note that for a unit cell typical of a virus crystal, a mosaic spread of more than about 0.2°

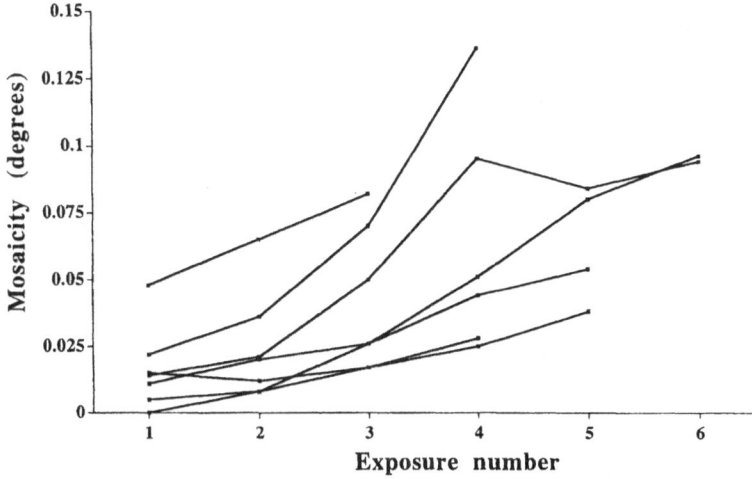

Fig. 17.1. Mosaic spread vs image number for several crystals of bovine enterovirus (each line represents a different crystal; Tate et al., unpublished). Note that the starting mosaic spread for a number of crystals is no greater than 0.02°.

renders the diffraction data hopelessly overlapped. This is an area that needs more careful experimentation, but is of great potential in reducing the number of crystals that need to be examined in order to solve a virus structure.

17.1.3 Data collection—beamline

The beam geometry, intensity and wavelength tunability of synchrotron radiation commend its use with large unit cells. Most synchrotron protein crystallography stations need no modification for virus data collection; however, the alignment of the various components should be carefully checked and a satisfactory way of precisely establishing the direct beam position established. For second generation synchrotrons where a station may take a wide angular fan of radiation, it is often necessary to reduce the width of the fan incident upon horizontal focusing elements so that the convergence (or divergence) of the beam is not excessive. Wavelengths of 1 Å or less can provide dramatic improvements in the signal-to-noise ratio and crystal lifetime. The measurement of low resolution reflections is often overlooked in protein crystallography. For virus crystallography these reflections can be very important, for instance in using a low resolution cryo-electron microscopy reconstruction as a phasing model, or for visualizing less well-ordered structure. By using a carefully designed back stop it is relatively easy to collect data down to 100 Å spacing (Grimes et al. 1998).

17.1.4 Data collection—detectors

In recent years, imaging plates have been the method of choice for data collection, combined with the oscillation method, so that each image records a small angular range. The Mar345, built by Mar research, is one of the most popular detectors and has a circular active area of 345 mm diameter and a useful pixel size of 150 μm (Fig. 17.2).

On a high brightness beamline such as ID2 at the European Synchrotron Radiation Facility (ESRF), Grenoble, France, the scan/erase cycle time of such a device exceeds the exposure time for a typical virus crystal by about an order of magnitude. There are two alternatives at present. Firstly, image plates can be scanned offline. This opens the way for using larger detectors, and hence improving the signal-to-noise ratio (Stuart and Jones 1993). Such an approach has been used, in combination with the Weissenberg method, by Sakabe at the Photon Factory (Sakabe 1991). There is no doubt that this approach can produce excellent data; however, it is more labour-intensive than the online detectors favoured elsewhere. An automated offline image plate system is being made at ID14 of the ESRF (S. Watasuki, personal communication). A radically different approach has been pioneered at synchrotrons in the USA: CCD detectors. Our preliminary experience with an ADSC device consisting of four CCD detectors arranged in a square to provide an 18.8 cm square active area indicates that the number of diffraction orders that can be resolved across the detector face is no less than for the 345 mm imaging plate system mentioned above. Since the readout time for this device is almost an order of magnitude less than the

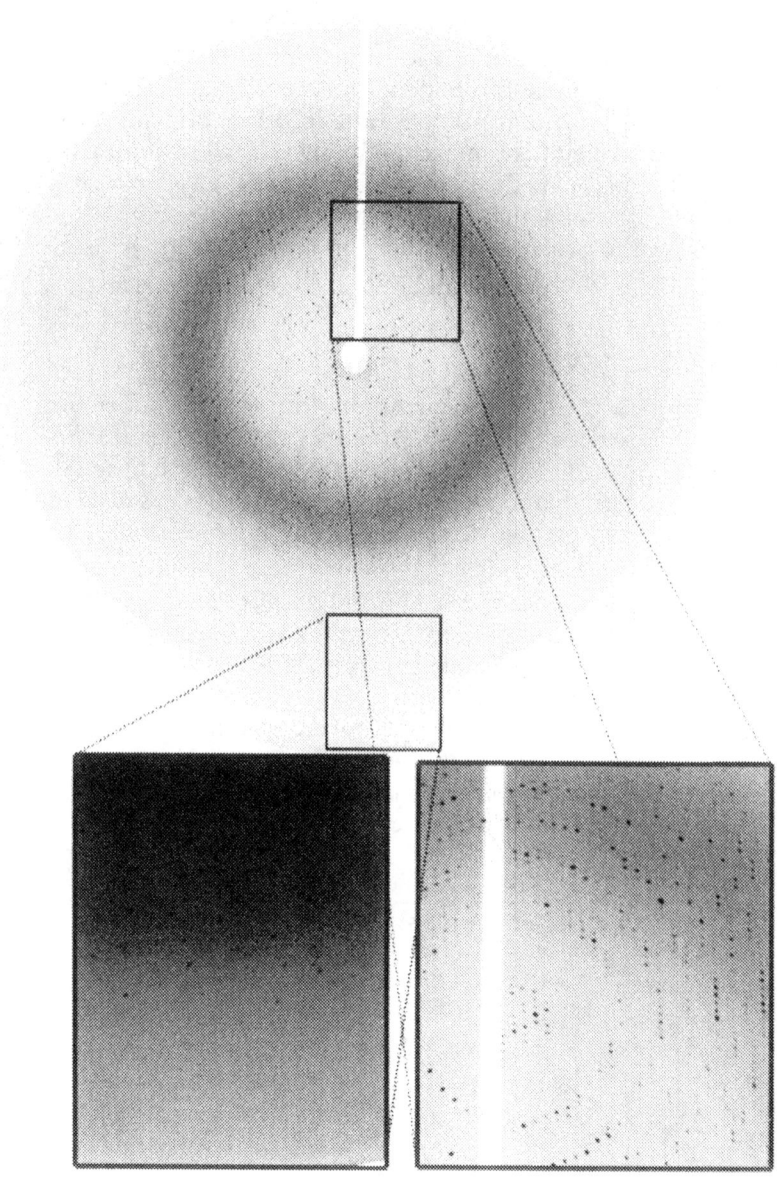

Fig. 17.2. Diffraction image for a crystal of foot-and-mouth disease virus (space group I23). Station ID2 ESRF, wavelength 1 Å, beam size 80 μm, Mar345 scanner, 0.2° oscillation. Crystal cooled to 100 K, resolution at edge about 1.7 Å.

scan/erase time for the imaging plate, the attraction of this new generation of detectors is obvious.

17.1.5 The Laue method

The Laue method has been used with virus crystals (Hadfield et al. 1995). The use of 'white' radiation leads to dramatic reduction in exposure times and in spite of the enormous intensity of incident radiation it has been claimed that some virus crystals may yield significantly more information with white radiation than with monochromatic illumination. In practice, the technical problems are very considerable and the method has not been used for a *de novo* virus structure determination. The availability of detectors capable of resolving more diffraction orders across their surface may make it worth revisiting this method, particularly for the study of dynamic events.

17.1.6 Data collection strategy

Since virus crystals are very sensitive to radiation, they are normally only aligned optically before data collection (Rossmann and Erickson 1983). Such a method is often supposed to be inefficient; we have analysed the effectiveness of a strategy based on deliberate randomization of the crystal orientation (Table 17.1). The method is in fact extremely effective. For instance, a data set

Table 17.1. Data from randomly oriented images. The figures shown are the percentage completeness to 3 Å resolution. These were calculated by predicting reflections for 90 images with randomly generated mis-setting angles, arbitrary cell, wavelength of 0.9 Å and oscillation range of 0.4°. Only reflections more than 50% recorded were accepted. These reflections were then reduced to the unique subset for each point group, merged, duplicates removed and the percentage completeness calculated. To a first approximation, these figures scale directly according to the oscillation range used. The figures are essentially independent of resolution.

Point group	No. of images (°)					
	5 (2°)	10 (4°)	20 (8°)	40 (16°)	60 (24°)	90 (36°)
1	2.2	4.4	8.5	16.3	23.4	33.0
2	4.3	8.4	16.2	29.4	40.8	54.4
222	8.2	15.7	29.0	48.9	63.5	77.8
4	8.4	16.1	29.6	50.3	64.9	79.2
422	15.1	28.0	48.4	72.6	85.5	94.2
23	22.7	40.2	64.0	86.2	94.6	98.6
432	37.6	60.9	83.8	96.8	99.2	99.8
3	6.5	12.6	23.5	41.3	54.8	69.5
321	12.3	22.8	40.5	63.5	77.8	89.4
6	12.5	23.1	41.0	64.7	78.9	90.2
622	22.4	39.1	62.6	84.6	93.5	98.1

64% complete can be collected from only 8° of data for space group I23. Since the virion transform of an icosahedral virus is inevitably oversampled such a level of completeness is often more than adequate.

It is usually possible to collect more than one photograph per position when using image plates, and with careful sizing of the beam, more than one position per crystal, so that in favourable cases one only needs to examine a few crystals. However, for more challenging analyses a very large number of crystals can be required, such as the work on BTV where over 1000 crystals were examined.

For the collection of, say, 3 Å resolution data for small viruses, the oscillation range is usually between 0.3° and 0.5° depending on the lattice properties. The often very small mosaic spread of virus crystals, coupled with a small beam divergence (undulator beamlines tend to have tiny divergence) means that it is possible to achieve a good yield of reflections fully recorded on a photograph with an oscillation range as small as 0.3°.

17.1.7 Space group determination

Space group determination may be problematic for virus crystals. Quite often the special zones of reciprocal space that give crucial information about the nature of the crystal symmetry will be largely missing from the partial data set available. The data processing programs are now good at suggesting point groups, although these should be carefully checked by analysis of the agreement of symmetry-related reflections. It is not infrequent for the final assignment of space group to depend upon molecular replacement methods, although careful analysis of the possible packing modes can sometimes be decisive.

17.1.8 Data processing

Data processing is now relatively standard for crystals with large unit cells. For instance, we have processed data from crystals with a primitive unit cell with dimensions of 1100, 1100 and 1600 Å with the DENZO system (Otwinowski and Minor 1997). With such a high density of spots on the detector, it is worth taking care to establish the direct beam position to an accuracy of within a few pixels. If not, the modern autoindexing algorithms will mis-index the diffraction pattern. Note that the R-factors tend to be higher than for those for crystals with smaller cells. However, a high degree of oversampling of the molecular transform can compensate, so that even a shell of data for which the merging R-factor is, say, 50% can produce a good contribution to the electron density map. It is often still necessary to specially configure the software to cope with the large number of data (frequently millions of reflections).

17.1.9 Phase determination

This divides into two problems: (1) initial phase determination, (2) phase refinement and extension. Oversampling of the molecular transform allows very poor initial phases to be refined to great accuracy, given a reasonable set of observed structure factor amplitudes.

Initial phase determination It is necessary to determine the precise orientation and position of the virus particle(s) in the unit cell. In some cases these can be defined trivially because the virus with its very high symmetry, can lie at a variety of special positions in different space groups. Further information on some of the parameters can often be directly derived from packing considerations, greatly simplifying later stages of the structure determination. In general, however, a self-rotation function will provide the precise particle orientation(s). Calculated in polar coordinates, plots of sections at constant κ will show the direction of non-crystallographic symmetry axes and the $\kappa = 72°$ is usually the easiest to interpret. If the parameters are correctly chosen, rotation functions give beautiful results for viruses. Translational searches can be performed using standard protocols to determine the position of the particle. Running on a fast computer, XPLOR (Brunger *et al.* 1987) provides a simple, flexible package for these procedures. A convenient program for the visualization of the various outputs of XPLOR is GROPAT (Esnouf, unpublished).

Once the position and orientation of the particle are known, phase refinement and extension can proceed. Starting phases are required and these may be derived from conventional heavy atom methods. Due to the large number of symmetry-related heavy atom sites and the constraints imposed by the non-crystallographic symmetry, the heavy atom data can be useful even if grossly incomplete [for instance, 5% completeness was sufficient in the study of CPV (Tsao *et al.* 1992)]. If one or more structurally similar viruses are available to provide a starting point for phase determination, phases are calculated for the virus or viruses placed correctly in the cell of the unknown structure. It is our experience that summing two or more homologous structures can provide improved phase estimates. A cryo-electron microscopy reconstruction can also provide an adequate starting point for phase determination (as discussed below).

Phase refinement and extension Iterative non-crystallographic symmetry averaging is used to improve the starting phases. In outline, the procedure used is to average the electron density for the non-crystallographically related subunits within the viral envelope, back-transform the average electron density, recombine the resultant calculated phases with the original F_{obs}, with suitable weighting, and compute a new electron density map. If phases are to be extended, the map is back-transformed to a slightly higher resolution than it was calculated at, which provides new phase estimates (albeit crudely) to this new limit, which can then be fed back into the cyclical procedure. The region outside the molecular envelope and thus beyond the limits of applicability of the local non-crystallographic symmetry is flattened to represent the disordered solvent. There are now a number of efficient computer programs to perform the real space part of this procedure (Kleywegt and Read 1997). There are a number of quirks to the procedure, and these are discussed in detail in Fry *et al.* (1993), and with particular discussion of optimization in Grimes and Stuart (1994). In tricky cases, attention to detail at this stage can make the difference between success and failure.

Carefully applied, phase extension from low resolution should be capable, even in the presence of the minimal five-fold non-crystallographic symmetry, of providing ultimately a set of very accurate phases to near atomic resolution (Fry et al. 1993). The process of phase refinement and extension is normally monitored by reference to averaging R-factors and correlation coefficients which should be analysed as a function of resolution. For one very satisfactory study these converged to values of 11.8% and 0.96%, respectively (Morgunova et al. 1994). The correlation coefficient is a more trustworthy indicator of success than the R-factor.

17.1.10   Map interpretation—model building—refinement

These aspects of the analysis are essentially the same as for a small protein. There is more to do, but if you are lucky enough to be working with an averaged map, it is likely to be of high quality. Beware not to throw away this benefit by relaxing the non-crystallographic constraints too early. Also the presence of non-crystallographic symmetry means that the R-free (the R-factor calculated for a subset of reflections not included in the target function of the refinement procedure) behaves in a quite different way (due to correlations between the working and test set of data). It can in fact be misleading if it is used as a guide to weighting schemes in reciprocal space (Stuart and Fry, unpublished).

17.2   Bluetongue virus

In collaboration with the group of Dr P. Mertens (Institute for Animal Health, Pirbright) we have worked on BTV. The work has built upon wider collaborations (P. Roy: recombinant proteins and B.V.V. Prasad: electron microscopy). The atomic structure of the core has now been determined (Grimes et al. 1998) and ordered RNA has been detected within the particle (Gouet et al. 1998). BTV belongs to the family *Reoviridae*, members of which cause plant and animal diseases, and are responsible for significant levels of child mortality in developing countries. The *Reoviridae* are characterized by their genome of 10–12 segments of linear dsRNA (Holmes et al. 1995). They are non-lipid containing icosahedral viruses with an outer capsid layer and a core, which contains the genome made up of a single copy of each dsRNA segment. Upon cell entry, the outer capsid is removed and the core released into the cytoplasm of the target cell. The dsRNA is retained within the core, which contains active polymerases (transcriptases) and capping enzymes. This core particle acts as a molecular engine producing exactly full length, capped, mRNA copies simultaneously from each of the genome segments. The BTV core has a diameter of 700 Å and is composed of two principal structural proteins. Each particle contains 780 copies of the protein T13 (38 kDa, 349 residues) arranged as trimers on a $T=13$ quasi-equivalent lattice, which form the 'bristly' core surface (Grimes et al. 1997). This outer T13 layer cloaks a thin, rather featureless 'subcore' shell constructed from 120 copies of the protein T2 (100 kDa, 901 residues). The icosahedral building block of the core is composed of 13 subunits of T13 attached to two molecules of T2, such that there must be 13 distinct sets of interactions between these molecules in the assembled core. An icosahedral shell built of 120 subunits is inexplicable within the conceptual framework of quasi-equivalence. It was also unclear whether the T13 layer is assembled using the conformational switching mechanism seen in other viruses [see e.g. Liddington et al. (1991)].

17.2.1 BTV structure determination

The structures of two serotypes, BTV-1 (SA) and BTV-10 (USA), have been solved by the method of X-ray crystallography. An atomic model of BTV-1 (SA) has been derived using data to about 3.5 Å resolution. Information from a 22 Å resolution cryo-electron microscopy reconstruction and a high resolution atomic structure for the T13 trimer was combined to give a model which provided phase information to higher resolution than the electron microscopy reconstruction. The crystals of BTV-1 (unit cell dimensions $a = 796$ Å, $b = 822$ Å and $c = 753$ Å), contain half a particle in the asymmetric unit. Over 21 000 000 data were collected, giving a unique data set of some 3 300 000 reflections to 3.5 Å resolution ($R_{merge} = 22.9\%$). A self-rotation function provided the orientation of the virus and a translation search, using the cryo-electron microscopy derived model, located the particle. The resolution was extended in reciprocal space from 12 to 6 Å. Averaging and solvent flattening then refined the phases, and an averaged map at 3.8 Å allowed unambiguous chain tracing and sequence alignment of both molecules of T2 (reciprocal space R-factor and correlation coefficient, 16.9% and 0.89, respectively). It was possible to refine the atomic structure in XPLOR despite very incomplete data at 3.5 Å (Brunger et al. 1987). The structure of BTV-10 was solved using BTV-1 as a search model. BTV-10 crystals have cell parameters $a = b = 1115$ Å, $c = 1584$ Å and contain one core particle in each asymmetric unit. As with BTV-1, data were collected at the high brilliance undulator beamline ID2 (BL4) at the ESRF. The highly parallel beam, in conjunction with limiting apertures on the incident beam of 50–80 μm and a 30 cm MAResearch imaging plate at 1.05 m from the crystal, allowed the diffraction orders to be resolved at a wavelength of some 1 Å. Careful experimental design allowed relatively low angle diffraction (60 Å resolution) to be collected with reasonable background noise. A unique data set of 1 034 180 reflections was obtained in the range 60–6.5 Å, with a completeness of 54%. The particle orientation was determined from inspection of a self-rotation function. The space group and particle position were determined by factoring the space group symmetry operators to reduce the problem to one- and two-dimensional searches. Rigid body refinement and a grid search refinement of the unit cell dimensions provided model structure factors whose phases formed the basis for cyclic averaging and solvent flattening. Cyclic averaging was performed using both BTV-1 and BTV-10 data to dissect out disordered structural features of the interior conserved between the two serotypes (data were collected to a low resolution limit of 100 Å for BTV-1). The electron density maps are strikingly similar for both viruses and indeed cross-averaging the two viruses leads to a similar consensus structure.

17.2.2 Overall architecture of the BTV core

The core obeys exact icosahedral symmetry. The overall features of the structure, a lattice of T13 trimers coating a thin inner shell of T2 monomers, enclosing the genome, are shown in Fig. 17.3. The inner surface of the T2 layer is relatively bland with some shallow sculpted grooves and very few charged residues.

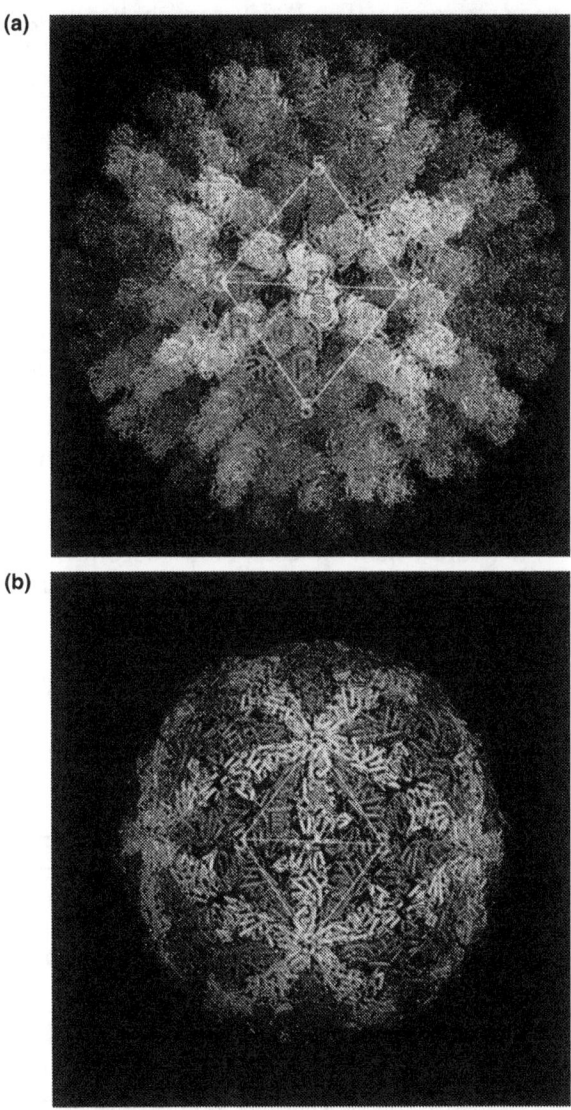

Fig. 17.3. (a) The outer layer of the core showing the T13 trimers P, Q, R, S and T that make up the viral icosahedral unit. (b) The inner scaffold of 120 copies of T2, showing the A molecules spanning the five-fold axes and the B molecules that plug the gaps.

17.2.3 A layer showing non-classical quasi-equivalence

The T2 subunits are arranged with icosahedral symmetry within the subcore layer, forming two sets (A and B) of 60 subunits each. The arrangement of the 120 T2 molecules on an icosahedrally symmetric lattice is shown in Fig. 17.3(a). Pairs of A subunits form a continuous scaffold defining the size of the subcore. The B subunits plug the holes in this scaffold. This unique architecture is likely to be a characteristic of dsRNA viruses and yet it does not fit within the dominant hypothesis for virus assembly. The T2 layer might be thought of as $T = 2$ lattice, however this T number is excluded by the quasi-equivalence formalism, described above since the two sub-triangles in a $T = 2$ lattice do not have the same shape. The core ameliorates this problem and achieves a closely packed shell by distortion of the triangular T2 building blocks. The molecule pivots about one end and is bent at the other, swinging the tip of the molecule by 35 Å. This new type of structure has been termed quasi-quasi-equivalence.

17.2.4 A layer showing classic quasi-equivalence

The outer layer of the core is built up of 13 icosahedrally independent copies of the T13 protein, in the form of trimers [Fig. 17.3(a)]. These copies are extremely similar to each other (with maximum rms deviations of 0.30 Å in Cα positions). Thus the $T = 13$ lattice follows the rules of classical quasi-equivalence to an unprecedented degree. The quasi-equivalent trimers [named P, Q, R, S and T, Fig. 17.3(a)] are arranged such that they relate to their nearest neighbours by two-fold axes. A distortion of the contacts between subunits is inherent in a quasi-equivalent structure where molecules must accommodate to both icosahedral five-fold and local six-fold symmetry axes. Usually, conformational switching occurs using alternative stabilizing contacts to provide controlled, path-dependent, assembly of the particle. BTV overcomes this by concentrating the interactions between the T13 trimers in a thin ring, facilitating the rolling of trimers against each other.

17.2.5 Multiple interactions between layers of different symmetry

The symmetry mismatch means that there are 13 different sets of contacts between the T13 and T2 layers. Since T13 subunits trimerize in solution, we may think of the core being made from the crystallization of 260 trimers onto the 120 T2 subunits, the driving force being the interaction of each trimer with the underlying layer. This two-dimensional crystallization model avoids the need for conformational switching.

17.2.6 Ordered RNA

Electron density maps for both BTV-1 and BTV-10 showed tubes of density within the protein capsid (Gouet *et al.* 1998). These tubes are the correct size to correspond to dsRNA. The electron density was sufficiently clear that a model

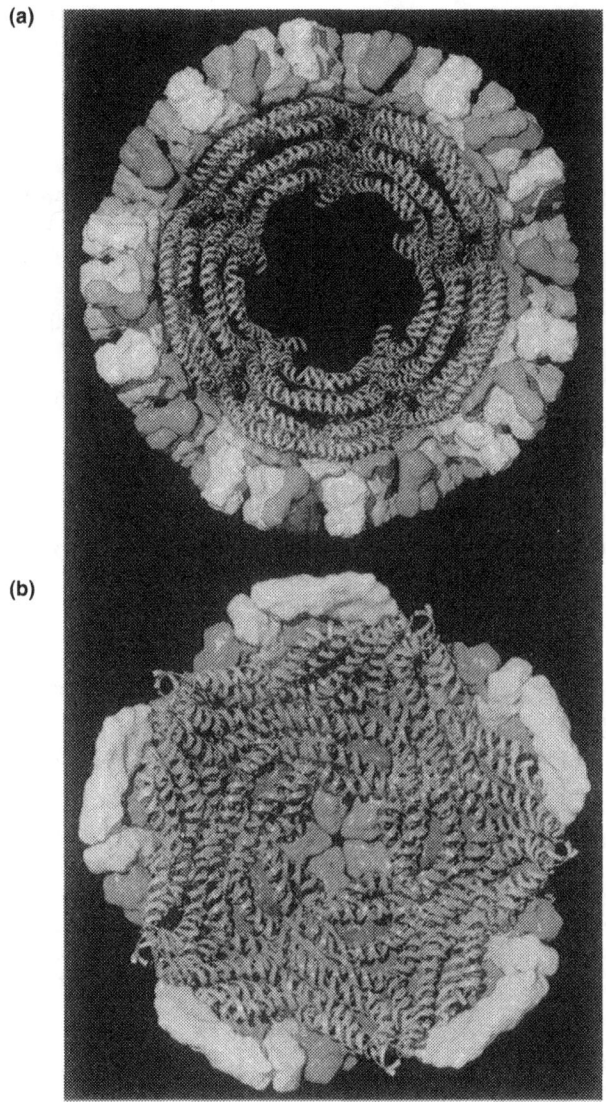

Fig. 17.4. (a) The model for the layer of dsRNA proximal to the T2 scaffold. (b) The view of the four layers of dsRNA modelled into the averaged electron density maps, orthogonal to the view in (a).

accounting for the dsRNA genome could be built (Fig. 17.4). In addition, there is diffuse density at the five-fold axes inside the protein shell, which is thought to represent the transcriptional complex. It is clear that there is very little free space within the capsid and the electron density shows that the RNA assumes an extraordinary degree of order, presumably organized by the peculiar T2 protein layer. The RNA appears to exist in a highly fluid liquid crystal form which allows it to writhe freely during the simultaneous transcription of the 10 gene segments.

References

Arnold, E *et al.* (1987). *Acta Crystallographica*, **A43**, 346–361.
Brunger, AT, Kuriyan, J and Karplus, M (1987). *Science*, **235**, 458–460.
Böttcher, B, Wynne, SA and Crowther, RA (1997). *Nature*, **386**, 88–91.
Caspar, DLD and Klug, A (1962). *Cold Spring Harbor Symp. Quant. Biol.*, **27**, 1–24.
Chen, Z *et al.* (1989). *Science*, **245**, 154–159.
Chiu, W, Burnett, RM and Garcea, RL, eds (1997). *Structural biology of viruses*, Oxford University Press, New York.
Crick, FHC and Watson, JD (1956). *Nature*, **177**, 473–475.
Fry, E, Acharya, R and Stuart, D (1993). *Acta Crystallographica*, **A49**, 45–55.
Fry, E, Logan, D and Stuart, D (1996) *Methods in molecular biology*, Vol. 56, *Crystallographic methods and protocols*, Jones, Mulloy, Sanderson eds, HUMANA Press, pp. 319–363.
Gouet, P *et al.* (1998). Submitted.
Grimes, J and Stuart, D (1994). *The use of the free R-factor as a guide in parameter optimization for density modification* in Daresbury Study Weekend, Bailey, Hubbard and Waller, eds, 67–76.
Grimes, J *et al.* (1997). *Structure*, **5**, 885–893.
Grimes, J *et al.* (1998). *Nature*, **395**, 470–478.
Hadfield, AT *et al.* (1995). *Acta Crystallographica*, **D51**, 859–870.
Harrison, SC, Strong, RK, Schlesinger, S and Schlesinger, MJ (1992). *Journal of Molecular Biology*, **226**, 277–280.
Holmes, IH *et al.* (1995). *Virus taxonomy: classification and nomenclature of viruses*, Murphy, FA *et al.*, eds, Vol. 10, pp. 208–239.
Kleygwegt, GJ and Read, R (1997). *Structure*, **5**, 1557–1569.
Liddington, RC *et al.* (1991). *Nature*, **354**, 278–284.
Luo, M, Vriend, G, Kamer, G and Rossmann, MG (1989). *Acta Crystallographica*, **B45**, 85–92.
Morgunova, E *et al.* (1994). *FEBS Letters*, **338**, 267–271.
Namba, K, Pattanyek, R and Stubbs, G (1989). *Journal of Molecular Biology*, **208**, 307–325.
Otwinowski, ZO and Minor, W (1997). *Methods in enzymology*, Vol. 276, Carter, CW and Sweet, RM, eds, Academic Press, pp. 307–326.
Rossmann, MG (1989). *Acta Crystallographica*, **A46**, 73–82.
Rossmann, MG and Erickson, JW (1983). *Journal of Applied Crystallography*, **16**, 629–636.
Sakabe, N (1991). *Nuclear Instruments and Methods in Physics Research*, **A303**, 448–463.
Stuart, DI and Jones, EY (1993). *Current Opinion in Structural Biology*, **3**, 737–740.
Tsao, J, Chapman, MS, Wu, H, Agbandje, M, Keller, W and Rossmann, MG (1992). *Acta Crystallographica*, **B48**, 75–88.

18

The quest for high resolution phasing for large macromolecular assemblies exhibiting severe non-isomorphism, extreme beam sensitivity and no internal symmetry

Ada Yonath

18.1 Introduction

The translation of the genetic code into polypeptide chains is a fundamental life process. In rapidly growing bacterial cells the biosynthetic machinery constitutes about half of the dry weight of the cell, and the biosynthetic process consumes up to 80% of the cell's energy. Among the over 100 different compounds participating in the biosynthetic process is a giant riboprotein complex which has been studied crystallographically for quite some time: the ribosome, an unstable 2350 kDa assembly of many proteins and RNA of diverse structures. This universal organelle mediates the translation step of the biosynthetic process by catalysing the sequential polymerization of amino acids according to the blueprint, encoded in the mRNA [for recent reviews see Wilson and Noller (1998), Yonath and Franceschi (1998)].

A typical bacterial ribosome (called 70S) has a molecular weight of 2.3×10^6 Da. About one-third of its mass is comprised of some 58–73 different proteins, depending on its source. The remaining two-thirds is made of three chains of rRNA, with a total of about 4500 nucleotides. These are arranged in two independent subunits of unequal size which associate upon the initiation of protein biosynthesis. The large subunit has a mass of 1.45×10^6 Da. It contains 36–48 different proteins and two RNA chains (with a total of some 3000 nucleotides). The small subunit has a mass of 0.85×10^6 Da and it contains 21 proteins and an RNA chain of about 1500 nucleotides. Each of the two ribosomal subunits carries out different tasks and displays different properties. The large subunit catalyses the chemical reaction of the formation of the peptide bond and provides the path along which the nascent protein progresses. The range of functional activities of the small subunit is larger. It provides the site for the initiation of the translation step, facilitates the decoding of the genetic information and creates the fundamental feature of the *in vivo* initiation selection mechanism. Most of the aminoglycoside antibiotics that cause misreading of mRNA codons during translation, interact with it.

The observations that ribosomes can pack periodically and the hypothesis that these ordered forms are the physiological mechanism for temporary storage, aimed at preserving the integrity and activity of the ribosomes for an expected better future, stimulated us to attempt crystallization. Ribosomes from prokaryotes were chosen, as they are smaller and have been characterized biochemically in much greater detail than those from eukaryotes. They provide systems for *in vitro* crystallization independent of *in vivo* events influenced by environmental influences (e.g. stress, cold shocks, wrong diet) which might be difficult or impossible to control and to reproduce. They can also be produced in high purity and large quantities, essential for effective crystallographic studies. The key for obtaining diffracting crystals from ribosomal particles was the choice of the organism: halophilic or thermophilic bacteria; presumably because these ribosomes are more stable than those from eubacteria. A strong correlation between the activity of the ribosomes and the quality of their crystals was found. Furthermore, in all cases, except for some fragmentation of rRNA, the crystallized material retains its integrity and biological activity for long periods, in contrast to the short lifetime of isolated ribosomes in solution.

Crystals have been grown from ribosomes, their complexes mimicking defined stages in protein biosynthesis and their natural, mutated, selectively depleted and chemically modified subunits (Berkovitch-Yellin *et al.* 1992). Far beyond the initial expectations, two of these crystal forms, of the large subunits from *Haloarcula marismortui* (H50S) and of the small subunit from *Thermus thermophilus* (T30S), diffract currently to around 3 Å (Fig. 18.1). Although this resolution range may seem to be inferior to what is obtained from crystals of other large macromolecular complexes, for ribosomal crystals it should be considered rather high in view of their enormous size, which does not contain any internal symmetry, and the high level of their complexity.

It was found that the ribosomes are tough subjects for crystallographic analysis, primarily because they are composed of highly degradable RNA along with proteins, some of which may be loosely held. Table 18.1 shows that, in contrast to the common observations in macromolecular crystallography, the high resolution obtained from the ribosomal crystals is not necessarily linked to the diffraction of a high quality. Thus, the crystal-type diffracting to the highest resolution (H50S) yields the most problematic diffraction data. The efforts towards the elucidation of the structure of the ribosome and the problems (solved as well as unsolved) encountered over the years, are the subjects of this article.

18.1.1 Crystals of the whole 70S ribosome

These crystals, diffract to low resolution, presumably due to the inherent conformational heterogeneity of their preparations, as they are extracted directly from cells during their growth phase. However, crystals of a complex mimicking a defined functional state, containing T70S ribosomes, an oligomer of about 35 uridines and two charged tRNA molecules, diffract to a higher resolution than that obtained from purified 70S ribosomes [see Table 18.1 and Hansen *et al.* (1990)]. A further improvement of the resolution is expected from

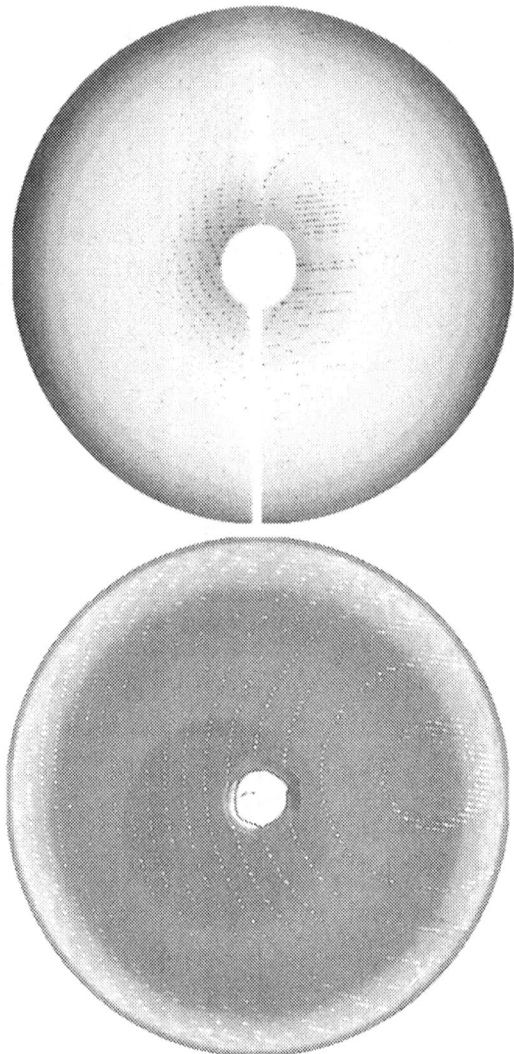

Fig. 18.1. *Top*: a 1° rotation diffraction pattern obtained from a crystal of treated small subunit from *Thermus thermophilus* (T30S), obtained at ID2/ESRF. *Bottom*: a 0.5° rotation diffraction pattern of a fresh crystal of H50S soaked in solution with 0.5 mM of W30, obtained in 20 s at the microfocus beamline (ID13) at ESRF (see Plate Section).

crystals programmed with mRNA of selected sequences. These complexes should eventually allow the mapping of the different conformations adopted by the ribosome while performing its function.

18.1.2 *The readiness of the large ribosomal subunit (50S) to crystallize*

This can be linked to its relative stability. Thus, 18 crystal forms have been grown from these particles, four of which were found to be suitable for crystallographic

Table 18.1. The quality of the ribosomal crystals

Source	Cell parameters (Å)	Resolution (Å)[a]	Isomorphism[b]	Beam sensitivity[c]	Spot shape[d]	Mosaicity[e]	Crystal shape and rigidity[f]
T70S[C]	524 × 524 × 306 P4₁2₁2	12/15	ok	U	ok	High	ok
T30S	407 × 407 × 176 P4₁2₁2	3.0/3.5	ok	High	ok	Rather high	Thin soft
T50S	495 × 495 × 196 P4₁2₁2	3.4/4.0	ok	U	ok	ok	ok
H50S	211 × 300 × 567 C222₁	2.7/3.2	Hardly any	Very high	Deformed elongated non-uniform	Very high	Problematic

T70S, T50S, T30S = the whole ribosome from *Thermus thermophilus* and its two subunits; H50S = the large subunit from *Haloarcula marismortui*.
[C] A complex of T70S ribosomes, two phe-tRNA^Phe molecules and an oligomer of 35 uridines (as mRNA). The crystals of the pure T70S ribosome pack in the same crystal form but diffract only to 18/14 Å.
[a] The highest detectable/useful resolution. 'Useful' means 75% completeness (or higher) in the last shell.
[b] Isomorphism refers to native crystals grown from the same preparation. An 'ok' indicates reasonable isomorphism for above 50% of the crystals. 'Very bad': less than 10% of the crystals are isomorphous (defined by cell dimensions and/or the distribution of the average $\Delta F/F$ values vs resolution).
[c] Beam sensitivity at cryo-temperature, using bright synchrotron radiation beam.
[d] An 'ok' corresponds to 0.2–0.5° mosaic spread. 'Very high' may reach up to 3°.
[e] An 'ok' spot shape means that the reflections have a defined shape which corresponds to the shape of the crystals and to the cross-section of the beam. Elongated and undefined shape indicates, in addition to these properties, a high chance for interpenetration.
[f] An 'ok' means crystals of fairly isotropic dimensions which, when handled carefully, do not develop severe deformations. 'Problematic' means extremely thin crystals, built from readily sliding layers, which, even upon extremely careful handling may suffer from resolution loss and fragmentation.
U: Unknown or not well determined, since these crystals do not diffract to the resolution showing fast decay.

analysis at various levels of detail. The crystals of the halophilic large subunits (H50S) have been the target of extensive crystallographic analysis, since they diffract to almost atomic resolution, 2.5–2.7 Å (von Böhlen et al. 1991). However, although data collected carefully at intermediate resolution led to MIR phasing (Yonath et al. 1998), it was found that the undesired properties of this crystal form (Table 18.1) become more problematic and less tolerable with the increase of resolution. For instance, at the higher resolution ranges the crystal decay is expressed not only by loss of resolution, which can be monitored visually, but also in an invisible, albeit substantial, growth of the longest unit cell axis (Fig. 18.3). It is conceivable that each of the problems encountered could have been tolerated, but the combination of severe non-isomorphism, high radiation sensitivity, non-stable cell constants, non-uniform mosaic spread, uneven reflection shape and high fragility led in many cases to extreme difficulties even in the mere production of reliable data sets, let alone the construction and the interpretation of high resolution difference Patterson maps.

It should be mentioned that in each preparation there are some crystals that lead to reasonable mosaic spread (0.1–0.2°), non-deformed reflection spot shape and no additional patterns (resulting from layer sliding). Experience showed that the probability of detecting such crystals decreases with the increase of resolution (from 3–4% for those diffracting to 6–7 Å to 1–2% of the crystals yield high quality diffraction to higher limits, namely 2.7–3.3 Å). Combined with the very low level of isomorphism and the extreme sensitivity to irradiation, the above rather poor statistics hamper smooth and efficient high resolution phasing.

The 10–12 Å map (Fig. 18.2) is based primarily on the contribution of a strong derivative, Ta_6Br_{14}. This derivative led to a well defined, thus readily interpretable difference Patterson map at 7.5 Å (Fig. 18.2), even when all data with resolution below 12.5 Å were omitted. Consequently, its difference Fourier map enabled the positioning of the sites of two weaker derivatives, W12 and W17 (Table 18.2). The resolution of this map is currently being extended to 7 Å. The currently available intermediate map (at 8 Å) shows a higher connectivity and can be partially interpreted (to be published). Indeed, the Ta_6Br_{14} derivatized crystals diffract well to a resolution much higher than the limits currently set by us, owing to the above-described obstacles.

This MIR map may shed light on the odd combination of the properties of H50S. Thus, the extensive interparticle contacts that are concentrated in parts of the unit cell may account for the high resolution. However, in contrast to this dense packing, only a relatively small region, surrounded by a sizeable volume of solvent with dimensions that may reach over 200 Å in their longest direction, is involved in contacts between the two halves of the unit cell along the very long c-axis (567 Å) direction. Although still uncertain, it seems that the isolated contact network is made by two symmetry-related particles via RNA chains (Harms et al. 1999). As these interparticle contacts appear to be rather loose, we assume that they are partially mediated by the solvent. Hence, the influence of Mg and Cd ions on the rigidity of the crystals and their thickness may also be understood.

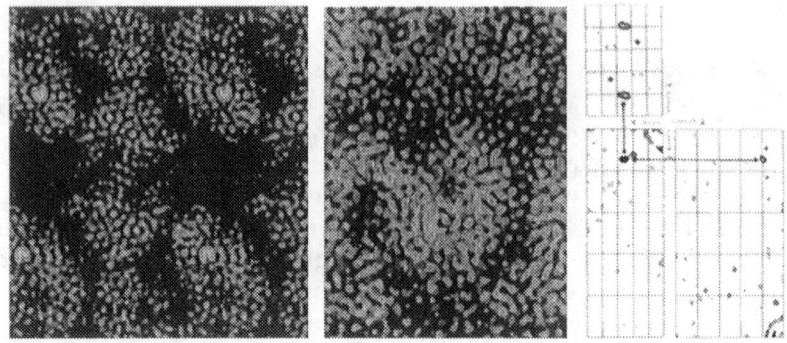

Fig. 18.2. *Left*: a part of the current 10–12 Å MIR map of H50S, showing the compact packing regions (around $z = 1/4$ and 3/4) as well as the isolated contact area along the z-axis. For clarity, two unit cells are shown along the y-direction (horizontal). The dense areas represent the position of the most occupied heavy atom site (at the interface between two subunits). *Middle*: the Ta_6Br_{14} difference Patterson map of H50S, including the data of the 7.5–12.5 Å resolution shell. The corresponding Harker peaks are shown by arrows. *Right*: a part of the map oriented to show the entrance to the main ribosomal internal tunnel (Yonath *et al.* 1987). More than 11 000 reflections were measured, and a total of 15 heavy atom sites of the three derivatives (Ta_6Br_{14}, W12 and W17, Table 18.2) were included. The positioning of the heavy atom sites was performed by a combination of difference Patterson and Fourier methods, based on the major position of Ta_6Br_{14}, found to be stable and consistent in all resolution ranges until 7.5 Å. Each heavy atom position was cross-verified and refined by MLPHARE with maximum likelihood. Since the contribution of the two W clusters was negligible beyond 10 Å, their scattering curve could be approximated by spherical averages of their corresponding radii (W18: 10 Å and W12: 8–9 Å). The Ta_6Br_{14}, however, was treated as in Knäblein *et al.* (1997) owing to its potential contribution to the higher resolution shells. Mean figure of merit: 0.32 (0.57 for centric); R_{cullis}: 0.76–0.97; phasing power: 0.98–1.15. The map was solvent flattened: one cycle, assuming 54% solvent (see Plate Section).

Table 18.2. Metal clusters

PIP = di-iododiplatinum (II) diethyleneamine
TAMM = tetrakis (acetoxymercuri)-methane
Ta_6Cl_{14}
$Ta_6Br_{14} \cdot 2H_2O$
Nb_6Cl_{14}
$Ir_4(CO)_8R'_2R''$ **
$C_{22}H_{280}N_{24}O_{38}P_7Au_{11}$ **
W12Rh = $CS_5H_xSiW_{11}O_{39}Rh^{III}CH_3COO(H)$ **
W30 = $K_{14}(NaP_5W_{30}O_{110})31H_2O$
W12 = $K_5H(PW_{12}O_{40})nH_2O$
W18 = $(NH_4)_6(P_2W_{18}O_{62})14H_2O$
W17Co = CoWLi17 = $Cs_7(P_2W_{17}O_{61}Co(NC_5H_5))nH_2O$
BuSnW17 = $K_7[(buSn)(P_2W_{17}O_{61})]nH_2O$
PhSnW15 = $K_5H_4[(phSn)_3(P_2W_{15}O_{59})]nH_2O$
BuSnW15 = $K_5H_4[(buSn)_3(P_2W_{15}O_{59})]nH_2O$
$Na_{16}[(O_3PCH_2PO_3)_4W_{12}O_{36}]nH_2O$

**Should be used for covalent binding. R = $CH_2CH_2CONH_2$; R' = $CH_2CH_2CONHCH_2$-CH_2CONH_2; bu = butyl; ph = phenyl.
For references see Weinstein *et al.* (1989), Thygesen *et al.* (1996), Knäblein *et al.* (1997), Lunger *et al.* (1997).

Such an unusual packing arrangement may be the reason for the low isomorphism of this crystal form, for the problematic morphology (plates reaching up to 0.5 × 0.5 mm² with a typical thickness of a few microns in the direction of the c-axis), for the layer structure of the crystals, for the high tendency of these layers to slide relative to each other (causing multilattice diffraction patterns), for the changes in the c-axis that are introduced by irradiation (Fig. 18.3)

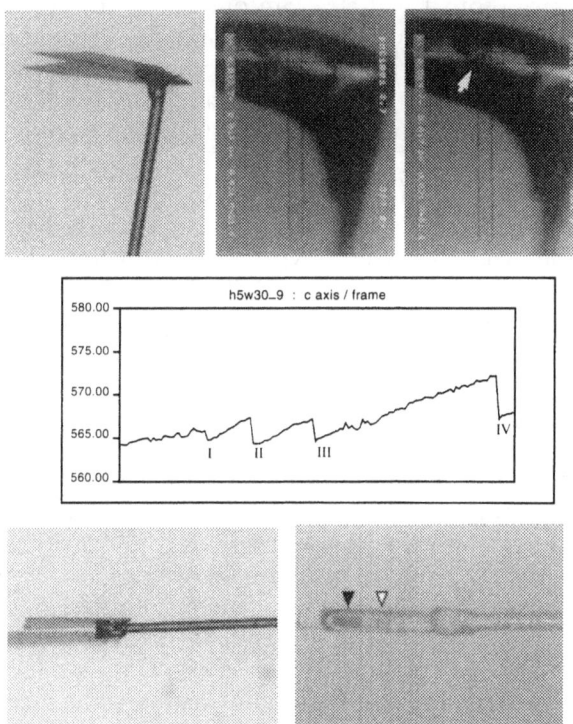

Fig. 18.3. *Top row*: (left) a perpendicular double spatula used for mounting H50S crystals; (Middle and Right) two photographs of the irradiated crystal, at the beginning of the experiment (the square shows the area being exposed) and after the decay of this position. Note that the irradiated region became dark (indicated by the arrow). *Middle row*: the 'fluctuating' c-axis. Data were collected sequentially from a crystal of H50S, mounted as shown on the right. The crystal size was 400 × 380 × 8 μm and the beam cross-section 100 × 100 μm. The initial resolution was higher than 3.2 Å (at the edge of the MAR detector). The loss of resolution was monitored by visual inspection, and when it reached 6–7 Å (points I, II, III), the crystal was translated to a new position. At the position marked IV the resolution limit is 9.5 Å. The region exposed last suffered from the decay of its neighbour even before its own exposure. While evaluating the data it was found that the crystal decay was accompanied by an increase in the c-dimension, from 564 to 572 Å. *Bottom row*: (left) a flat spatula, used for mounting T30S crystals; (right) a crystal of 300 × 50 × 30 μm, placed in a spatula similar to that shown on the left, was irradiated by a beam with a cross-section of 65 μm at ID19/APS. The first position was at the far end of the spatula (black arrow) and translated once decayed (total 4°, 20 rotations 0.2° each) to the middle of the spatula (white arrow). There it was exposed for 15 s. The intact crystal was transparent. Note that the intensities of the 'burns' of the crystals are proportional to the exposure time (see Plate Section).

and for the penetration of very large clusters (e.g. W30, W18, W17, see Table 18.2) into the crystals. It should be mentioned that a similar packing arrangement was obtained independently by molecular replacement, exploiting the image reconstructed from T50S (Yonath et al. 1998; Harms et al. 1999). However, although this solution yields reasonable scores (i.e. 93% correlation with an R-factor of 27% for the region > 60 Å; and 48% correlation with an R-factor 42% for the region 30–90 Å), it is comparable to several other solutions. The use of the envelope of a ribosomal particle from one bacterium for determining the packing diagram of crystals grown from the same particle, but from a different source, is based on the assumption that at low resolution the gross structural features of prokaryotic ribosomes are rather similar. However, at higher resolution the validity of such studies is questionable and the loss in scores with the advance of resolution may be accounted for by the difference between T50S and H50S (the latter contains 12–14 additional proteins).

Three additional packing arrangements have previously been suggested for H50S crystals. The first was based on MIR and anomalous phase information, initially determined at 15 Å (Schlünzen et al. 1995). The second (Roth et al. 1996), the third (Ban et al. 1998) and the current (Yonath et al. 1998) seem to be rather similar not only in their packing scheme but also because both originate from non-MIR information (direct methods and molecular replacement, respectively) at 30 Å, and so far both could be validated only to low resolution limits, 10–12 Å.

Apart from the general concern regarding the domination of the phases determined by molecular replacement or similar procedures, relying solely on phase information originating from very low resolution information may be rather misleading. Thus, the extension of procedures developed for systems possessing extensive levels of internal non-crystallographic symmetry and/or for the use of heavy atom derivatives showing (potentially or in reality) phasing power extending near-molecular resolution (Jack et al. 1975), is not always justified. This is so especially when the non-MIR information is the main source for the determination of the heavy atom sites (using difference Fourier), and when the phasing power of these heavy atoms extends only to low resolution (i.e. 10–14 Å). In these resolution ranges the contribution of the non-crystalline material (the solvent) to the structure factors may reach the same order of magnitude as the contribution of the crystalline material. Therefore, the diffraction may be strongly influenced by the contribution of the solvent. Consequently, the chances that the heavy atom sites so determined may indeed represent changes in the solvent rather than in the structured material are non-negligible, even when they display acceptable phasing statistics (e.g. Schlünzen et al. 1995) and/or apparent anomalous signals.

This risk is especially high for the combination of ribosomal crystals with multitungsten clusters. In recent studies we have found that a significant amount of W clusters, in quantities much higher than those directly incorporated in the phasing procedure (i.e. detected in difference Patterson or Fourier maps) remain within the crystal environment even after applying an extensive washing procedure (12 times during 40–50 h). Thus, the number of W atoms found in the washed crystals by inductively coupled plasma mass spectrometry as well as by atomic emission spectrometry, are equivalent to 18–20 clusters of W30 or to 25–27 clusters of W18 per ribosomal particle. Such large amounts of 'floating' W clusters are sufficient to generate measurable anomalous signals by contributing to the structure factors, as well as to introduce subtle non-isomorphism, which may not be detected as such in routinely treated diffraction data. It therefore remains to be seen whether the bases for these three structures (Schlünzen et al. 1995; Roth et al. 1996; Ban et al. 1998), i.e. the signals produced by these clusters at 12–15 Å, were a consequence of real derivatization, or of inherent or induced non-isomorphism or of the mere modification of the density of the crystal solvent.

In contrast to the marked tendency of large ribosomal subunits to crystallize, only one crystal form has so far been obtained from the *small ribosomal subunit* (Trakhanov et al. 1987; Yonath et al. 1988). For almost a decade this crystal form (T30S) yielded satisfactory data only to 12–15 Å (Schlünzen et al. 1995), although reflections were observed up to 7.3 Å. The low internal order of the crystals of the small ribosomal subunits was correlated with their marked instability, which reaches a higher level than that observed for the large subunits. For example, by exposing 70S ribosomes to a potent proteolytic mixture, the 50S subunits remained intact, whereas the 30S subunits were completely digested. Similarly, large differences in the integrity of the two subunits were observed while attempting the crystallization of functionally active 70S ribosomes, constructed from purified large and small subunits, then combined to form active ribosomes. It was found that the crystals obtained from these preparations consisted only of 50S subunits (Berkovitch-Yellin et al. 1992). This indicates that the self-affinity of the large subunits overcame their interactions with the small subunits to produce 70S particles not engaged in protein biosynthesis. It is noteworthy that at the end of this experiment, the supernatant of the crystallization drop did not contain intact small subunits, but instead their proteins and their fragmented RNA chain. Thus, while the large subunits crystallized, the small ones dissociated into their individual components.

Subtle modifications in the procedures of bacterial growth and crystal treatment led recently to diffraction of high quality at much higher resolution, 3–3.5 Å. Against all odds, the crystals of T30S are of a higher quality than any obtained from the large subunits, and display reasonable (though far from perfect) isomorphism (Table 18.1). Furthermore, data collected from T30S crystals under different conditions (stations and detectors): IP(MAR 345) at ID2-(ESRF), CCD at F1/CHESS and IP(MAR 300) at BW6/DESY, merged very well (R_{merge} 8–11%, compared to individual R_{merge} values of 6–9%). Large, medium-size and smaller metal compounds are being exploited for MIR and MAD phasing, leading typically to multisite binding which imposes extensive cross-verifications. This approach led to sufficient phasing power up to 6–6.3 Å resolution (limits currently dictated by the derivatized crystals) and allowed the construction of an interpretable electron density map (to be published).

18.2 Synchrotron radiation and crystal decay

The large unit cell dimensions and the extremely weak diffraction power do not permit any crystallographic preparative work, including crystal screening, to be performed on home generators, thus dictating absolute dependence of synchrotron radiation for all stages of structure determination. At ambient temperature, ribosomal crystals decay upon the first instance of irradiation. To overcome this unusual sensitivity, the concept of data collection at cryo-temperature was pioneered (Hope et al. 1989). Consequently, flash frozen ribosomal crystals can be irradiated by synchrotron radiation of moderate intensities (i.e. the bending magnet stations at DESY) at 85–95 K, with no observable radiation damage for periods sufficient for the collection of a complete data set at medium or low resolution (6–9 Å) and in exceptional cases even at 5 Å (Hope et al. 1989). However, even when the decay was not manifested in resolution loss, in many cases prominent damage has been observed at the outer resolution shells. This 'hidden' decay is detectable in the data quality, which becomes poorer with the progression of the irradiation, and is expressed in lower signal-to-noise ratios (e.g. from the original 9–10 to 2–3), higher R_{merge} values (e.g. from 0.06 to 0.17), fluctuations in the intraframe scaling factors between successive frames and frequent changes in the unit cell dimensions.

Because of the outstanding experimental demands of ribosomal crystallography, even the pre-freezing treatment must be performed with special care, although it introduces a higher level of complication. In attempts to maximize the useful resolution and to minimize the variation between crystals, procedures were developed for careful mounting of the crystals in a protective miniature double-layer thin glass spatula, and plunging them rapidly into liquid propane at its melting temperature (about 85 K). These double spatulas were found to be superior to the popular loops (Teng et al. 1994), although diffraction patterns of lower quality could also be collected from loop-mounted crystals. Thus, the double spatula accommodates the delicate properties of the ribosomal crystals: the extreme anisotropic morphology (one very thin dimension, of about 2–5 µm and two rather normal ones, of 200–400 µm), the notable softness and the high pliability, whereas crystals mounted in loops may float on the surface of the solvent bubble caught by the loop and bend around its concave shape. To address the hypothesis that the pre-freezing procedures may introduce apparent non-isomorphism, relatively large crystals were halved and each of its halves was flash-frozen separately. The two halves were positioned in the beam in a similar orientation, and data were collected around the cell axes of each part. Differences of magnitude similar to the experimental errors in the determination of all cell axes were observed.

It was firmly established that for collecting the higher resolution (2.5–6 Å) diffraction, high brilliance SR radiation (such as ID2 and ID13/ESRF or F1/CHESS) is essential. Unfortunately, such high brightness causes substantial radiation damage even at 15–95 K, within a period sufficient for the collection of a few rotations. The need to merge data from many crystals, coupled with the low level of isomorphism of the H50S crystals, makes it almost impossible to construct complete data sets even from native crystals. Since no improvement was obtained by the addition of free-radical absorbers, or by using a He stream (at 15–20 K) during data collection, a procedure was designed for the irradiation of the crystals in parts, using a beam of a cross-section smaller than the crystal. This procedure led to useful medium resolution data, but the high resolution shells are still rather problematic, since, as mentioned above, the decay is expressed not only in loss of resolution, but also in gradual, albeit substantial increase of the C unit cell axis (Fig. 18.3), which occurs mainly at the higher resolution shells.

18.3 Elucidation of phases

The assignment of phases to the structure factor amplitudes is the most crucial, albeit most complicated step in structure determination. Clearly, for large macromolecular assemblies that cannot be subdivided by internal symmetry, the magnitude and the complexity of phasing is greatly increased and involves outstanding experimental demands.

18.3.1 *The derivatization agents: soaking experiments*

The commonly used methods for phase determination in biological crystallography, MIR, SIR and MAD, require the preparation of derivatives, usually by introducing electron-dense compounds into the crystalline lattice at a limited number of distinct locations while keeping the crystal parameters isomorphous with those of the native molecule. The most common procedure is to soak the crystals in solutions containing heavy atom compounds. Single heavy atoms have yielded useful high resolution phases for several large complexes. Among these are the viruses [e.g. Jack *et al.* (1995), Rossmann (1995)], the 371 K ATPase (Abrahams *et al.* 1994) and the 250 K tRNA^{phe} synthetase and its complex with its cognate tRNA (Goldgur *et al.* 1997).

Owing to the enormous size of the ribosome, a large number of sites is required for generating accurately measured signals. Such multisite derivatives should be extremely difficult to locate in the unit cell. Alternatively, advantage can be taken of compact and dense compounds containing several heavy atoms linked directly to each other, or arranged in close proximity. However, in contrast to the availability of numerous single-atom agents, there are only a few stable water-soluble polymetallic compounds that may be suitable for the derivatization. Examples are heteropolyanions and multicoordination compounds usable for soaking experiments and monofunctional reagents of dense metal clusters, designed for covalent binding at specific sites prior to the crystallization (Table 18.2 and in Thygessen *et al.* 1995).

With the increase of resolution of the ribosomal crystals, medium-size compounds were tested. Among these is TAMM, which proved suitable for phasing data from crystals of rather large particles, such as the photosynthetic reaction centre (Deisenhofer *et al.* 1984), the nucleosome-core-particle (Luger *et al.* 1997), an iodotype–anti-idiotype complex (Bentley *et al.* 1990) and glutathione transferase (Reinemer *et al.* 1991), but could not be exploited in ribosomal crystallography either because of low solubility (H50S) or because it introduces severe non-isomorphism (T30S and T50S). The situation with PIP, which was also used for phasing in some of the studies mentioned above (e.g. Luger et al. 1997) is unclear. It obviously did not introduce substantial non-isomorphism, but at the same time its phasing power was found to be lower than that obtained from smaller compounds, showing presumably that it decomposes in an uncontrolled fashion during the course of the experiment.

Ta_6Cl_{14} has recently become rather popular in macromolecular crystallography as it was shown to phase at different resolution ranges over a wide pH range. Thus, it was used for structure determination of ribulose-1,5-phosphate carboxylase/oxygenase (rubisco) and transketolase at 5.5 Å (Schneider and Lindquist 1994), as well as of the proteosome at 3.4 Å (Löwe et al. 1995) and at atomic resolution (Knäblein et al. 1997). This compound was found to be useful to various extents, for H50S, T50S and T30S crystal forms (see above and in Yonath et al. 1998).

18.3.2 Quantitative attachment of heavy atom compounds

Quantitative attachment to predetermined sites prior to crystallization ensures high occupancy. This approach requires complicated and time-consuming procedures, but is bound to yield indispensable information not only for phasing but also at later stages of map interpretation. Examples of compounds that may be bound are the clusters of undecatungsten (Wei et al. 1997), undecagold and tetra-iridium (Jahn 1989a,b), shown in Table 18.2.

The feasibility of phasing by specifically bound heavy atoms has been proven in several cases, including the nucleosome-core-particle (Luger et al. 1997), as well as for low resolution derivatization of B50S, performed with a monofunctional reagent of an undecagold cluster (Weinstein et al. 1989; Bartels et al. 1995).

The studies on the structure of the nucleosome-core-particle are illuminating. This 206 kDa particle consists of 146 base pairs of DNA wrapped around an internal core, composed of an octamer made of two copies of four histones. Since the fine characteristics of the structure of each individual nucleosome-core is dictated by the sequence of the incorporated DNA, which varies as a function of its position on the genome, crystals obtained from naturally occurring nucleosome-core-particles diffract at best to only 7 Å (Richmond et al. 1984). In order to decrease the natural crystal variability, a semi-artificial nucleosome-core-particle was designed, consisting of genetically produced wild-type or mutated histones together with a fragment of 146 base pairs, synthesized with a defined sequence. Thus, in addition to the provision of a homogenous population, the use of recombinant nucleosome-cores facilitated the insertion of exposed cysteines at selected sites on the surface of the histone proteins.

This elegant and logical approach cannot be adopted for the derivatization of the ribosomes, since so far all fully reconstructed particles did not yield crystals that diffracted well. This was rather unexpected since functionally active ribosomes can be reconstituted *in vitro* from isolated ribosomal components, indicating that the information required to obtain the active quaternary structure resides within them. It is conceivable that the *in vitro* reconstitution pathways lead to slight deviations from the natural conformation, since the conditions under which the reconstitution is performed *in vitro* are dramatically different from the physiological events. For example, the *in vitro* assembly of the ribosome requires 90 min, whereas the *in vivo* process is completed within 3 min. These slight conformational differences can be tolerated and induced to form the correct active sites by their substrates, but are sufficiently severe to prohibit quality crystallization. Therefore the pre-crystallization heavy atom binding is limited either to the available sites on the native particles (there is one exposed cysteine on T30S and none on H50S) or to the genetic creation of potential binding sites on selected ribosomal proteins, those that can be quantitatively and reversibly detached from the ribosome under mild conditions [there are four such proteins in H50S and eight in T30S (Sagi *et al.* 1995)].

An example is in the studies exploiting a mutant of *Bacillus stearothermophilus*, lacking one ribosomal protein, BL11. The mutated 70S ribosomes and 50S subunits formed three-dimensional crystals and two-dimensional sheets under the conditions used for crystallizing the wild-type particles. Thus, the absence of BL11 does not cause gross conformational changes in the ribosomal particles, and it is not crucial for the interparticle interactions forming the crystallographic network. Protein BL11 is a ribosomal component that undergoes conformational changes upon isolation from the particle. Thus, when incorporated into the ribosome, its single free –SH group is exposed and chemically reactive. In isolation, however, this group is not reactive unless the protein is denatured. The gold cluster was quantitatively bound to isolated BL11 under denaturing conditions, and the modified protein was incorporated in the core particles, to form a specific and quantitative derivative. To place the size of the clusters in perspective, we note that the molecular weight of the undecagold cluster (6200 Da) is more than a third of the molecular weight of BL11 (15 500 Da) and its diameter, 22 Å, approaches three-fourths of the end-to-end dimension of lysozyme.

An obvious target for attachment of heavy atom clusters to ribosomal particles is tRNA. To facilitate co-crystallization of tRNA with ribosomal particles, we have determined conditions for stoichiometric binding of tRNA to ribosomal particles. One of these complexes mimics a defined state in the process of protein biosynthesis, composed of the 70S ribosome, a short segment of mRNA and two molecules of charged tRNA (Hansen *et al.* 1990). The undecagold cluster was covalently attached to tRNA^{phe} at base 47. The modified tRNA molecule binds to the ribosome and can be aminoacylated by its cognate synthetase. Furthermore, it seems that the undecagold cluster did not introduce large perturbations into the ribosomal structure, as crystals of the complex containing the modified tRNA^{phe} diffracted to resolution comparable to that obtained from the complex containing the native tRNA molecule.

In conducting proper isomorphous replacement experiments, the low level of isomorphism of H50S crystals dictated the exposure of many crystals in order to construct complete either native or derivatized data sets. A larger number is needed for the selection of rather isomorphous pairs. Thus, 27 data sets were constructed from data collected from 76 native and derivatized crystals, until four of them were found to exhibit reasonable isomorphism at about 7 Å. As described above, similar studies performed on T30S progressed somewhat more smoothly. In this case, large and medium-size compounds alongside smaller salts were exploited. In both cases, some of the difference Patterson maps were too complicated to be interpreted in the conventional way. Extensive alternations between difference Patterson and Fourier maps were performed to allow heavy atom site location. This sophisticated approach was later combined with careful analysis and led to preliminary indications for sufficient phasing power.

18.3.3 Phasing by MAD

MAD phasing should eliminate the dependence on isomorphism, provided that all the data can be collected at a few wavelengths from a single crystal. Owing to the severe radiation decay of the ribosomal crystals, this requirement cannot be fulfilled, but it is anticipated that even partial high resolution phase information obtained from individual crystals should be more useful than that expected to be obtained from difference Patterson maps. The suitability of this method for ribosomal crystallography is currently being assessed and a word of caution is due, since the anticipated anomalous signals may be of the same order of magnitude as the changes in structure factors induced by the decay of the crystals. MAD studies exploiting selenium recently gained a lot of popularity in protein crystallography. For obtaining selenated halophilic ribosomal particles, a methionine-dependent strain was constructed (M. Mevarech, private communication). The 50S subunits of this strain yield crystals that may be of a higher quality than those grown from the problematic H50S wild-type (Fig. 18.4). The exact numbers of methionines in T30S and H50S are still to be determined, since only a part of the sequences of the ribosomal proteins from these sources is known. Their estimated numbers (25 and 55, respectively) may not be sufficient to provide measurable signals, therefore efforts are being made to increase their amounts by genetic techniques (Franceschi *et al.* 1993).

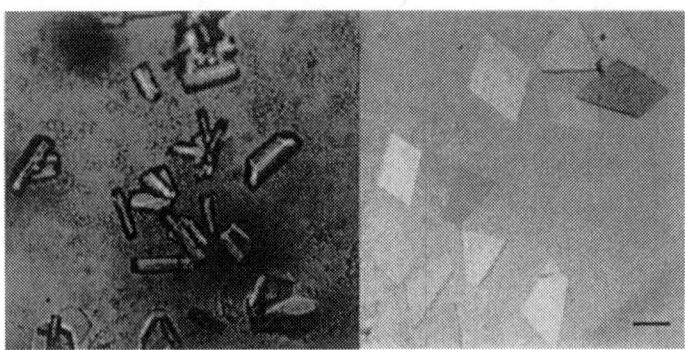

Fig. 18.4. *Left*: Crystals of large ribosomal subunits from strain H2 grown under conditions somewhat milder than those used for H50S (1.45 instead of 1.6 M KCl); *right*: those of H50S. Bars: 0.2 mm (see Plate Section).

18.3.4 Molecular replacement

Molecular replacement is based on manoeuvring the positioning of a known model in the unit cell of the unknown structure until the calculated structure factors match best the observed ones. The rotation–translation searches have been performed, initially by exploiting models obtained from electron micrographs of tilt series of negatively stained crystalline arrays of 70S ribosomes and 50S subunits from *B. stearothermophilus* (Yonath *et al.* 1987; Arad *et al.* 1987). These were used despite their rather low (28–40 Å) resolution, since key functional features, such as the site of protein synthesis and the path of the nascent chain, were observed in them for the first time.

Higher resolution images (16–25 Å) have recently been obtained by angular reconstructions of single ribosomal particles embedded in vitreous ice. As similar images, resembling the views observed by traditional electron microscopy and containing the features revealed in the crystalline arrays, were obtained independently by two groups (Stark *et al.* 1995; Frank *et al.* 1995), they seem to be reliable. To enhance the chances for elucidating the correct packing arrangements of the ribosomal crystals, particles of the same preparations which yielded the best T50S and T70S crystals (Table 18.1) have been subjected to reconstruction at 18 and 26 Å resolution, respectively (Harms *et al.* 1999). In both cases, a unique solution was obtained in the rotation–translation studies showing no collisions or short contacts, with $R_{\text{merge}}(I)$ of 42% and 46% and correlations of 75% and 79%, respectively (Figs 18.5 and 18.6).

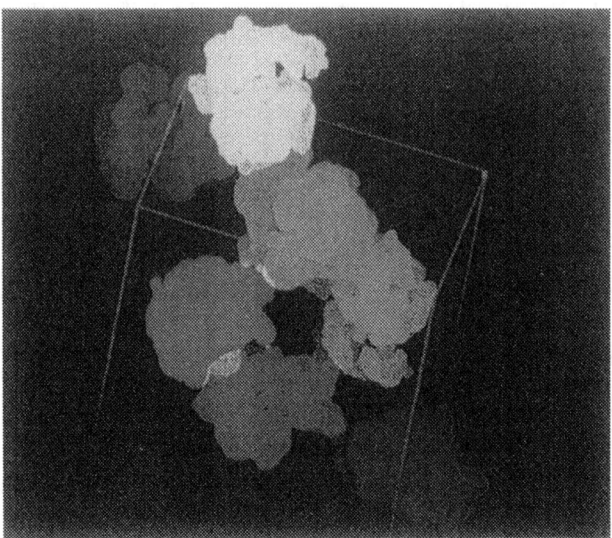

Fig. 18.5. The packing diagram of the crystallographic unit cell of whole ribosome (T70S), assembled by positioning the 26 Å electron microscope model in the crystallographic unit cell according to the most prominent result of the molecular replacement search, and applying the eight symmetry operations. Data were collected at BW6/DESY to 17 Å resolution (see Plate Section).

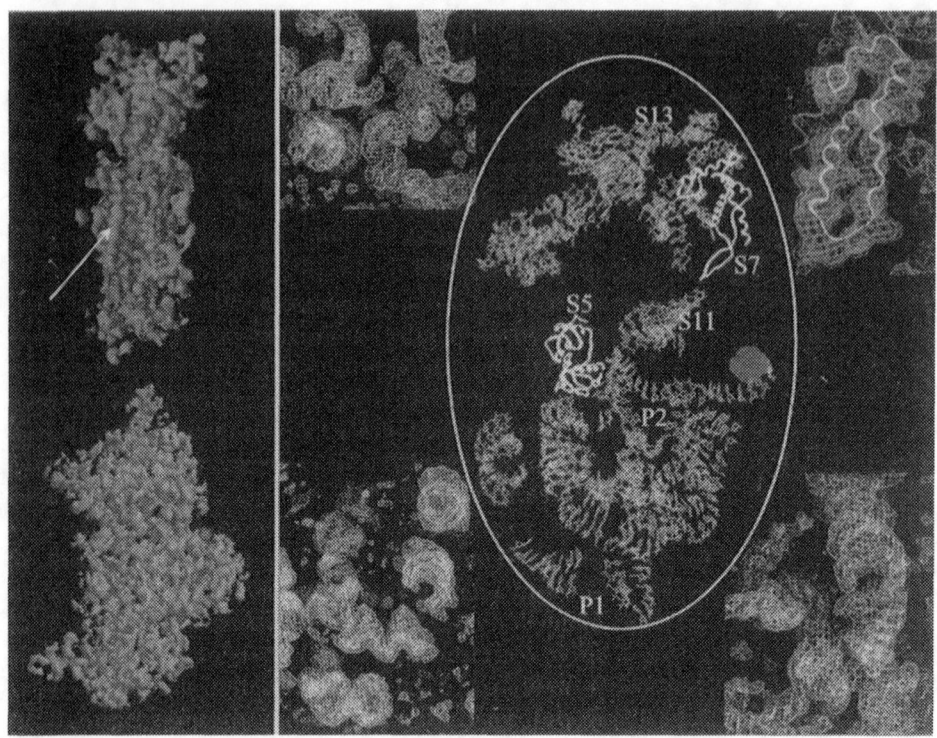

Fig. 18.6. *Left*: Two orthogonal views of the overall structure of the small subunit, as extracted from the 4.5 Å map. The arrow points to an exceptionally long dense region, suitable for hosting a double-helical RNA chain that may be interpreted as helix 44 in the model of the 16S RNA (Müller and Brimacombe 1997). For orientation, the location of the tail of the cDNA complementary to the 3' end of the 16S RNA is shown in red on both sides (here and on the *Right* side). *Right*: Part of the 16S RNA chain so far traced in the 4.5 Å map of T30S is shown within the white ellipse. The position of the centre of mass of the TAMM molecules that were bound to the mRNA analog is shown as a red sphere (artificially enlarged). The locations of proteins TS5, TS7 are represented by their backbone structure, as determined crystallographically. Two tentative locations for protein TS15 are marked P1 and P2. The positions of the exposed cysteines of proteins TS11 and TS13 are marked by their numbers (see Plate Section).

Electron density maps were constructed from the observed crystallographic structure factor amplitudes and calculated phases, as well as from the observed amplitudes and the combination of the rotation search phases with those determined by SIRAS from crystals soaked in a solution of Ta_6Br_{14}, performed since a reasonable correlation was found between them (Fig. 18.6). An examination of the relation between the packing arrangement and the positions of the two most prominent Ta_6Br_{14} sites showed that one of them is located between two particles and the second in a small 'nest' within the particle (Fig. 18.6).

18.3.5 Isolated ribosomal components and in situ complexes

The immense difficulties anticipated (and encountered) in the determination of the structure of intact ribosomal particles has led to a parallel approach, focusing on isolated ribosomal components. For over a decade the yield of this approach was rather poor, but the substitution of the E. coli ribosome, which used to be the favoured research object, by more robust particles (e.g. from thermophilic bacteria), the employment of genetic techniques and the introduction of three-dimensional NMR spectroscopy, have resulted in major accomplishments in the determination of the molecular structure of an impressive number of ribosomal proteins (reviewed in Liljas and Al-Karadaghi 1997; Hosaka et al. 1997; Wimberly et al. 1997) as well as RNA fragments (Betzel et al. 1994; Puglisi et al. 1997; Dallas and Moore 1997; Correll et al. 1997). It remains to be seen whether the structures determined in isolation bear resemblance to the *in situ* situation. It is widely assumed that the inherent conformation is maintained despite changes in the environment. However, recent results challenge this assumption, as a significant discrepancy has been observed between the structure of a crystalline ribosomal protein and its NMR solution (Chlemons et al. 1998). Clearly, some of the structures of individual ribosomal components should not differ from their *in situ* state. These should be instrumental for the interpretation of the electron density map of the entire ribosomal particles, as they may provide useful markers.

Conformational readjustments have been predicted to be associated with the creation of the *in situ* microenvironment within the ribosome. It was widely assumed that the ribosomal components which undergo the main conformational changes are the rRNA molecules. The observation that flexible loops in ribosomal proteins become ordered upon binding to rRNA, indicates that changes in the protein conformation may also be essential for the assembly of the ribosome. At the same time it is anticipated that, provided the association with the *in situ* closest neighbours is maintained, isolated internal ribosomal complexes are likely to keep their natural conformations in solution. Assuming that some of the structures of the isolated complexes and single components indeed reflect the *in situ* situation, the crystal structures of these complexes may provide phase information in molecular replacement studies. They should also be instrumental in the interpretation of the electron difference maps, providing 'flags' and 'markers'.

18.4 Discussion, conclusions and perspectives

This is an exciting time in biological crystallography as projects that were considered beyond our reach until not too long ago, are currently being carried out. One of the most striking examples is ribosomal crystallography, which underwent dramatic progress since the submission of this chapter. As mentioned above, these studies required the pioneering of revolutionary concepts and sophisticated techniques, not only because ribosomes are giant assemblies with no internal symmetry, but also because their crystals are extremely delicate, radiation sensitive, of rather low isomorphism and frequent polymorphism (Makowski et al. 1997; von Böhlen et al. 1991; Ban et al. 1999). Nevertheless, the approaches developed to minimize the harm caused by these negative properties, together with the increasing availability of bright synchrotron radiation beams coupled with reliable area and CCD detectors, led to spectacular results. Among those are (in decreasing order of resolution) a 7.5 Å map of functional complexes of 70S ribosome (Cate et al. 1999), a 5 Å map of the halophilic large subunit (Ban et al. 1999), a 5.5 Å map of the thermophilic small subunit (Clemons et al. 1999), and a 4.5 Å map of the small subunit, containing either chemical markers or functional analogs (Tocilj et al. 1999).

The 4.5 Å study of the small subunits is of great importance because it shows clearly that the border of 5 Å can be crossed. This is a major breakthrough since below 5 Å resolution complete data sets can be collected from single crystals, whereas the bright beam that is essential for collecting the higher resolution shells causes severe radiation decay. Consequently the data are more problematic and several crystals are required in order to produce complete data sets, a task that was found to be extremely demanding because of the severe non-isomorphism of some of the ribosomal crystal systems (i.e. H50S).

The strategy that proved suitable for phasing of all crystal systems is based on the determination of an initial phase set at very low resolution, followed by its extension by experimental and/or computational methods. For this aim, molecular replacement exploiting cryo EM reconstructions was performed successfully for H50S (Ban et al. 1999), T50S (Yonath and Franceschi 1998) and for the whole ribosome from *T. thermophilus*, T70S (Cate et al. 1999; Harms et al. 1999): however, for T30S these attempts were found not suitable, presumably because of the high conformational variability (Stark et al. 1995; Frank et al. 1995; Gabashvili et al. 1999; Harms et al. 1999; Wang et al. 1999). In this case the initial set of phases was obtained by using heavy atom clusters (Clemons et al. 1999; Tocilj et al. 1999).

So far, all attempts at interpreting the ribosomal electron density maps are based fully or partially on placements of structures of ribosomal components or of similar molecules, guided by the available non-crystallographic structural information obtained by electron microscopy, neutron scattering, footprinting, modelling and biochemical experiments [for a review, see Müller and Brimacombe (1997)]. The building blocks used for tracing the map were constructed from known RNA motifs or from the ribosomal proteins whose structures have been determined crystallographically or by NMR.

Placement of structures determined at high resolution in medium resolution maps requires special concern, as considerable uncertainties are associated with such attempts. In the case of the ribosome there are further potential ambiguities, since most of its individual components are built from common motifs (Liljas and Al-Karadaghi 1997; Ramakrishnan and White 1998). Also, at low or medium resolution, molecular mimicries (Nyborg et al. 1996) may mislead the differentiation between proteins and RNA regions. In addition, most of the ribosomal components possess non-negligible conformational variability that may lead to misinterpretations, since their structures are likely to be influenced by the *in situ* ribosomal environment.

An exceptional case is the 4.5 Å map of small subunit (Tocilj *et al.* 1999) since heavy atom markers were used for its unbiased interpretation. These allowed independent positioning of ribosomal components as they attached to known chemically active ribosomal moieties or to carriers with a high affinity to specific locations in the 30S subunit, such as antibiotics or DNA oligomers, complementary to exposed single stranded RNA. Thus, post-crystallization activation by controlled heating led to higher proportions of satisfactorily diffracting crystals and enabled almost quantitative binding of compounds participating in protein biosynthesis or their analogs. In this way, close to stoichiometric hybridization with mercurated cDNA oligomers was achieved, despite their large-size, which may reach 70 Å in length.

Targeting the 16S RNA region, where mRNA docks to allow the formation of the initiation complex by a mercurated mRNA analog, led to the characterization of its vicinity (Weinstein *et al.* 1999; Auerbach *et al.* 1999; Bartels *et al.* 1999; Bashan *et al.* 1999; Tocilj *et al.* 1999). This region of the 16S chain is known to be rather flexible and may adopt several conformations and it is likely that the high quality diffraction obtained from the crystals derivatized by this oligomer results from stabilization of the flexible 3′ arm of the 16S RNA in a fashion similar to its binding to mRNA.

Similarly, heavy atom clusters (a tetrairidium and a tetramercury compound) covalently bound to the exposed sulphydryls of two ribosomal proteins, S11 and S13, were used to reveal their position in difference Fourier maps (Weinstein *et al.* 1999; Auerbach *et al.* 1999). Interestingly, the location of one of the two proteins, S13, found this way is in reasonable agreement with that suggested by neutron scattering (Moore *et al.* 1985), immunoelectron microscopy (Stöffler and Stöffler-Meilicke, 1986) and modelling based on cross-linking and enzymatic data (Müller and Brimacombe 1997). For protein S11 the situation is somewhat different. Its position in the electron density map is in accord with that proposed by electron microscopy and by modelling, but differs from that obtained by neutron scattering, by a distance larger than the expected diameter of this protein.

Despite the difficulties with molecular replacement, the overall structure of the small ribosomal subunit, as seen at 4.5 Å in the map, is remarkably similar to most of the electron microscopy reconstructions of this particle at its functionally active conformation. It shows the recognizable small subunit features, including the traditional division into three main parts: a rather large head, a short neck and a bulky lower body (Stark *et al.* 1995; Frank *et al.* 1995; Gabashvili *et al.* 1999). It contains elongated dense features as well as lower-density globular regions. In the latter, proteins S5 and S7 were placed visually. Suitable host regions for the fold of protein S15 were detected in several positions, all at a significant distance from the location of this protein in the neutron scattering map (Fig. 18.6).

The level of detail of some of the ribosomal electron-density maps, the ability to insert specific markers and the availability of crystals of functional complexes that diffract to rather high resolution (3–3.5 Å) indicate that the elucidation of the molecular structure of the ribosomes is no longer so far away.

Acknowledgements

We express our exceptional gratitude to the late Prof. H.G. Wittmann with whom these studies were initiated, to Drs M. Pope, W. Preetz and W. Jahn who gave us generous gifts of heavy atom clusters and to the team working with us on this project. The studies presented here were performed at the Weizmann Institute, Rehovot, the Max-Planck Research Unit in Hamburg and the Max-Planck Institute for Molecular Genetics in Berlin. Data were collected at EMBL and MPG lines at DESY; F1/CHESS, Ithaca, NY; ID2, ID13, D2AM/ESRF/Grenoble; ID19/APS/ARGONNE,IL; PF/KEK, Tsukuba, Japan. Support was provided by the Max-Planck Society, the US National Institute of Health (NIH GM 34360), the German Ministry for Science and Technology (BMBF 05-641EA) and the Kimmelman Center for Macromolecular Assembly at the Weizmann Institute. A.Y. holds the Martin S. Kimmel Professorial Chair.

Abbreviations

tRNA, rRNA and mRNA stand for transfer, ribosomal and messenger RNA, respectively. r-proteins are ribosomal proteins. 70S, 50S, 30S: the whole ribosome and its two subunits. A letter as a prefix represents the bacterial source: E: *E. coli*; B: *Bacillus stearothermophilus*; T: *Thermus thermophilus*; H: *Haloarcula marismortui*. 5S RNA, the shortest RNA chain in the large ribosomal subunit.

References

Abrahams, JP, Leslie, AGW, Lutter, R and Walker, JE (1994). *Nature*, **370**, 621–628.
Arad, T, Piefke, J, Weinstein, S, Gewitz, HS, Yonath, A and Wittmann, HG (1987) *Biochimie*, **69**, 1001–1006.
Auerbach, T, Pioletti, M, Avila, H, Anagnostopoulos, K, Weinstein S, Franceschi, F and Yonath, A (1999). *Biomolecular Structure and Dynamics* (in press).
Ban, N, Freeborn, B, Nissen, P, Penczek, P, Grassucci, RA, Sweet, R, Frank, F, Moore, P and Steitz, T (1998). *Cell*, **93**, 1105–1115.
Ban, N, Nissen, P, Capel, M, Moore P and Steitz, T (1999). *Nature*, **400**, 841–847.
Bartels, H, Bennett, WS, Hansen, HAS, Eisenstein, M, Weinstein, S, Müssig, J, Volkmann, N, Schlünzen, F, Agmon, I, Franceschi, F and Yonath, A (1995). *Biopolymers*, **37**, 411–419.
Bartels, H, Glühmann, M, Janell, D, Schlünzen, F, Tocilj, A, Bashan, A, Levin, I, Hansen, HAS, Harms, J, Kessler, M, Pioletti, M, Auerbach, T, Agmon, I, Avila, H, Simitsopoulou, M, Weinstein, S, Peretz, M, Bennett, WS, Franceschi, F and Yonath, A (1999). *Cellular and Molecular Biology* (in press).
Bashan, A, Pioletti, M, Bartels, H, Janell, D, Schlünzen, F, Glühmann, M, Levin, I, Harms, J, Hansen, HAS, Tocilj, A, Auerbach, T, Avila, H, Anagnostopoulos, K, Simitsopoulou, M, Peretz, M, Bennett, WS, Agmon, I, Kessler, M, Weinstein, S, Franceschi, F and Yonath, A (1999). *ASM publications* (in press).

Bentley, GA, Boulot, G, Riottot, MM and Poljak, RJ (1990). *Nature*, **348**, 254–257.
Berkovitch-Yellin, Z, Bennett, WS and Yonath, A (1992). *Critical Reviews in Biochemistry and Molecular Biology*, **27**, 403–444.
Betzel, C, Lorenz, S, Fürste, JP, Bald, R, Zhang, M, Schneider, R, Wilson, KS and Erdmann, VA (1994). *FEBS Letters*, **351**, 159–164.
von Böhlen, K, Makowski, I, Hansen, HAS, Bartels, H, Berkovitch-Yellin, Z, Zaytzev-Bashan, A, Meyer, S, Paulke, C, Franceschi, F and Yonath, A (1991). *Journal of Molecular Biology*, **222**, 11–15.
Cate, JH, Yusupov, MM, Yusupova, GZ, Earnest, TN and Noller, HF (1999). *Science*, **285**, 2095–2104.
Clemons, WM, Davies, C, White, S and Ramakrishnan, V (1998). *Structure*, **6**, 429–438.
Clemons, WM, May, JLC, Wimberly, BT, McCutcheon, JP, Capel, MS and Ramakrishnan, V (1999). *Nature*, **400**, 833–840.
Correll, CC, Freeborn, B, Moore, PB and Steitz, TA (1997). *Cell*, **91**, 705–712.
Dallas, A and Moore, PB (1997). *Structure*, **5**, 1639–1653.
Deisenhofer, J, Epp, O, Miki, K, Huber, R and Michel, H (1984). *Journal of Molecular Biology*, **180**, 385–398.
Frank, J, Zhu, J, Penczek, P, Li, Y, Srivastava, S, Verschoor, A, Radamacher, M, Grassucci, R, Lata, AK and Agrawal, RK (1995). *Nature*, **376**, 441–444.
Franceschi, F, Weinstein, S, Evers, U, Arndt, E, Jahn, J, Hansen, HAS, von Böhlen, K, Berkovitch-Yellin, Z, Eisenstein, M, Agmon, I, Thygesen, J, Volkmann, N, Bartels, H, Schlünzen, F, Bashan, A, Sharon, R, Levin, I, Dribin, A, Sagi, I, Choli-Papadopoulou, T, Tsiboly, P Kryger, G, Bennett WS and Yonath, A (1993). *The Translational Apparatus*, ed. Nierhaus, K, Plenum Press. pp. 397–406.
Gabashivili, IS, Agrawal, RK, Grassucci, R and Frank, J (1999). *J. Mol. Biol.*, **286**, 1285–1291.
Goldgur, Y, Mosyak, L, Reshetnikova, L, Ankilova, V, Lavrik, O, Khodyreva, S and Safro, M (1997). *Structure*, **5**, 59–68.
Hansen, HAS, Volkmann, N, Piefke, J, Glotz, C, Weinstein, S, Makowski, I, Meyer, S, Wittmann, HG and Yonath, A (1990). *Biochemical and Biophysical Acta*, **1050**, 1–5.
Harms, J, Tocilj, A, Levin, I, Agmon, I, Kölln, I, Stark, H, van Heel, M, Cuff, M, Schlünzen, F, Bashan, A, Franceschi, F and Yonath, A (1999). *Structure*, **7**, 931–941.
Hope, H, Frolow, F, von Böhlen, K, Makowski, I, Kratky, C, Halfon Y, Danz, H, Webster, P, Bartels, K, Wittmann, HG and Yonath, A (1989). *Acta Crystallographica*, **B45**, 190–199.
Hosaka, H, Nakagawa, A, Tanaka, I, Harada, N, Sano, K, Kimura, M, Yao, M and Wakatsuki, S (1997). *Structure*, **5**, 1199–1208.
Jack, A, Harrison, SC and Crowther, RA (1975). *Journal of Molecular Biology*, **97**, 163–172.
Jahn, W (1989a). *Zeitschrift für Naturforschung*, **44b**, 79–82.
Jahn, W (1989b). *Zeitschrift für Naturforschung*, **44b**, 1313–1322.
Knäblein, J, Neuefeind, T, Schneider, F, Bergner, A, Messerschmidt, A, Löwe, J, Steipe, B and Huber, R (1997). *Journal of Molecular Biology*, **270**, 1–7.
Liljas, A and Al-Karadaghi, S (1997). *Nature Structural Biology*, **4**, 767–771.
Löwe, J, Stock, D, Jap, B, Zwickl, P, Baumeister, W and Huber, R (1995). *Science*, **268**, 533–539.
Luger, K, Mäder, AW, Richmond, RK, Sargent, DF and Richmond, TJ (1997). *Nature*, **389**, 251–260.
Makowski, I, Frolow, F, Saper, MA, Shoham, M, Wittmann, HG and Yonath, A (1987) *J. Mol. Biol.*, **193**, 819–821.

Moore, PB, Capel, MS, Kjeldgaard, M and Engelman, DM (1985). in *Structure, Function & Genetics of Ribosomes*, (Hardesty, B and Kramer, G, eds) Springer Verlag, Heidelberg & NY pp. 87–100.
Müller, F and Brimacombe, R (1997). *J. Mol. Biol.*, **271**, 524–544.
Nyborg, J, Nissen, P, Kjeldgaard, M, Thirup, S, Polekhina, G, Clark, BFC and Reshetnikova, L (1996). *Trends in Biochemical Sciences*, **21**, 81–82.
Puglisi, EV, Green, R, Noller, HF and Puglisi, JD (1997). *Nature Structural Biology*, **4**, 775–778.
Ramakrishnan, V and White, SW (1998). *Trends in Biochemical Sciences*, **3**, 208–212.
Reinemer, P, Dirr, HW, Ladenstein, R, Schäffer, J, Gallay, O and Huber, R (1991). *EMBO Journal*, **10**, 1997–2005.
Richmond, TJ, Finch, JT, Rushton, B, Rhodes, D and Klug, A (1984). *Nature*, **311**, 532–537.
Rossmann, MG (1995). *Current Opinion in Structural Biology*, **5**, 650–659.
Roth, M, Pebay-Peyroula, E, Bashan, A, Berkovitch-Yellin, Z, Agmon, I, Franceschi, F, Lewit-Bentley, A and Yonath, A. (1996). *Biological Structure and Dynamics, Proceedings of the 9th Conversation*, Sarma, RH and Sarma, MH, eds, p. 15.
Sagi, I, Weinrich, V, Levin, I, Glotz, C, Laschever, M, Melamud, M, Franceschi, F, Weinstein, S and Yonath, A (1995). *Biophys. Chem.*, **55**, 31–41.
Schlünzen, F, Hansen, HAS, Thygesen, J, Bennett, WS, Volkmann, N, Levin, I, Harms, J, Bartels, H, Zaytzev-Bashan, A, Berkovitch-Yellin, Z, Sagi, I, Franceschi, F, Krumbholz, S, Geva, M, Weinstein, S, Agmon, I, Böddeker, N, Morlang, S, Sharon, R, Dribin, A, Maltz, E, Peretz, M, Weinrich, V and Yonath, A (1995). *Journal of Biochemistry and Cell Biology*, **73**, 739–749.
Schlünzen, F, Kölln, I, Janell, D, Glühmann, M, Levin, I, Bashan, A, Harms, J, Bartels, H, Auerbach, T, Pioletti, T, Avila, H, Anagnostopoulos, K, Hansen, HAS, Bennett, WS, Agmon, I, Kessler, M, Tocilj, A, Peretz, M, Weinstein, S, Franceschi, F and Yonath, A (1999). *J. Syn. Radiation*, **6**, 928–941.
Schneider, G and Lindquist, Y (1994). *Acta Crystallographica*, **D50**, 186–192.
Stark, H, Mueller, F, Orlova, EV, Schatz, M, Dube, P, Erdemir, T, Zenin, F, Brimacombe, R and Van Heel, M (1995). *Structure*, **3**, 815–914.
Stöffler, G and Stöffler-Meilicke, M (1986). In *Structure, Function and Genetics of Ribosomes*, (Hardesty, B and Kramer, G, eds) Springer Verlag, Heidelberg and NY. pp.28–46.
Tocilj, A, Schlünzen, F, Janell, D, Glühmann, M, Hansen, HAS, Harms, J, Bashan, A, Bartels, H, Agmon, I, Franceschi, F and Yonath, A (1999). *PNAS* (in press).
Trakhanov, SD, Yusupove, MM, Agalarov, SC, Garber, MB, Ryazantsev, SN, Tichenko, SV and Shirokov, VA (1987). *FEBS Letters*, **220**, 319–323.
Teng, TY, Schildkamp, W, Dolmer, P and Moffat, K (1994). *Journal of Applied Crystallography*, **27**, 133–137.
Thygesen, J, Weinstein, S, Franceschi, F and Yonath, A (1996). *Structure*, **4**, 513.
Wang, R, Alexander, RW, VanLoock, M, Vladimirov, S, Bukhtiyarov, Y, Harvey, SC and Cooperman, BS (1999). *J. Mol. Biol.*, **286**, 521–40.
Wei, X, Dickman, MH and Pope MT (1997). *Inorg. Chem.*, **36**, 130–131.
Weinstein, S, Jahn, W, Hansen, HAS, Wittmann, HG and Yonath, A (1989). *Journal of Biological Chemistry*, **264**, 19 138–19 142.
Weinstein, S, Jahn, W, Glotz, C, Schlünzen, F, Levin, I, Janell, D, Harms, J, Kölln, I, Hansen, HAS, Glühmann, M, Bennett, WS, Bartels, H, Bashan, A, Agmon, I, Kessler, M, Pioletti, M, Avila, H, Anagnostopoulos, K, Peretz, M, Auerbach, T, Franceschi, F and Yonath, A (1999). *J. Struct. Biol.*, **127**, 141–151.

Wilson, KE and Noller, HF (1998). *Cell*, **92**, 337–349.
Wimberly, BT, White, SW and Ramakrishnan, V (1997). *Structure*, **5**, 1187–1198.
Yonath, A and Francschi, F (1998). *Structure*, **6**, 678–684.
Yonath, A, Harms, J, Hansen, HAS, Bashan, A, Schlünzen, F, Levin, I, Kölln, I, Tocilj, A, Agmon, I, Peretz, M, Bartels, H, Bennett, WS, Krumbholz, S, Janell, D, Weinstein, S, Auerbach, T, Piolleti, M, Morlang, S, Bhanumoorthy, P and Franceschi, F (1998). *Acta Crystallographica*, **A54**, 945–955.
Yonath, A, Leonard, KR and Wittmann, HG (1987). *Science*, **236**, 813–817.
Yonath, A, Glotz, C, Gewitz, HS, Bartels, S, von Böhlen, K, Makowski, I and Wittmann, HG (1988). *Journal of Molecular Biology*, **203**, 831–834.
Yonath, A, Harms, J, Hansen, HAS, Bashan, A, Peretz, M, Bartels, H, Schlünzen, F, Kölln, I, Bennett, WS, Levin, I, Krumbholz, S, Tocilj, A, Weinstein, S, Agmon, I, Piolleti, M, Janell, D, Auerbach, T and Franceschi, F (1998). *Acta Cryst.*, **54A**, 945–955.

19

Nuclear spin contrast variation studies on macromolecular complexes

Heinrich B. Stuhrmann

19.1 From isotopic substitution to nuclear spin polarization

Isotopic substitution of the hydrogen isotope ^{1}H by its heavier isotope ^{2}H (deuterium) is used in most experiments of neutron scattering from hydrogenous matter. Polarized neutron scattering from polarized nuclear spins adds another dimension of contrast variation.

19.1.1 *The scattering length of hydrogen in the absence of spin polarization*

The concept of contrast was developed very early in neutron small-angle scattering of particles in mixtures of H_2O and D_2O. The scattering length density ρ changes with the fraction X of heavy water:

$$\rho_{\text{solvent}} = \rho_{H_2O} + X(\rho_{D_2O} - \rho_{H_2O}). \tag{19.1}$$

Similarly, we introduce the effective scattering length b at the site of a hydrogen atom,

$$b = b_H + X(b_D - b_H), \tag{19.2}$$

where b_H and b_D are the coherent scattering lengths from the isotopes ^{1}H and ^{2}H = D, respectively. Thus, b is an average, as are many scattering lengths of coherent neutron scattering; b_H and b_D are spin dependent and result from averages over spin states, as we will now explain.

Let us consider the scattering length of light hydrogen, b_H. Both the incident neutron and the nucleus of the hydrogen atom (= proton) have a spin $\frac{1}{2}$. In the scattering of the neutron by the proton there are two different channels (Fig. 19.1): one for the total spin $\frac{1}{2} + \frac{1}{2} = 1$ with a scattering length b^+, and one for the total spin $\frac{1}{2} - \frac{1}{2} = 0$ with a scattering length b^-. The spin 1 has three substates $1, 0, -1$, whereas there is only one state for the total spin 0. The effective length b of coherent scattering is the weighted average of b^+ and b^-. With $b^+ = 1.083 \times 10^{-12}$ cm and $b^- = -4.74 \times 10^{-12}$ cm, we obtain

$$b_H = \frac{3}{4}b^+ + \frac{1}{4}b^- = -0.374 \times 10^{-12} \text{ cm}. \tag{19.3}$$

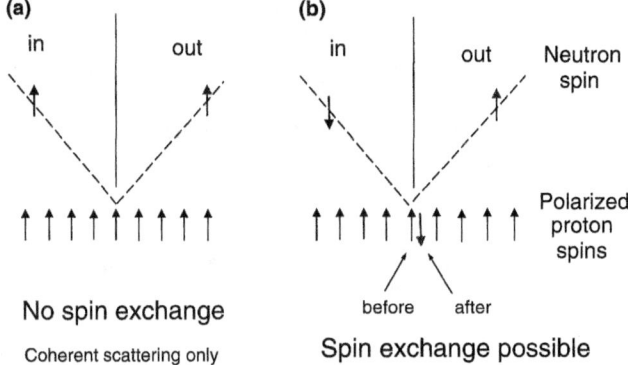

Fig. 19.1. Scattering of a neutron by polarized proton spins. (a) All proton spins point in the same direction as the spin of the incident neutron. There is no spin flip. (b) Proton spin polarization before scattering as in (a) but the spin of the incident neutron points in the opposite direction. Spin exchange is possible. In that case one of the proton spins changed its direction (marked by 'after'). The scattering centre could be identified—at least in principle.

This is the well-known negative coherent scattering length of hydrogen. In this chapter, it will be shown how the probability for the occurrence of b^+ and b^- can be changed by polarization.

19.1.2 The polarization of spins. Definitions

The polarization P is the ratio of the average projection of the spins along the static magnetic field to its maximum value. For spin $=\frac{1}{2}$ we have

$$P = \frac{n_+ - n_-}{n_+ + n_-}, \qquad (19.4)$$

where n_+ and n_- denote the numbers of spins in the states $+\frac{1}{2}$ and $-\frac{1}{2}$, respectively.

Example 1 Out of a very large number of proton spins, 75% are 'spin up' and 25% are 'spin down' with respect to an external magnetic field. The proton spin polarization is $(0.75 - 0.25)/(0.75 + 0.25) = 0.5 = 50\%$. A negative polarization is found if the majority of spins is in the 'down' state, i.e. pointing in the direction opposite to the magnetic field.

Example 2 Out of a small number of proton spins, the probability of finding spins in the state 'up' is $w = 0.75$. It has been determined in an independent experiment as described by Example 1. The actual number of 'spin up' (k) and 'spin down' ($n - k$) for n protons will vary according to the binomial distribution

$$W_n(k) = \binom{n}{k} w^k (1-w)^{n-k}.$$

For $n=1000$ protons this distribution gives a variance $\sigma = (nw(1-w))^{1/2} = 13.7$ around the mean value of $nw = 750$. This argument also applies to an ensemble of unpolarized nuclear spins ($p=0.5$). For $n=1000$ we have $\sigma = 15.8$. The variance vanishes for a completely polarized spin system ($w=1$ or $w=0$). These considerations may be important, for instance, in small-angle scattering from dilute solutions of macromolecules when the contrast is very low (see Section 19.3).

19.1.3 The coherent scattering length of polarized neutrons scattered from polarized nuclei

When the polarization of both the incident neutron beam p and that of the proton spins P_H is complete and in the same direction, the scattering process is described by b^+, i.e. $b_H = b^+$ [Fig. 19.1(a)]. In analogy to (19.2) we write

$$b_H = b_0 \pm P_H(b^+ - b_0) = [-0.374 \pm 1.456 P_H] \times 10^{-12} \text{ cm}, \qquad (19.5)$$

where $-1 \leq P_H \leq 1$. b_0 is the scattering length in the absence of proton spin polarization. This equation holds for a completely polarized beam of incident neutrons. The direction of neutron spin may point in the direction of the external magnet field (the $+$ sign of $\pm$ holds) or it may point in the opposite direction (the $-$ sign of $\pm$ holds). There is a huge variation of the scattering length with the polarization of protons/neutrons from -1.83×10^{-12} to $+1.082 \times 10^{-12}$ cm. A considerably smaller variation of the scattering length is observed with polarized deuterons (see Fig. 19.2),

$$b_D = [0.667 \pm 0.28 P_D] \times 10^{-12} \text{ cm}. \qquad (19.6)$$

19.1.4 Coherent and incoherent scattering from polarized nuclei

When the polarization p of the incident neutron beam is not complete ($|p| < 1$) the eqns (19.5) and (19.6), in terms of amplitudes, are no longer valid. They have to be replaced by the coherent scattering cross-sections (Abragam and Goldman 1982; Glättli and Goldman 1987):

$$\frac{\sigma_{coh}}{4\pi} = b_0^2 + b_0 b_N I p P + \frac{1}{4} b_N^2 I^2 P^2, \qquad (19.7)$$

where I is the nuclear spin, $b_0 = [(I+1)b^+ + Ib^-]/(2I+1)$, $b_N = 2(b^+ - b^-)/(2I+1)$. Note that the cross-section σ_{coh} is *not* the square of $b_0 + \frac{1}{2} b_N I p P$, as one would expect with isotopic substitution described by (19.2).

The incoherent scattering cross-section is

$$\frac{\sigma_{inc}}{4\pi} = \frac{1}{4} b_N^2 [I(I+1) - I p P - I^2 P^2]. \qquad (19.8)$$

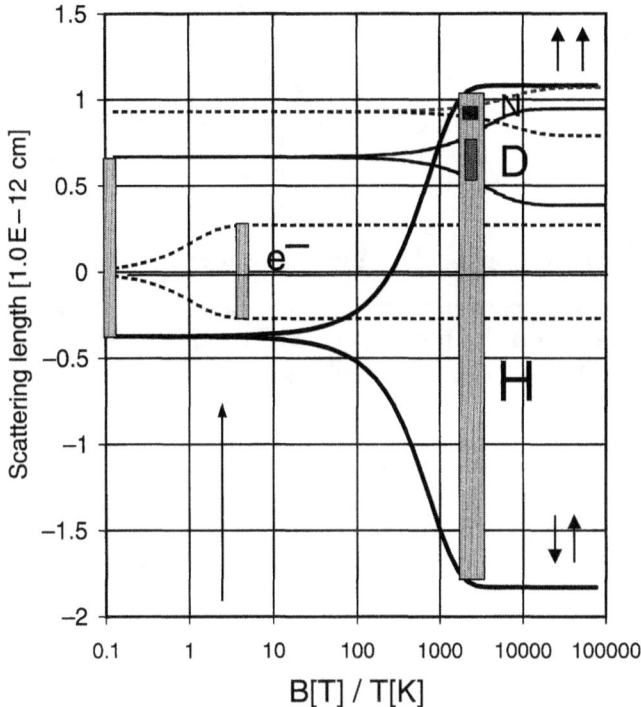

Fig. 19.2. With increasing magnetic field B and decreasing temperature T, the polarization increases according to the magnetic momentum of the particles. In a completely polarized neutron beam the scattering lengths develop as described in (19.5) and (19.6). The magnetic neutron scattering amplitude of an unpaired electron is given for comparison. The standard magnetic field of 2.5 T at 1 K is indicated by a long vertical arrow.

Table 19.1. Some spin-dependent scattering lengths (10^{-12} cm)

Isotope	Spin I	b_0	$b^+ - b^-$
^{1}H	1/2	−0.374	5.824
^{2}H (D)	1	0.667	0.855
^{14}N	1	0.93	0.42

For additional values, see Glättli and Goldman (1987).

We see that σ_{inc} vanishes for $pP = 1$, but not for $pP = -1$. This is easy to understand (Fig. 19.1): the states $p = 1$, $P = 1$ or $p = -1$, $P = -1$ give rise to the total spin $I + \frac{1}{2}$. All individual scattering lengths being equal, there is no incoherent scattering. By contrast, the states with $pP = -1$ are mixtures of the states $I + \frac{1}{2}$ and $I - \frac{1}{2}$ having different scattering lengths. This leads to the occurrence of incoherent scattering (Glättli and Goldman 1987) (Table 19.1).

Example Incoherent scattering from ^{1}H. Using (19.8) one obtains

$$\sigma_{\text{inc}} = 4\pi \frac{1}{4}[2(b^+ - b^-)/2]^2 \left[\frac{1}{2}\left(\frac{1}{2}+1\right) - \frac{1}{2}pP - \frac{1}{4}P^2\right]$$

$$= 106.56 \left[\frac{3}{4} - \frac{1}{2}pP - \frac{1}{4}P^2\right] 10^{-24} \text{ cm}^2.$$

In the absence of proton spin polarization ($P = 0$) we have $\sigma_{\text{inc}} = 79.9$ barn.[1]

There is a strong dependence of neutron beam transmission with polarized proton spin targets[2] on the proton spin polarization as can be deduced from Fig. 19.3. Hence, polarized hydrogenous materials are excellent *neutron spin filters* (Masuda *et al.* 1988).

Low incoherent scattering as achieved by proton spin polarization would be highly beneficial to structural studies, e.g. in protein crystallography. This method presents an alternative to the suppression of incoherent scattering from protons by perdeuteration of proteins. It has not been tried yet.

Note also that no spin analysis of the scattered neutrons is required. It is the polarization *before* scattering that matters.

Note the non-linear variation of b_D with P_H in Fig. 19.4. This is due to the assumed thermal equilibrium between the proton spins and the deuteron spins.

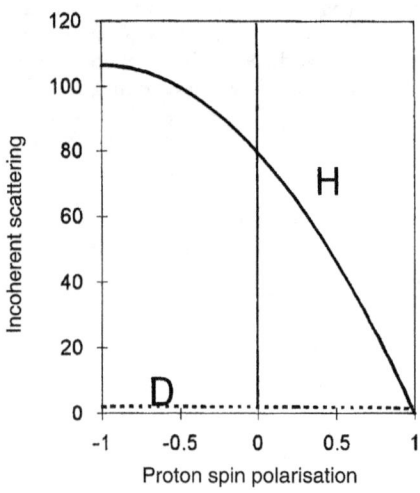
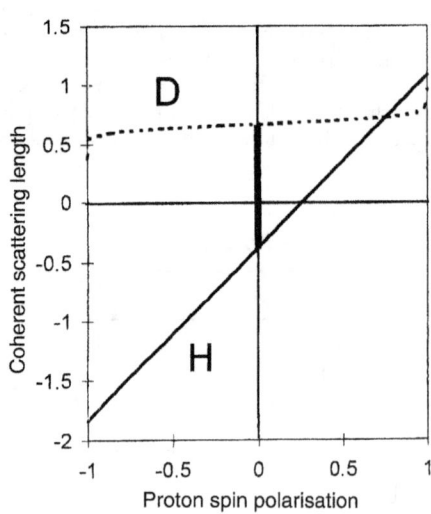

Fig. 19.3. Incoherent scattering cross-section in units of 10^{-24} cm^2 vs pP_H.

Fig. 19.4. Coherent scattering lengths in units of 10^{-12} cm vs pP_H.

---

[1] barn = 10^{-24} cm^2.

[2] Following the usage in nuclear and particle physics we call samples with polarized nuclear spins *polarized targets*. A sample containing protons and deuterons is called a *proton spin target* if only its proton spins are polarized, and it is called a *deuteron spin target* if only its deuteron spins are polarized.

As the nuclear magnetic moment of deuterons is much smaller than that of the protons, deuterons are relatively less polarized. For $|P_H| < 0.8$, $P_D = P_H/(4.8 - 1.8P_H^2)$ is a good empirical approximation.

The largest contrasts between H and D are encountered with negative pP_H where incoherent scattering is strongest. Clearly, this is a nuisance in structural studies. The way to minimize incoherent scattering from protons is to replace them by deuterons except in those components of the sample that are of interest (cf. Section 19.3).

19.1.5 Nuclear spin contrast variation

For a lattice with more than one atom per unit cell the coherent cross-section is obtained as follows. Let $U(\mathbf{Q})$ be the structure factor of the unpolarized sample and $V(\mathbf{Q})$ the amplitude due to the polarization of all nuclei having spin, then the coherent scattering cross-section of polarized neutrons is (Abragam and Goldman 1982)

$$\left(\frac{d\sigma}{d\Omega}\right)_{coh} = |U(\mathbf{Q})|^2 + 2p\,\mathrm{Re}\{U(\mathbf{Q})V^*(\mathbf{Q})\} + |V(\mathbf{Q})|^2. \qquad (19.9)$$

Only the cross-term $2p\,\mathrm{Re}\{UV^*\}$ depends on the neutron spin polarization p. The other terms, $|U|^2$ and $|V|^2$, are insensitive to the polarization of the incident neutron beam, i.e. they are also obtained with an unpolarized neutron beam. In a deuterated sample the polarization of both the proton spins and the deuteron spins will have to be taken into account. Thus

$$V(\mathbf{Q}) = P_H V_H(\mathbf{Q}) + P_D V_D(\mathbf{Q}). \qquad (19.10)$$

For random orientation of dissolved particles the basic scattering functions $|U(\mathbf{Q})|^2$, $2\,\mathrm{Re}\{U(\mathbf{Q})V^*(\mathbf{Q})\}$ and $|V(\mathbf{Q})|^2$ in (19.9) need to be replaced by the corresponding averaged intensities $I_U(Q)$, $I_{UV}(Q)$ and $I_V(Q)$.

As an example, we give the explicit form of the cross-term

$$I_{UV}(Q) = \int 2\,\mathrm{Re}\{U(\mathbf{Q})V'(\mathbf{Q})\}\,d\Omega$$

$$= \int\int u(\mathbf{r})v(\mathbf{r}')\frac{\sin(Q|\mathbf{r}-\mathbf{r}'|)}{Q|\mathbf{r}-\mathbf{r}'|}d^3r\,d^3r' \qquad (19.11)$$

which results from integration over the solid angle Ω. Both $u(\mathbf{r})$ and $v(\mathbf{r})$ are scattering length densities. For a dilute solution the extrapolated zero-angle scattering $I_{UV}(0)$ of N particles each having the same volume V is

$$I_{UV}(Q) = N\int\int u(\mathbf{r})v(\mathbf{r}')d^3r\,d^3r' = NV\rho_u\rho_v, \qquad (19.11')$$

where ρ_U is the contrast of the dissolved particle at $P=0$ with respect to the solvent, ρ_V is the contrast with respect to the solvent added by the nuclear polarization $P=1$. ρ_U and ρ_V may have different signs yielding a negative $I_{UV}(0)$. The solution of a macromolecule in a deuterated solvent is an example of that kind.

Similarly, $I_U(Q)$ and $I_V(Q)$ can be constructed. For proton spin targets we introduce $I_{UH}(Q)$ and $I_H(Q)$ and similarly for deuteron targets $I_{UD}(Q)$ and $I_D(Q)$. Then the variation of the intensity $I(Q)$ of small-angle scattering with P_H and P_D is given by

$$I(Q) = \begin{cases} I_U(Q) + pP_H I_{UH}(Q) + P_H^2 I_H(Q), \\ I_U(Q) + pP_D I_{UD}(Q) + P_D^2 I_D(Q). \end{cases} \quad (19.12)$$

19.2 Polarized neutron scattering from polarized nuclear spins

A diffractometer suitable for polarized neutron scattering from polarized nuclei in solids is shown in Fig. 19.5. The main extension is the facility for nuclear spin polarization.

Nuclear spins get polarized in strong magnetic fields B at very low temperatures T (Fig. 19.2). At $T = 1$ K and $B = 2.5$ T, the proton spin polarization in thermal equilibrium is $P_H = 0.0025$. Much more drastic conditions would be needed to achieve a higher nuclear polarization, e.g. $P_H = 0.7$ could be reached at $T = 1$ mK and $B = 10$ T. This brute-force method would have very little success for the class of substances we are discussing here, as the nuclear spin system is only weakly coupled to the system of lattice vibrations, i.e. the spin–lattice relaxation times T_1 become extremely long. A completely different approach has to be used, which is described next.

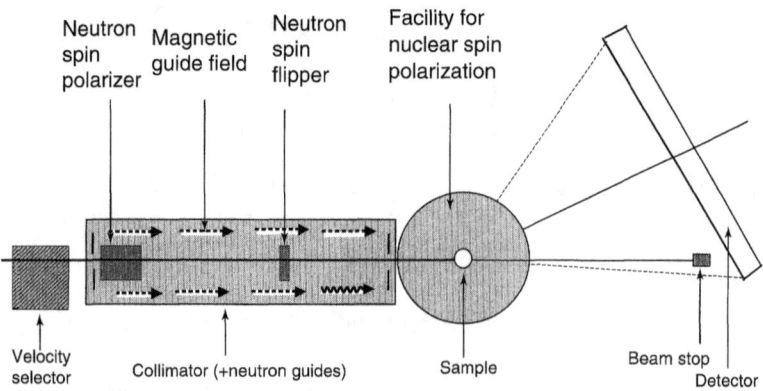

Fig. 19.5. An instrument for polarized neutron diffraction from polarized nuclei.

19.2.1 Dynamic nuclear spin polarization

The method of dynamic nuclear spin polarization (DNP) makes it possible to impart to the nuclear spins a polarization whose magnitude is comparable with the thermal equilibrium polarization of the electrons and whose orientation can be chosen at will, parallel or antiparallel to the external field (Abragam and Goldman 1982; Glättli and Goldman 1987). We outline the practical procedure used with solutions: the sample is doped by a small amount of an organic radical (a paramagnetic centre having one unpaired electron) with concentrations of 10^{19}–10^{20} cm^{-3}. Very often this is done by adding a stable compound of Cr(v), such as EHBA-Cr(v),[3] Na(C$_{12}$H$_{20}$O$_7$Cr) · H$_2$O (Krumpolc and Rocek 1979). Alternatively, radicals can be created by irradiation of the frozen sample with high energy particles or by UV light. The solvent must solidify as a glass under conditions of rapid freezing.[4] For biological macromolecules, a mixture of glycerol and water (11/9) turned out to be most useful.

At $T = 1$ K and $B = 2.5$ T the spins of the unpaired electrons belonging to the radical are nearly completely polarized whereas those of the nuclear spins are not (Fig. 19.2). Irradiation by 4 mm microwaves slightly below the Larmor resonance frequency of the unpaired electrons of EHBA-Cr(v) ($\sim 69.1 - 0.15$ GHz) aligns the nuclear spins in the direction of the external magnetic field. Using microwaves slightly above the electronic Larmor frequency ($\sim 69.1 + 0.15$ GHz) aligns the nuclear spins in the direction *opposite* to the external field (see Fig. 19.6). For proton spins, under favourable conditions a polarization of

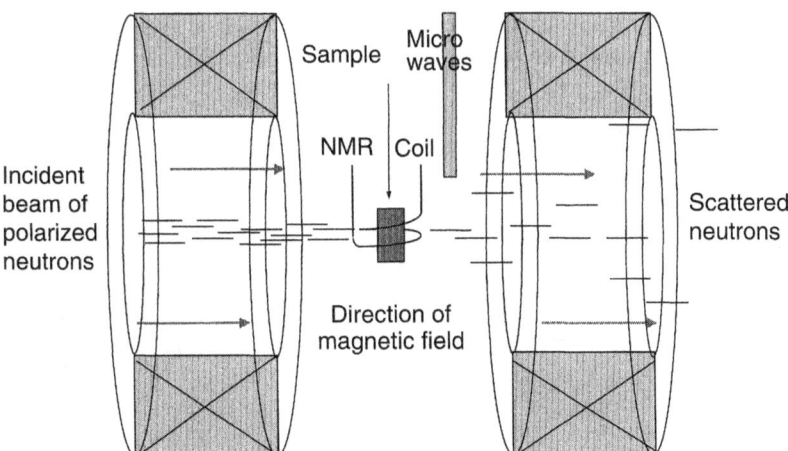

Fig. 19.6 The facility for dynamic nuclear spin polarization. A strong magnetic field is created by two superconducting magnets (cross-section of coils in each corner). The sample is kept at temperatures below 1 K. The refrigerator is not shown. Irradiation by microwaves polarizes the proton spins (see text). The beam of polarized neutrons is entering the target station on the left side.

---

[3] 2-Ethyl 2-hydroxy butyric acid is the ingredient for the synthesis of the **Cr(v)** compound.
[4] The liquid is injected into a copper mould kept at liquid nitrogen temperature. A sample of 3 mm thickness will have reached the final temperature in less than 10 s.

more than 95% can be obtained after a fraction of an hour to several days. For deuterons, because of the smaller magnetic momentum, the present practical limit is $P_D = 45\%$ (cf. legend of Fig. 19.4).

19.2.2 Measurement of nuclear polarization

In a magnetic field B, a nucleus with a spin $I = \frac{1}{2}$ has two energy levels. If $\hbar\omega$ denotes the energy difference between the two levels, then the condition for nuclear magnetic resonance (NMR) absorption is given by $\omega = \gamma B$, where the magnetogyric ratio is a constant depending on the kind of spin [see e.g. Kittel (1966)]. A proton NMR signal is shown in Fig. 19.7. For deuterons with spin $I = 1$ the signal is more complicated (Fig. 19.8).

The proton spin polarization is determined by comparing the enhanced NMR signal with the thermal equilibrium signal (Figs 19.7 and 19.8). The deuteron spin polarizations are calculated from the asymmetry of the resonance profile.

19.2.3 Selective nuclear spin depolarization

This method starts from a *dynamically* polarized target. All nuclei having spin are polarized, though to a different extent (Fig. 19.2). In order to obtain a target with only one isotopic spin polarized, the spins of the other isotopes need to be depolarized. In the case of specifically deuterated biomolecules or polymers two options may be envisaged:

- selective depolarization of the deuteron spins ⇒ proton spin contrast variation
- selective depolarization of the proton spins ⇒ deuteron spin contrast variation

Selective depolarization is achieved by a radio frequency sweep across the NMR peak at a power rate that is ∼100 times higher than in the measurement of nuclear polarization (Fig. 19.7).

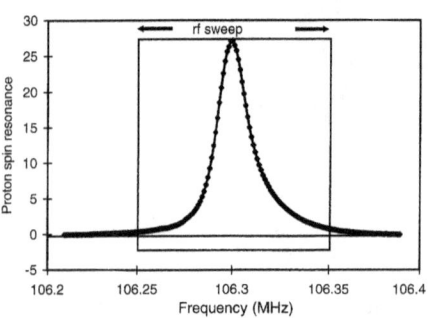

Fig. 19.7. Proton NMR. The integral of the peak is a measure of the proton spin polarization.

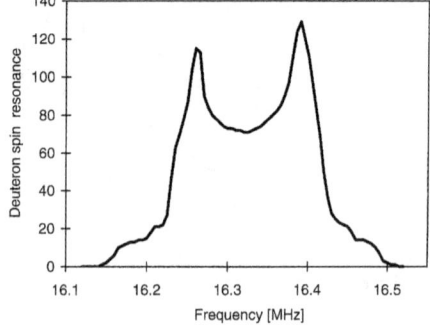

Fig. 19.8. Deuteron NMR. The asymmetry of the signal is a measure of the deuteron spin polarization.

19.2.4 Polarization reversal by adiabatic fast passage

Spin polarization reversal by using the adiabatic fast passage (AFP) mechanism is a method frequently used in NMR spectroscopy. Goldman *et al.* (1968) gave an extensive description of the AFP mechanism. The efficiency of this method is quite high if certain criteria are obeyed (Hautle *et al.* 1992). After a radio frequency sweep (or a magnetic field sweep) of about 2 s duration across the NMR profile (Fig. 19.7), typically more than 80% of the nuclear spins can be reversed.

19.3 Nuclear spin contrast is complementary to solvent contrast

This is easily seen from the variation of the scattering density of proteins and ribonucleic acid (RNA) with proton spin polarization (Fig. 19.9). With increasing proton spin polarization the scattering density of both RNA and protein increase. At P_H close to 0.65 both become equal to the scattering density of the deuterated solvent (deuterated glycerol/D_2O, 11:9 w/w). Nuclear spin contrast variation hardly discriminates between RNA and proteins, whereas (external) contrast variation in H_2O/D_2O mixtures does.

19.3.1 Proton spin contrast variation

Proton spin contrast variation is the method of choice in *internal contrast variation*: contrast is given to a selected region of a particle. Usually this is done by deuteration of a major part of an assembly, while leaving the region of interest protiated.[5] A deuterated solvent ensures low incoherent scattering. Polarization of the proton spins varies the contrast of the labelled region of the particle. The gain in contrast with respect to mere isotopic exchange is impressive as is shown in Fig. 19.9. It is the prerequisite for the *in situ* structure determination of relatively small macromolecular components in large complex particles. Labelled regions of less than 1% of the total volume of the particle can still be studied.

This also applies to copolymers, where proton spin polarization of selectively deuterated particles reveals the conformation of the components (Glättli *et al.* 1989; des Cloizeaux and Jannink 1987).

19.3.2 Deuteron spin contrast variation

As the solvent is deuterated, polarization of the deuteron spins will have an influence on the intensity of neutron scattering that, in principle, is similar to that of solvent contrast variation using H_2O/D_2O mixtures (*external* contrast variation). Contrary to the latter, deuteron spin contrast variation of the solvent is much less efficient. The scattering densities of RNA and proteins lie outside the range of scattering densities of the deuterated solvent even if P_D varies from -1 to $+1$ (Fig. 19.10). In practice, $|P_D|$ in glycerol–water mixtures rarely

---
[5] The expression 'protiated' has been adopted instead of 'protonated'.

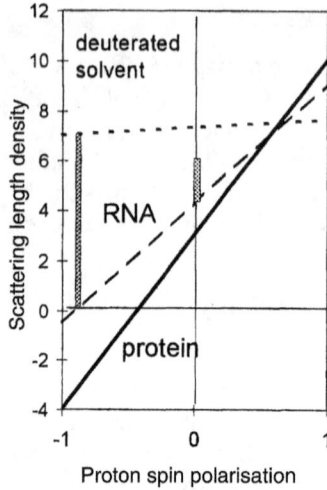 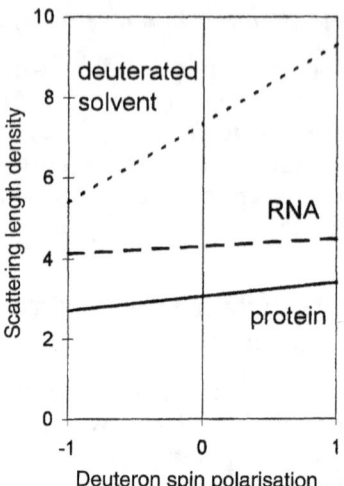

Fig. 19.9. The scattering density of RNA, protein and a mixture of deuterated glycerol and D$_2$O (11:9) in units of 10^{10} cm^{-2} vs the proton spin polarization P_H. The bar at $P_H = 0$ is the contrast of RNA in an H$_2$O/D$_2$O mixture containing 95% D$_2$O. A much larger contrast is obtained by proton spin polarization of -90%.

Fig. 19.10. The scattering density of RNA, protein and the same deuterated solvent as in Fig. 19.9 vs the deuteron spin polarization. The slope of the straight lines is a measure of the deuteron concentration. Usually deuteron polarizations of around 20% are achieved.

exceeds 0.2. Nevertheless, deuteron spin contrast variation plays a certain role with deuterated particles in deuterated solvents, where the contrast is low and its sign ambiguous. As the concentration of deuterons in the solvent is always higher than in the solute (even at full deuteration) the deuteron spin contrast must be negative: $V_D(0) < 0$. This facilitates the appraisal of small-angle scattering data at very low contrast of the solute.

19.4 *In situ* structure of the mRNA[tRNA]$_2$ complex in the *E. coli* ribosome

For large structures like ribosomes (see Chapter 18 of this volume), the separate structure determination of their components *in situ*, i.e. in its native environment, is an approach that is often used in neutron scattering. Two conditions are important:

1. The contrast of the ribosome should be low,[6] i.e. the ribosome should be transparent to coherent neutron scattering. This is achieved remarkably well for deuterated ribosomes in a mixture of deuterated glycerol and heavy water. Moreover, incoherent scattering is low.

---

[6] This is a case where statistical fluctuations of proton spin polarization (see Section 19.1.2) might be significant. They are not taken into account in the data analysis.

2. The contrast of the labelled region, e.g. of the bound tRNAs, should be high. This is achieved by the presence of ^{1}H in the label, i.e. in the absence of deuteration. Proton spin polarization considerably enhances the contrast of the label (Fig. 19.9). The studies on the functional complex of the ribosome are meant as an introduction to the practice of nuclear spin contrast variation.

19.4.1 The sample

A solution (0.6 ml) of the functional complex of the ribosome (15 mg/ml) in D_2O consisting of the deuterated 70S particles, two tRNAs and a short mRNA fragment (both protonated), in a pretranslocational state (Nierhaus et al. 1998). Addition of deuterated glycerol containing a small amount of the Cr(v) complex decreases the concentration of ribosomes to about 6 mg/ml. The final concentration of the Cr(v) complex is 7 mg/ml. The sample is frozen immediately after preparation in a liquid nitrogen cooled copper mould. Size of the frozen platelet: $2.8 \times 17 \times 17$ mm^3.

19.4.2 The experiment

After transfer into a dilution refrigerator, the sample is cooled to 1 K. The proton NMR signal from the sample in a magnetic field of 2.5 T is calibrated. Then the sample is cooled to 0.15 K. Irradiation by 4 mm microwaves for several hours yields a proton spin polarization of 70% and deuteron spin polarization of 15%. Then, in the absence of microwaves, the deuteron spins are selectively depolarized, as explained in Section 19.2.3. The proton spins remain polarized. A typical measurement of polarized neutron scattering from the proton spin target takes several days.

The sample is then again irradiated by microwaves. This time the proton spins of the dynamic polarized sample are selectively depolarized. The measurement with deuteron spin contrast variation takes another couple of days. Several hours are needed to unload the dilution refrigerator. This is a routine currently used at the GKSS Research Centre, Geesthacht. It has been established in cooperation with CERN, Geneva.

A similar procedure holds for the polarization facility at the Orphée reactor, Saclay (Glättli et al. 1989). Facilities working at $T = 1$ K are based on helium evaporators. They have much shorter turn-around times.

19.4.3 Data analysis

The data are corrected for absorption and background scattering. The amount of incoherent scattering is estimated from the scattering at $Q = 0.25$ Å^{-1}, where coherent small-angle scattering has been assumed to be negligibly small. It is subtracted as a constant intensity in the whole Q-range. The following scattering curves are needed for the calculation of the basic scattering functions: $I(Q)$ of the unpolarized sample, and two scattering functions from the proton spin target, $I_{\uparrow\uparrow}(Q)$ and $I_{\downarrow\uparrow}(Q)$, differing in the direction of neutron spin polarization,

$$I_U(Q) = I(Q)_{(P=0)},$$
$$I_{UH}(Q) = \frac{I_{\uparrow\uparrow}(Q) - I_{\downarrow\uparrow}(Q)}{2P_H}, \quad (19.13)$$
$$I_H(Q) = \frac{I_{\uparrow\uparrow}(Q) + I_{\downarrow\uparrow}(Q) - 2I_U(Q)}{2P_H^2}.$$

These basic scattering functions and the cross-term $I_{UD}(Q)$ from the deuteron spin target enter into the *in situ* structure determination of the tRNA complex.

1. *Where is the centre of mass of the tRNA complex in the coordinate system of the ribosome?*

An answer can be given if the low resolution structure of the ribosome is known. We take a model from electron microscopy studies (Frank *et al.* 1995). The basic scattering functions are calculated from the ribosome model with its tentatively located tRNA complex (sphere), and compared with the basic scattering functions from neutron scattering.

In a search for the best fit (minimum of least-squares) the position of the tRNA complex is varied systematically (Stuhrmann and Nierhaus 1996). A unique site is found (Fig. 19.11).

We outline the calculation briefly. Both $U(\mathbf{Q})$ and $V(\mathbf{Q})$ are sums of the amplitudes of the ribosomal model $M(\mathbf{Q})$ and the label $L(\mathbf{Q})$:

$$U(\mathbf{Q}) = c_M M(\mathbf{Q}) + c_L L(\mathbf{Q}),$$
$$V(\mathbf{Q}) = k_M M(\mathbf{Q}) + k_L L(\mathbf{Q}), \quad (19.14)$$

where c_M and k_M are the rather low average contrasts of the ribosome, c_L and k_L are the contrasts of the label, which are strongly enhanced by isotopic substitution and/or nuclear spin polarization.

A very handy and CPU-time-saving expression is obtained by expanding both $M(\mathbf{Q})$ and $L(\mathbf{Q})$ as a series of spherical harmonics $Y_{lm}(\Omega)$:

$$M(\mathbf{Q}) = \sum_{l=0}^{\infty} \sum_{m=-l}^{l} M_{l,m}(Q) Y_{l,m}(\Omega), \quad (19.15)$$

where Ω is a unit vector; $M_{lm}(Q)$ contains the information specific to $M(\mathbf{Q})$. The ribosomal model is described in a three-dimensional cubic lattice with a spacing of 5 Å. The scattering density ρ_n at the nth cube at $\mathbf{r}_n = (r_n, \theta_n, \phi_n)$ is taken from the electron microscopic model. Then the radial function $M_{lm}(Q)$ is

$$M_{l,m}(Q) = \sqrt{\frac{2}{\pi}} i^l \sum_{n=1}^{N} \rho_n j_l(Qr_n) Y_{l,m}^*(\theta_n, \phi_n). \quad (19.16)$$

As N may be of the order of some thousand, the calculation of the multipole components $M_{lm}(Q)$ up to $l = 20$ at 40 values of Q may take half an hour on a

PC. This calculation has to be done only once. The $L_{lm}(Q)$ of the label are subject to frequent change in an iterative process. It is therefore fortunate that their calculation is fast:

$$L_{l,m}(Q) = \sqrt{\frac{2}{\pi}} i^l j_l(Qr_L) Y_{l,m}^*(\theta_L, \phi_L). \tag{19.17}$$

The centre of mass of the label has the polar coordinates (r_L, θ_L, ϕ_L) in the coordinate system of the ribosome. Starting from multipoles, the calculation of the basic scattering functions becomes very easy. We give the cross-term as an example ($V = H$ or $V = D$):

$$I_{UV}(Q) = 4\pi^2 \sum_l \sum_{m=-l}^{l} U_{l,m}(Q) V_{l,m}^*(Q). \tag{19.18}$$

The expression to be minimized is

$$\text{Min} = \sum_i \frac{[I_{\exp}(Q_i) - I_{\text{calc}}(Q_i)]^2}{\sigma_i^2}, \tag{19.19}$$

where σ_i is the standard deviation of the intensity measured at Q_i.

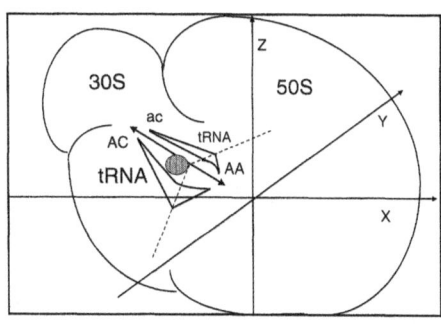

Fig. 19.11. The position and orientation of the two tRNA molecules in the coordinate system of the 70S ribosome [schematic drawing of the electron microscopy model from Frank et al. (1995)]. AC and ac are the anticodons of the two tRNAs. AA is the aminoacyl residue of one of the tRNAs. For further explanation see A. Yonath (this book).

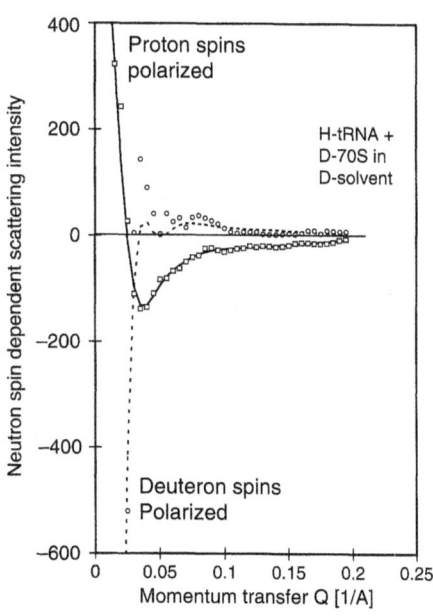

Fig. 19.12. The cross-terms of polarized neutron scattering, $I_{UH}(Q)$ (□), and $I_{UD}(Q)$ (○). The lines are not a guide for the eye. They are calculated from the model. Intensity in units of n/channel/10^3 s (sample–detector distance: 0.7 m, $\lambda = 8.5$ Å, $\Delta\lambda/\lambda = 0.1$) (Nierhaus et al. 1998).

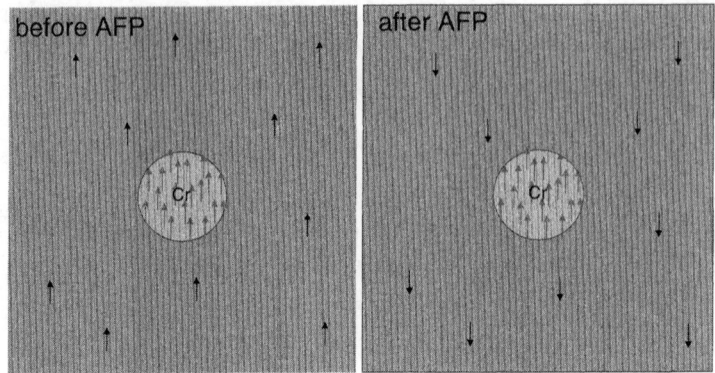

Fig. 19.13. The spatial distribution of the polarized proton spins in a dilute solution of EHBA-Cr(v) dissolved in a deuterated solvent. The schematic drawing shows the cluster of proton spins aligned in the same direction as the very few protons of the deuterated solvent. This is the situation after DNP and before AFP (right side). The method of AFP will preferably reverse the direction of the proton spins of the solvent. A gradient of proton spin polarization exists at the boundary of the EHBA molecule. The gradient will disappear by the mechanism of spin diffusion at a rate that depends on the concentration of the proton spins in the solvent.

2. How are the tRNA molecules oriented with respect to the 70S ribosome?

The answer relies on both the quality of the 70S model and the structure of the tRNA molecule. The latter is known to atomic resolution, which is by far too good for the present low resolution study. The tRNA a roughly L-shaped particle is approximated by four spheres. Restrictions for the mutual arrangements of the two tRNA molecules are given by the close proximity of their anticodons (AC and ac in Fig. 19.11) and by the site of protein elongation (AA in Fig. 19.11), which must be close to the end of both tRNA chains. A solution is given in Figs 19.11 and 19.12. The planes of the tRNA molecules form an angle of $110° \pm 10°$.[7] The data do not tell to which side the angle is open. Although the result is ambiguous, it does exclude some other models. That is the way spin contrast variation from solutions contributes to the knowledge of the *in situ* structure of ribosomal components.

Similarly, the post-translocational state of the functional complex has been studied. A displacement of the centre of the tRNA complex by 12 ± 4 Å and a slight change of the orientation during translocation have been observed (Nierhaus *et al.* 1998a,b).[8]

This example shows that the proton spin contrast is most convenient for visualizing small labels in large complex particles. In another experiment, a protiated mRNA fragment (tRNAs and the 70S ribosome were deuterated) which

---

[7] This result appears to be rather precise. Indeed, it is not, as systematic errors due to the poor knowledge of the total ribosome structure may influence the result. Using different models of the ribosome, different angles varying between 50° and 150° were obtained. In all cases there was hardly any change of the angle during the transition from the pre-translocational to the post-translocational state of the ribosome.

[8] The error in the displacement of the centre of the tRNA complex has been estimated from a map showing the deviations of the expression (19.19).

fills 0.5% of the ribosomal volume has been localized, even at the low occupancy of only 40%. The method of triple isotopic substitution (TIS, Chapter 12) is about an order of magnitude less sensitive (Serdyuk 1997).

19.5 Nuclear spin diffusion

During the process of DNP by microwave irradiation gradients of proton spin polarization occur around paramagnetic centres in lanthanum magnesium nitrate (Leslie *et al.* 1980). Similar phenomena were observed in dilute solutions of EHBA-Cr(v) with selective proton spin depolarization (Zhao *et al.* 1995) and with the reversal of proton spin polarization by the method of AFP (Stuhrmann *et al.* 1997).

These findings are explained by the creation of selectively polarized proton spin domains near the unpaired electrons (Fig. 19.13). The lifetime of these domains strongly depends on the concentration of the proton spins of the solvent. It is of the order of seconds in the absence of deuteration, and it may increase to hours in highly deuterated samples. The decay of the polarization gradients by spin diffusion (Cox *et al.* 1977) needs further investigation by both NMR and polarized neutron scattering.

Preliminary measurements of polarized neutron scattering from a dilute solution of EHBA-Cr(v) indicate that the size of the proton spin domains comprises about 20 protons, i.e. those of the EHBA molecule. The scattering amplitude of these 20 protons is two orders of magnitude larger than that of magnetic neutron scattering from the single unpaired electron of the EHBA-Cr(v) molecule. A new method for the investigation of dilute paramagnets is at hand. This development is triggered by the possibility of establishing radical density maps of oxidoreductases from polarized neutron scattering of selectively proton spin reversed targets.

References

Abragam, A and Goldman, M (1982). *Nuclear magnetism: order and disorder*, Clarendon Press, Oxford.
des Cloizeaux, J and Jannink, G (1987). *Les Polymères en Solution: leur Modélisation et leur Structure*, Les édition de physique, Les Ulis, France.
Cox, SFJ, Read, SFJ and Wenckebach, Th (1977). *Journal of Physics C: Solid State Physics*, **10**, 2917–2936.
Frank, J, Zhu, J, Penczek, P, Li, Y, Srivastava, S, Verschoor, A, Rademacher, M, Grasucci, R, Lata, RK and Agarwal, RK (1995). *Nature*, **376**, 441.
Glättli, H and Goldman, M (1987). *Methods of Experimental Physics*, **23C**, 241–286.
Glättli, H, Fermon, C and Eisenkremer, M (1989). *Journal de Physique (Paris)*, **50**, 2375–2388.
Goldman, M, Chapellier, M and Vu Hoang Chau (1968). *Physical Review*, **168**, 301.
Hautle, P, Gruebler, W, van den Brandt, B, Konter, JA, Mango, S and Wessler, M (1992). *Physical Review Letters*, **33**, 696–699.
Kittel, C (1966). *Introduction to solid state physics*, John Wiley, New York.

Krumpolc, M and Rocek, J (1979). *Journal of the American Chemical Society*, **101**, 3206–3209.
Leslie, M, Jenkins, GT, White, JW, Cox, S and Warner, G (1980). *Philosophical Transactions of the Royal Society London*, **B290**, 497–503.
Masuda, Y, Ishimoto, S, Ishida, M, Ishikawa, Y, Koghi, M and Masaike, A (1988). *Nuclear Instruments and Methods in Physics Research*, **A264**, 169–172.
Nierhaus, K-H, Stuhrmann, HB and Svergun, D (1998a). *Progress in nucleic acid research and molecular biology*, ed, K. Moldave, Vol. **59**, pp. 177–204.
Nierhaus, K-H, Wadzack, J, Burkhardt, N, Jünemann, R, Meerwinck, W, Willumeit, R and Stuhrmann, HB (1998b). *Proceedings of the National Academy of Sciences USA*, **95**, 945–950.
Niinikoski, TO (1995). *Nuclear Instruments and Methods in Physics Research*, **A356**, 62–73.
Serdyuk, IN (1997). *Physica*, **B234**, 188–192 (and citations therein).
Stuhrmann, HB and Nierhaus, K-H (1996). *Neutrons in Biology*, Schoenborn and Knott, eds, Plenum Press, New York.
Stuhrmann, HB, van den Brandt, B, Hautle, P, Konter, JA, Niinikoski, TO, Schmitt, M, Willumeit, R, Zhao, J and Mango, S (1997). *Journal of Applied Crystallography*, **30**, 839–843.
Zhao, J, Meerwinck, W, Niinikoski, TO, Rijllart, A, Schmitt, M, Willumeit, R and Stuhrmann, HB (1995). *Nuclear Instruments and Methods in Physics Research*, **A356**, 133–137.

INDEX

acetanilide, vibrational dynamics 169–70
acoustic scattering (TDS) 185–6
actin filaments
  generation of Fourier transform 288
  structure and diffraction 286–90
L-alanine, lattice vibrations 170–2
Alcaligenes xyloxidans, azurins 131
alchemical perturbation 156–7
alkanes, diffusion in urea compounds 172–3
allosteric transition, *E. coli* aspartate transcarbamylase (ATCase) 205, 224–32
amide group, orbitals and transitions 328
annealing, simulations 157–8
anomalous dispersion 45–51
  accessible wavelength range 49
  AS of C, N, O, H vs heavy metals 82
  data measurement and processing 63–8
    area detector 65–6
    instrumentation 63–4
    X-ray optics 65
    X-ray sources 64
  defined 44
  diffraction experiment 67–8
  normal vs anomalous scattering 45
  properties of f' and f'' 46–8
  structure phases 45–51
Archaeoglobus fulgidus, genome 4
area detector, data measurement and processing, anomalous dispersion 65–6
Argand diagram 109
argon ion pumped dye laser, characteristics 313
aspartate transcarbamylase, *E. coli*
  allosteric transition 224–32
  Guinier plots 205
asymptotic regime 207
atomic models, small-angle X-ray scattering (SAXS) 213–14
ATP synthase *E. coli*, stoichiometry 215–16
autoindexing
  parameters known 21–3
  parameters unknown 25
automated refinement protocol (wARP) 92
azurins 131

Bacillus stearothermophilus, tryptophan, fluorescence in LDH 315, 322
Bacillus subtilis, genome 4
bacterial ribosomes 367–75
bacteriorhodopsin 259–69
  charge-controlled conformational changes 267

crystal structure, crystallization of bR 267–9
EM structure, refinement by neutron and X-ray diffraction 263
ground state crystal structure 269
M splits into two states M1 and M2 265–7
molecules and α-helices in the PM arrangement 261–3
mutant Asp 96 Asn, X-ray diffraction, time resolved 264–5
photocycle intermediates using PMs, investigations 263–7
photocycle and proton translocation 259–61
trapping M state
  mutant Asp 96 Asn 264
  wild-type bR 263–4
Bayesian approach 40, 57
beam intensity monitor 66–7
Bessel functions 278–9
beta-octyl glycoside
  ring structure 111
  scattering length densities 108
Bijvoet differences 52
  ratios 57
  SIRAS 62
Bijvoet pairs
  data 68
  mirror-related reflections 51
biocrystallography
  phase problem 36–40
  see also multiple wavelength; neutron crystallography; structure factor phases
bluetongue virus (BTV) 361–6
Boltzmann's constant 149
Born-Oppenheimer approximation, calculation of potential energy 142
Borrelia burgdorferi, genome 4
bovine enterovirus crystals 355
bovine pancreatic trypsin inhibitor (BPTI) 175–6
Bragg angle 26, 30
Bragg law 285
  diffraction 185
  mono- and polychromatic diffraction 120
Bragg peak 168
Bragg reflections 15, 181, 185
  elastic intensity 194
Bravais lattice vectors 25
  diffuse X-ray scattering 187
Brillouin zone 168, 170
bromo-5cytosine-2 derivative, MAD phasing 72
buoyancy term, small-angle neutron scattering (SANS) 243

Index

calcium, calculated spectra 47
calcium hydroxyapatite, absorption spectrum 348
calcium oxide, absorption spectrum, X-ray microscopy 345
calmodulin, fluorescence anisotropy 323
carbamyl phosphate (CP), formation via aspartate transcarbamylase 224–32
carbon dioxide, absorption spectrum, X-ray microscopy 345
cardiotoxin, *Naja nigricollis* 316
alpha-carrageenan, SRCD 334
CATH, URL 5
charge coupled devices (CCD) detectors 19–20
circular dichroism 324–36
   applications in structural biology 333–6
      instrumental developments and future prospects 335–6
   comparison of absorbance spectra 331
   electric vector, snapshots 326
   idealized chiral molecule 327
   measurements using SR 328–32
      data analysis 333
      instrumentation 332–3
cobalt hexamine, bromo-5-cytosine-2 derivative, MAD phasing 72
collagen, absorption spectrum 348
collimation, beam intensity monitor 66–7
collimation devices, X-ray optics 15–16
condensed phase systems 146
confocal microscopy 317
   compared with conventional microscopies 339
conjugate gradient algorithm 147
convolution theorem 273–5
critical assessment of protein structure (CASP), homology 11–12
cryocooling
   crystal mounting 17
   data collection 121
CRYSOL, applications and program, ideal monodisperse solution 213–15
crystal mounting 16–18
   cryocooling 17
   data collection and reduction methods 16–18
   glass capillaries 17–18, 355
crystalline enzymes 117–21
crystallography, *see* biocrystallography
crystals
   non-crystallographic symmetry (NCS) 87–92
   orientation
      definition 22
      integration of images 28–30
   and solution, quaternary structure, comparison 226–8
Cu–Zn superoxide dismutase, structure 133–5
cupredoxins, EXAFS, structure 130–3

DALI, URL 5
data analysis, solutions 207–9
data collection and reduction methods 14–34
   crystal mounting 16–18
   detectors 18–20
   goniostat 16

   X-ray optics 14–16
data processing 21–32
   anomalous dispersion 63–8
   autoindexing 21–2
      cell parameters, known/unknown 22–5
      crystal orientation, definition 22
      refinement 25–7
   data reduction 30–2
      data merging 31
      scaling 30–1
   image integration 28–30
      peak/background mask definition 28–9
      reflection position integration 28
      standard deviation estimation 30
      summation and profile fitting 29, 34
   profile fitting equations 32–4
   steps 21
databases and URL addresses 5
Debye–Waller factor 168
density functional theory 143
density modification
   electron density maps 84–92
   Sayre's equation 88–9
derivatization agents, soaking experiments, high resolution phasing 377–8
detectors
   characteristics 19–20
   ideal 19
detergent, beta-octyl glycoside 108, 111
deuterium
   $D_2O:H_2O$ exchange, and SANS 241–6
   label triangulation 244
   see also isotopic substitution
deuterium flash radiation, characteristics 313
deuteron spin polarization 400
Dictyostelium discoideum, NDPK protein, mercury phasing 73
difference vectors 24
diffraction
   Bragg diffraction 185
   experiment, anomalous dispersion 67–8
   low energy electron diffraction (LEED) 124
   protein diffraction, diffuse X-ray scattering 192–4
   time-resolved X-ray diffraction 298
   see also anomalous dispersion; fibre diffraction; helical diffraction; multiple wavelength anomalous diffraction; single anomalous diffraction; X-ray diffraction
diffraction pattern, resonant scattering 49–51
diffractometer, polarized neutron scattering 396
diffuse X-ray scattering 181–94
   disordered diffuse scattering 186–92
   experimental considerations 191–2
   molecular Fourier transform 188–90
   protein studies 192–4
   random orientational disorder 190–1
   thermal diffuse scattering (TDS), acoustic/phonic 183–6
diffusion
   nuclear spin diffusion 405
   reaction initiation, time-resolved biocrystallography 119

Index 409

diffusive motion, protein dynamics, incoherent quasielastic neutron scattering 177–9
disordered states, static vs dynamic 182–3
distance-dependent dielectric method 146
DNA
 fibre diffraction 284–6
 and protein, sperm, measurement and distribution 350
dynamic structure factor
 25K incoherent, spectrometry 170
 simulation relationship, neutron scattering 161–3

eigenvalues 147
Einstein crystal 189
elastic incoherent structure factor (EISF) 165–6
electron beam lithography, X-ray microscopy 343
electron density 79–100
 interpretation 92–100
  density strategy 92–3
  first model building (rough) 97
  model improvement 98–100
  molecular boundary, NCS 94–5
  protein model
   common errors 100
   good/bad features 97
  real-space correlation coefficient (RSCC) 98
  Richards Box 92–3
  rotameter side-chain score (RSC) 98
  sequence factor placement 95–6
  skeletal improvement 96–7
  skeleton 93–4
electron density distribution, Fourier transform 200
electron density maps 79–92
 calculations 81–2
 difference maps 82–4
 electron density equation 79–81
 fast Fourier transform 81–2
 non-crystallographic symmetry (NCS) 87–92
 phase combination 82
 phase improvement by density modification 84–92
  DM program 88–9
  molecular envelope 89
  NCS operator determination 90–1
  non-crystallographic averaging 89–92
  phase extension 91–2
  solvent flattening 85–6
  solvent flipping 87–8
  truncation of density 86
  wARP—automated refinement protocol 92
 σ_A weighted maps 83–4
electron microscopy
 compared with other microscopies 339
 electron probe microanalysis 339–40
 STEM and TEM 339–40
electron radius 45
electron transfer, cupredoxins 130–5
elongation factors, nucleic acid–proteins, functional systems in solution 246–8
ENTREZ, URL 5
enzyme reactions in crystals 116–21
 data collection
  experimental set-up 121

mono- and polychromatic diffraction 120
 initiation by diffusion 120
 initiation by photolysis 119
 intermediates 116–17
 kinetic vs time-resolved crystallography 117–18
epidermal growth factor, circular dichroism 325
equations, profile fitting 32–4
erbium, anomalous scattering factor 48
ergodic principle, trajectory analysis 154
Escherichia coli
 ATCase 205, 224–32
 ATP synthase stoichiometry 215–16
 genome 4–5
 LacR phasing with mercury 69
 methionyl–tRNA synthase 246–8
 Ompf protein 111–12
 see also ribosome
2-ethyl 2-hydroxybutyric acid (EHBA-Cr(v)), nuclear spin diffusion 404–5
ethyl–mercury–thiosalicylate (EMTS) 190
European Synchrotron Radiation Facility (ESRF) 64–5
Ewald sphere
 geometry 276
 lattice points 26
 Lorentz factor 30
Ewald summation, evaluation method 146
extended X-ray absorption fine structure (EXAFS) 124, 345–7
 single and multiple scattering 126–8

f' and f''
 anomalous dispersion, MAD 46–8
 Kramers–Kronig transformation 67–8
fibre diffraction 272–86
 experimental geometry 285–6
 future developments 297–9
  high resolution studies 297–8
  new synchrotron sources and detectors 298–9
 helical diffraction theory 273–84
 synthetic polymers and DNA 284–6
 see also muscle diffraction
fluorescence, process, stages 306–7
fluorescence anisotropy, see time-resolved fluorescence
fluorescence lifetime analysis, site-directed mutagenesis 316
fluorescence lifetime measurement 308–12
 microvolume 317–19
 pulse coincidence vs phase modulation 308–12
fluorescence probes
 amino acids 307–8
 DPH 323
fluorescence spectroscopy 305–14
 fluorescence decay analysis 314
 single photon counting 311
 time-resolved 305–6
  pulsed sources, review 313
  time-to-amplitude converter (TAC) 311–12
fold recognition (threading) 11–12
folding studies 232–3

folds
  families, structures, Protein Data Bank (PDB) 9–10
  superfold types 10
foot-and-mouth (FMV) disease crystal 357
force fields, empirical 143–5
Fourier synthesis 36
Fourier transform
  convolution product 218
  diffuse X-ray scattering 188–90
  electron density distribution 200
  Internet information 81
  phase problem of biocrystallography 36–40
  scattering density F_{hkl} 104
free energies
  calculation 155–7
  thermodynamic integration method 155
  thermodynamic perturbation method 155
Fresnel zone plate, X-ray lens diagram 340
Friedel pairs
  data 68
  mirror-related refections 51
  SAD 62

genomics 3–13
  3-D structures 8–10
  complete sequences 4
  gene sequencing 4–6
  homology modelling 11–12
  homology studies 7–10
  major databases and URL addresses 5
glass capillaries, crystal mounting 17–18, 355
glycogen, diffuse X-ray scattering 194
GNOM visual search program 208
goniometer 66–7
goniostat 16, 66–7
  data collection and reduction methods 16
guanidinium chloride 245
Guinier approximation, SAS 240
Guinier relation, gyration radius 204–6

Haemophilus influenzae, genome 4
Haloarcula marismortui 259, 368
Hartree–Fock molecular orbital method 143
helical diffraction theory 273–84
  continuous helix transform appearance 279–80
  crystalline arrays of helical structures 282–4
  discontinuous helix of points transforms 280–1
  helices of atoms and molecules 281–2
  repeated helical turns 277–8
  single helical turn 275–7
Helicobacter pylori, genome 4
hen egg white lysozyme (HEWL)
  folding studies 232–4
  SAD 72
  see also lysozyme
high resolution phasing for large macromolecular assemblies 367–87
  H50S 10–12 Angstrom MIR map 372
  50S ribosomal subunit crystallization 369–75
  crystals of whole 70S ribosome 368–9
  H50S 10–12 Angstrom MIR map 372

H50S crystal mounting and photographs of irradiated crystal 373
H50S diffraction pattern 369
metal clusters 372
phases 377–83
  derivatization agents: soaking experiments 377–8
  H2 crystals grown under mild conditions compared to H50S 380
  heavy atom compounds, quantitative attachment 378–80
  isolated ribosomal components and in situ complexes 383
  MAD phasing 380–1
  molecular replacement 381–2
  T50S, T70S ribosome, packing diagrams 381–3
  T70S ribosome, packing diagram 381
ribosomal crystal quality 370
synchrotron radiation and crystal decay 376–7
Thermus thermophilus (T30S) diffraction pattern 369
HIV-1 RT, shape determination 211
hydration
  $D_2O : H_2O$ exchange, and SANS 245–6
  nucleic acids and proteins 245–6
hydrodynamics
  buoyancy term 243
  buoyancy term and SANS, molecular weights and solvent interactions 243
hydrogen, scattering length 390–1

image collection 20–1
  data collection strategy 21
  fine vs coarse phi slicing 20–1
  software 21
image intensifier 16
image plate detectors 18
inelastic/quasielastic neutron scattering, *see* neutron scattering
infra-red microscopy, compared with conventional microscopies 339
insect, myosin filament nets 290–2
Institut Laue Large in (ILL) 105–7, 112, 113, 257–8
Internet information, Fourier transform 81
insulin, diffuse X-ray scattering 193
isotopic substitution, triple 244
isotopic substitution to nuclear spin polarization
  $D_2O : H_2O$ exchange 241–2
  nuclear spin contrast variation studies 390–6

Jablonski diagram, fluorescence process 306–7

Kendrew models 92
kinetic crystallography 116–21
Kramers–Kronig dispersion ratio 45
Kramers–Kronig transformation, f' and f' 67
Kratky plot 217–18
krypton, SIRAS phasing 73

LacR, MAD phasing with mercury 69–70, 73
lactate dehydrogenase, *Bacillus stearothermophilus*, time-resolved fluorescence 315, 322

lanthanides
  MAD phasing 70-2
  nuclear spin diffusion 405
lasers, radiation characteristics 313
lattice vibrations, in L-alanine 170-2
Laue diffractometer (LADI), near-atomic resolution neutron crystallography 105-7
Laue method
  data collection, virus crystallography 358
  quasi-Laue neutron diffraction, near-atomic resolution, neutron crystallography 105-7
  X-ray diffraction, data collection, mono- and polychromatic diffraction 120
LEED, see low energy electron diffraction
Lennard-Jones energy 144
ligands and effectors, structural transition, equilibrium study 228-9
light microscopy, compared with other microscopies 339
light-driven proton pumps, bacteriorhodopsin 259-69
lipid bilayers 251-9
  diffraction experiments
    orientated films multilamellar stacks 252-4
    powders of multilamellar stacks 254
  lipid polymorphism 254-6
  phospholipid molecule conformation in bilayer, neutron diffraction studies 256-9
Lorentz factor, Ewald sphere 30
low energy electron diffraction (LEED) 124
LURE, wire chamber detector 18
lysozyme
  diffuse X-ray scattering 193
    data analysis 209
  folding studies 232-4
  quasi-Laue neutron diffraction 105-7
  SAD 72

macromolecular assemblies, see high resolution
macromolecules
  complexes, structure definition 239
  solutions 238-9
  very large, crystallography 353-66
MAD, see multiple wavelength anomalous diffraction
malaria, life cycle, human red blood cells 343
mannose-binding protein (MBP), MAD phasing with lanthanides 70-2
MASC (multiple wavelength anomalous solvent contrast) 52, 60-2, 67
Maxwell-Boltzmann law 152
membranes 251-69
  bacteriorhodopsin 259-69
  lipid bilayers 251-9
  proteins and complexes, structure and interactions 243-5
mercury
  MAD phasing 69-70
  SIRAS 72-3
metal clusters 372
metalloprotein structure 124-35
  applications

Cu-Zn superoxide dismutase 133-5
cupredoxin structure 130-3
domain closure in transferrins 135
data analysis 128
experimental requirements 128-9
multiple scattering (EXAFS and XANES) 127-35
single scattering 126-7
Methanobacterium thermoautotrophicum, genome 4
Methanococcus jannaschii, genome 4
methionine, MAD phasing with sulphur 59
methionyl-tRNA synthase, *E. coli* 246-8
microfluorimetry, time-resolved real time 317-19
microscopy, comparisons of microscopies 339
microvolume fluorescence lifetime measurement 317-19
MIR, see multiple isomorphous replacement
mirror-related refections
  Bijvoet pairs 51
  Friedel pairs 51
mirrors, X-ray optics 15-16
molecular dynamics simulations 148-58
  methodology 149-53
  trajectory analysis 153-5
molluscs, myosin filament nets 290-2
monochromatic diffraction, enzyme reactions in crystals 120
monochromator
  for X-ray optics 15
  X-ray optics 15, 65
monochromator-mirror camera, for SAXS 221
Monte Carlo techniques, protein dynamics 158
mosaic spread vs image number, bovine enterovirus crystals 355
Mössbauer radiation, myoglobin 194
multiple isomorphous replacement (MIR) 39, 81
  formalism 55
  method 41-2
multiple wavelength anomalous diffraction (MAD) 40, 125
  anomalous ratios 57-8
  choice of anomalous scattering species 58-9
  methods 51-62
    data collection strategy 57-8
    MASC method 60-2
    MIR formalism 55
    normal and anomalous scattering 45
    properties of f' and f'' 46-8
    resonant scattering on diffraction pattern 49-51
  phasing, high resolution for large macromolecular assemblies 380-1
  phasing with bromine 58, 72
  phasing with lanthanides 59, 70-2
  phasing main structure 55-7
  phasing with mercury 69-70
  phasing with selenium 59, 69
  phasing with sulphur 59
  and SIRAS, comparison 63
  solution of substructure of anomalous scatterers 52-4
multiple wavelength anomalous solvent contrast (MASC) 52, 60-2, 67

Index

multiwire proportional chambers (MWPCs) 18–20, 66, 298–9
muscle structure and diffraction 286–97
  actin filaments structure 286–90
  future developments
    high resolution diffration studies 297–8
    new synchrotron sources and detectors 298–9
    time-resolved X-ray diffration from dynamic systems 298
  hierarchy 287
  modelling, myosin head array in fish muscle 294–7
  myosin filaments 290–2
  whole muscle diffraction patterns 292–4
Mycoplasma genitalium, genome 4
Mycoplasma pneumoniae, genome 4
myelin, deep sea fish 323–4
myoglobin
  CD spectrum 330
  diffuse X-ray scattering 194
myosin filaments, structure and diffraction 290–2
myosin head array, fish muscle 294–7
myosin S1 fragment, Guinier plots 205
N-myristoyl transferase, MAD phasing with selenium 69

Naja nigricollis, cardiotoxin 316
near-edge X-ray absorption fine structure (NEXAFS) 344–7
  molecules with covalent bonds 346–7
  molecules with ionic bonds 346
  origin and use 346–7
neutron crystallography 102–14
  absolute phase determination 110–11
  contrast variation method 107–8
  detergent in membrane protein crystals 111–12
  experimental set-up 113
  low resolution 107–11
  phase problem 109–10
  viruses 112–13
  near-atomic resolution 103–7
    methodology 104–5
    quasi-Laue neutron diffraction 105–7
neutron scattering
  by polarized proton spins 391
  dynamics in small molecule condensed phases 169–73
    diffusive motions in molecular crystals 172–3
    lattice vibrations in L-alanine 170–2
    vibrational dynamics 169–70
  inelastic/quasielastic 161–79
    inelastic incoherent 166–7
    protein dynamics 173–9
      diffusive motion, incoherent quasielastic neutron scattering 177–9
      vibrations 174–6
        density of states 175–6
        high frequency 176–7
      simulation relationship 162–8
        coherent inelastic neutron scattering 168
        dynamic structure factor 161–3

elastic incoherent structure factor 165–6
incoherent neutron scattering 163–6
inelastic incoherent scattering 166–7
quasielastic incoherent scattering 164–5
various elements 103
see also small-angle neutron scattering (SANS)
neutrons and X-rays 238
nickel filters 15
nitrogen flash radiation, characteristics 313
non-crystallographic symmetry (NCS) 87–92
  averaging 89–92
  operators 90–1
  phase extension 91–2
nuclear spin contrast variation studies 390–405
  isotopic substitution 390–6
    coherent cross-section calculation 395–6
    coherent and incoherent scattering from polarized nuclei 392–5
    coherent scattering length of polarized neutrons from polarized nuclei 392
    magnetic momentum and polarization 393
    neutron scattering by polarized proton spins 391
    nuclear spin contrast variation 399–400
    polarization of spins 391–2
    scattering length of hydrogen in absence of spin polarization 390–1
  nuclear spin contrast is complementary to solvent contrast 399–400
    deuteron spin contrast variation 399–400
    proton spin contrast variation 399
  nuclear spin diffusion 405
  polarized neutron scattering 396–9
    cross-terms 403
    dynamic nuclear spin polarization 397–8
    instrumentation 396
    nuclear polarization measurement 398
    polarization reversal by adiabatic fast passage 399
    selective nuclear spin depolarization 398
    proton and deuteron NMR measurements 398
    proton and deuteron spin polarization 400
    proton spin contrast variation 399
    spin-dependent scattering lengths 393
    structure of mRNA[tRNA]2 complex, *E. coli* ribosome 400–5
      data analysis 401–5
      orientation in 70S coordinate system 403
      sample and experiment 401
nuclear spin diffusion 405
nucleic acids
  functional systems in solution 246–8
  hydration 245–6
nucleosome core particle 377–8
  glass capillary mounting 17–18
  quantitative attachment of heavy atom compounds 378–9

beta-octyl glycoside ring structure 108, 111
Omnilyser, time-resolved fluorescence anisotropy 324
Ompf protein, *Escherichia coli* 111–12
open reading frames (ORFs) 6–7

Index 413

Opsanus tau, parvalbumin, MAD phasing with lanthanides 70–2
optical microscopy, vs single crystal X-ray diffraction 39
optical parametric amplifier laser, characteristics 313
optics, *see also* X-ray optics
ORTOGNOM program 209
osteoporosis, collagen, mineral and hydrated volume, measurement and distribution 348

pancreatic elastase (PPE), SIRAS 73
pancreatic trypsin inhibitor (BPTI) 175–6
particle mesh Ewald method 146
particles
   asymptotic regime 207
   Guinier relation, gyration radius 204–6
   hydration 206–7
   Shannon sampling theorem 210
parvalbumin, MAD phasing with lanthanides 70–2
Patterson maps 43, 52, 71, 73
peptides, SRCD 334–5
phase combination 82
phase probability curve 82
phasing methods
   principles 40–1
   solvent electron density, variations 43–4
phi slicing, fine vs coarse 20–1
phonon scattering (TDS) 183–5
photolysis, reaction initiation, time-resolved biocrystallography 119
photon counting 29
   multiwire proportional chambers (MWPCs) 18–20, 66, 298–9
photosynthetic reaction centres (PRCs), *Rhodopseudomonas* and *Rhodobacter* 111–12
pinhole collimation, X-ray optics 15
platelets, Guinier relation 206
polychromatic diffraction, enzyme reactions in crystals 120
polyethylene, diffraction pattern 284–6
porcine pancreatic elastase (PPE), SIRAS 73
Porod invariant 206
potential energy, calculation 141–6
potential energy surfaces, local exploration 146–8
potential of mean force (PMF) 157
profile fitting
   equations 32–4
   intensity
      strong reflections 33
      weak reflections 33–4
   and summation integration 29
protein crystals 117
Protein Data Bank (PDB) 5–13
   structures, families, folds 9–10
protein denaturation–renaturation 223–33
protein diffraction, diffuse X-ray scattering 192–4
protein dynamics 141–58
   diffuse X-ray scattering 181–94
   diffusive motion, incoherent quasielastic neutron scattering 177–9
   high frequency vibrations 176–7

Monte Carlo techniques 158
neutron scattering, inelastic/quasielastic 161–79
potential energy calculation 141–6
   Born-Oppenheimer approximation 142
   empirical force fields 142–5
potential energy surfaces, local exploration 146–8
simulations 148–58
   annealing 157
   free energies, calculation 155–7
   methodology 149–53
   trajectory analysis 153–5
vibrational density of states 175–6
vibrations, incoherent inelastic neutron scattering 175
protein structure 3–140
   critical assessment of protein structure (CASP), homology 11–12
   data collection and reduction methods 14–34
   electron density 79–100
   enzyme reactions 116–21
   genomics 3–13
   metalloproteins 124–35
   models, common errors 100
   neutron crystallography 102–14
   structure factor phases 36–78
   time-resolved crystallography 116–21
   see also structure factor phases
proteins
   concentrated solutions, interactions 218–20, 223–4
   hydration 245–6
   hydration layer, $D_2O : H_2O$ exchange, and SANS 245–6
   neutron crystallography, near-atomic resolution 103–7
   secondary structure, VUV CD 329
proton pumps, bacteriorhodopsin 259–69
purple membranes, bacteriorhodopsin 259–69

quantum mechanics methods 142–6
quartz capillaries, crystal mounting 17–18, 355
quasi-elastic incoherent scattering 164–5
quasi-Laue neutron diffraction, HEWL, near-atomic resolution neutron crystallography 105–7

radiation damage 67, 376–7
rat mannose-binding protein (MBP), MAD phasing with lanthanides 70–2
real-space correlation coefficient (RSCC) 98
real-space R factor (RSRF) 98
real-time time-resolved microfluorimetry 317–19
reflections, and profile fitting 32–4
Rhodobacter rubrum, proton–translocating transhydrogenase 315
Rhodopseudomonas viridis, photosynthetic reaction centres (PRCs) 111–12
ribosome
   bacterial 367
   derivatization agents, high resolution phasing 377–8
   isolated ribosomal components and in situ complexes 383

ribosome (*Cont.*)
  structure of mRNA[tRNA]$_2$ complex 400–5
    centre of mass 402
    orientation 403
  subunit 50S 369–75
  subunit 70S 368–9
Richards Box 92
RNA
  blue-tongue virus 364–6
  methionyl–tRNA synthase, *E. coli* 246–8
  scattering density vs deuteron spin polarization 400
  tRNA complex, *E. coli* ribosome 400–5
RNA reductase R1 protein, shape determination 212
RNAse, acoustic scattering 185–6
rods, Guinier relation 206
rotameter side-chain score (RSC) 98

Saccharomyces, genome 4–5
samarium, calculated spectra 47
SANS, SAXS, *see* small-angle neutron scattering; small-angle X-ray scattering
Sayre's equation, density modification 88–9
scattering
  acoustic scattering 185–6
  by ideal monodisperse solution 203–18
  phonon scattering 183–5
  polarized nuclei, coherent/ incoherent scattering 392–5
  resonant, effects on diffraction pattern 49–51
  thermal diffuse (TDS) 183–6
  very low scattering vector unit 240
  see also diffuse X-ray; small-angle neutron scattering
scattering density
  F_{hkl}, Fourier transform 104
  matching by $D_2O : H_2O$ exchange 241–2
scattering length, hydrogen, in absence of spin polarization 390–1
Schrödinger equation 142
selenium
  calculated spectra 47–8
  isomorphous effects 42
  MAD phasing 69
SHAKE algorithm 143
Shannon sampling theorem 210
simulated annealing 157
simulations
  free energies, calculation 155–7
  molecular dynamics 148–58
  techniques, increased computational power 158
single anomalous diffraction (SAD), single wavelength method 62–3
single crystal X-ray diffraction, vs optical microscopy 39
single isomorphous replacement and anomalous dispersion (SIRAS) 62–3
  phasing with mercury 73
  phasing with noble gases 73–4
single isomorphous replacement (SIR) 81
singular value decomposition analysis, solutions 215

site-directed mutagenesis, and SR fluorescence lifetime analysis 316
small-angle neutron scattering (SANS) 238–48
  biological solutions 239–41, 246–8
    examples 246–8
  buoyancy term 243
  contrast variation 241
  scattering density matching by $D_2O : H_2O$ exchange 241–2
  solvent interactions 244
  see also small-angle X-ray scattering (SAXS)
small-angle scattering (SAS) 238
  low scattering vector limit 240
small-angle X-ray scattering (SAXS) 199–233, 238
  ideal monodisperse solution 203–18
    atomic models 213–14
    by extended chain 217–18
    data analysis 207–9
    global structural parameters 203–7
    singular value decomposition 215
  instrumentation 220–22
  interactions, proteins in concentrated solutions 218–20, 223–4
  see also diffuse X-ray: small-angle neutron scattering
SOLOMON program 56, 87
solutions
  atomic models 213–14
  data analysis 207–9
  fibre and muscle diffraction 272–303
  global structural parameters 203–7
  interactions, concentrated solutions 218–20, 223–4
  membrane structure and dynamics 251–69
  small-angle neutron scattering (SANS) 238–48
  small-angle X-ray scattering 199–233
    ideal monodisperse solution 203–18
solvent exchange, contrast variation 243–4
solvent flattening
  electron density maps 85–6
  molecular envelope 89
  Wang-type 88
solvents
  interactions in SANS 244
  phasing, electron density, variations 43–4
  special studies 245
  X-ray scattering 201
sound velocities, L-alanine 171
spectroscopy
  inelastic/quasielastic neutron scattering 161–79
  using low energy (VUV/UV SR) 305–37
  see also fluorescence spectroscopy
spermatozoa, DNA and protein, X-ray microscopy 348–50
Structural Classification of Proteins (SCOP) 5, 8–9
Staphylococcus spp. nuclease
  SRCD 335
  time-focusing crystal analyser spectrometer (TXFA) 177
structural biology
  SRCD applications 333–6

instrumental developments and future prospects 335–6
time-resolved fluorescence spectroscopy 315–24
structure factor
  dynamic, simulations in neutron scattering 161–3
  $F(h)$ 44, 109
  F_{hkl} 104
structure factor phases
  anomalous dispersion 45–51
  chemical variation methods 41–5
    based on partial structures, ordered heavy atoms 41–3
    chemical to physical contrast variation 44–5
    solvent electron sensity 43–4
  data, accessible spectral range 49
  experimental determination 36–74
  measurement and processing 63–8
    area detector 65–6
    goniometry 66–7
    instrumentation 63–4
    X-ray optics 65
  sample mounting to structure amplitudes 67–8
  single wavelength methods
    anomalous diffraction (SAD) 62–3, 73
    phasing with mercury 72–3
    phasing with noble gases 73–4
    SIRAS 62–3
  X-ray sources 64
  see also protein structure
sulphur, calculated spectra 47
summation integration, and profile fitting 29, 34
superfolds, types 10
superoxide dismutases, structure 133–5
Swiss-Prot, URL 5
synchrotron
  ESRF 64–5
  new sources and detectors, future developments 298–9
  X-ray sources 64
synchrotron low energy VUV/UV radiation 305–37
  characteristics 313
  sources 310
  time-resolved fluorescence anisotropy 319–24
  time-resolved fluorescence spectroscopy 305–6
synchrotron radiation, and crystal decay 376–7
Synechocystis, genome 4
synthetic polymers, fibre diffraction 284–6

tapered glass capillaries, crystal mounting 17–18, 355
thermal diffuse scattering (TDS) 183–6
  acoustic scattering 185–6
  phonon scattering 183–5
thermodynamic integration method 155–7
thermodynamic perturbation method 155–6
Thermus aquaticus, electron density map 84
Thermus thermophilus
  diffraction pattern 369
  ribosome crystals 368
threading, fold recognition 11–12
time correlation functions 155
time-dependent Schrödinger equation 142

time-focusing crystal analyser spectrometer (TXFA) 177
time-resolved crystallography 116–21
time-resolved fluorescence anisotropy 319–24
  applications 322–4
  decay theory 320–1
  instrumental developments 324
  SR sources 321–2
  time domain measurements 324
time-resolved fluorescence spectroscopy 305–19
  applications 315–16
  decay analysis 314
  instrumental developments 317–19, 324
  simultaneous spectral and lifetime collection 317
time-resolved microfluorimetry, real time 317–19
time-resolved microvolume fluorescence lifetime measurements 317
time-resolved X-ray diffraction
  muscle structure and diffraction 298
  striated muscle 298
time-to-amplitude converter (TAC) 311–12
titanium sapphire solid state laser, characteristics 313
trajectory analysis, molecular dynamics simulations 153–5
transferrins, domain closure 135
triple isotopic substitution (TIS) 244
tRNA, and SANS 246–8
tropomyosin
  diffuse X-ray scattering 192–3
  structure and diffraction 287–90
tryptophan fluorescence
  decay analysis 315
  as probe 307–8

umbrella sampling method 155, 157
uranium, calculated spectra 47
urea compounds, alkane diffusion 172–3

vacuum UV/UV synchrotron radiation 305–37
valence bond algorithm 143
van der Waals interactions 144
van Hove correlation functions 162
Verlet integration algorithm 143
vibrational density of states 167
virus and v.l. macromolecule crystallography 353–66
  bluetongue virus (BTV) 357, 361–66
    A layer, classical/ non-classical quasi-equivalence 364
    core architecture 363
    core outer layers showing T13 trimers 363
    model for dsRNA layer proximal to T2 scaffold 365
    multiple interactions between layers of different symmetry 364
    ordered RNA 364–6
    structure 362
  crystallizaton and crystal preparation 355–6
  data collection
    beamline 356
    detectors 356–8
    initial phase determination 360

virus and v.l. macromolecule crystallography (*Cont.*)
  Laue method 358
  phase determination 359–61
  phase refinement and extension 360–1
  processing 359
  randomly oriented images 358
  space group determination 359
  strategy 358–9
  foot-and-mouth disease crystal 357
  icosahedral symmetry 354
  map interpretation, model building, refinement 361
  mosaic spread vs image number of bovine enterovirus crystals 355
viruses, neutron crystallography 112–13

X-ray absorption fine structure (XAFS) 124–35
  applications 129–35
    extended X-ray absorption (EXAFS) 124–5, 130–5
    X-ray absorption near edge (XANES) 124–5, 129–30
  metalloprotein structure, experimental requirements 128–9
X-ray detectors
  characteristics 19–20
  charge coupled devices (CCD) 19
  data collection and reduction methods 18–20
  ideal detector, characteristics 19–20
  image plate detectors 18
  MWPCs 18
  technologies 18–20
X-ray 'eye' 16
X-ray microscopy 338–52
  absorbtion spectra
    CaO 345
    gaseous CO_2 345
  analytical microscopies using native contrast 339–40
  chemical state microscopy via the transmission signal 348–9
  collagen, mineral and hydrated volume in osteoporotic bone, measurement and distribution 348
  compared with conventional microscopies 339
  DNA and BSA absorption spectrum 350

  DNA and protein in sperm, measurement and distribution 348–9
  elemental and chemical state contrast, X-ray probes via X-ray absorption fine structure 344–5
  elemental and chemical state maps, absorption contrast differences 347–8
  embedding medium map of mineralizing tissue 349
  Fresnel zone plate X-ray lens diagram 340
  future directions 349–51
  malaria life cycle in human red blood cells 343
  microscope choice
    factors affecting 338–9
    types 340–3
  near-edge X-ray absorption fine structure (NEXAFS) 344–7
  photoemission 341
  schematic potentials of carbon atoms and carbon based diatomic molecule 344
  transmission, scanning/non-scanning 341–3
  X-ray-induced fluorescence 341
  zone plate by electron beam lithography 343
X-ray optics
  data collection and reduction methods 14–16
  data measurement and processing
    anomalous dispersion 65
    structure factor phases 65
  double mirror system 15–16
  monochromator 15, 65
  novel collimation devices 16
  pinhole collimation 15
X-ray scattering
  by matter 200–3
  diffuse 181–94
  principles 200–3
  small angle
    instrumentation 220–22
    questions and systems 222–24
  see also small-angle X-ray scattering (SAXS)
X-ray scattering, *see* diffuse X-ray scattering
xenon
  isomorphous effects 42–3
  SIRAS phasing 74

zwitterions, L-alanine 170–2